A TEXTBOOK

OF

APPLIED MECHANICS

A TEXTBOOK
OF
APPLIED MECHANICS

(Including Laboratory Practicals)

S.I. UNITS

By

R.K. RAJPUT

M.E. (Heat Power Engg.) *Hons.–Gold Medallist* ; Grad. (*Mech. Engg. & Elect. Engg.*) ;
M.I.E. (*India*) ; M.S.E.S.I. ; M.I.S.T.E. ; C.E. (*India*)

Principal (*Formerly*),

Punjab College of Information Technology

PATIALA (Punjab)

LAXMI PUBLICATIONS (P) LTD

BANGALORE • CHENNAI • COCHIN • GUWAHATI • HYDERABAD
JALANDHAR • KOLKATA • LUCKNOW • MUMBAI • PATNA
RANCHI • NEW DELHI

Published by :

LAXMI PUBLICATIONS (P) LTD
113, Golden House, Daryaganj,
New Delhi-110002
Phone : 011-43 53 25 00
Fax : 011-43 53 25 28
www.laxmipublications.com
info@laxmipublications.com

Third Edition: 2011

OFFICES

✆ **Bangalore**	080-26 75 69 30	✆ **Kolkata**	033-22 27 43 84
✆ **Chennai**	044-24 34 47 26	✆ **Lucknow**	0522-220 99 16
✆ **Cochin**	0484-237 70 04, 405 13 03	✆ **Mumbai**	022-24 91 54 15, 24 92 78 69
✆ **Guwahati**	0361-251 36 69, 251 38 81	✆ **Patna**	0612-230 00 97
✆ **Hyderabad**	040-24 65 23 33	✆ **Ranchi**	0651-221 47 64
✆ **Jalandhar**	0181-222 12 72		

EAM-0791-250-ATB APPLIED MECH-RAJ **C—689/09/09**

Typeset at : Goswami Associates, Delhi

Printed at : Repro Knowledgecast Limited, Than

CONTENTS

PREFACE TO THE THIRD EDITION

This 2011-edition of the book **"Applied Mechanics"** has the following features:

The book has been thoroughly revised and the following additions have been incorporated to make it still more useful for the readers, from examination point of view:

- It has been completely converted into S.I. Units.
- **"Selected Questions from recent Examination Papers—With Solutions"** have been added at the end of each chapter.

Any suggestions for the improvement of this book will be thankfully acknowledged and incorporated in the next edition.

—Author

PREFACE TO THE FIRST EDITION

Several books on the subject of **"Applied Mechanics"** are available in the market, but only a couple of them strictly covers the syllabus at a diploma and draughtsmanship level ; but even these books do not properly cater to the need of the students in respect of subject matter—its bulk, presentation and placement. This gap has been judiciously bridged by this book.

This book covers the syllabi of this subject of almost all polytechnics in the country in various disciplines, *e.g.,* Civil, Electrical, Mechanical, Textile, Automobile etc. The book, however, is also suitable for other competitive examinations including studentship examination conducted by Institute of Engineers (India).

The book is divided into two parts : Part I and Part II. Part I assimilates 12 chapters and each chapter comprises of the following : (*i*) Text, (*ii*) Worked examples (both in M.K.S. and S.I. units), (*iii*) Highlights, (*iv*) Objective type questions and (*v*) Exercises (including theoretical questions and unsolved problems). Part II contains **'Laboratory Practicals'.**

The language of the book is direct, simple, lucid and book is profousely illustrated.

The author's heartiest thanks are due to his wife Ramesh Rajput for rendering all assistance during preparation and proof reading of the book. The author is thankful to Mr. R.K. Syal for drawing beautiful and well proportioned figures for the book.

The author is also highly grateful to M/s Laxmi Publications (Pvt.) Ltd. for bringing out the book in time and pricing it so moderately inspite of heavy cost of paper and printing.

—Author

PREFACE TO THE THIRD EDITION

- ... Set of Questions from ... Examinations ... have been included at the end of each chapter.

To My Reversed Parents
Late Smt. & Shri Hans Raj Rajput

Introduction to S.I. Units and Conversion Factors

A. INTRODUCTION TO S.I. UNITS

S.I., the international system of units are divided into three classes :

1. Base units
2. Derived units
3. Supplementary units.

From the scientific point of view division of S.I. units into these classes is to a certain extent arbitrary, because it is not essential to the physics of the subject. Nevertheless the General Conference, considering the advantages of a single, practical, world-wide system for international relations, for teaching and for scientific work, decided to base the international system on a choice of six well-defined units given in Table 1 below :

Table 1. S.I. Base Units

Quantity	*Name*	*Symbol*
length	metre	m
mass	kilogram	kg
time	second	s
electric current	ampere	A
thermodynamic temperature	kelvin	K
luminous intensity	candela	cd
amount of substance	mole	mol

The second class of S.I. units contains derived units, *i.e.,* units which can be formed by combining base units according to the algebraic relations linking the corresponding quantities. Several of these algebraic expressions in terms of base units can be replaced by special names and symbols can themselves be used to form other derived units.

Derived units may, therefore, be classified under three headings. Some of them are given in Tables 2, 3 and 4.

Table 2. Examples of S.I. Derived Units Expressed in terms of Base Units

Quantity	S.I. Units	
	Name	Symbol
area	square metre	m^2
volume	cubic metre	m^3
speed, velocity	metre per second	m/s
acceleration	metre per second squared	m/s^2
wave number	1 per metre	m^{-1}
density, mass density	kilogram per cubic metre	kg/m^3
concentration (of amount of substance)	mole per cubic metre	mol/m^3
activity (radioactive)	1 per second	s^{-1}
specific volume	cubic metre per kilogram	m^3/kg
luminance	candela per square metre	cd/m^2

Table 3. S.I. Derived Units with Special Names

Quantity	S.I. Units			
	Name	Symbol	Expression in terms of other units	Expression in terms of S.I. base units
frequency	hertz	Hz	—	s^{-1}
force	newton	N	—	$m.kg.s^{-2}$
pressure	pascal	Pa	N/m^2	$m^{-1}.kg.s^{-2}$
energy, work, quantity of heat power	joule	J	N.m	$m^2.kg.s^{-2}$
radiant flux quantity of electricity	watt	W	J/s	$m^2.kg.s^{-3}$
electric charge	coulomb	C	A.s	s.A
electric tension, electric potential	volt	V	W/A	$m^2.kg.s^{-3}.A^{-1}$
capacitance	farad	F	C/V	$m^{-2}.kg^{-1}.s^4$
electric resistance	ohm	Ω	V/A	$m^2.kg.s^{-3}.A^{-2}$
conductance	siemens	S	A/V	$m^{-2}.kg^{-1}.s^3.A^2$
magnetic flux	weber	Wb	V.S.	$m^2.kg.s^{-2}.A^{-1}$
magnetic flux density	tesla	T	Wb/m^2	$kg.s^{-2}.A^{-1}$
inductance	henry	H	Wb/A	$m^2.kg.s^{-2}.A^{-2}$
luminous flux	lumen	lm	—	cd.sr
illuminance	lux	lx	—	$m^{-2}.cd.sr$

Table 4. Examples of S.I. Derived Units Expressed by means of Special Names

Quantity	*S.I. Units*		
	Name	*Symbol*	*Expression in terms of S.I. base units*
dynamic viscosity	pascal second	Pa-s	$m^{-1}.kg.s^{-1}$
moment of force	metre newton	N.m	$m^2.kg.s^{-2}$
surface tension	newton per metre	N/m	$kg.s^{-2}$
heat flux density, irradiance	watt per square metre	W/m^2	$kg.s^{-2}$
heat capacity, entropy	joule per kelvin	J/K	$m^2.kg.s^{-2}.K^{-1}$
specific heat capacity, specific entropy	joule per kilogram kelvin	J/(kg.K)	$m^2.s^{-2}.K^{-1}$
specific energy	joule per kilogram	J/kg	$m^2.s^{-2}$
thermal conductivity	watt per metre kelvin	W/(m.K)	$m.kg.s^{-3}.K^{-1}$
energy density	joule per cubic metre	J/m^3	$m^{-1}.kg.s^{-2}$
electric field strength	volt per metre	V/m	$m.kg.s^{-3}.A^{-1}$
electric charge density	coulomb per cubic metre	C/m^3	$m^{-3}.s.A$
electric flux density	coulomb per square metre	C/m^2	$m^{-2}.s.A$
permittivity	farad per metre	F/m	$m^{-3}.kg^{-1}.s^4.A^4$
current density	ampere per square metre	A/m^2	—
magnetic field strength	ampere per metre	A/m	—
permeability	henry per metre	H/m	$m.kg.s^{-2}.A^{-2}$
molar energy	joule per mole	J/mol	$m^2.kg.s^{-2}mol^{-1}$
molar heat capacity	joule per mole kelvin	J/(mol.K)	$m^2.kg.s^{-2}.K^{-1}mol^{-1}$

The S.I. units assigned to third class called "Supplementary units" may be regarded either as base units or as derived units. Refer Table 5 and Table 6.

Table 5. S.I. Supplementary Units

Quantity	*S.I. Units*	
	Name	*Symbol*
plane angle	radian	rad
solid angle	steradian	sr

Table 6. Examples of S.I. Derived Units Formed by Using Supplementary Units

Quantity	S.I. Units	
	Name	Symbol
angular velocity	radian per second	rad/s
angular acceleration	radian per second squared	rad/s^2
radiant intensity	watt per steradian	W/sr
radiance	watt per square metre steradian	$W\text{-}m^{-2}.sr^{-1}$

Table 7. S.I. Prefixes

Factor	Prefix	Symbol	Factor	Prefix	Symbol
10^{12}	tera	T	10^{-1}	deci	d
10^{9}	giga	G	10^{-2}	centi	c
10^{6}	mega	M	10^{-3}	milli	m
10^{3}	kilo	k	10^{-6}	micro	μ
10^{2}	hecto	h	10^{-9}	nano	n
10^{1}	deca	da	10^{-12}	pico	p
			10^{-15}	fasnto	f
			10^{-18}	atto	a

B. CONVERSION FACTORS

1. Force :

$$1 \text{ newton} = \text{kg-m/sec}^2 = 0.012 \text{ kgf}$$
$$1 \text{ kgf} = 9.81 \text{ N}$$

2. Pressure :

$$1 \text{ bar} = 750.06 \text{ mm Hg} = 0.9869 \text{ atm} = 10^5 \text{ N/m}^2 = 10^3 \text{ kg/m-sec}^2$$
$$1 \text{ N/m}^2 = 1 \text{ pascal} = 10^{-5} \text{ bar} = 10^{-2} \text{ kg/m-sec}^2$$
$$1 \text{ atm} = 760 \text{ mm Hg} = 1.03 \text{ kgf/cm}^2 = 1.01325 \text{ bar}$$
$$= 1.01325 \times 10^5 \text{ N/m}^2$$

3. Work, Energy or Heat :

$$1 \text{ joule} = 1 \text{ newton metre} = 1 \text{ watt-sec}$$
$$= 2.7778 \times 10^{-7} \text{ kWh} = 0.239 \text{ cal}$$
$$= 0.239 \times 10^{-3} \text{ kcal}$$
$$1 \text{ cal} = 4.184 \text{ joule} = 1.1622 \times 10^{-6} \text{ kWh}$$
$$1 \text{ kcal} = 4.184 \times 10^3 \text{ joule} = 427 \text{ kgf m}$$
$$= 1.1622 \times 10^{-3} \text{ kWh}$$
$$1 \text{ kWh} = 8.6042 \times 10^5 \text{ cal} = 860.42 \text{ kcal} = 3.6 \times 10^6 \text{ joule}$$
$$1 \text{ kgf-m} = \left(\frac{1}{427}\right) \text{ kcal} = 9.81 \text{ joules}$$

4. Power :

1 watt = 1 joule/sec = 0.86 kcal/h

1 h.p. = 75 m kgf/sec = 0.1757 kcal/sec = 735.3 watt

1 kW = 1000 watts = 860 kcal/h

5. Specific heat :

1 kcal/kg-°K = 0.4184 joules/kg-K

6. Thermal conductivity :

1 watt/m-K = 0.8598 kcal/h-m-°C

1 kcal/h-m-°C = 1.16123 watt/m-K = 1.16123 joules/s-m-K.

7. Heat transfer co-efficient :

1 watt/m^2-K = 0.86 kcal/m^2-h-°C

1 kcal/m^2-h-°C = 1.163 watt/m^2-K.

C. IMPORTANT ENGINEERING CONSTANTS AND EXPRESSIONS IN S.I. UNITS

Engineering constants and expressions	*M.K.S. system*	*S.I. Units*
1. Value of g_0	9.81 kg-m/kgf-sec^2	1 kg-m/N-sec^2
2. Universal gas constant	848 kgf-m/kg mole-°K	848 × 9.81 = 8314 J/kg-mole-°K ($\because$ 1 kgf-m = 9.81 joules)
3. Gas constant (R)	29.27 kgf-m/kg-°K for air	$\frac{8314}{29}$ = 287 joules/kg-K for air
4. Specific heats (for air)	c_v = 0.17 kcal/kg-°K c_p = 0.24 kcal/kg-°K	c_v = 0.17 × 4.184 = 0.71128 kJ/kg-K c_p = 0.24 × 4.184 = 1 kJ/kg-K
5. Flow through nozzle-Exit velocity (C_2)	$91.5\sqrt{U}$, where U is in kcal	$44.7\sqrt{U}$, where U is in kJ
6. Refrigeration 1 ton	= 50 kcal/min	= 210 kJ/min
7. **Heat transfer** The Stefan Boltzman Law is given by :	$Q = \sigma T^4$ kcal/m^2-h where $\sigma = 4.9 \times 10^{-8}$ kcal/h-m^2-°K^4	$Q = \sigma T^4$ watts/m^2-h where $\sigma = 5.67 \times 10^{-8}$ W/m^2K^4

Part I

APPLIED MECHANICS—THEORY

1

Basic Concepts

1.1. Introduction to mechanics. 1.2. Basic definitions. 1.3. Rigid body. 1.4. Scalar and vector quantities. 1.5. Fundamental units and derived units. 1.6. Systems of units—Highlights—Objective Type Questions—Exercises.

1.1. INTRODUCTION TO MECHANICS

In any field the importance of a thorough knowledge of fundamentals cannot be over emphasised. Fundamentals have always been stressed in the learning of new skills. Similarly, the mechanics branch of science which deals with the forces and their effect on bodies on which they act is founded on basic concepts and forms the ground-work for further study in the design and analysis of machines and structures. Mechanics can be divided into two parts. (Fig. 1.1)

(*i*) *'Statics'* which relates to the *bodies at rest* and (*ii*) *'dynamics'* which deals with *bodies in motion.* (In mechanics the term strength of materials refers to the ability of the individual parts of a machine or structure to resist loads. It also permits the determination of dimensions to ensure sufficient strength of the various parts). Dynamics may be further into the following two groups :

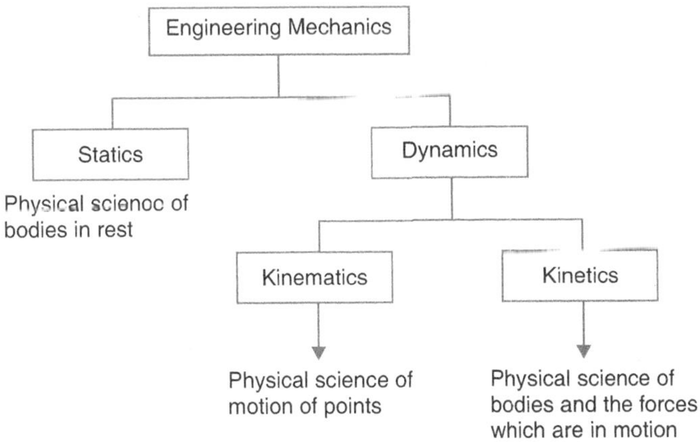

Fig. 1.1

(*a*) *'Kinematics'* which deals with the motion of bodies without any reference to the cause of motion.

(*b*) *'Kinetics'* which deals with the relationship between forces and the resulting motion of bodies on which they act.

The *branch of science which deals with the study of different laws of mechanics as applied to solution of engineering problems is called* ***Applied Mechanics.***

1.2. BASIC DEFINITIONS

Length. This term is applied to the linear dimensions of a straight or curved line. For example, the diameter of circle is the length of a straight line which divides the circle into two equal parts ; the circumference is the length of its curved perimeter.

Area. The two dimensional size of a shape or a surface is its **area.** The shape may be flat (lie in a plane) or curved, for example, the size of a plot of land, the surface of a fluorescent bulb, or the cross-sectional size of a shaft.

Volume. The three dimensional or cubic measure of the space occupied by a substance is known as its **volume.**

Force. This term is applied to any action on the body which tends to make it move, change its motion, or change its size and shape. A **force** is usually thought of a push or pull, such as a hand pushing against a wall or the pull of a rope fastened to a body.

Pressure. The external force per unit area, or the total force divided by the total area on which it acts, is known as **pressure.** Water pressure against the face of a dam, steam pressure in a boiler, or earth pressure against a retaining wall are some examples.

Mass. The amount of matter contained in a body is called its mass, and for most problems in mechanics, mass may be considered **constant.**

Weight. The force with which a body is attracted towards the centre of earth by the gravitational pull is called its **weight.**

The relation between mass (m) and weight (W) of a body is given by the equation

$$W = m \times g \qquad ...(1.1)$$

The value of g is taken as 9.81 m/sec^2 (usually 9.80 m/sec^2 to make the calculation work easier) in M.K.S. system as well as in S.I. units.

Density. The weight of a unit volume of a body or substance is the **density.** This term is sometimes called weight density, to distinguish it from a similary definition (mass density) made in terms of mass.

Moment. The tendency of a force to cause rotation about some point is known as a **moment.**

Table 1.1. Difference between Mass and Weight

Mass	*Weight*
1. It is the quantity of matter contained in a body.	1. It is the force with which the body is attracted towards the centre of earth.
2. It is constant at all places.	2. It is different at different places.
3. It resists motion in the body.	3. It produces motion in the body.
4. It is a scalar quantity since it has magnitude only.	4. It is a vector quantity since it has magnitude as well as direction.
5. It can be measured by an ordinary balance.	5. It is measured by a spring balance.
6. It is never zero.	6. It is zero at the centre of earth.
7. It is measured in kilogram (kg) in M.K.S. system of units as well as in S.I. units.	7. It is measured in kilogram weight (kg wt. or kgf) in M.K.S. system of units and in newton (N) in S.I. units.

Torque. The action of a force which causes rotation to take place is known as torque. The action of a belt on a pulley causes the pulley to rotate because of **torque.** Also if you grasp a piece of chalk near each end and twist your hands in opposite directions, it is the developed torque that causes the chalk to twist and, perhaps, snap.

Work. *The energy developed by a force acting through a distance against resistance is known as work.* The distance may be along a straight line or along a curved path. When the distance is linear, the work can be found from *work = force × distance.* When the distance is along a circular path the work can be found from *work = toque × angle*. Common forms of work include a weight lifted through a height, a pressure pushing a volume of substance, and torque causing a shaft to rotate.

Power. *The rate of doing work, or work done per unit time is called power*. For example, a certain amount of work is required to raise an elevator to the top of its shaft. A 5 HP motor can raise the elevator, but a 20 HP motor can do the same job four times faster.

1.3. RIGID BODY

A *rigid body is one which does not change its shape and size under the effect of forces acting over it.* It differs from an elastic body in the sense that the latter undergoes deformation under the effect of forces acting on it and returns to its original shape and size on removal of the forces acting on the body. The rigidity of a body depends upon the fact that how far it undergoes deformation under the effect of forces acting on it.

In real sense no solid body is perfectly rigid because everybody changes it size and shape under the effect of forces acting on it. It actual practice, the deformation (*i.e.*, change in shape and size of a body under the effect of forces acting on it) is very small and therefore it may be considered as a rigid body.

1.4. SCALAR AND VECTOR QUANTITIES

Scalar quantity. A scalar quantity is one that has *magnitude only.*

Examples. Mass, volume, time and density.

Vector quantity. A vector quantity is one that has *magnitude as well as direction.*

Examples. Force, velocity, acceleration and moment etc.

A vector quantity is represented by a line carrying an arrow head at one end. The length of the line (to convenient scale) equals the *magnitude* of the vector. The line, together with its arrow head, defines the *direction* of the vector. Suppose a force of 60 N is applied to point *A* in Fig. 1.3 at an angle of 45° to the horizontal. The vector *AB* represents this force since its length equals 60 N (to scale) and its direction is proper. If the vector *BA* is drawn to same scale (Fig. 1.2) it represents a 60 N force having a direction exactly opposite to vector *AB*.

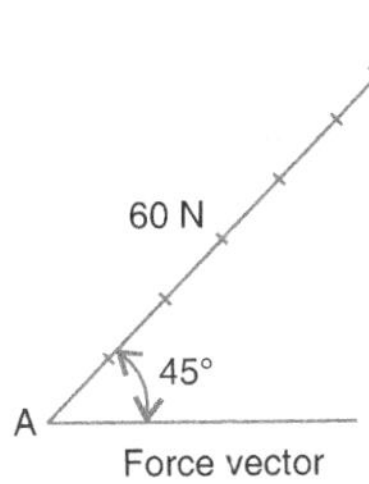

Fig. 1.2

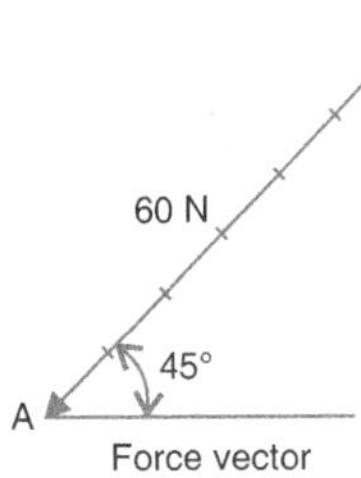

Fig. 1.3

1.5. FUNDAMENTAL UNITS AND DERIVED UNITS

The basic quantities or fundamental quantities of mechanics are those quantities which cannot be expressed in terms of one another. Length, mass, time are usually considered as basic or fundamental quantities. The units of these quantities are called fundamental units and are developed by L, M and T respectively. The units of all other quantities except above are derived with the help of fundamental units and thus they are known as *derived units.* For example, units of velocity, acceleration, density etc. are derived units as explained below :

$$\text{Velocity} = \frac{\text{Displacement}}{\text{Time}} = \frac{L}{T} = LT^{-1}$$

$$\text{Acceleration} = \frac{\text{Velocity}}{\text{Time}} = \frac{LT^{-1}}{T} = LT^{-2}$$

$$\text{Density} = \frac{\text{Mass}}{\text{Volume}} = \frac{M}{L^3} = ML^{-3}.$$

1.6. SYSTEMS OF UNITS

The four systems of units in use are given below :

1. Foot-Pound-Second system (F.P.S. system)
2. Centimetre-Gram-Second system (C.G.S. system)
3. Metre-Kilogram-Second system (M.K.S. system)
4. International system of units (S.I. units).

Foot-Pound-Second system. In this system the units of fundamental quantities *i.e.*, length, mass and time are foot, pound and second respectively.

Centimetre-Gram-Second system. In this system the value of length, mass and time are expressed as centimetre, gram and second respectively.

Metre-Kilogram-Second system. In this system units of length mass and time are metre, kilogram and second respectively.

International system of units. This system considers three more fundamental units of electric current, temperature and luminous intensity in addition to the fundamental units of **length, mass** and **time.**

Important M.K.S. and S.I. system units are given in Table 1.2.

Table 1.2. Important M.K.S. and S.I. system units

S.No. Quantities	*M.K.S. System*		*S.I. System*	
	Units	*Symbols*	*Units*	*Symbols*
1. Length	metre	m	metre	m
2. Mass	kilogram	kg	kilogram	kg
3. Time	second	sec or s	second	s
4. Temperature	celsius or kelvin	°C or °K	kelvin	K
5. Plane angle	radians	rad	radians	
6. Area	square metre	m^2	square metre	m^2
7. Volume	cubic metre	m^3	cubic metre	m^3
8. Density	kilogram per cubic metre	kg/m^3	kilogram per cubic metre	kg/m^3 or $kg\ m^{-3}$
9. Velocity	metre per second	m/sec	metre per second	ms^{-1}
10. Acceleration	metre per second per second	m/sec^2	metre per second per second	m/s^2 or ms^{-2}

11. Angular acceleration	radians per second per second	rad/sec^2	radians per second per second	rad/s^2 or rad s^{-2}
12. Force	kilogram weight	kgf	newton	N
13. Moment of force	kilogram weight metre	kgf m	newton metre	Nm
14. Momentum	kilogram metre per second	kg m/sec	kilogram metre per second	kg m s^{-1}
15. Angular momentum	-do-	kgm/sec	-do-	kg m s^{-1}
16. Moment of inertia	kilogram metre squared	kg m^2	kilogram metre squared	kg m^2
17. Work 'W'	kilogram metre	kgf m	joule or newton metre	J or Nm
18. Torque 'T'	-do-	kgf m	-do-	J or Nm
19. Power	kilogram metre per second	kg m/sec	watts or joules per second	W or J/s
20. Energy 'E'	kilogram metre	kgf m	joule or newton metre	J or Nm
21. Pressure intensity	kilogram per square metre	kgf/m^2	newton per square metre	N/m^2 or Nm^{-2}
22. Specific volume	cubic metre per kilogram mass	m^3/kg	cubic metre per kilogram mass	m^3/kg or m^3 kg^{-1}
23. Dynamic viscosity	kilogram per metre per sec	kg/m-sec	newton second per square metre	Ns/m^2 or Nsm^{-2}
24. Kinematic viscosity	metre per second per second	m/sec^2	metre per second per second	m/s^2 or ms^{-2}
25. Specific heat	kilocalorie per unit mass per degree celsius	kcal/kg °C	joule per unit mass per degree kelvin	J/kg K or J kg^{-1} K^{-1}

HIGHLIGHTS

1. '*Statics*' is the branch of mechanics which relates to bodies at rest.
2. '*Dynamics*' is the branch of mechanics which deals with bodies in motion. Kinematics deals with the motion of bodies without any reference to the cause of motion.
   Kinetics deals with the relationship between forces and the resulting motion of bodies on which they act.
3. *Mass* is the amount of matter contained in a body.
4. *Weight* is the force with which a body is attracted towards the centre of the earth by the gravitational pull.
5. *Density* is the weight of unit volume of a body or substance.
6. *Power* is the rate of doing work.
7. A rigid body is one which does not change its shape and size under the effect of forces acting over it.
8. A *scalar quantity* is one that has magnitude only.
9. A *vector quantity* is one that has magnitude as well as direction.
10. The basic quantities or fundamental quantities are those quantities which cannot be expressed in terms of one another.
11. *System of units*. The four system of units in use are :
    (*i*) Foot-Pound-Second system (F.P.S. system)
    (*ii*) Centimetre-Gram-Second system (C.G.S. system)

(*iii*) Metre-Kilogram-Second system (M.K.S. system)

(*iv*) International system of units (S.I. units).

OBJECTIVE TYPE QUESTIONS

1. Fill in the Blanks :

(*i*) ......is the branch of mechanics which relates to bodies at rest.

(*ii*) ......is the branch of mechanics which deals with bodies in motion.

(*iii*) ......deals with the motion of the bodies without any reference to the cause of motion.

(*iv*) ......deals with relationship between forces and the resulting motion of bodies on which they act.

(*v*) ......is the term applied to the linear dimensions of a straight or curved line.

(*vi*) The external force per unit area, or the total force divided by the total area on which it acts, is known as......

(*vii*) The amount of matter contained in a body is called its......

(*viii*) ......= torque × angle.

(*ix*) The rate of doing work is called......

(*x*) A scalar quantity is one that has......only.

2. Say 'Yes' or 'No' :

(*i*) The two dimensional size of a shape or a surface is its area.

(*ii*) Mass helps motion.

(*iii*) Velocity is a vector quantity.

(*iv*) A rigid body is one which undergoes change in its shape and size under the effect of forces acting over it.

(*v*) Acceleration is a scalar quantity.

(*vi*) In real sense, no solid body is perfectly rigid.

(*vii*) There is no difference between mass and weight.

(*viii*) The tendency of a force to cause rotation about some point is known as a moment.

(*ix*) Mass can be measured by a spring balance.

(*x*) Weight resists motion in the body.

Answers

1. (*i*) Statics (*ii*) Dynamics (*iii*) Kinematics (*iv*) Kinetics (*v*) Length
(*vi*) pressure (*vii*) mass (*viii*) Work (*ix*) power (*x*) magnitude.

2. (*i*) Yes (*ii*) No (*iii*) Yes (*iv*) No (*v*) No (*vi*) Yes
(*vii*) No (*viii*) Yes (*ix*) No (*x*) No.

EXERCISES

1. (*a*) Define the terms :

(*i*) Statics

(*ii*) Dynamics.

(*b*) Differentiate between 'Kinematics' and 'Kinetics'.

2. Define the following terms :
Force, volume, pressure, work and power.

3. Write a short note on 'Rigid body'.

4. (*a*) Explain the terms 'Mass' and 'Weight'.

(*b*) How does mass differ from weight ?

5. (*a*) Define 'Scalar' and 'Vector' quantities.

(*b*) Which of the following quantities are scalar ?
Mass, force, volume, velocity, time, acceleration.

6. Describe the various systems of units. Which system of units is being followed these days and why ?

2

Laws of Forces

2.1. FORCE

Force is some thing which *changes or tends to change the state of rest or of uniform motion of a body in a straight line.* Force is the direct or indirect action of one body on another. The bodies may be in direct contact with each other causing direct motion or separated by distance but subjected to gravitational effects.

There are different kinds of forces such as gravitational, frictional, magnetic, inertia or those cause by mass and acceleration. A static force is the one which is caused without relative acceleration of the bodies in question.

The force has a magnitude and direction, therefore, it is **vector**. While the directions of the force is measured in absolute terms of angle relative to a co-ordinate system, the magnitude is measured in different units depending on the situation.

When a force acts on a body, the following effects may be produced in that body : (*i*) *It may bring a change in the motion of the body i.e., the motion may be accelerated or retarded ; (ii) it may balance the forces already acting on the body thus bringing the body to a state of rest or of equilibrium, and (iii) it may change the size or shape of the body i.e., the body may be twisted, bent, stretched, compressed or otherwise distorted by the action of the force.*

2.2. UNITS OF FORCE

The two commonly used units of force are :

1. Absolute units
2. Gravitational units.

Absolute units. Because the mass and acceleration are measured differently in different systems of units, so the units of force are also different in the various systems as given below :

In the F.P.S. (Foot-Pound-Second) system the absolute unit of force is a *poundal* which is that much force as produces an acceleration of 1 ft/sec^2 in a mass of one pound.

In the C.G.S. (Centimetre-Gram-Second) system the absolute unit of force is *dyne* which is that much force as produces an acceleration of 1 cm/sec^2 in a mass of one gram.

In the M.K.S. (Metre-Kilogram-Second) system the absolute unit of force is a *newton* which is that much force as produces an acceleration of 1 m/sec^2 in a mass of the kilogram.

Obviously, 1 newton = 10^5 dynes.

Gravitational units of force. Gravitational units are the units which are used by engineers for all practical purposes. These units depend upon the weight of a body (*i.e.*, the force with which the body is attracted towards the centre of the earth). Now the weight of a body of mass m (*i.e.*, the quantity of matter contained in a body). $= mg$, where g is the acceleration due to gravity.

So the gravitational units of force in the three systems of units *i.e.*, F.P.S., C.G.S. and M.K.S. are Pound weight, Gram weight and kilogram weight.

The relationship of units of force is given as under :

1 lb wt. (or lbf) = g poundal = 32.2 poundals (app.)

1 gm wt. (or gmf) = g dynes = 981 dynes (app.)

1 kg wt. (or kgf) = g newtons = 9.81 newtons (app.)

which means

Gravitational unit of force = 'g' times the corresponding absolute units of force.

Usually, kg, wt (or kgf) is written simply as kg.

2.3. CHARACTERISTICS OF A FORCE

The characteristics or elements of the force are the quantities by which a force is fully represented. These are :

1. Magnitude (*i.e.*, 50 N, 100 N, etc.)
2. Direction or line of action (angle relative to a co-ordinate system).
3. Sense or nature (push or pull).
4. Point of application.

2.4. REPRESENTATION OF FORCES

Forces may be represented in the following two ways :

1. Vector representation
2. Bow's notation.

Vector representation. A force can be represented graphically by a vector as shown in Figs. 1.2 and 1.3.

Bow's notation. It is a method of designating a force by writing two capital letters one on either side of the force a shown in Fig. 2.1, where force P_1 (200 N) is represented by AB and force P_2 (100 N) by CD.

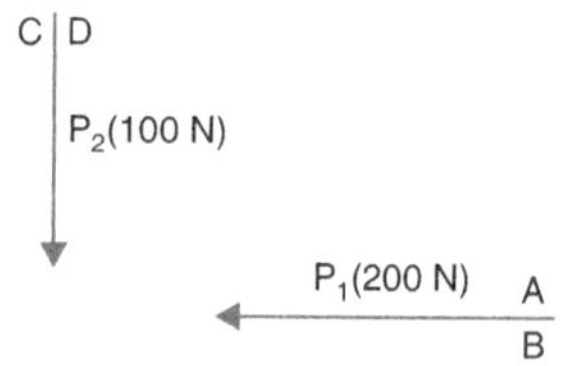

Fig. 2.1

2.5. CLASSIFICATION OF FORCES

There are several ways in which forces can be classified. Some of the important classifications are given as under :

1. According to the effect produced by the force :

(*i*) **External force.** When a force is applied external to a body it is called **external force.**

(*ii*) **Internal force.** The resistance to deformation, or change of shape, exerted by the material of a body is called an **internal force.**

(*iii*) **Active force.** An **active force** is one which causes a body to move or change its shape.

(*iv*) **Passive force.** A force which prevents the motion, deformation of a body is called a **passive force.**

2. According to nature of the force :

(*i*) **Action and reaction.** Whenever there are two bodies in contact, each exerts a force on the other. Out of these forces one is called action and other is called reaction. *Action and reaction are equal and opposite.*

(*ii*) **Attraction and repulsion.** These are actually non-contacting forces exerted by one body or another without any visible medium transmission such as magnetic forces.

(*iii*) **Tension and thrust.** When a body is dragged with a string the force communicated to the body by the string is called the *tension* while, if we push the body with a rod, the force exerted on the body is called a *thrust*.

3. According to whether the force acts at a point or is distributed over a large area.

(*i*) **Concentrated force.** The force whose point of application is so small that it may be considered as a point is called a concentrated force.

(*ii*) **Distributed force.** A **distributed force** is one whose place of application is area.

4. According to whether the force acts at a distance or by contact.

(*i*) **Non-contacting forces or forces at a distance.** Magnetic, electrical and gravitational forces are examples of non-contacting forces or forces at a distance.

(*ii*) **Contacting forces or forces by contact.** The pressure of steam in a cylinder and that of the wheels of a locomotive on the supporting rails are examples of contacting forces.

2.6. FORCE SYSTEMS

A *force system* is a collection of forces acting on a body in one or more planes.

According to the relative positions of the lines of action of the forces, the forces may be classified as follows :

1. Coplanar concurrent collinear force system. It is the simplest force system and includes those forces whose vectors lie along the same straight line (refer Fig. 2.2.).

Fig. 2.2

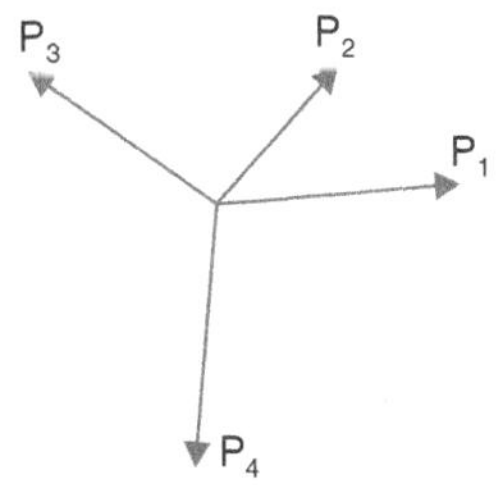

Fig. 2.3

2. Coplanar concurrent non-parallel force system. Forces whose lines of action pass through a common point are called **concurrent forces.** In this system lines of action of all the forces meet at a point but have different directions in the same plane as shown in Fig. 2.3.

3. Coplanar non-concurrent parallel force system. In this system, the lines of action of all the forces lie in the same plane and are parallel to each other but may not have same direction as shown in Fig. 2.4.

4. Coplanar non-concurrent non-parallel force system. Such a system exists where the lines of action of all forces lie in the same plane but do not pass through a common point. Fig. 2.5 shows such a force system.

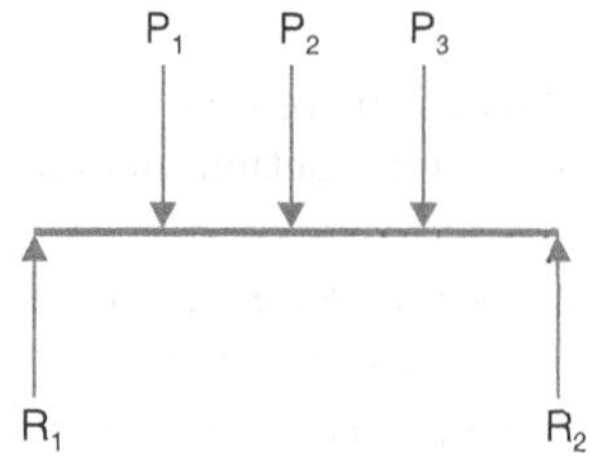

Fig. 2.4

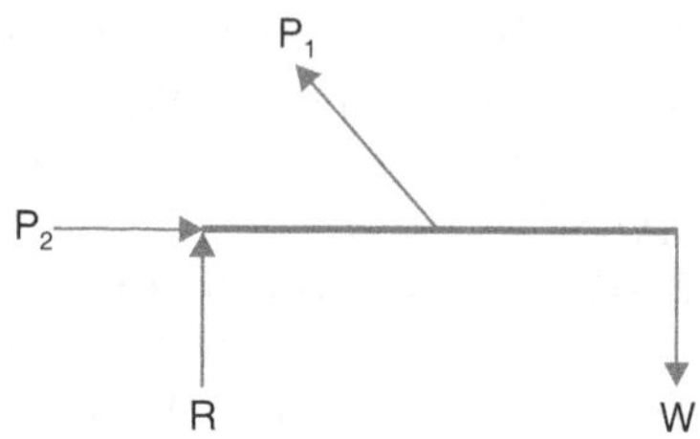

Fig. 2.5

5. Non-coplanar concurrent force system. This system is evident where the lines of action of all forces do not lie in the same plane but do pass through a common point. An example of this force system is the forces in the legs of tripod support for camera (Fig. 2.6).

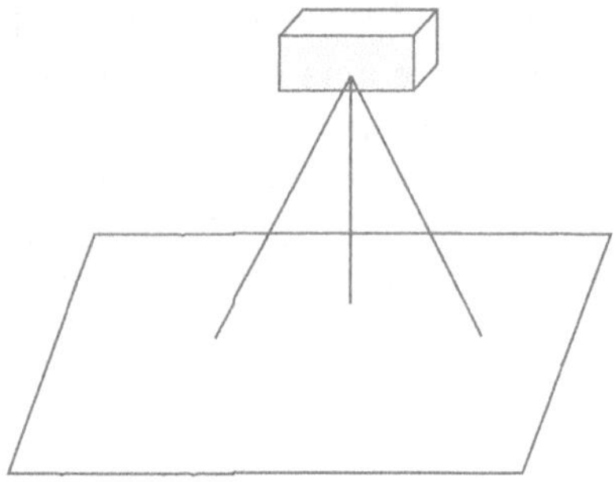

Fig. 2.6

6. Non-coplanar non-concurrent force system. Where the lines of action of all forces do not lie in the same plane and do not pass through a common point, a non-coplanar non-concurrent system is present.

2.7. FREE BODY DIAGRAMS

A body may consist of more than one element and supports. Each element or support can be isolated from the rest of the system by incorporating the net effect of the remaining system through a set of forces. This *diagram of the isolated element or a portion of the body along with the net effects of the system on it is called a 'free-body diagram'.* Free-body diagrams are useful in solving the forces and deformations of the system.

In case of a body in Fig. 2.7, we remove the supporting springs and replace it by the reactive force R equal to W in magnitude. The Fig. 2.7 (*c*) in which the body is completely isolated from its support and in which all forces acting on it are shown by vectors is called a *free body diagram.*

Let us consider another case of a beam shown in Fig. 2.8 (*a*). The beam is supported on a hinge at the left end and on a roller at the right end. The hinge offers vertical and horizontal reaction whereas the roller offers vertical reaction. The beam can be isolated from the supports by setting equivalent forces of the supports. Fig. 2.8 (*b*) illustrates the free body diagram of the beam in which R_1 and R_2 are reactions of the hinge support and R_3 the reaction of the roller support. Similarly, the free body diagrams of hinge and roller supports are shown in Figs. 2.8 (*c*) and 2.8 (*d*) respectively.

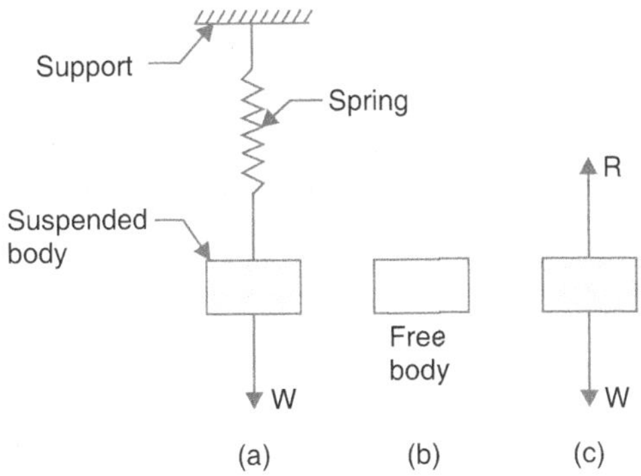

Fig. 2.7

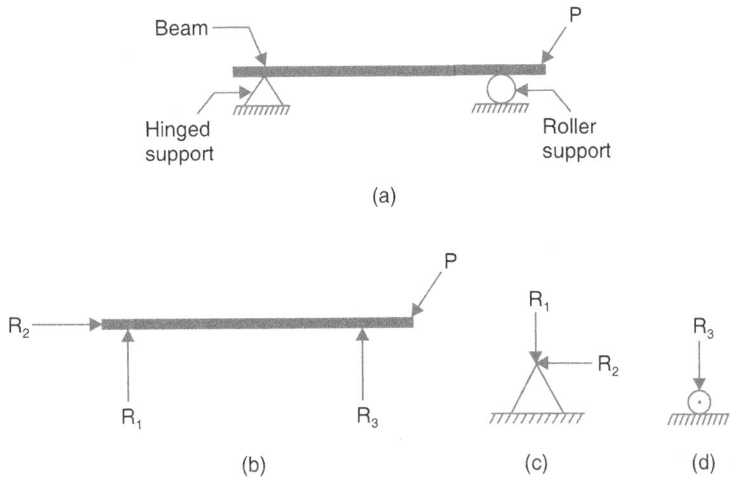

Fig. 2.8

2.8. TRANSMISSIBILITY OF A FORCE

The principle of transmissibility of forces states that *when a force acts upon a body, its effect is the same whatever point in its line of action is taken as the point of the application provided that the point is connected with the rest of the body in the same invariable manner.*

A force may be considered as acting at any point on its line of action so long as the direction and magnitude are not changed. Suppose a body (Fig. 2.9) is to be moved by a horizontal force P applied by hooking a rope to some point on the body. The force P will have the same effect if it is applied at 1, 2, 3 (Fig. 2.10) or any point on its line of action. This property of force is called *transmissibility*.

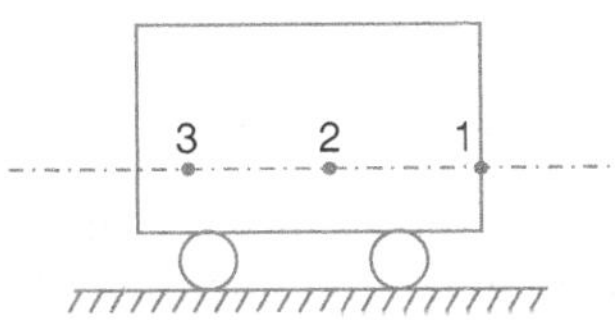

Fig. 2.9

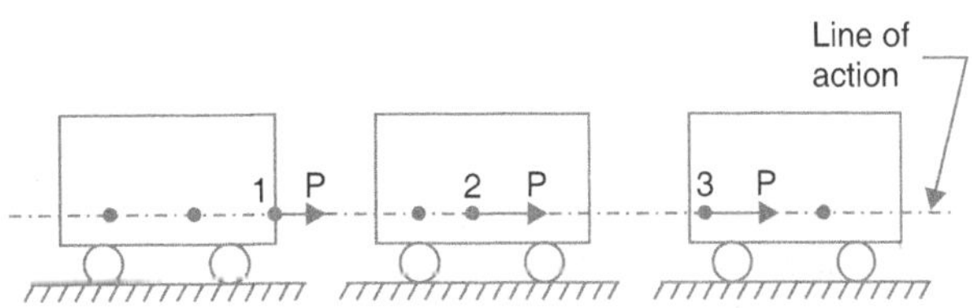

Fig. 2.10

2.9. PARTICLE

A body whose dimensions are practically negligible is called a *particle*. In any problem of mechanics, when the applied forces have no tendency to rotate the body on which they act, the body may be considered as a particle. Forces acting on the particle are concurrent, the point through which they pass being the point representing the particle.

2.10. RESULTANT FORCE

A *resultant force is a single force which can replace two or more forces and produce the same effect on the body as the forces.* It is fundamental principle of mechanics, demonstrated by experiment, that when a force acts on a body which is free to move, the motion of the body is in the direction of the force, and the distance travelled in a unit time depends on the magnitude of the force. Then, for a system of concurrent forces acting on a body, the body will move in the direction of the resultant of that system, and the distance travelled in a unit time will depend on the magnitude of the *resultant*.

2.11. COMPONENT OF A FORCE

As two forces acting simultaneously on a particle acting along directions inclined to each other can be replaced by a single force which produces the same effect as the given force, similarly, a single force can be replaced by two forces acting in directions which will produce the same effect as the given force. This breaking up of a force into two parts is called the *resolution of a force.* The force which is broken into two parts is called the *resolved force* and the parts are called *component forces* or the *resolutes.*

Generally, a force is resolved into the following two types of components :

1. Mutually perpendicular components
2. Non-perpendicular components.

1. Mutually perpendicular components. Let the force P to be resolved is represented in magnitude and direction by oc in Fig. 2.11. Let P_x is the component of force P in the direction oa making an angle α with the direction oc of the force. Complete the rectangle $oacb$. Then the other component P_y at right angle to P_x will be represented by ob which is also equal to ac.

From the right-angled triangle oac

$$P_x = oa = P \cos \alpha$$

$$P_y = ac = P \sin \alpha.$$

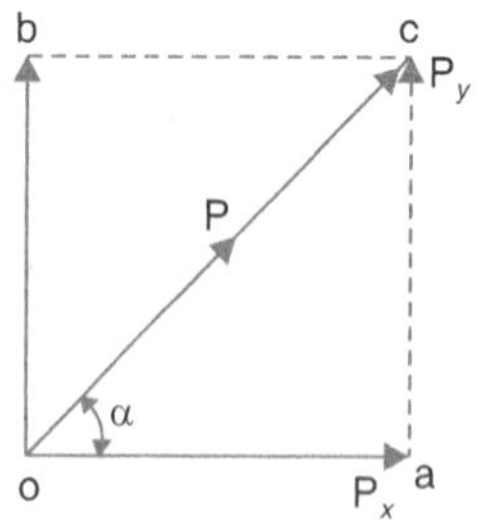

Fig. 2.11

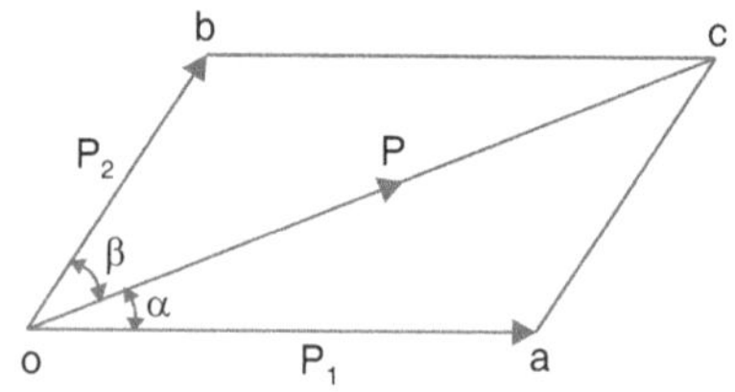

Fig. 2.12

2. Non-perpendicular components. Refer Fig. 2.12. Let oc represents the given force P in magnitude and direction to some scale. Draw oa and ob making angle α and β with oc. Through c draw ca parallel to ob and cb parallel to oa to complete the parallelogram $oacb$. Then the vectors oa and ob represent in magnitude and direction (to the same scale) the components P_1 and P_2 respectively.

Now from the triangle *oac*, by applying sine rule,

$$\frac{oa}{\sin \beta} = \frac{oc}{\sin [180 - (\alpha + \beta)]} = \frac{ac}{\sin \alpha}$$

or
$$\frac{P_1}{\sin \beta} = \frac{P}{\sin (\alpha + \beta)} = \frac{P_2}{\sin \alpha}$$

∴
$$P_1 = P \cdot \frac{\sin \beta}{\sin (\alpha + \beta)} \quad ...(2.1)$$

and
$$P_2 = P \cdot \frac{\sin \alpha}{\sin (\alpha + \beta)} \quad ...(2.2)$$

2.12. PRINCIPLE OF RESOLVED PARTS

The principle of resolved parts states : "*The sum of the resolved parts of two forces acting at a point in any given direction is equal to the resolved parts of their resultant in that direction.*

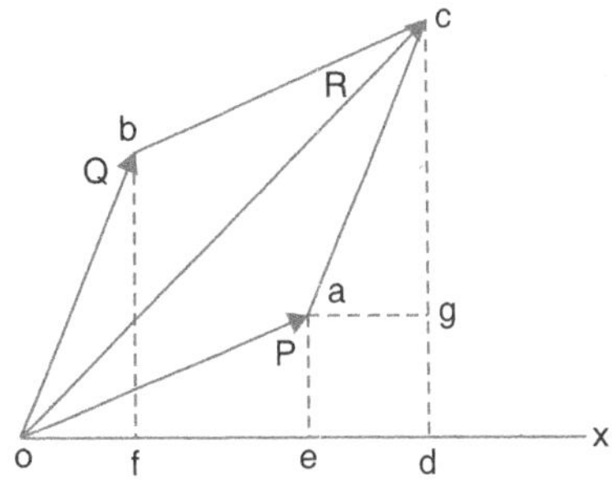

Fig. 2.13

Refer Fig. 2.13. Let the two forces P and Q be represented by the sides *oa* and *ob* of the parallelogram *oacb* and the resultant R of these two forces is given by the diagonal *oc* in magnitude and direction. Let *ox* is the given direction. Draw *bf*, *ae*, *cd* and *ag* perpendicular to *cd*.

Now from the two triangles *obf* and *acg* which are same in all respects, we get

$$of = ag = ed$$

∴
$$od = oe + ed = oe + of$$

But *oe*, *of* and *od* represent the resolved components or parts of the forces P, Q and R respectively in the direction of *ox*.

It may be noted that this principle *holds good for any number of forces.*

2.13. LAWS OF FORCES

The method of determination of the resultant of some forces acting simultaneously on a particle is called *composition of forces.* The various laws used for the composition of forces are given as under :

1. Parallelogram law of forces
2. Triangle law of forces
3. Polygon law of forces.

1. Parallelogram law of forces. It states as under :

"If two forces, acting simultaneously on a particle, be represented in magnitude and direction by the two adjacent sides of a parallelogram then their resultant may be represented in magnitude and direction by the diagonal of the parallelogram which passes through their point of intersection."

Refer Fig. 2.14. Let two forces P and Q acting simultaneously on a particle be represented in magnitude and direction by the adjacent sides *oa* and *ob* of a parallelogram *oacb* drawn from a point *o*, their resultant R will be represented in magnitude and direction by the diagonal *oc* of the parallelogram.

The value of R can be determined either graphically or analytically as explained below :

Graphical method. Draw vectors *oa* and *ob* to represent to some convenient scale the forces *P* and *Q* in magnitude and direction. Complete the parallelogram *oacb* by drawing *ac* parallel to *ob* and *bc* parallel to *oa*. The *vector oc* measured to the same scale will represent the resultant force *R*.

Analytical method. As shown in Fig. 2.14, in the parallelogram *oacb*, from *c* drop a perpendicular *cd* to *oa* at *d* when produced. Now from the geometry of the figure.

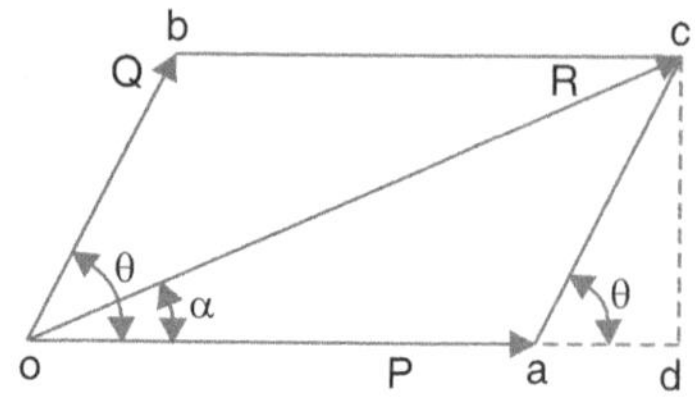

Fig. 2.14

$$\angle cad = \theta,\ ac = Q$$

$\therefore$ $$cd = Q \sin\theta$$

and $$ad = Q\cos\theta$$

From right-angled triangle, *odc*

$$oc = \sqrt{(od)^2 + (cd)^2}$$

$$= \sqrt{(oa + ad)^2 + (cd)^2}$$

or $$R = \sqrt{(P + Q\cos\theta)^2 + (Q\sin\theta)^2}$$

$$= \sqrt{P^2 + Q^2\cos^2\theta + 2PQ\cos\theta + Q^2\sin^2\theta}$$

$$= \sqrt{P^2 + Q^2(\sin^2\theta + \cos^2\theta) + 2PQ\cos\theta}$$

$$= \sqrt{P^2 + Q^2 + 2PQ\cos\theta} \qquad (\because \sin^2\theta + \cos^2\theta = 1)$$

$\therefore$ $$R = \sqrt{P^2 + Q^2 + 2PQ\cos\theta} \qquad ...(2.3)$$

Let the resultant makes an angle α with *P* as shown in figure.

Then $$\tan\alpha = \frac{cd}{od} = \frac{cd}{oa + ad}$$

$$= \frac{Q\sin\theta}{P + Q\cos\theta} \qquad ...(2.4)$$

Case 1. If θ = 0°, *i.e.*, when the forces *P* and *Q* act along the same straight line then equation (2.3) reduces to

$$R = P + Q \qquad (\because \cos 0° = 1)$$

Case 2. If θ = 90°, *i.e.*, when the forces *P* and *Q* act at right angles to each other, then

$$R = \sqrt{P^2 + Q^2} \qquad (\because \cos 90° = 0)$$

Case 3. If θ = 180°, *i.e.*, the forces *P* and *Q* act along the same straight line but in opposite directions, then

$$R = P - Q \qquad (\because \cos 180° = -1)$$

The resultant will act in the direction of the greater force.

2. Triangle law of forces. It states as under :

"If two forces acting simultaneously on a body are represented in magnitude and direction by the two sides of triangle taken in order then their resultant may be represented in magnitude and direction by the third side taken in opposite order."

Let *P* and *Q* be the two coplanar concurrent forces. The resultant force *R* in this case can be obtained with the help of the triangle law of forces both graphically and analytically as given below :

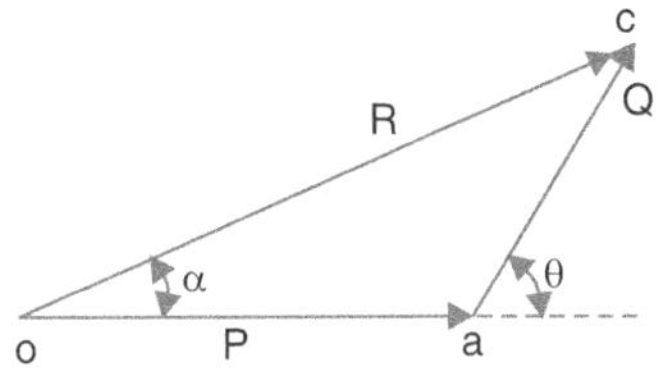

Fig. 2.15

Graphical method. Refer Fig. 2.15. Draw vectors *oa* and *ac* to represent the forces *P* and *Q* to some convenient scale in magnitude and direction. Join *oc* which will represent the resultant force *R* in magnitude and direction to the same scale.

Analytical method. From the geometry of triangle *oac* (Fig. 2.15).

$$\angle coa = \alpha, \angle oca = \theta - \alpha, \angle cao = 180° - \theta$$

$$\therefore \quad \frac{oa}{\sin(\theta - \alpha)} = \frac{ac}{\sin \alpha} = \frac{oc}{\sin(180° - \theta)}$$

or
$$\frac{P}{\sin(\theta - \alpha)} = \frac{Q}{\sin \alpha} = \frac{R}{\sin(180° - \theta)}$$

or
$$\frac{P}{\sin(\theta - \alpha)} = \frac{Q}{\sin \alpha} = \frac{R}{\sin \theta} \quad ...(2.5)$$

3. Polygon law of forces. It states as under :

"If a number of coplanar concurrent forces, acting simultaneously on a body are represented in magnitude and direction by the sides of a polygon taken in order, then their resultant may be represented in magnitude and direction by the closing side of a polygon, taken in the opposite order".

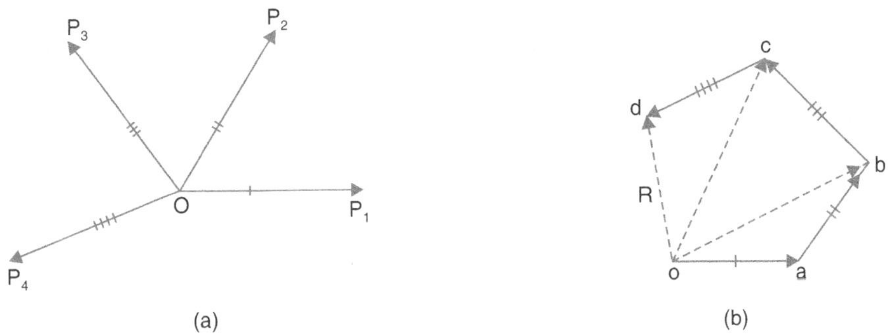

Fig. 2.16

If the forces P_1, P_2, P_3, and P_4 acting simultaneously on a particle be represented in magnitude and direction by the sides *oa, ab, bc* and *cd* of a polygon respectively, their resultant is represented by the closing side *do* in the opposite direction as shown in Fig. 2.16 (*b*).

The law is actually an extension of triangle law of forces. This is so because *ob* is the resultant of *oa* and *ab* and therefore *oc* which is resultant of *ob* and *bc* is also the resultant of *oa, ab* and *bc*. Similarly, *od* is the resultant of *oc* and *cd* and therefore of *ob, bc* and *cd* and finally of *oa, ab, bc* and *cd*.

2.14. RESULTANT OF SEVERAL COPLANAR CONCURRENT FORCES

To determine the resultant of a number of coplanar concurrent forces any of the following two methods may be used :

1. Graphical method (Polygon law of forces)
2. Analytical method (Principle of resolved parts).

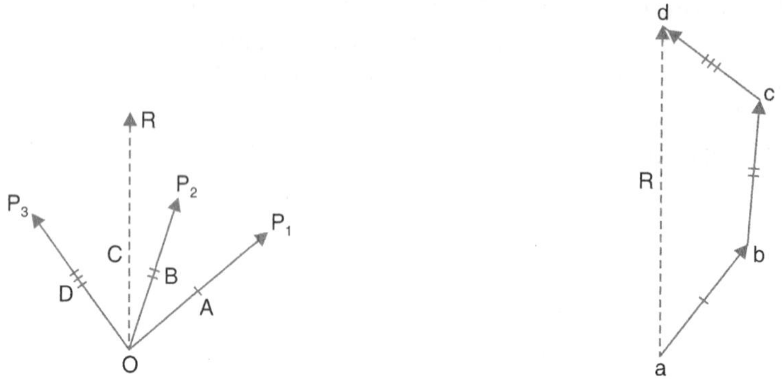

(a)
Space diagram

(b)
Vector diagram

Fig. 2.17

Resultant by graphical method. Fig. 2.17 (*a*) shows the forces P_1, P_2 and P_3 simultaneously acting at a particle O.

Draw a vector *ab* equal to force P_1 to some suitable scale and parallel to the line of action of P_1.

From '*b*' draw vector *bc* to represent force P_3 in magnitude and direction.

Now from '*c*' draw vector *cd* equal and parallel to force P_3. Join *ad* which gives the required resultant in magnitude and direction, the direction being *a* to *d* as shown in the vector diagram.

Resultant by analytical method. Refer Fig. 2.18.

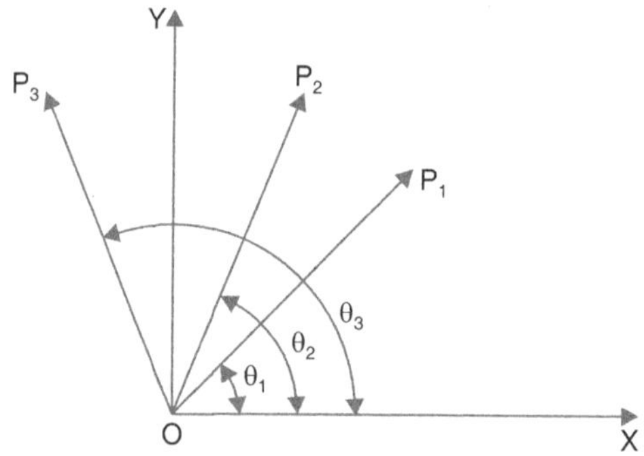

Fig. 2.18

The resolved parts in the direction OX and OY of

P_1 are $P_1 \cos \theta_1$ and $P_1 \sin \theta_1$, respectively,

P_2 are $P_2 \cos \theta_2$ and $P_2 \sin \theta_2$ respectively

and P_3 and $P_3 \cos \theta_3$ and $P_3 \sin \theta_3$ respectively.

If the resultant R makes an angle θ with OX then by the principle of resolved parts :

$$R \cos \theta = P_1 \cos \theta_1 + P_2 \cos \theta_2 + P_3 \cos \theta_3 = \Sigma H \quad ...(i)$$

and

$$R \sin \theta = P_1 \sin \theta_1 + P_2 \sin \theta_2 + P_3 \sin \theta_3 = \Sigma V \quad ...(ii)$$

Now, by squaring and adding eqns. (*i*) and (*ii*), we get

$$R = \sqrt{(\Sigma H)^2 + (\Sigma V)^2} \qquad ...(2.6)$$

and by dividing eqn. (*ii*) by eqn. (*i*), we get

$$\frac{R \sin \theta}{R \cos \theta} = \tan \theta = \frac{\Sigma V}{\Sigma H}$$

$$\theta = \tan^{-1}\left(\frac{\Sigma V}{\Sigma H}\right) \qquad ...(2.7)$$

It may be noted that while solving problems proper care must be taken about the signs (+ve or –ve) of the resolved parts. Following sign conventions may be kept in view :

Vertical components

Upward direction ↑ Positive (+)

Downward direction ↓ Negative (–)

Horizontal components

Directions :
: From left to right → Positive (+)
: From right to left ← Negative (–)

Example 2.1. *Find the magnitude and direction of the resultant of two forces 40 N and 60 N acting at a point with an included angle of 40° between them. The force of 60 N being horizontal.*

Sol. Refer to Fig. 2.14,

$$P = 60 \text{ N}, Q = 40 \text{ N}, \theta = 40°$$

Using the relation,

$$R = \sqrt{P^2 + Q^2 + 2PQ \cos \theta}$$

$$= \sqrt{(60)^2 + (40)^2 + 2 \times 60 \times 40 \times \cos 40°}$$

$$= \sqrt{3600 + 1600 + 4800 \times 0.766}$$

$$= \mathbf{94.22 \text{ N}.}$$

Hence **magnitude** of the resultant force **= 94.22 N. (Ans.)**

$$\tan \alpha = \frac{Q \sin \theta}{P + Q \cos \theta} = \frac{40 \sin 40°}{60 + 40 \cos 40°}$$

$$= \frac{25.71}{60 + 30.64} = 0.284$$

$$\therefore \qquad \alpha = 15.85° \text{ or } 15° \, 51'$$

Hence the direction of the resultant force

= 15° 51′ with the 60 N force. (Ans.)

Example 2.2. *The angle between the two forces of magnitude 20 N and 15 N is 60° ; the 20 N force being horizontal. Determine the resultant in magnitude and direction, if*

(*a*) *the forces are pulls ; and*

(*b*) *the 15 N force is a push and 20 N force is a pull.*

Sol. Case (*a*). Refer to Fig. 2.19.

$P = 20$ N, $Q = 15$ N,

$\theta = 60°$

Using the relation,

$$R = \sqrt{P^2 + Q^2 + 2PQ \cos \theta}$$

$$= \sqrt{(20)^2 + (15)^2 + 2 \times 20 \times 15 \times \cos 60°}$$

$$= \sqrt{400 + 225 + 600 \times 0.5}$$

$$= \mathbf{30.4\ N.\ (Ans.)}$$

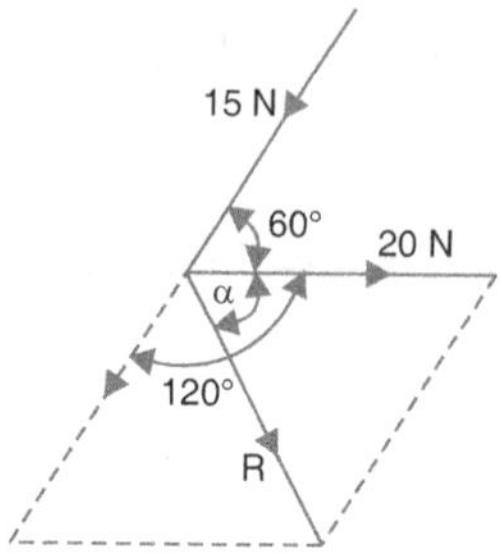

Fig. 2.19

$$\tan \alpha = \frac{Q \sin \theta}{P + Q \cos \theta} = \frac{15 \times \sin 60°}{20 + 15 \cos 60°}.$$

or $$\tan \alpha = \frac{12.99}{20 + 7.5}$$

$$\alpha = \tan^{-1}\left(\frac{12.99}{20 + 7.5}\right)$$

or $$\alpha = \mathbf{32.05° = 25°\ 3'\ with\ 20\ N\ force.\ (Ans.)}$$

(*b*) Refer to Fig. 2.19

$$\boldsymbol{R} = \sqrt{(20)^2 + (15)^2 + 2 \times 20 \times 15 \cos 120°}$$

$$= \sqrt{400 + 225 - 300} = \mathbf{18\ N.\ (Ans.)}$$

$$\tan \alpha = \frac{15 \sin 120°}{20 + 15 \cos 120°} = \frac{12.99}{20 - 15 \times 0.5}$$

$$= 1.039$$

or $$\alpha = \tan^{-1} (1.039)$$

or $$\alpha = \mathbf{46.1°\ or\ 46°\ 6'\ with\ 20\ N\ force.\ (Ans.)}$$

Example 2.3. *The resultant of two forces P and 30 N is 40 N inclined at 60° to the 30 N force. Find the magnitude and direction of P.*

Sol. Refer to Fig. 2.20.

From the knowledge of trigonometry, we know that in ΔOAB,

$$AB^2 = OA^2 + OB^2 - 2.OA\ .\ OB \cos 60°$$

$$P^2 = (30)^2 + (40)^2 - 2 \times 30 \times 40 \cos 60°$$

$$\boldsymbol{P} = \sqrt{900 + 1600 - 2400 \times 0.5}$$

$$= \mathbf{36.06\ N.\ (Ans.)}$$

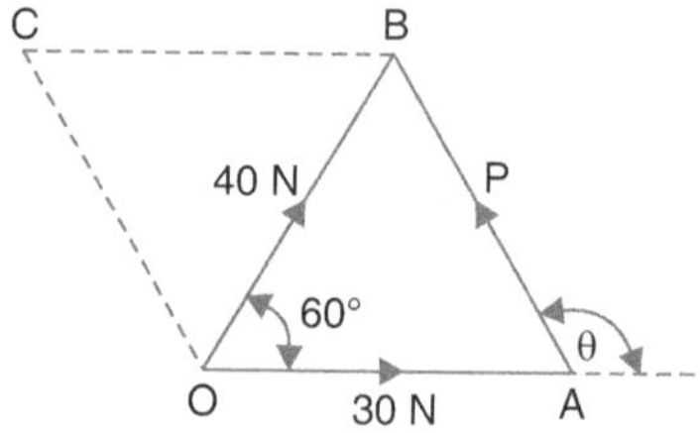

Fig. 2.20

Now applying sine rule,

$$\frac{P}{\sin 60°} = \frac{40}{\sin (180° - \theta)}$$

$$\frac{36.06}{0.866} = \frac{40}{\sin (180° - \theta)}$$

$$\therefore \quad \sin(180° - \theta) = \frac{40 \times 0.866}{36.06} = 0.96$$

$$180 - \theta = 73.74° \quad \text{or} \quad 73° 44'$$

or $$= \mathbf{106° 16'. (Ans.)}$$

Example 2.4. *Find the magnitude of two forces such that, if they act at right angles, their resultant is 5 N whilst when they act at an angle of 60°, their resultant is $\sqrt{37}$ N.*

Sol. Let the two forces be P and Q, when they act at right angles to each other then the resultant

$$R = \sqrt{P^2 + Q^2}$$

or $$R^2 = P^2 + Q^2$$

or $$25 = P^2 + Q^2 \quad ...(i)$$

In the second case, when the forces act at an angle of 60° then the resultant

$$R = \sqrt{P^2 + Q^2 + 2PQ\cos 60°}$$

or $$R^2 = P^2 + Q^2 + 2PQ\cos 60°$$

or $$(\sqrt{37})^2 = P^2 + Q^2 + 2PQ \times \frac{1}{2}$$

or $$37 = P^2 + Q^2 + PQ \quad ...(ii)$$

Substracting eqn. (*i*) from eqn. (*ii*), we get

$$12 = PQ \quad ...(iii)$$

Multiply eqn. (*iii*) by 2 and add to eqn. (*i*),

$$25 + 12 \times 2 = P^2 + Q^2 + 2PQ$$

or $$25 + 24 = P^2 + Q^2 + 2PQ$$

or $$49 = (P + Q)^2$$

or $$7 = P + Q \quad ...(iv)$$

From eqns. (*iii*) and (*iv*), we get

$$\boldsymbol{P = 3\ \text{N}}$$

$$\boldsymbol{Q = 4\ \text{N. (Ans.)}}$$

Example 2.5. *Two forces of magnitudes 3P, 2P respectively acting at a point have a resultant R. If the first force is doubled, the magnitude of the resultant is doubled. Find the angle between the forces.*

Sol. Let θ = angle between the two forces ($3P$, $2P$).

We know that,

$$R^2 = (3P)^2 + (2P)^2 + 2(3P)(2P)\cos\theta$$

$$= 9P^2 + 4P^2 + 12P^2\cos\theta$$

$$= 13P^2 + 12P^2\cos\theta \quad ...(i)$$

On doubling the first force and doubling the resultant force, we have

$$(2R)^2 = (6P)^2 + (2P)^2 + 2(6P)(2P)\cos\theta$$

$$4R^2 = 36P^2 + 4P^2 + 24P^2\cos\theta$$

$$= 40P^2 + 24P^2\cos\theta \quad ...(ii)$$

Substituting the value of R^2 from eqn. (*i*) in eqn. (*iii*), we get

$$4(13P^2 + 12P^2\cos\theta) - 40P^2 + 24P^2\cos\theta$$

$$52P^2 + 48P^2\cos\theta = 40^2 + 24P^2\cos\theta$$

$$24P^2 \cos \theta = -12P^2$$

$$\cos \theta = -\frac{12}{24} = -\frac{1}{2}$$

$\therefore$ $\theta = \mathbf{120°}$. **(Ans.)**

Example 2.6. *Two forces P and Q acting at a point have a resultant R. If Q be doubled, R is doubled. Again if the direction of Q is reversed, then R is doubled, show that*

$$P : Q : R = \sqrt{2} : \sqrt{3} : \sqrt{2}.$$

Sol. We know that

$$R^2 = P^2 + Q^2 + 2PQ \cos \theta \quad ...(i)$$

If Q is doubled the resultant is also doubled,

$$(2R)^2 = P^2 + (2Q)^2 + 2P(2Q) \cos \theta$$

$$4R^2 = P^2 + 4Q^2 + 4PQ \cos \theta \quad ...(ii)$$

If the direction of Q is reversed, the resultant is again doubled,

$$(2R)^2 = P^2 + (-Q)^2 + 2P(-Q) \cos \theta$$

$$4R^2 = P^2 + Q^2 - 2PQ \cos \theta \quad ...(iii)$$

Adding eqns. (*i*) and (*iii*), we get

$$5R^2 = 2P^2 + 2Q^2 \quad ...(iv)$$

Multiplying eqn. (*iii*) by 2 and adding to eqn. (*ii*), we get

$$12R^2 = 3P^2 + 6Q^2$$

$$4R^2 = P^2 + 2Q^2 \quad ...(v)$$

Subtracting eqn. (*v*) from eqn. (*iv*), we get

$$R^2 = P^2$$

or $$R = P$$

Substituting the value of R in eqn. (*v*),

$$4P^2 = P^2 + 2Q^2$$

or $$3P^2 = 2Q^2$$

or $$\frac{3}{2}P^2 = Q^2$$

or $$\sqrt{\frac{3}{2}}P = Q$$

Now $$P : Q : R$$

or $$P : \sqrt{\frac{3}{2}}P : P$$

or $$1 : \sqrt{\frac{3}{2}} : 1$$

or $$\sqrt{2} : \sqrt{3} : \sqrt{2}. \textbf{ Proved.}$$

Example 2.7. *Two equal weights of 10 N are attached to the ends of a thin string which passes over three smooth pegs in a wall arranged in the form of an equilateral triangle with one side horizontal. Find the pressure on each peg.*

Sol. Refer to Fig. 2.21.

Let *A*, *B* and *C* be three pegs.

Let R_A, R_B and R_C be the reactions at *A*, *B* and *C* and *T* be the tension.

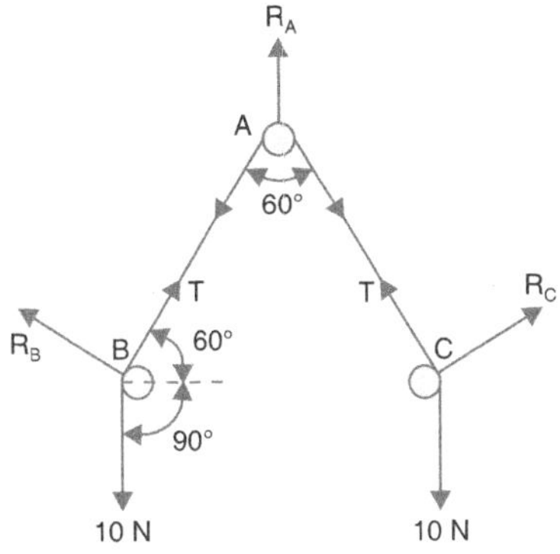

Fig. 2.21

$$T = 10 \text{ N}$$

We know that $R^2 = P^2 + Q^2 + 2PQ \cos \theta$

$$\therefore \quad R_B^2 = (10)^2 + (10)^2 + 2 \times 10 \times 10 \cos (90° + 60°)$$

$$= 100 + 100 - 200 \sin 60°$$

$$= 200 - 200 \times \frac{\sqrt{3}}{2}$$

$$= 200 \left(1 - \frac{\sqrt{3}}{2}\right) = 200 (1 - 0.866) = 26.8$$

$$\boldsymbol{R_B} = \sqrt{26.8} = \mathbf{5.17 \text{ N. (Ans.)}}$$

Hence $R_B = R_C = 5.17$ N

Again, $R_A^2 = (10)^2 + (10)^2 + 2 \times 10 \times 10 \times \cos 60°$

$$= 100 + 100 + 200 \times \frac{1}{2} = 300$$

$$\boldsymbol{R_A} = \sqrt{300}$$

$$= \mathbf{17.32 \text{ N. (Ans.)}}$$

Example 2.8. *Find the components of a force of 150 N into two directions inclined at angle of 45° and 30° with the force.*

Sol. Refer to Fig. 2.22.

$$P = 150 \text{ N}, \angle \alpha = 45°, \angle \beta = 30°$$

$$P_1 = P \cdot \frac{\sin \beta}{\sin (\alpha + \beta)} = 150 \times \frac{\sin 30°}{\sin (45° + 30°)}$$

$$= 150 \times \frac{0.5}{0.966}$$

$$= \mathbf{77.63 \text{ N. (Ans.)}}$$

$$P_2 = P \cdot \frac{\sin \alpha}{\sin (\alpha + \beta)}$$

$$= 150 \frac{\sin 45°}{\sin (45° + 30°)} = \frac{150 \times 0.707}{0.966}$$

$$= \mathbf{109.78 \text{ N. (Ans.)}}$$

Example 2.9. *A particle is acted upon by the following forces :*

(*i*) *A pull of 8 N due North East ;*

(*ii*) *A pull of 10 N due North ;*

(*iii*) *A pull of 12 N due East ;*

(*iv*) *A pull of 4 N in a direction inclined 60° South of West ;*

(*v*) *A pull of 6 N in a direction inclined 30° East of South.*

Find graphically the magnitude and direction of the resultant force.

Sol. Draw space diagram as in Fig. 2.22 (*a*) showing relative positions of the lines of action of the various forces acting on point *O*.

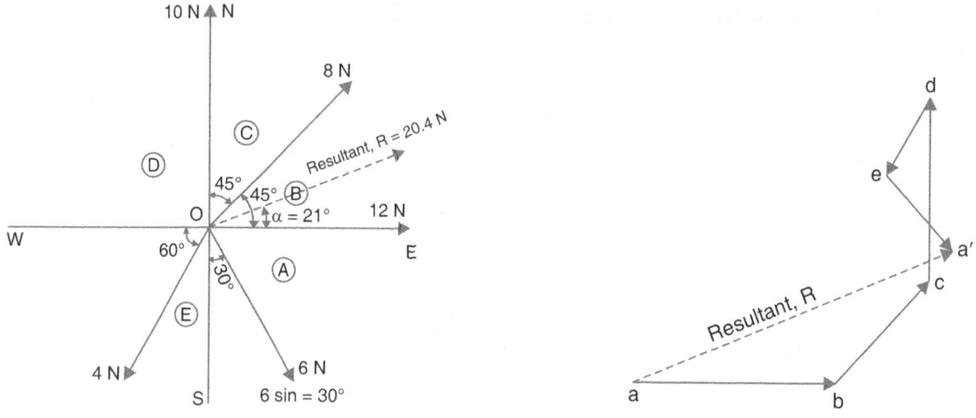

(a) Space diagram (b) Vector diagram

Fig. 2.22

Draw vectors *ab, bc, cd, de, ea* to represent to some scale, the forces 12 N, 8 N, 10 N, 4 N and 6 N respectively [Fig. 2.22 (*b*)].

Join *aa′* which represents the resultant in magnitude and direction

$$\boldsymbol{R = 20.4 \text{ N}}$$

$$\alpha = \mathbf{21°}.$$

So the resultant is a **20.4 N pull acting at 21° North of East. (Ans.)**

Example 2.10. *Determine analytically the magnitude and direction of the resultant of the following four forces acting at a point :*

(*i*) *10 N pull N 30° E ;*

(*ii*) *12.5 N push S 45° W ;*

(*iii*) *5 N push N 60° W ;*

(*iv*) *15 N push S 60° E.*

Sol. The various forces acting at a point are shown in Fig. 2.23.

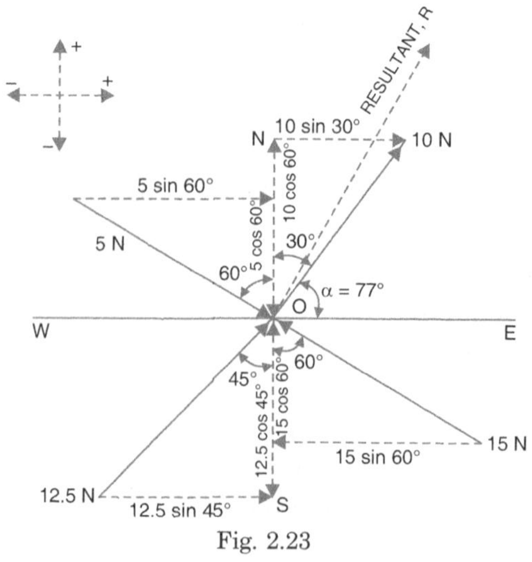

Fig. 2.23

Resolving the forces horizontally, we get

$$\Sigma H = 10 \sin 30° + 5 \sin 60° + 12.5 \sin 45° - 15 \sin 60°$$
$$= 10 \times 0.5 + 5 \times 0.866 + 12.5 \times 0.707 - 15 \times 0.866$$
$$= 5 + 4.33 + 8.84 - 12.99$$
$$= 5.18 \text{ N}.$$

Similarly, resolving forces vertically, we get

$$\Sigma V = 10 \cos 30° - 5 \cos 60° + 12.5 \cos 45° + 15 \cos 60°$$
$$= 10 \times 0.866 - 5 \times 0.5 + 12.5 \times 0.707 + 15 \times 0.5$$
$$= 8.66 - 2.5 + 8.84 + 7.5 = 22.5 \text{ N}.$$

∴ Resultant,

$$\boldsymbol{R} = \sqrt{(\Sigma H)^2 + (\Sigma V)^2}$$
$$= \sqrt{(5.18)^2 + (22.5)^2}$$
$$= \sqrt{26.83 + 506.25}$$
$$= \mathbf{23.09\ N.\ (Ans.)}$$

$$\tan \alpha = \frac{\Sigma V}{\Sigma H} = \frac{22.5}{5.18} = 4.34$$

∴ α = **77°. (Ans.)**

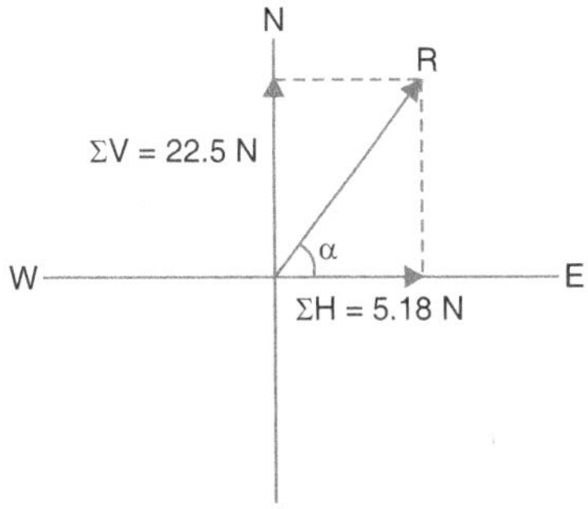

Fig. 2.24

Example 2.11. *The following forces (all pull) act at a point :*

(i) 25 N due North ; *(ii) 10 N North-East ;*

(iii) 15 N due East ; *(iv) 20 N 30° East of South ;*

(v) 30 N 60° South of West.

Find the resultant force. What angle does it make with East ?

Sol. The various forces acting at a point are shown in Fig. 2.25.

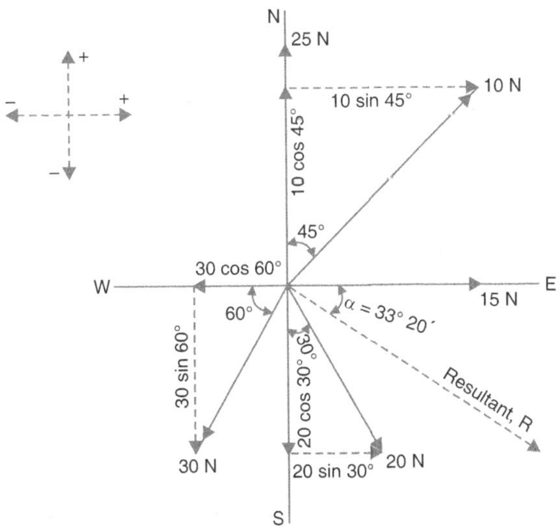

Fig. 2.25

Resolving the forces horizontally, we get

$$\Sigma H = 10 \sin 45° - 30 \cos 60° + 20 \sin 30° + 15$$
$$= 10 \times 0.707 - 30 \times 0.5 + 20 \times 0.5 + 15$$
$$= 7.07 - 15 + 10 + 15$$
$$= 17.07 \text{ N.}$$

Similarly, resolving forces *vertically*, we get

$$\Sigma V = 10 \cos 45° + 25 - 30 \sin 60° - 20 \cos 30°$$
$$= 10 \times 0.707 + 25 - 30 \times 0.866 - 20 \times 0.866$$
$$= 7.07 + 25 - 25.98 - 17.32$$
$$= -11.23 \text{ N.}$$

Resultant, $\boldsymbol{R} = \sqrt{(\Sigma H)^2 + (\Sigma V)^2}$

$$= \sqrt{(17.07)^2 + (-11.23)^2}$$
$$= \sqrt{291.38 + 126.11}$$
$$= \mathbf{20.43\ N\ .\ (Ans.)}$$

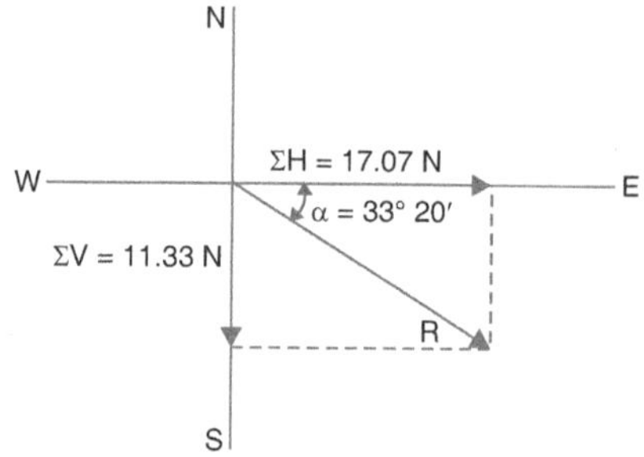

Fig. 2.26

$$\tan \alpha = \frac{\Sigma V}{\Sigma H} = \frac{11.23}{17.07} = 0.6578$$

or $\alpha =$ **33° 20′ South of East. (Ans.)**

Example 2.12. *ABCDEF is a regular hexagon. Forces of magnitudes 2, $4\sqrt{3}$, 8, $2\sqrt{3}$ and 4 N act at A in the directions of AB, AC, AD, AE and AF respectively. Determine the resultant completely.*

Sol. Refer to Fig. 2.27.

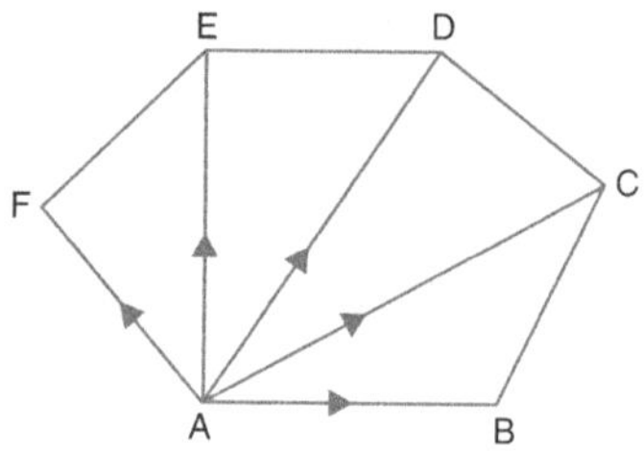

Fig. 2.27

In *ABCDEF* regular hexagon *AE* and *AB* are perpendicular to each other.

Let us resolve the forces along *AB* and *AE*.

Forces along *AB*

$$= 2 + 4\sqrt{3} \cos 30° + 8 \cos 60° + 4 \cos 120°$$
$$= 2 + 6 + 4 - 2 = 10 \text{ N}$$

Forces along $AE = 2\sqrt{3} + 4\sqrt{3} \sin 30° + 8 \sin 60° + 4 \sin 120°$

$$= 2\sqrt{3} + 2\sqrt{3} + 4\sqrt{3} + 2\sqrt{3} = 10\sqrt{3} \text{ N.}$$

Now, 10 N act along *AB* and $10\sqrt{3}$ N act along *AE*, and their resultant force

$$= \sqrt{(10)^2 + (10\sqrt{3})^2} = \sqrt{100 + 300}$$
$$= \mathbf{20\ N.\ (Ans.)}$$

Let α be the angle between the resultant force and the horizontal direction *AB*.

Then, $\tan \alpha = \dfrac{10\sqrt{3}}{10} = \sqrt{3}$; $\tan \alpha = \tan 60°$

$\therefore$ $\alpha =$ **60°. (Ans.)**

2.15. EQUILIBRIUM CONDITIONS FOR COPLANAR CONCURRENT FORCES

When several forces act on a particle, the particle is said to be in *equilibrium* if there *is no unbalanced* forces acting on it, *i.e.*, the resultant of all the forces acting on the particle is *zero*.

Analytical and graphical conditions of equilibrium of coplanar concurrent forces are given as under :

Analytical conditions :

1. The algebraic sum of components of all the forces in any direction which may be taken as horizontal, in their plane must be zero. Mathematically, $\Sigma H = 0$.

2. The algebraic sum of components of all the forces in a direction perpendicular to the first direction, which may be taken as *vertical*, in their plane, must be zero. Mathematically, $\Sigma V = 0$.

Graphical conditions. The force polygon, *i.e.*, force or vector diagram *must close.*

2.16. LAMI'S THEOREM

It states as under :

"If three coplanar forces acting on a point in a body keep it in equilibrium, then each force is proportional to the sine of the angle between the other two forces."

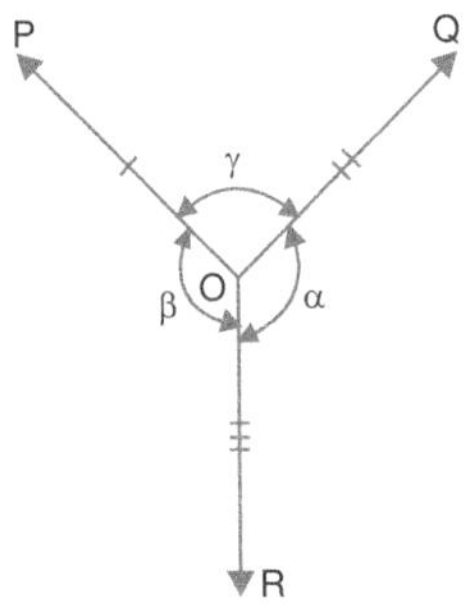

Fig. 2.28

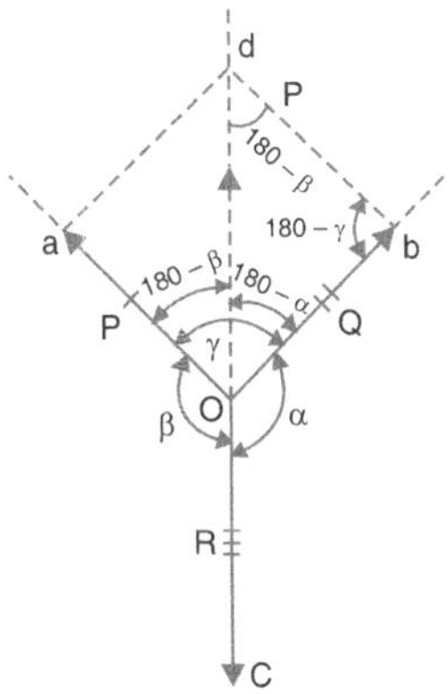

Fig. 2.29

Fig. 2.28 shows three forces P, Q and R acting at a point O. Let the angle between P and Q be γ, between Q and R be α and between R and P be β. If these forces are in equilibrium then according to Lami's theorem :

$$\frac{P}{\sin \alpha} = \frac{Q}{\sin \beta} = \frac{R}{\sin \gamma} \qquad ...(2.8)$$

Proof. Let us first consider the two forces P and Q which are represented by the two sides *oa* and *ob* of a parallelogram *oadb* as shown in Fig. 2.29. Then the resultant of these two *forces* will be given by *od* (the diagonal of the parallelogram) in magnitude and direction. This means *od* should be equal to R in magnitude but opposite in direction to *oc* as P, Q and R are in equilibrium.

From geometry of parallelogram *oadb*, we find

$$bd = P \text{ and } ad = Q$$

$$\angle bod = (180 - \alpha)$$

and $$\angle bdo = \angle aod = \angle(180 - \beta)$$

Now, from triangle *obd*,

$$\angle obd = 180° - \angle bod - \angle bdo$$

($\because$ Sum of all the angles of triangle = 180°)

$$= 180 - (180 - \alpha) - (180 - \beta) = \alpha + \beta - 180°$$

But $\alpha + \beta + \gamma = 360°$ ($\because$ Sum of all the angles at a point = 360°)

Subtracting 180° from both the sides, we get

$$\alpha + \beta + \gamma - 180° = 360° - 180°$$

$$(\alpha + \beta - 180°) + \gamma = 180°$$

$$\angle obd + \gamma = 180°$$

$$\angle obd = 180 - \gamma$$

Now, applying sine formula to triangle *obd*,

$$\frac{bd}{\sin(180 - \alpha)} = \frac{ob}{\sin(180 - \beta)} = \frac{od}{\sin(180 - \gamma)}$$

$$\frac{P}{\sin(180 - \alpha)} = \frac{Q}{\sin(180 - \beta)} = \frac{R}{\sin(180 - \gamma)}$$

Hence $$\frac{P}{\sin\alpha} = \frac{Q}{\sin\beta} = \frac{R}{\sin\gamma}.$$

Example 2.13. *A wheel has five equally spaced radial spokes, all in tension. If the tensions of three consecutive spokes are 500 N, 700 N and 600 N respectively, find the tensions in the other two spokes.*

Sol. Refer to Fig. 2.30.

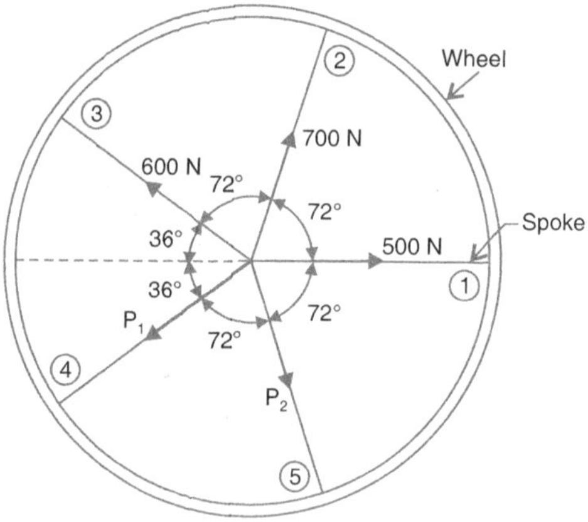

Fig. 2.30

Let P_1 and P_2 be the tensions in the spokes (4) and (5) respectively.

Resolving the forces horizontally, we get

$$\Sigma H = 500 + 700\cos 72° - 600\cos 36° - P_1\cos 36° + P_2\cos 72°$$

But $\Sigma H = 0$ since the *spokes are in equilibrium* (horizontally)

$\therefore$ $500 + 700\cos 72° - 600\cos 36° - P_1\cos 36° + P_2\cos 72° = 0$

or $500 + 700 \times 0.309 - 600 \times 0.809 - P_1 \times 0.809 + P_2 \times 0.309 = 0$

or $500 + 216.3 - 485.4 - 0.809\,P_1 + 0.309\,P_2 = 0$

or $0.809\,P_1 - 0.309\,P_2 = 230.9$

or $P_1 - 0.38\,P_2 = 285.4$ ...(*i*) (Dividing both sides by 0.809)

Similarly, resolving the forces vertically, we get

$$\Sigma V = 700 \sin 72° + 600 \sin 36° - P_1 \sin 36° - P_2 \sin 72°$$

But $\Sigma V = 0$, since the *spokes are in equilibrium* (vertically)

$\therefore$ $700 \sin 72° + 600 \sin 36° - P_1 \sin 36° - P_2 \sin 72° = 0$

or $700 \times 0.951 + 600 \times 0.588 - P_1 \times 0.588 - P_2 \times 0.951 = 0$

or $665.7 + 352.8 - 0.588\, P_1 - 0.951\, P_2 = 0$

or $0.588\, P_1 + 0.951\, P_2 = 1018.5$

or $P_1 + 1.62\, P_2 = 1732.1$ ...(*ii*)

(Dividing both sides by 0.588)

Subtracting (*i*) from (*ii*), we get

$$2.0\, P_2 = 1446.7$$

$$\boldsymbol{P_2 = 723.3 \text{ N. (Ans.)}}$$

Putting the value of P_2 in (*i*), we get

$$P_1 - 0.38 \times 723.3 = 285.4$$

or $\boldsymbol{P_1 = 560.2 \text{ N. (Ans.)}}$

Example 2.14. *Three forces keep a particle in equilibrium. One acts towards east, another towards north-west and the third towards south. If the first be 5 N, find the other two.*

Sol. Refer to Fig. 2.31.

Let force P act towards north-west and Q towards south. On applying Lami's theorem, we get

$$\frac{P}{\sin 90°} = \frac{Q}{\sin 135°} = \frac{5}{\sin 135°}$$

$$\boldsymbol{P} = \frac{5 \sin 90°}{\sin 135°} = \boldsymbol{5\sqrt{2} \text{ N. (Ans.)}}$$

$$\boldsymbol{Q} = \frac{5 \sin 135°}{\sin 135°} = \boldsymbol{5 \text{ N. (Ans.)}}$$

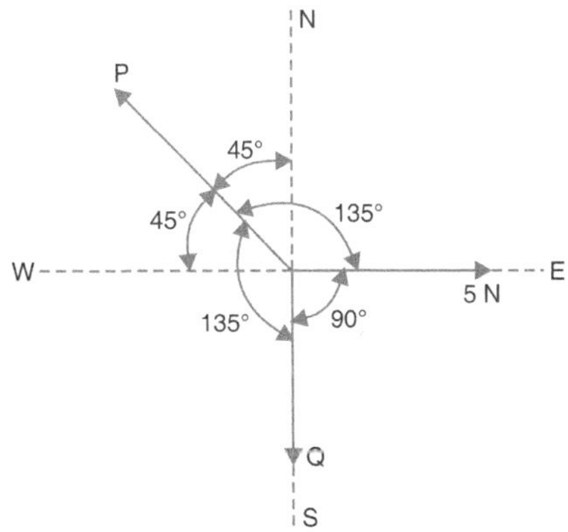

Fig. 2.31

Example 2.15. *A machine weighing 1500 N is supported by two chains attached to some point on the machine. One of these ropes goes to the eye bolts in the wall and is inclined 30° to the horizontal and other goes to the hook in ceiling and is inclined at 45° to the horizontal. Find the tensions in the two chains.*

Sol. The machine is in equilibrium under the following forces :

(*i*) W (weight of the machine) acting vertically down ;

(*ii*) Tension T_1 in the chain OA ;

(*iii*) Tension T_2 in the chain OB.

Now, applying Lami's theorem at O, we get

$$\frac{T_1}{\sin (90° + 45°)} = \frac{T_2}{\sin (90° + 30°)} = \frac{W}{\sin 105°}$$

$$\frac{T_1}{\sin 135°} = \frac{T_2}{\sin 120°} = \frac{1500}{\sin 105°}$$

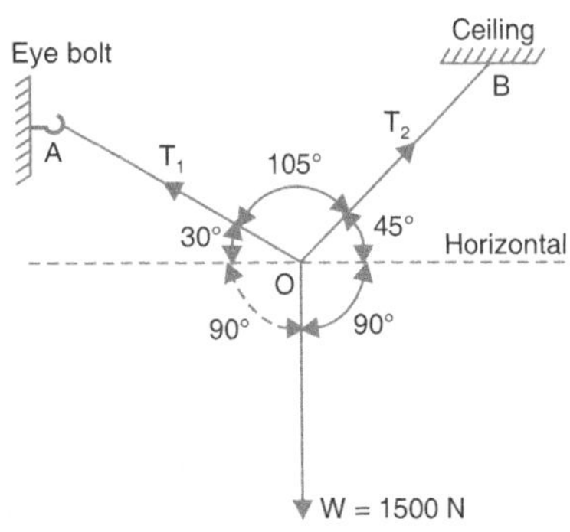

Fig. 2.32

$$\therefore \quad T_1 = \frac{1500 \sin 135°}{\sin 105°} = \frac{1500 \times 0.707}{0.965}$$

$$= \textbf{1098.96 N. (Ans.)}$$

and $$T_2 = \frac{1500 \times \sin 120°}{\sin 105°} = \frac{1500 \times 0.866}{0.965}$$

$$= \textbf{1346.11 N. (Ans.)}$$

Example 2.16. *Fig. 2.33 represents a weight of 20 kN supported by two cords, one 3 m long and the other 4 m long with points of support 5 m apart. Find the tensions T_1 and T_2 in kN in the cords.*

Sol. Refer to Fig. 2.33.

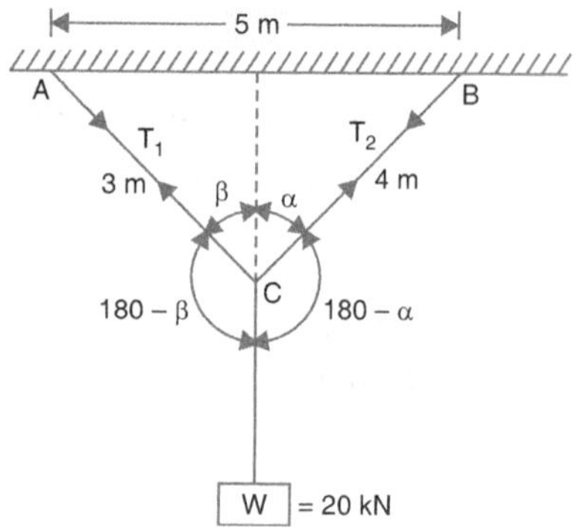

Fig. 2.33

In ΔABC, $\angle C$ in a right angle because $AB^2 = AC^2 + BC^2$

i.e., $$\begin{bmatrix} (5)^2 = (3)^2 + (4)^2 \\ 25 = 9 + 16 = 25 \end{bmatrix}$$ or

Also by sine rule :

$$\frac{3}{\sin \angle B} = \frac{4}{\sin \angle A} = \frac{5}{\sin \angle C}$$

$$\frac{3}{\sin \angle B} = \frac{4}{\sin \angle A}$$

$$= \frac{5}{\sin 90°}$$

$$\therefore \quad \sin \angle B = 3/5 = 0.6$$

or $$\angle B = 36° \ 52'$$

and $$\sin \angle A = 4/5 = 0.8$$

or $$\angle A = 53° \ 8'$$

Now, $$\angle \alpha = 90 - \angle B = 90 - 36° \ 52' \text{ or } 53° \ 8'$$

and $$\angle \beta = 90° - \angle A = 90 - 53° \ 8' \text{ or } 36° \ 52'$$

Using Lami's theorem at C, we get

$$\frac{T_1}{\sin (180 - \alpha)} = \frac{T_2}{\sin (180 - \beta)} = \frac{W}{\sin (\alpha + \beta)}$$

$$\frac{T_1}{\sin \alpha} = \frac{T_2}{\sin \beta} = \frac{W}{\sin (\alpha + B)}$$

$$\frac{T_1}{\sin 53° 8'} = \frac{T_2}{\sin 36° 52'} = \frac{20}{\sin 90°}$$

$$\therefore \quad T_1 = \frac{20}{\sin 90°} \times \sin 53° \ 8'$$

$$= \textbf{16 kN. (Ans.)}$$

and $$T_2 = \frac{20}{\sin 90°} \times \sin 36° \ 52'$$

$$= \textbf{12 kN. (Ans.)}$$

Example 2.17. *A smooth sphere of weight 'W' is supported in contact with a smooth vertical wall by a string fastened to a point on its surface, the end being attached to a point on the wall. If the length of the string is equal to the radius of sphere, find tensions in the string and reaction on the wall.*

Sol. Refer to Fig. 2.34.

The sphere is in equilibrium under the action of following forces :

(*i*) Self weight '*W*' acting vertically downwards through the centre of sphere.

(*ii*) Tension *T* in the string *AC*.

(*iii*) Reaction 'R_B' of the wall at the point of contact *B* of the sphere, acting perpendicular to surface of wall as shown because the wall is smooth.

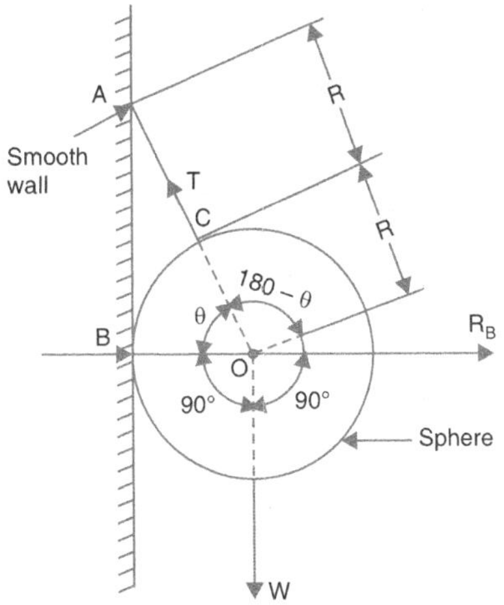

Fig. 2.34

Now, applying Lami's theorem at *O*, we get

$$\frac{T}{\sin 90^\circ} = \frac{R_B}{\sin (90^\circ + \theta)} = \frac{W}{\sin (180^\circ - \theta)}$$

$$\left(\begin{array}{c} \because \text{ In } \Delta AOB, \\ \cos\theta = \dfrac{R}{2R} = \dfrac{1}{2} \\ \text{or } \theta = 60^\circ \end{array}\right)$$

or $$\frac{T}{1} = \frac{R_B}{\sin (90^\circ + 60^\circ)} = \frac{W}{\sin (180^\circ - 60^\circ)}$$

or $$\frac{T}{1} = \frac{R_B}{\sin 150^\circ} = \frac{W}{\sin 120^\circ}$$

$\therefore$ $$\boldsymbol{T} = \frac{W}{\sin 120^\circ} = \frac{W}{\sqrt{3}/2} = \frac{\mathbf{2}}{\sqrt{\mathbf{3}}}\ \mathbf{W.\ (Ans.)}$$

and $$\boldsymbol{R_B} = \frac{W \sin 150^\circ}{\sin 120^\circ} = \frac{W \times 1/2}{\sqrt{3}/2} = \mathbf{W/\sqrt{3}\ .\ (Ans.)}$$

Example 2.18. *A string is tied to two point at the same level and a smooth ring of weight W, which can slide freely along the string, is pulled by horizontal force P. If, in the position of equilibrium the portions of the string are inclined at 60° and 30° to the vertical. Find the value of P and the tension in the string.*

Sol. Refer to Fig. 2.35.

Let *C* be the position of the ring. Considering horizontal equilibrium,

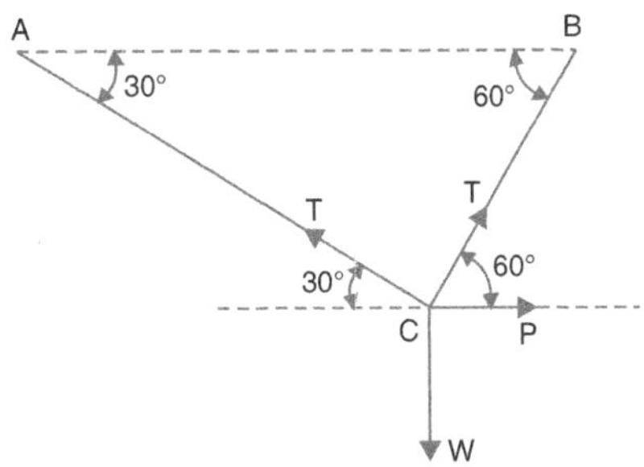

Fig. 2.35

$$T \cos 30^\circ = T \cos 60^\circ + P$$

$$\frac{\sqrt{3}}{2} T = \frac{T}{2} + P$$

$$P = \frac{\sqrt{3} - 1}{2} T \qquad ...(i)$$

Considering vertical equilibrium

$$T \sin 30^\circ + T \sin 60^\circ = W$$

$$\frac{T}{2} + \frac{\sqrt{3}}{2} T = W$$

$$\frac{(1+\sqrt{3})}{2} T = W$$

or $$\boldsymbol{T = \frac{2}{\sqrt{3}+1} W.} \quad \textbf{(Ans.)} \qquad ...(ii)$$

Substituting the value of T from Eqn. (*ii*) in eqn. (*i*), we get

$$\boldsymbol{P} = \frac{\sqrt{3}-1}{2} \times \frac{2}{\sqrt{3}+1} W$$

$$= \frac{\sqrt{3}-1}{\sqrt{3}+1} W$$

$$= \frac{(\sqrt{3}-1)(\sqrt{3}-1)}{(\sqrt{3}+1)(\sqrt{3}-1)} W = \frac{4-2\sqrt{3}}{3-1} W$$

$$= \boldsymbol{(2-\sqrt{3})\, W.} \quad \textbf{(Ans.)}$$

Example 2.19. *A body of weight 20 N is suspended by two strings 5 m and 12 m long and other ends being fastened to the extremities of a rod of length 13 m. If the rod be so held that the body hangs immediately below its middle point, find out the tensions in the strings.*

Sol. Refer to Fig. 2.36.

The body is in equilibrium under the action of following forces :

(*i*) W (weight of the rod) acting vertically downwards.

(*ii*) Tension 'T_1' in the string OA.

(*iii*) Tension 'T_2' in the string OB.

Now, in ΔAOB, $AB = 13$ m,

$$OA = 5 \text{ m}, OB = 12 \text{ m},$$

$$\angle ABO = \theta$$

and $$\angle AOB = 90° \qquad (\because \; AB^2 = OA^2 + OB^2)$$

then $$\sin\theta = \frac{OA}{AB} = \frac{5}{13}$$

and $$\cos\theta = \frac{OB}{AB} = \frac{12}{13}$$

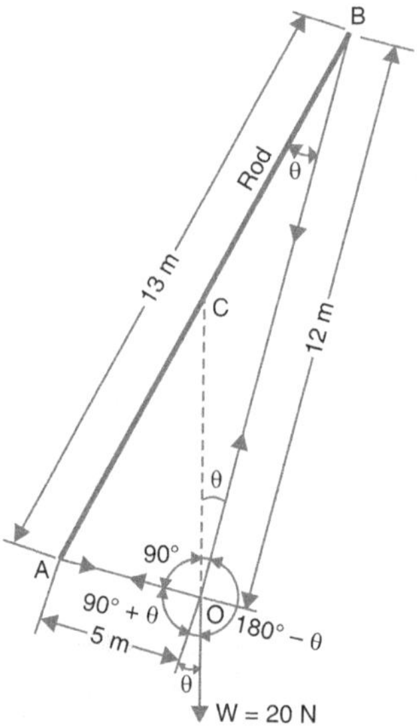

Fig. 2.36

Applying Lami's theorem, at 'O', we have

$$\frac{T_1}{\sin(180° - \theta)} = \frac{T_2}{\sin(90° + \theta)} = \frac{W}{\sin 90°}$$

or $$\frac{T_1}{\sin\theta} = \frac{T_2}{\cos\theta} = \frac{20}{1}$$

or $$\frac{T_1}{5/13} = \frac{T_2}{12/13} = \frac{20}{1}$$

$\therefore$ $$\boldsymbol{T_1} = \frac{20 \times 5}{13} = \textbf{7.69 N.} \quad \textbf{(Ans.)}$$

and $$\boldsymbol{T_2} = \frac{20 \times 12}{13} = \textbf{18.46 N.} \quad \textbf{(Ans.)}$$

Example 2.20. *What axial forces does the vertical load W = 800 N induce in the tie rod and the jib of the jib crane shown in Fig. 2.37 ? Neglect the self-weight of the members.*

Sol. Let P_1 and P_2 be the forces induced in the tie rod and jib respectively. The *tie rod will be under tension* and *jib will be compression* as shown in Fig. 2.37.

Let us now consider equilibrium of the point *C* and apply Lami's theorem to it.

$$\frac{P_1}{\sin(180° - 30°)} = \frac{P_2}{\sin(30° + 15°)} = \frac{W}{\sin(180° - 15°)}$$

or
$$\frac{P_1}{\sin 30°} = \frac{P_2}{\sin 45°} = \frac{800}{\sin 15°}$$

∴
$$\boldsymbol{P_1} = \frac{800 \sin 30°}{\sin 15°} = \frac{800 \times 0.5}{0.2588}$$

$$= \mathbf{1545.6\ N.\ (Ans.)}$$

and
$$\boldsymbol{P_2} = \frac{800 \sin 45°}{\sin 15°} = \frac{800 \times 0.707}{0.2588}$$

$$= \mathbf{2185.5\ N.\ (Ans.)}$$

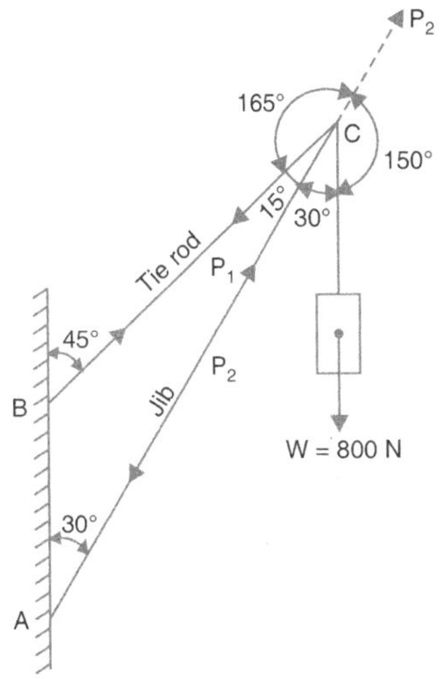

Fig. 2.37

Example 2.21. *A string ABCD hangs from fixed point A and D carrying a weight of 12 N at B and W at C. AB is inclined at 60° to the horizontal, CD is inclined at 30° to the horizontal and BC is horizontal, find W.*

Sol. Refer to Fig. 2.38.

Let *T* be the tension in *BC*.

By Lami's theorem,

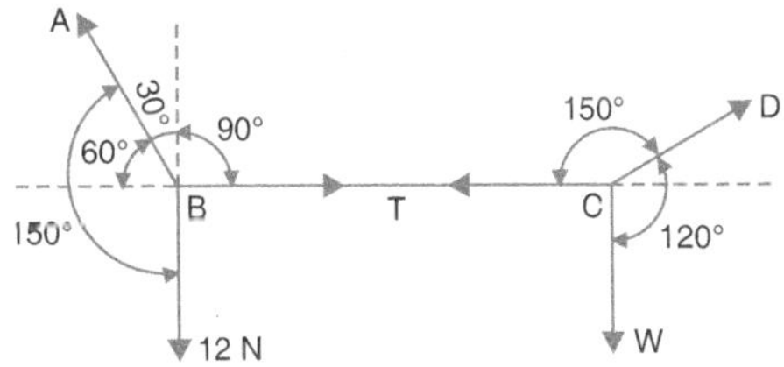

Fig. 2.38

$$\frac{T}{\sin 150°} = \frac{12}{\sin 120°}$$

$$T = \frac{12 \sin 150°}{\sin 120°} = \frac{12 \times \frac{1}{2}}{\frac{\sqrt{3}}{2}} = \frac{12}{\sqrt{3}}$$

Again by Lami's theorem,

$$\frac{W}{\sin 150°} = \frac{T}{\sin 120°}$$

$$\frac{W}{\sin 150°} = \frac{12}{\sqrt{3} \times \frac{\sqrt{3}}{2}}$$

$$\frac{W}{\frac{1}{2}} = \frac{24}{3}$$

$$\mathbf{W = 4\ N.\ (Ans.)}$$

Example 2.22. *The extremities A and D of a light inextensible string ABCD are tied to two points in the same horizontal line. Weights W and 3 W are tied to the string at the points B and C respectively. If AB and CD are inclined to the vertical at angles 60° and 30° respectively, show that BC is horizontal and find the tensions in the various parts of the string.*

Sol. Refer to Fig. 2.39.

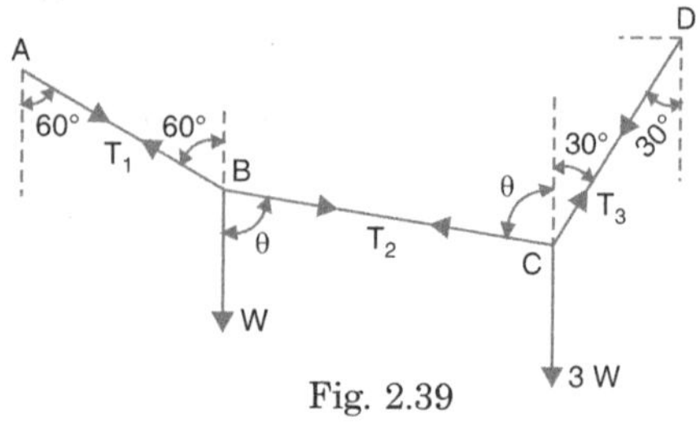

Fig. 2.39

Let BC makes an angle θ with the vertical and let T_1, T_2 and T_3 are tensions in the three parts AB, BC and CD of the string as shown in Fig. 2.39. At B, the forces in equilibrium are W, T_1 and T_2.

Applying Lami's theorem, we get

$$\frac{T_1}{\sin\theta} = \frac{T_2}{\sin 120°} = \frac{W}{\sin(240° - \theta)} \qquad ...(i)$$

Similarly, applying Lami's theorem to the point C,

$$\frac{T_2}{\sin 150°} = \frac{T_3}{\sin(180° - \theta)} = \frac{3W}{\sin(30° + \theta)} \qquad ...(ii)$$

From eqns. (*i*) and (*ii*),

$$T_2 = \frac{W\sin 120°}{\sin(240° - \theta)} = \frac{3W\sin 150°}{\sin(30° + \theta)}$$

or $$\frac{\sin 60°}{-\sin(60° - \theta)} = \frac{3\sin 30°}{\sin(30° + \theta)}$$

or $$\frac{\sqrt{3}}{2}\sin(30° + \theta) = -3 \times \frac{1}{2}\sin(60° - \theta)$$

or $$\sin 30°\cos\theta + \cos 30°\sin\theta$$
$$= -\sqrt{3}\,(\sin 60°\cos\theta - \cos 60°\sin\theta)$$

or $$\frac{1}{2}\cos\theta + \frac{\sqrt{3}}{2}\sin\theta = -\frac{3}{2}\cos\theta + \frac{\sqrt{3}}{2}\sin\theta$$

or $$2\cos\theta = 0$$

or $$\theta = 90°.$$

So BC makes 90° with the vertical *i.e.*, BC is horizontal.

Putting this value of θ in eqns. (*i*) and (*ii*), we get

$$\boldsymbol{T_1} = \frac{W\sin 90°}{W\sin 150°} = \frac{W}{\frac{1}{2}} = \textbf{2 W. (Ans.)}$$

$$\boldsymbol{T_2} = \frac{W\sin 120°}{\sin 150°} = \frac{\sqrt{3}/2\,W}{\frac{1}{2}} = \boldsymbol{\sqrt{3}}\ \textbf{W. (Ans.)}$$

$$\boldsymbol{T_3} = \frac{3W\sin 90°}{\sin 120°} = \frac{3W}{\sqrt{3}/2} = \boldsymbol{2\sqrt{3}}\ \textbf{W. (Ans.)}$$

Example 2.23. *A weight W is supported by two strings at the right angle to one another and attached to two points in the same horizontal line. Prove that their tensions are inversely proportional to their lengths.*

Sol. Refer to Fig. 2.40.

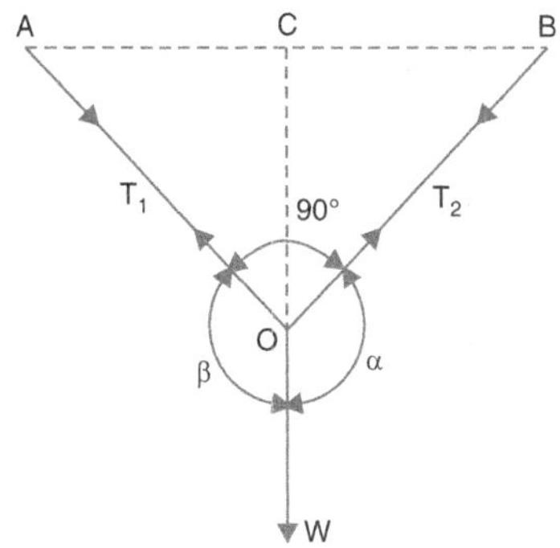

Fig. 2.40

Let T_1 and T_2 be the tensions in the strings OA and OB respectively.

By Lami's theorem,

$$\frac{T_1}{\sin \alpha} = \frac{T_2}{\sin \beta} = \frac{W}{\sin 90°}$$

$$T_1 = W \sin \alpha = W \sin \angle BOC = W \sin A = W \cdot \frac{OB}{AB}$$

$$T_2 = W \sin \beta = W \sin \angle AOC = W \sin B = W \cdot \frac{OA}{AB}$$

$$\frac{T_1}{T_2} = \frac{W \times OB}{AB} \times \frac{AB}{W \times OA} = \frac{OB}{OA}$$

$$T_1 \propto \frac{1}{OA} \quad \text{or} \quad T_1 \propto \frac{1}{\text{length}}$$

$$T_2 \propto \frac{1}{OB} \quad \text{or} \quad T_2 \propto \frac{1}{\text{length}}$$ ...... **Proved.**

Example 2.24. *A spherical ball of weight W, rests in a triangular groove whose sides are inclined at angles α and β to the horizontal. Find the reactions at the surfaces of contact.*

Sol. Refer to Fig. 2.41.

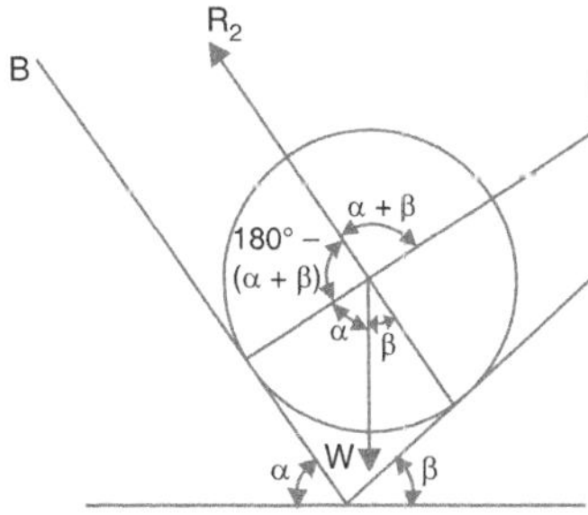

Fig. 2.41

Let AB and AC be the two inclined planes at an angle α, β with the horizontal respectively.

Let R_1, R_2 and W be the reaction on the inclined planes AB and AC and weight of the sphere respectively.

On applying Lami's theorem, we get

$$\frac{R_1}{\sin (180° - \beta)} = \frac{R_2}{\sin (180° - \alpha)} = \frac{W}{\sin (\alpha + \beta)}$$

$$\frac{R_1}{\sin \beta} = \frac{R_2}{\sin \alpha} = \frac{W}{\sin (\alpha + \beta)}$$

∴ $$\boldsymbol{R_1 = \frac{W \sin \beta}{\sin (\alpha + \beta)}}$$. **(Ans.)**

and $$\boldsymbol{R_2 = \frac{W \sin \alpha}{\sin (\alpha + \beta)}}$$. **(Ans.)**

Example 2.25. *A cast iron sphere 30 cm in diameter rests in 20 cm × 20 cm angle, one leg of which is at an angle of 30° with the horizontal as shown in Fig. 2.42. Assuming all surfaces smooth, compute the reactions on the spheres at A and B. Cast iron weighs 72 kN/m³.*

Sol. Refer to Fig. 2.42.

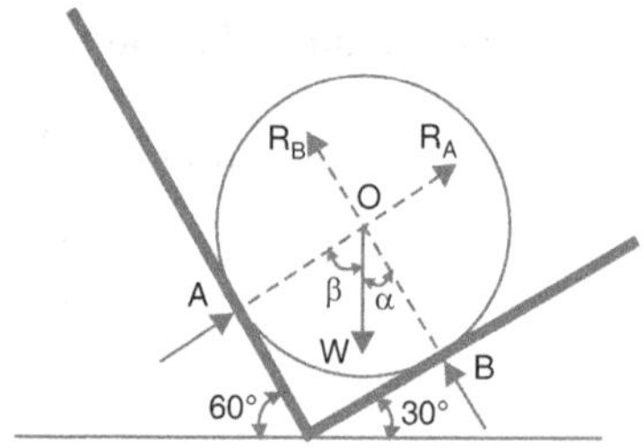

Fig. 2.42

Weight of sphere = volume of sphere × density of cast iron.

$= 4/3\ \pi r^3 \times$ density of cast iron

$= 4/3\ \pi \times (0.15)^3 \times (72 \times 1000)$

$= 1017.9$ N

Since, the surfaces are smooth, so the reactions at A and B will be normal at the points of contact of sphere and the angle. *If the system is in equilibrium the line of action of the weight must pass through O, the point of intersection of the lines of action of the two reactions R_A and R_B.*

Now, $\angle\alpha = 30°, \angle\beta = 60°$

Applying Lami's theorem, we get

$$\frac{R_A}{\sin(180° - \alpha)} = \frac{R_B}{\sin(180° - \beta)} = \frac{W}{\sin 90°}$$

$$\frac{R_A}{\sin\alpha} = \frac{R_B}{\sin\beta} = \frac{1017.9}{1}$$

$$\frac{R_A}{\sin 30°} = \frac{R_B}{\sin 60°} = \frac{1017.9}{1}$$

∴ $\boldsymbol{R_A} = 1017.9 \times \sin 30° =$ **508.9 N. (Ans.)**

$\boldsymbol{R_B} = 1017.9 \times \sin 60° =$ **881.5 N. (Ans.)**

Example 2.26. *A uniform wheel 40 cm in diameter rests against a rigid rectangular block 10 cm thick as shown in Fig. 2.43. Find the least pull through the centre of the wheel to just turn it over the corner of the block. All surfaces are smooth. Find also the reaction of the block. The wheel weighs 800 N.*

Sol. Refer to Fig. 2.43.

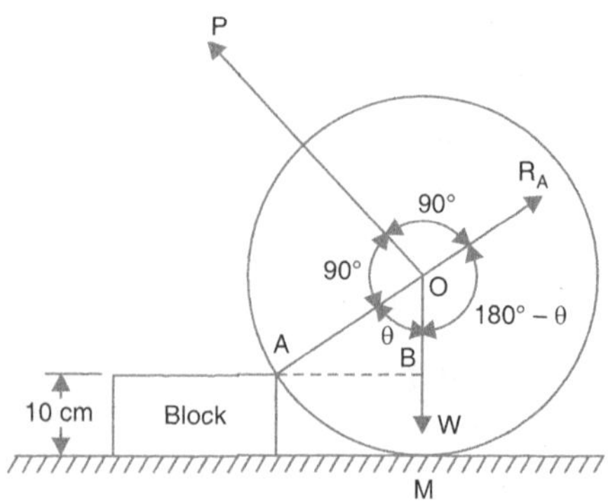

Fig. 2.43

Consider the equilibrium of the wheel when it is just about to turnover the block. In this position, the following forces shall act on the wheel :

(*i*) The pull P ;

(*ii*) Weight of the wheel = 800 N ;

(*iii*) Reaction, R_A.

In this position, there will be no contact between the wheel and the floor, so $R_M = 0$.

If the pull is to be *minimum* it must be applied normal to AO.

Now, applying Lami's theorem, we get

$$\frac{P}{\sin (180° - \theta)} = \frac{R_A}{\sin (90° + \theta)} = \frac{W}{\sin 90°}$$

or $$\frac{P}{\sin \theta} = \frac{R_A}{\cos \theta} = \frac{800}{1}$$

$\therefore$ $$P = 800 \sin \theta$$

and $$R_A = 800 \cos \theta$$

But from right-angled Δ AOE,

$$\cos \theta = \frac{OB}{OA} = \frac{10}{20} = \frac{1}{2}$$

$\therefore$ $$\theta = 60°$$

$\therefore$ $$P = 800 \times \sin 60°$$
$$= \mathbf{692.8\ N.\ (Ans.)}$$

and $$\boldsymbol{R_A} = 800 \cos 60°$$
$$= \mathbf{400\ N.\ (Ans.)}$$

Example 2.27. *The cylinders shown in Fig. 2.44 have the same diameter but the cylinder '1' weighs 200 N and cylinder '2' weighs 150 N. Find the reactions at the supports.*

Sol. Refer to Fig. 2.44.

The following forces keep the cylinder '2' *in equilibrium*

(*i*) Weight of cylinder, W_2 (= 150 N) acting vertically downward ;

(*ii*) Reaction (R_s) acting at right angle to OQ ; and

(*iii*) Reaction of cylinder '1' (R_N) in the direction O_1O_2 (action and reaction are equal and opposite).

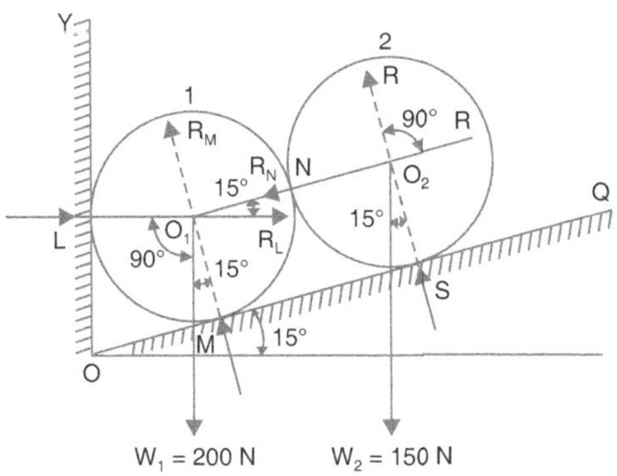

Fig. 2.44

Applying Lami's theorem :

$$\frac{R_N}{\sin (180° - 15°)} = \frac{R_S}{\sin (90° + 15°)} = \frac{W_2}{\sin 90°}$$

or
$$\frac{R_N}{\sin 15°} = \frac{R_S}{\sin 105°} = \frac{150}{1}$$

$\therefore$
$$R_N = 150 \sin 15° = 38.82 \text{ N}$$

and
$$R_S = 150 \sin 105° = 144.89 \text{ N}$$

The forces which keep the cylinder '1' in equilibrium are :

(*i*) Weight of the cylinder W_1 (= 200 N) acting vertically down ;

(*ii*) Reaction R_L acting at right angle to OY ;

(*iii*) Reaction R_M acting at right angle to OQ ;

(*iv*) R_N, the pressure of the cylinder '2' acting in the direction O_2O_1.

Since in this case, number of forces acting are four, so we *cannot apply Lami's theorem here which is applicable in a case where there are only three forces.* The unknown in this case can be determined by resolving the forces along O_1O_2 and in a direction *perpendicular to* O_1O_2.

$\therefore$
$$R_L \cos 15° - W_1 \sin 15° - R_N = 0 \quad ...(i)$$

and
$$R_M - R_L \sin 15° - W_1 \cos 15° = 0 \quad ...(ii)$$

From eqn. (*i*),

$$R_L \times 0.9659 - 200 \times 0.2588 - 38.82 = 0$$

or
$$R_L \times 0.9659 - 51.76 - 38.82$$

or
$$\mathbf{R_L = 93.78 \text{ N. (Ans.)}}$$

Putting the value of R_L in eqn. (*ii*), we get

$$R_M - 93.78 \times 0.2588 - 200 \times 0.9659 = 0$$

$$R_M - 24.27 - 193.18 = 0$$

or
$$\mathbf{R_M = 217.45 \text{ N. (Ans.)}}$$

Example 2.28. *Two cylinders '1' and '2' rest in a horizontal channel as shown in Fig. 2.45. The cylinder '1' has a weight of 500 N and radius of 180 mm. The cylinder '2' has a weight of 200 N and a radius of 100 mm. The channel is 360 mm wide at the bottom with one side vertical. The other side is inclined at an angle 60° with the horizontal. Find the reactions.*

Sol. Refer to Figs. 2.45 and 2.46.

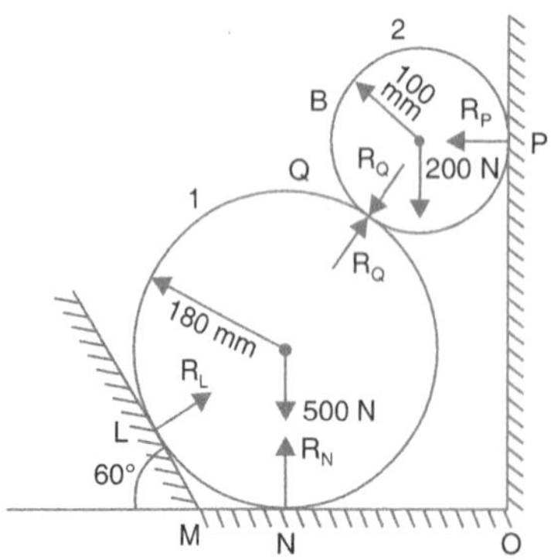

Fig. 2.45

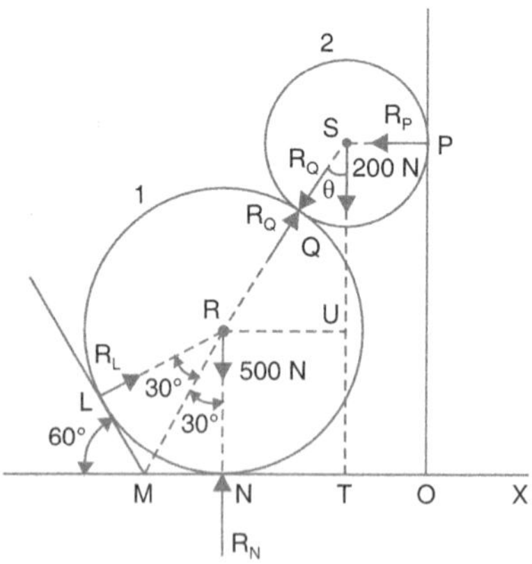

Fig. 2.46

The following forces keep the **cylinder '2'** *in equilibrium* :

(*i*) Weight of the cylinder '2' = 200 N, acting downward

(*ii*) Reaction of the channel wall at the point $P = R_P$

(*iii*) Reaction of cylinder '1' on the cylinder '2' = R_Q acting along the line *RS*, *R* and *S* being the centres of the cylinders.

The following forces keep the **cylinder '1'** in equilibrium :

(*i*) Weight of the cylinder '1' = 500 N acting downward.

(*ii*) Reaction of the cylinder '2' on the cylinder '1' = R_Q acting along the line *RS*. *R* and *S* being the centres of the cylinders.

(*iii*) Reaction of the bottom surface at the point *N* on the cylinder = R_N.

(*iv*) Reaction of the channel wall at the *L* on the cylinder = R_L.

Except for the reaction R_Q, the directions of all other forces are known. Let R_Q makes an angle θ with the vertical.

Then $$\sin\theta = \frac{RU}{RS}$$

Now $$RS = 180 + 100 = 280 \text{ mm}$$

$$RU = NT = MO - MN - TO$$
$$= 360 - MN - 100$$

But $$MN = RN \tan 30° = 180 \times \frac{1}{\sqrt{3}} = 104 \text{ mm}$$

∴ $$RU = 360 - 104 - 100 = 156 \text{ mm}$$

∴ $$\sin\theta = \frac{156}{280} = 0.5571$$

i.e., $$\theta = 33.86°$$

Equations of equilibrium for cylinder '2' :

$$\Sigma H = 0 : R_Q \sin 33.86° - R_P = 0 \quad ...(i)$$

$$\Sigma V = 0 : R_Q \cos 33.86° - 200 = 0 \quad ...(ii)$$

From eqn. (*ii*) $R_Q = \dfrac{200}{\cos 33.86°}$ = **240.84 N. (Ans.)**

From eqn. (*i*) $R_P = R_Q \sin 33.86°$

$= 240.84 \sin 33.86°$ = **134.18 N. (Ans.)**

Equations of equilibrium for cylinder '1' :

$$\Sigma H = 0 : R_L \sin 60° - R_Q \sin 33.86° = 0$$

∴ $$R_L = \frac{R_Q \sin 33.86°}{\sin 60°} = \frac{240.84 \times 0.5571}{0.866}$$

i.e., $\boldsymbol{R_L}$ **= 154.93 N. (Ans.)**

$$\Sigma V = 0 : R_N - 500 - R_Q \cos 33.86° + R_L \cos 60° = 0$$

∴ $$R_N = 500 + 240.84 \times \cos 33.86° - 154.93 \times 0.5$$

$= 500 + 200 - 77.46$ = **622.54 N. (Ans.)**

SELECTED QUESTIONS FROM EXAMINATION PAPERS

Example 2.29. *Two forces 400 N and 300 N act at a point as shown in Fig. 2.47. The resultant of the two forces is 600 N. Determine the angle θ between the forces and direction of the resultant.*

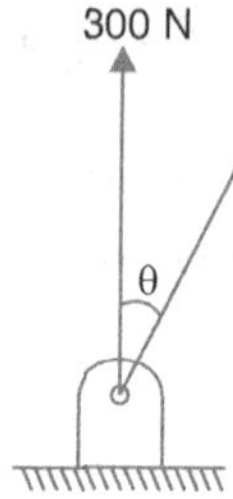

Fig. 2.47

Sol. Refer to Figs. 2.48 and 2.49

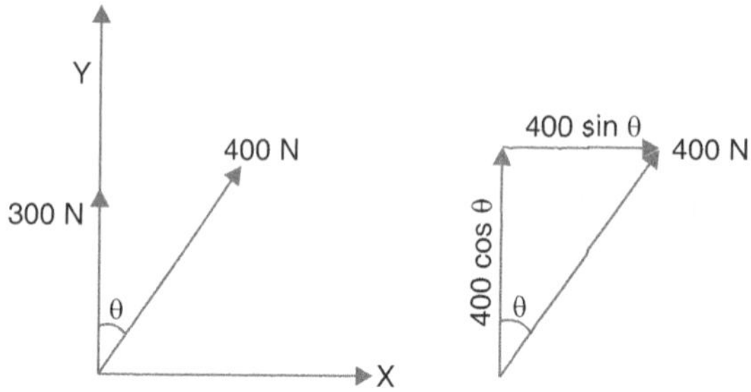

Fig. 2.48

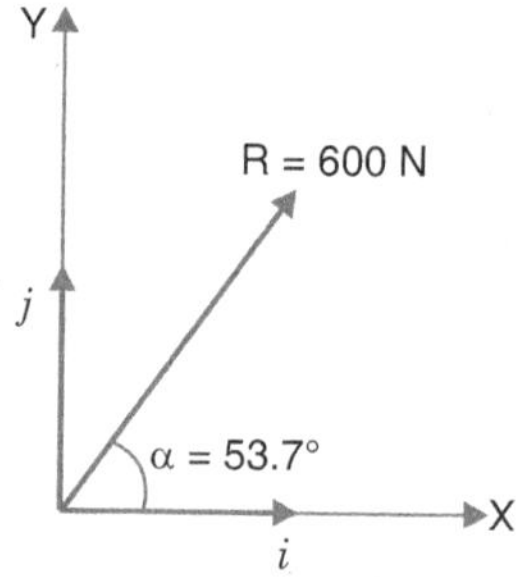

Fig. 2.49

$$\boldsymbol{R} = 400 \sin \theta\, i + 400 \cos \theta\, j + 300\, j$$
$$= 400 \sin \theta\, i + (400 \cos \theta + 300)\, j$$

Magnitude,
$$R = \sqrt{(\Sigma P_x)^2 + (P_y)^2}$$
$$600 = \sqrt{(400 \sin \theta)^2 + (400 \cos \theta + 300)^2}$$

Squaring both sides, we get

$$(600)^2 = (400 \sin \theta)^2 + (400 \cos \theta + 300)^2$$
$$(600)^2 = (400 \sin \theta)^2 + (400 \cos \theta)^2 + 2 \times 400 \times 300 \cos \theta + (300)^2$$
$$360000 = (400)^2 (\sin^2 \theta + \cos^2 \theta) + 240000 \cos \theta + 90000$$

or
$$360000 = 160000 + 240000 \cos \theta + 90000$$

or
$$110000 = 240000 \cos \theta$$

∴
$$\cos \theta = \frac{110000}{240000} = 0.4583$$

or
$$\cos \theta = \cos^{-1} (0.4583) = \mathbf{62.7°.\ (Ans.)}$$

Direction of the resultant (Fig. 2.49)

$$\tan \alpha = \frac{\Sigma P_y}{\Sigma P_x} = \frac{400 \cos 62.7° + 300}{400 \sin 62.7°}$$
$$= \frac{483.46}{355.45} = 1.36$$

∴
$$\alpha = \tan^{-1} (1.36) = \mathbf{53.7°.\quad (Ans.)}$$

Example 2.30. *The force system shown in Fig. 2.50 has resultant of 200 N towards the Y-axis. Compute the values of P and θ required to give this resultant.*

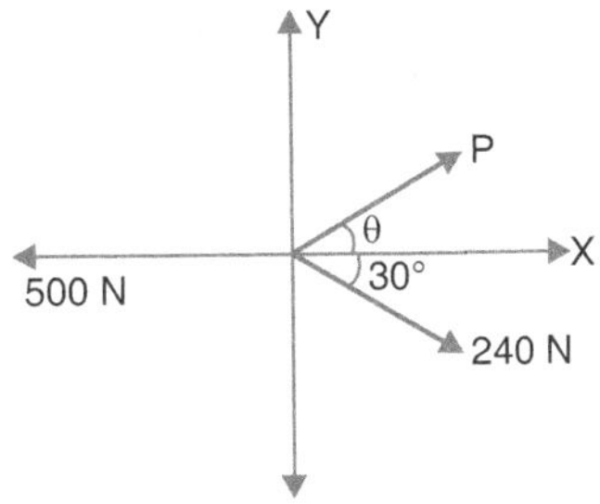

Fig. 2.50

Sol. As the resultant R is in upward direction, $\Sigma P_x = 0$

$\therefore \quad \Sigma P_x = -500 + 240 \cos 30° + P \cos \theta = 0$

$\therefore \quad P \cos \theta = 292.15 \text{ N} \quad ...(i)$

and $\quad \Sigma P_y = 200$

$\Sigma P_y = P \sin \theta - 240 \sin 30° = 200$

$\therefore \quad P \sin \theta = 320 \quad ...(ii)$

From eqns. (*i*) and (*ii*), we have

$$\frac{P \sin \theta}{P \cos \theta} = \frac{320}{292.15} = 1.095$$

or $\quad \tan \theta = 1.095$

or $\quad \theta = \tan^{-1} (1.095) = \mathbf{47.6°.}$ **(Ans.)**

Example 2.31. *The clock shown in Fig. 2.51 is acted on by its weight W = 400 N, a horizontal force F = 600 N and pressure P exerted by the inclined plane. The resultant R of these forces is parallel to the incline. Determine P and R. Does the block move up or down the plane?*

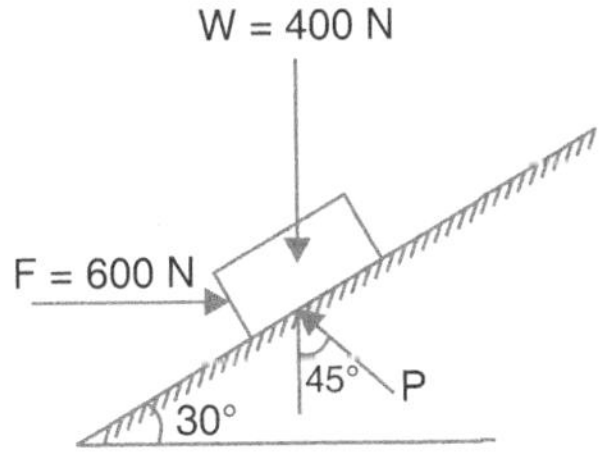

Fig. 2.51

Sol. Refer to Fig. 2.52

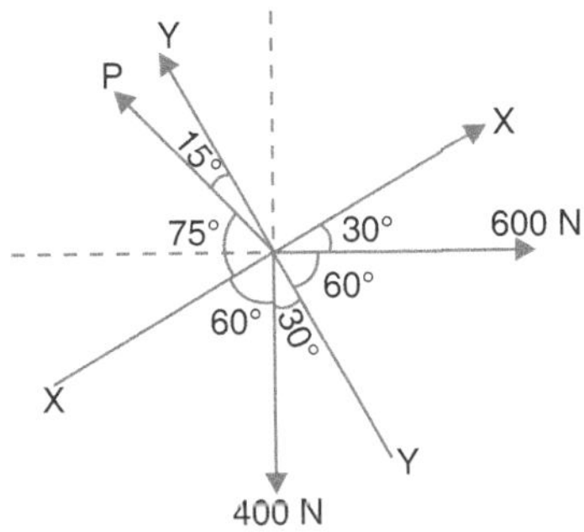

Fig. 2.52

As resultant is along the *X*-axis (new), hence

$$\Sigma F_y = 0$$

$\therefore$ $$\Sigma F_y = P \cos 15° - 400 \cos 30° - 600 \cos 60° = 0$$

or $$0.966\, P - 346.41 - 300 = 0$$

$\therefore$ $$\mathbf{P = 669.2\ N.}\quad \textbf{(Ans.)}$$

As *R* is along the incline,

$\therefore$ $$\boldsymbol{R} = \Sigma F_x = -P \cos 75° - 400 \cos 60° + 600 \cos 30°$$
$$= -669.2 \times 0.2588 - 400 \times 0.5 + 600 \times 0.866$$
$$= \mathbf{146.41\ N.}\quad \textbf{(Ans.)}$$

The block will move up the incline (**Ans.**)

Example 2.32. *Two spheres A and B weighing 100 N and 75 N respectively and with the corresponding radii 75 mm and 50 mm are placed in a container as shown in Fig. 2.53. Determine the support reactions.*

Sol. Join the centres O_1 and O_2 of the two spheres and drop a vertical line from point O_2 and a horizontal line from point O_1 to form a $\Delta O_1O_2\,T$ as shown in Fig. 2.53

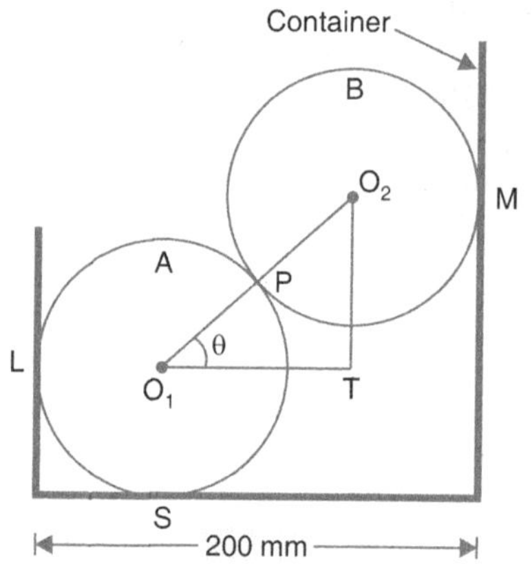

Fig. 2.53

$$O_1T = 200 - (75 + 50) = 75 \text{ mm}$$
$$O_1O_2 = 75 + 50 = 125 \text{ mm}$$
$$\cos \theta = \frac{O_1T}{O_1O_2} = \frac{75}{125} = 0.6$$

$\therefore$ $$\theta = 53°$$

Consider equilibrium of sphere B : Refer to Fig. 2.54 (*a*)

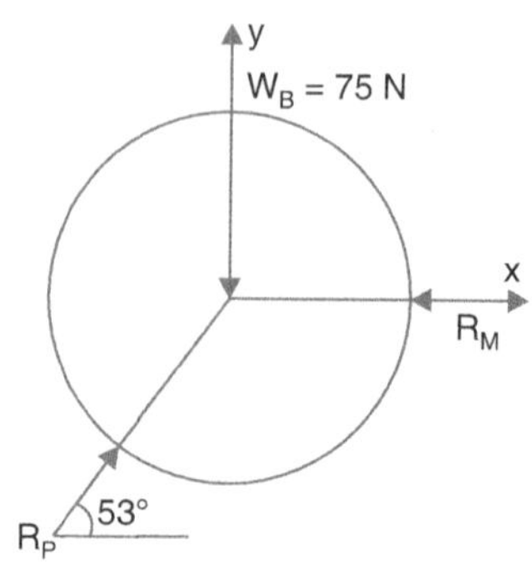

Fig. 2.54 (*a*)

$$-75\,j - R_N\,i + R_P \cos 53°\,i + R_P \sin 53°\,j = 0$$

The corresponding scalar equations are :

$$-R_M + R_P \cos 53° = 0$$

or $$R_M = R_P \cos 53° \quad ...(i)$$

and $$-75 + R_P \sin 53° \quad ...(ii)$$

or $$R_P = \mathbf{93.91\ N.}\quad \textbf{(Ans.)}$$

and $$R_M = 93.91 \times \cos 53° = \mathbf{56.52\ N.}\quad \textbf{(Ans.)} \quad \text{[From eqn. }(i)\text{]}$$

Consider equilibrium of sphere A : Refer to Fig. 2.54 (*b*)

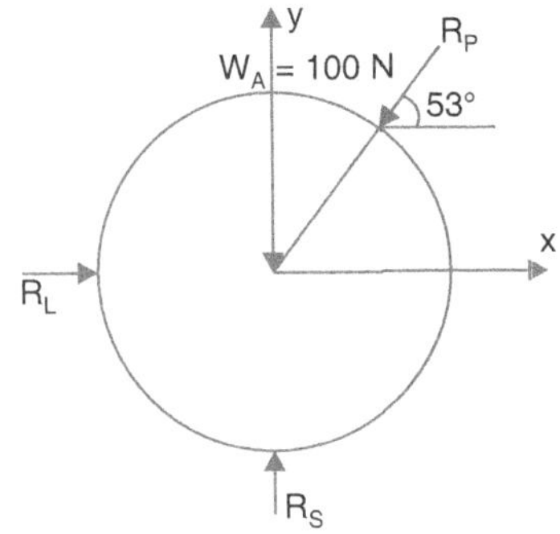

Fig. 2.54 (*b*) Free body diagram

$$-100\,j - R_P \cos 53°\,i - R_P \sin 53°\,j + R_S\,j + R_L\,i = 0$$

The corresponding scalar equations are :

$$-R_P \cos 53° + R_L = 0 \quad ...(iii)$$

$\therefore$ $$R_L = R_P \cos 53° = 93.91 \cos 53°$$

$$= \mathbf{56.52\ N.}\quad \textbf{(Ans.)}$$

and $$-100 - R_P \sin 53° + R_s = 0 \quad ...(iv)$$

$$-100 - 93.91 \sin 53° + R_s = 0$$

$\therefore$ $$R_0 = \mathbf{175\ N.}\quad \textbf{(Ans.)}$$

Example 2.33. *Fig. 2.55 shows the cylinders, A of weight 100 N and B of weight 50 N resting on smooth inclined planes. They are connected by a bar of negligible weight hinged to each cylinder at their geometric centres by smooth pins. Find the forces P, as shown that holds the system in the given position.*

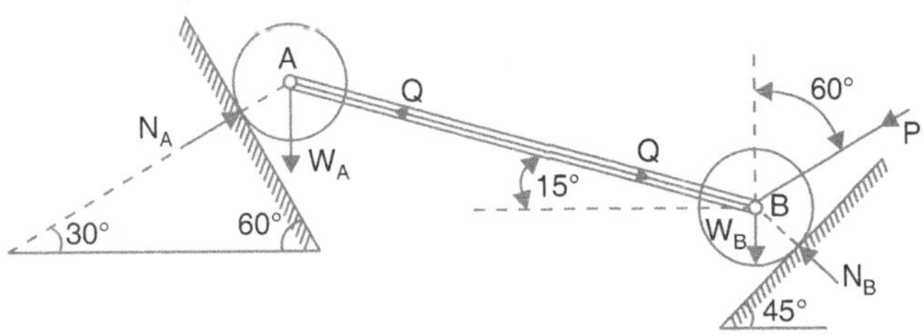

Fig. 2.55

Sol. Free body diagrams (FBDs) for cylinders A and B are shown in Fig. 2.56 (*a*) and (*b*) respectively. In these figures N_A and N_B represent the reactions (normal) on the cylinders A and B respectively and Q represents the force exerted by the connected bar. It may be noted here that the direction of the force Q will be along the line AB, joining the centres of the two cylinders, because the bar AB is a two force member and the two forces acting on it must be equal and opposite to keep it in equilibrium.

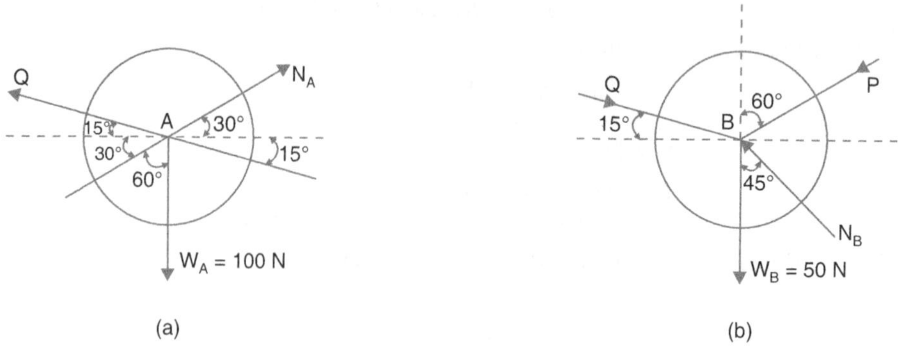

Fig. 2.56

Applying *Lami's theorem* for cylinder *A*, we get

$$\frac{Q}{\sin 120°} = \frac{W_A}{\sin 135°} = \sin \frac{N_A}{105°}$$

or
$$Q = \sin 120° \times \frac{W_A}{\sin 135°} = 0.866 \times \frac{100}{0.707}$$
$$= 122.49 \text{ N}$$

Considering the equilibrium of the cylinder *B* and resolving the forces in horizontal and vertical directions respectively, we get

$$Q \cos 15° - P \sin 60° - N_B \sin 45° = 0 \quad ...(i)$$

and
$$- Q \sin 15° - P \cos 60° + N_B \cos 45° - W_B = 0 \quad ...(ii)$$

Adding eqns. (*i*) and (*ii*), we get

$$Q (\cos 15° - \sin 15°) - P (\sin 60° + \cos 60°) - W_B = 0$$

or
$$P = \left[\frac{Q (\cos 15° - \sin 15°) - W_B}{\sin 60° + \cos 60°}\right]$$
$$= \left[\frac{122.49 (\cos 15° - \sin 15°) - 50}{\sin 60° + \cos 60°}\right]$$
$$= \frac{122.49 (0.966 - 0.258) - 50}{0.866 + 0.5}$$
$$= \frac{36.72}{1.366} \approx 26.88 \text{ N}$$

i.e., $P =$ **26.88 N. (Ans.)**

Coplanar Non-Concurrent Forces-Resultant Force

Example 2.34. *Determine the resultant of the forces acting on the board as shown in Fig. 2.57.*

Sol. Resolving the forces *horizontally* :

$$\Sigma P_x = 80 \cos 60° + 120 \cos 60° - 150 + 50$$
$$= 80 \times 0.5 + 120 \times 0.5 - 150 + 50 = 0$$

Resolving the forces *vertically* :

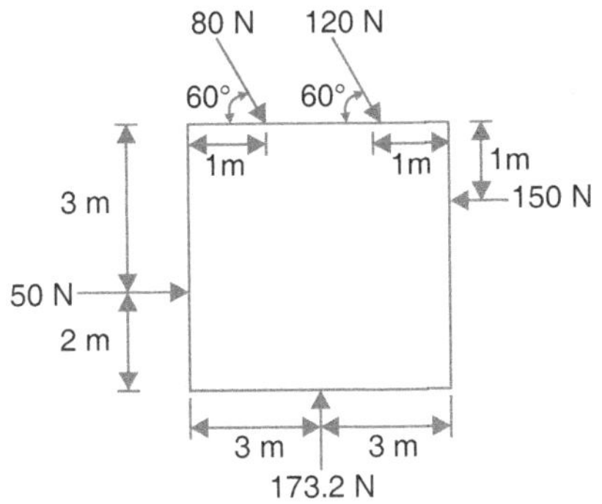

Fig. 2.57

$$\Sigma P_y = -80 \sin 60° - 120 \sin 60° + 173.2$$
$$= -80 \times 0.866 - 120 \times 0.866 + 173.2 = 0$$

∴ Resultant, $R = \sqrt{(\Sigma P_x)^2 + (\Sigma P_y)^2}$ = **0. (Ans.)**

Example 2.35. *Determine the resultant of the force system shown in Fig. 2.58. Assume that the co-ordinates are in metres.*

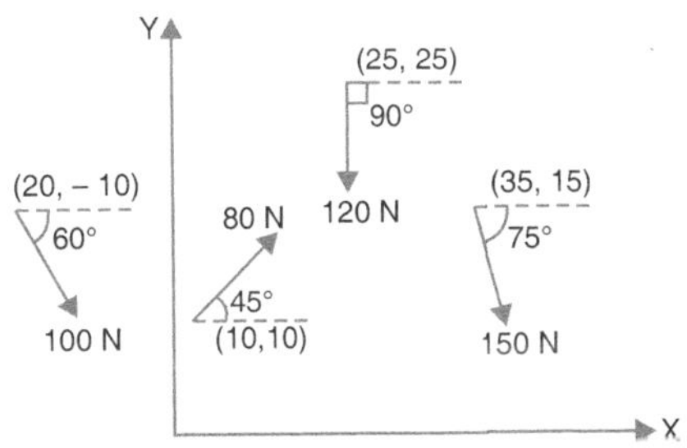

Fig. 2.58

Sol. Resolving the forces *horizontally* :

$$\Sigma P_x = 100 \cos 60° + 80 \cos 45° + 150 \cos 75°$$
$$= 100 \times 0.5 + 80 \times 0.707 + 150 \times 0.259$$
$$= 145.4 \text{ N}$$

Resolving the forces *vertically* :

$$\Sigma P_y = -100 \sin 60° + 80 \sin 45° - 120 - 150 \sin 75°$$
$$= -100 \times 0.866 + 80 \times 0.707 - 120 - 150 \times 0.966$$
$$= -86.6 + 56.56 - 120 - 144.9$$
$$= -294.94 \text{ N}$$

∴ Resultant force, $R = \sqrt{(\Sigma P_x)^2 + (\Sigma P_y)^2}$

$$= \sqrt{(145.4)^2 + (-294.94)^2}$$

= **328.83 N. (Ans.)**

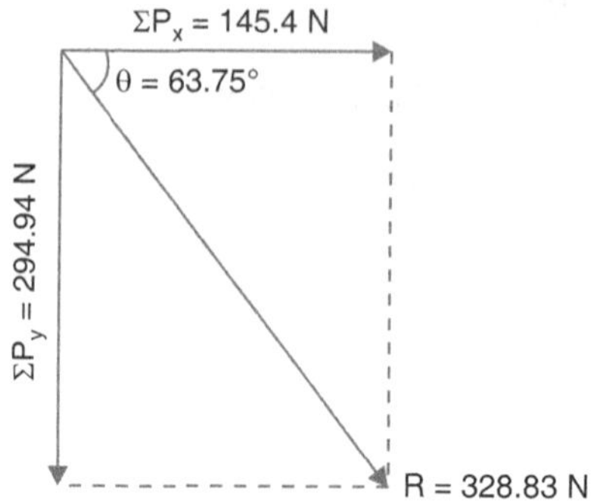

Fig. 2.59

Since ΣP_y is having (– ve) sign the resultant falls in the *4th quadrant*

and
$$\tan \theta = \frac{\Sigma P_y}{\Sigma P_x} = \frac{-294.94}{145.4} = -2.03$$

∴ θ = **63.75° with the horizontal. (Ans)**

Example 2.36. *Find the resultant of the system of the forces shown in Fig. 2.60.*

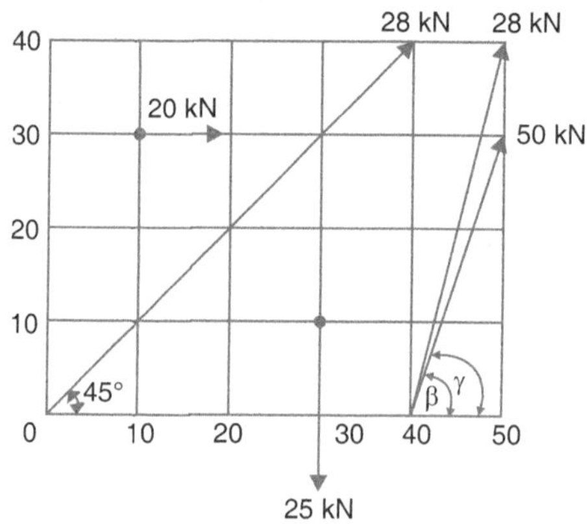

Fig. 2.60

Sol. Resolving the forces *horizontally,* we get

$$\Sigma P_x = 20 + 28 \cos \alpha + 28 \cos \beta + 50 \cos \gamma \quad ...(i)$$

Resolving the forces *vertically,* we get

$$\Sigma P_y = 28 \sin \alpha + 28 \sin \beta + 50 \sin \gamma - 25 \quad ...(ii)$$

From the given graph we can find out α, β and γ by geometrical properties as follows :

$$\tan \alpha = \frac{40}{40} = 1 \quad \therefore \quad \alpha = 45°$$

$$\tan \beta = \frac{40}{10} = 4 \quad \therefore \quad \beta = 75.96°$$

$$\tan \gamma = \frac{30}{10} = 3 \quad \therefore \quad \gamma = 71.56°$$

Substituting the values of α, β and γ in eqns. (*i*) and (*ii*), we get

$$\Sigma P_x = 20 + 28 \cos 45° + 28 \cos 75.96° + 50 \cos 71.56°$$
$$= 20 + 19.79 + 6.79 + 15.81$$
$$= 62.39 \text{ kN}$$

and
$$\Sigma P_y = 28 \times \sin 45° + 28 \sin 75.96° + 50 \sin 71.56° - 25$$
$$= 19.79 + 27.16 + 47.43 - 25$$
$$= 69.38 \text{ kN}$$

∴ Resultant
$$R = \sqrt{(\Sigma P_x)^2 + (\Sigma P_y)^2}$$
$$= \sqrt{(62.39)^2 + (69.38)^2}$$
$$= 93.3 \text{ kN}$$

and
$$\tan \theta = \frac{\Sigma P_y}{\Sigma P_x} = \frac{69.38}{62.39} = 1.112$$

∴
$$\theta = 48.03°$$

Hence, *magnitude of resultant force*, **R = 93.3 kN. (Ans.)**

and, *direction of the resultant force*, **R = 48.03° with the horizontal. (Ans.)**

HIGHLIGHTS

1. *Force* is something which changes or tends to change the state of rest or of uniform motion of a body in a straight line.
2. *Units of force.* The two commonly used units of force are :
   (*i*) Absolute units
   (*ii*) Gravitational units.
   Gravitational Units are used by *engineers* for all practical purposes. 1 kg wt (or kgf) = 9.81 N (app.)
3. Characteristics of a force are :
   (*i*) Magnitude
   (*ii*) Direction or line of action (angle relative to a co-ordinate system)
   (*iii*) Sense or nature
   (*iv*) Point of application.
4. Forces may be represented in two way :
   (*i*) Vector representation
   (*ii*) Bow's notation.
5. *Bow's notation* is a method of designating a force by writing two capital letters one on either side of the force.
6. *A force system* is a collection of forces acting on a body in one or more planes.
7. The diagram of the isolated element or a portion of the body along with the net effects of the system on it is called a '*free body diagram.*'
8. The *principle* of *transmissibility* of forces states that when a force acts upon a body its effect is the same whatever point in its line of action is taken as the point of application provided that the point is connected with the rest of the body in the same invariable manner.
9. A *resultant force* is a single force which can replace two or more forces and produce the same effect on the body as the forces.
10. *The principle of resolved parts* states. "The sum of the resolved parts of two forces acting at a point in any given direction is equal to the resolved parts of their resultant in that direction."

11. *Parallelogram law of forces.* It states : "If two foces acting simultaneously on a particle be represented in magnitude and direction by the two adjacent sides of a parallelogram, their resultant may be represented in magnitude and direction by the diagonal of the parallelogram which passes through their point of intersection."

Analytically, $$R = \sqrt{P^2 + Q^2 + 2PQ \cos \theta}$$

$$\tan \alpha = \frac{Q \sin \theta}{P + Q \cos \theta}.$$

12. *Triangle law of forces states :* "If two forces acting simultaneously on a body are represented in magnitude and direction by the two sides of a triangle taken in order, then their resultant may be represented in magnitude and direction by the third side taken in opposite order."

13. *Polygon law of forces states* : "If a number of coplanar concurrent forces acting simultaneously on a body are represented in magnitude and direction by the sides of a polygon taken in order, then the resultant may be represented in magnitude and direction by the closing side of the polygon, taken in opposite order.

14. Equilibrium conditions for coplanar concurrent forces :

Analytical : $\Sigma H = 0$

$\Sigma V = 0$

Graphical : The force polygon must close.

15. *Lami's theorem states* : "If three coplanar forces acting on a point in a body keep it in equilibrium, then each force is proportional to the sine of the angle between the other two forces."

OBJECTIVE TYPE QUESTIONS

1. Fill in the blanks :

(*i*) ...... is something which changes or tends to change the state of rest or of uniform motion of a body in a straight line.

(*ii*) Magnitude, direction, sense and ...... are the characteristics of a force.

(*iii*) The resistance to deformation, or change of shape, exerted by the material of body is called an ...... force.

(*iv*) A force which prevents the motion, deformation of body is called a ...... force.

(*v*) An ...... force is one which causes a body to move or change its shape.

(*vi*) ...... is a method of designating force by writing two capital letters one on either side of the force.

(*vii*) Forces whose lines of action pass through a common point are called ...... forces.

(*viii*) A ...... is a single force which can replace two or more forces and produce the same effect on the body as the forces.

(*ix*) A body whose dimensions are practically negligible is called a ......

(*x*) ...... units are used by engineers for all practical purposes.

2. Say 'Yes' or 'No' :

(*i*) A force system is a collection of forces acting on a body in one or more planes.

(*ii*) Coplanar concurrent collinear force system includes those forces whose vectors do not lie along the same straight line.

(*iii*) The force whose point of application is so small that it may be considered as a point is called a concentrated force.

(*iv*) A distributed force is one whose place of application is a point.

(*v*) The method of determination of the resultant of some forces acting simultaneously on a particle is called composition of forces.

Answers

1. (*i*) Force (*ii*) point of application (*iii*) internal (*iv*) passive
(*v*) active (*vi*) Bow's notation (*vii*) concurrent (*viii*) resultant force (*ix*) particle
(*x*) Gravitational.

2. (*i*) Yes (*ii*) No (*iii*) Yes (*iv*) No (*v*) Yes.

EXERCISES

Theoretical Questions

1. (*a*) Define a force. State the effects which a force may produce when it acts on the body.
   (*b*) Enumerate characteristics of a force.
2. (*a*) How are the forces classified ?
   (*b*) Define a 'force system'. Name the different force systems.
3. Write a short note on 'Free body diagrams'.
4. State the principle of 'Transmissibility of a force'.
5. Explain briefly the following :
   (*i*) Resultance force.
   (*ii*) Components of a force.
   (*iii*) Principle of resolved parts.
6. State the following law of forces :
   (*i*) Parallelogram law of forces ;
   (*ii*) Triangle law of forces ;
   (*iii*) Polygon law of forces.
7. Discuss graphical and analytical methods for finding resultant of several coplanar concurrent forces.
8. State equilibrium conditions for coplanar concurrent forces.
9. State and prove Lami's theorem.

Unsolved Examples

1. Find the components of a force of 100 N in directions, inclined to it at 30° and 40° on opposite sides. [**Ans.** 68.4 N ; 53.2 N]
2. Two forces equal to $2P$ and P respectively act on a particle ; if the first be doubled, and the second increased by 12 N the direction of the resultant is unaltered, find the value of P. [**Ans.** 12 N]
3. The angle between the two forces of magnitude 40 N and 30 N is 60° ; the 40 N force being horizontal. Determine the resultant in magnitude and direction : if (*i*) the forces are pulls and (*ii*) the 30 N force is a push and 40 N force is a pull. [**Ans.** (*i*) 60.8 N, 25° 16′ ; (*ii*) 36.05 N 46° with 40 N force]
4. The resultant of two forces P and 15 N is 20 N inclined at 60° to the 15 N force. Find the magnitude and direction of P. [**Ans.** P = 18.03 N, 106° 52′]
5. The following forces act at a point :
   (*i*) 50 N due North ;
   (*ii*) 20 N North-East ;
   (*iii*) 30 N due East ;
   (*iv*) 40 N in the direction inclined 30° East of South ;
   (*v*) 60 N in a direction inclined 60° South of West.
   Find the resultant force. What angle does it make with East ? [**Ans.** 40.87 N at 33° 20′ S of E]
6. A particle is acted upon by the following forces :
   (*i*) 20 N inclined 30° to North of East ;
   (*ii*) 25 N towards North ;
   (*iii*) 30 N towards North West ;
   (*iv*) 35 N inclined 40° South of West.
   Find the magnitude and direction of the resultant force. [**Ans.** 45.59 N, 132° 20′ with East]
7. A particle is acted upon by three forces, equal to 5 N, 10 N and 13 N along three sides of an equilibrium triangle, taken in order. Find graphically or otherwise the magnitude and direction of the resultant force. [**Ans.** 7 N, 20° 47′]
8. Forces of 2, 3, 5, $\sqrt{3}$ and 2 N respectively act at one of the angular points of regular hexagon towards the other five angular points, taken in order. Find the magnitude and direction of the resultant force. [**Ans.** 11.11 N, 56° 42′]

9. Forces of 7, 1, 1 and 3 N respectively act at one of the angular points of regular pentagon towards the four other points, taken in order. Find the magnitude and direction of their resultant force.

[**Ans.** 8.368 N, 31° 55′ with 7 N force]

10. A weight of 40 N is supported by two strings inclined 30° and 45° respectively to the horizontal. Find graphically or otherwise the tension in each cord. [**Ans.** 29.30 N, 35.8 N]

11. A machine weighing 3000 N is supported by two chains attached to some point on the machine. One of these chains goes to an eye bolt in the wall and is inclined 30° to horizontal and the other goes to the hook in ceiling and is inclined at 45° to the horizontal. Find tensions in the two chains.

[**Ans.** 2196 N, 2690 N]

12. A small ring is situated at the centre of a hexagon, and is supported by six strings drawn tight, all in the same plane and radiating from the centre of the ring, and each fastened to a different angular points of the hexagon. The tensions in four consecutive strings are 1 N, 3.5 N, 4.5 N and 3 N respectively. Find the tensions in the two remaining strings. [**Ans.** 1.5 N, 6.5 N]

13. A body of weight 45 N is suspended by two strings 18 cm and 61 cm long and other ends being fastened to the extremities of rod of length 63.6 cm. If the rod be so held that the body hangs immediately below its middle point, find out the tensions of the strings. [**Ans.** 43.2 N, 12.6 N]

14. In a simple jib crane, the vertical post is 2.4 m high, the jib is 3.9 metres long and the tie 2.1 metres long. Find the forces on the jib and tie rod when a weight of 2.5 kN is supported at the crane end.

[**Ans.** Jib = 4.05 kN (comp.) ; Tie = 2.9 kN (tension)]

15. A rope *AB*, 2.7 m long is connected at *A* and *B* to two points on the same level, 2.4 m apart. A load of 136 N is suspended from a point *C* on the rope 90 cm from *A*. What load connected to a point *D* on the rope, 60 cm from *B*, will be necessary to keep the portion *CD* level ? [**Ans.** 272 N]

16. A sphere of radius 200 mm and weight 10 N is suspended against a smooth wall by a string of length 200 mm. The string joins a point in the wall and a point on the surface of the sphere. Find the inclination and the tension of the string and the reaction of the wall. $\left[\textbf{Ans. } 60°, T = \frac{20}{\sqrt{3}}\text{ N} ; R = \frac{10}{\sqrt{3}}\text{ N}\right]$

17. Two weights *P* and *Q* are fixed to a string *ABCD* at the points *B* and *C*, the ends *A* and *B* of which are fixed. Portions *AB* and *CD* make angles 50° and 20° respectively with the horizontal and portion *BC* remains horizontal. If *Q* is 12 N, find the value of *P* and tension in portion *BC* of the string.

[**Ans.** 39.29 N ; 32.97 N]

18. Two equal heavy spheres of 5 cm radius are in equilibrium with a smooth cup of 15 cm radius. Show that reaction between the cup and one sphere is double than that between the two spheres.

19. Solve for the reactions at *L, M, N* and *S* on the two cylinders shown in Fig. 2.61, assuming all surfaces to be smooth. [**Ans.** R_L = 62 N ; R_M = 200 N ; R_N = 79.5 N ; R_S = 62 N]

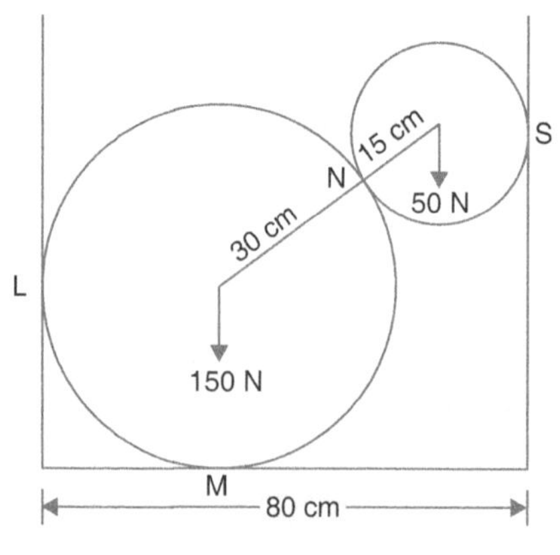

Fig. 2.61

3

Moments

3.1. MOMENTS

The tendency of forces is not only to move the body but also to rotate the body. *This rotational tendency of a force is called* ***moment.*** *The force multiplied by the perpendicular distance from the point to the line of action of the force is called moment about that point.* Unit of moment is equal to the force unit multiplied by the distance unit. It can be in kgfm or Nm etc.

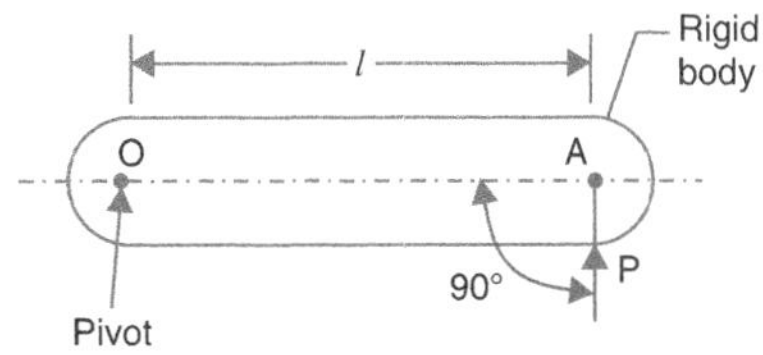

Fig. 3.1

Consider, a finite rigid body capable of rotation about point O as shown in Fig. 3.1. The diagram shows the section of the body in the plane of the paper. The axis of rotation is the line perpendicular to the paper and passing through the point O. Let us apply a force P directed along the paper and acting on the body at the point A. The direction PA is the line of action of the force which is perpendicular to OA. Then the moment (or torque) of the force P about the point O is given by the product of force P and the distance OA, *i.e.*

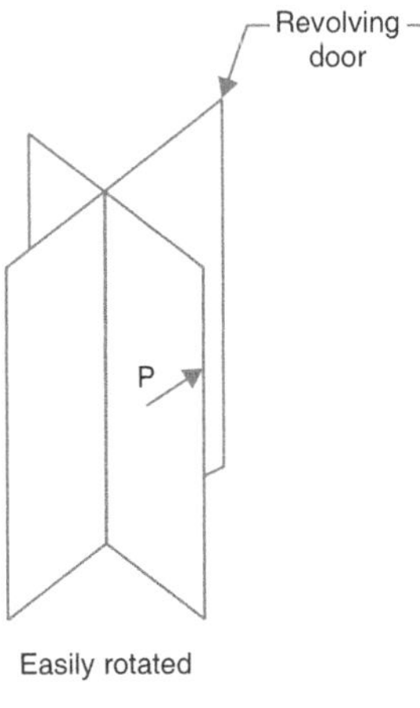

Fig. 3.2

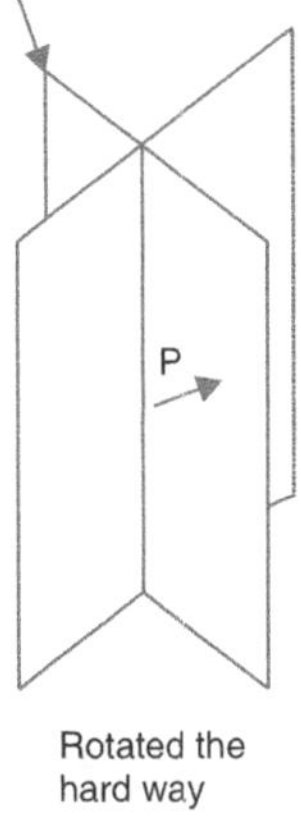

Fig. 3.3

Moment of force $= \textit{force} \times \textit{perpendicular distance}$

$= P \times OA$

$= P \times l$

Moment of a force is vector quantity as it has a magnitude as well as a direction.

It may be noted that the moment of force varies directly with its distance from the pivot. For example, it is much easier to turn a revolving door by pushing at the outer edge of the door, as in Fig. 3.2, than by pushing in the centre, as in Fig. 3.3.

3.2. CLOCKWISE AND ANTI-CLOCKWISE MOMENTS

If a force P is applied to a body in such a way that it tends to rotate the body in the clockwise sense as, shown in Fig. 3.4 (a), then the moment is said to be *clockwise*. If, on the other hand, the force P tends to rotate the body in the anti-clockwise sense, as shown in Fig. 3.4 (b), the moment is said to be anti-clockwise.

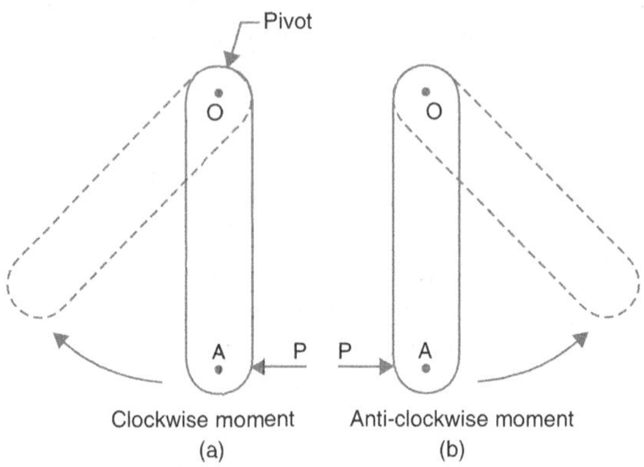

Fig. 3.4

3.3. PRINCIPLE OF MOMENTS

The *principle of moments* may be stated as follows :

"When a body acted upon by several forces is in rotational equilibrium, the sum of the clockwise moments of the forces about any point is equal to sum of the anti-clockwise moments of the forces about the same point."

This law can be verified as follows :

Let us consider, four coplanar forces P_1, P_2, P_3 and P_4 about P_4 acting simultaneously on a body and keeping the body in equilibrium (Fig. 3.5). AA is the axis about which the body can rotate l_1, l_2, l_3 and l_4 are the respective perpendicular distances from O (about which moments are to be taken) to the lines of action of these forces.

The moments of the forces P_1, P_2, P_3 and P_4 about O are given as under :

Moment of force $P_1 = P_1 \times l_1$ clockwise (–)

Moment of force $P_2 = P_2 \times l_2$ anti-clockwise (+)

Moment of force $P_3 = P_3 \times l_3$ clockwise (–)

Moment of force $P_4 = P_4 \times l_4$ anti-clockwise (+)

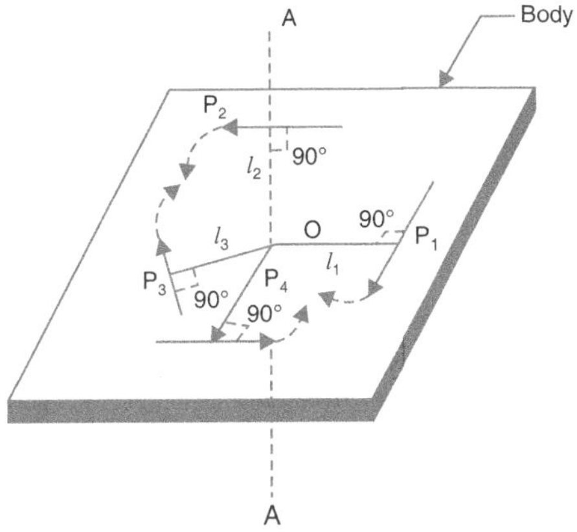

Fig. 3.5

The resultant of these moments will be equal to the algebraic sum of all the moments about O.

$\therefore$ The resultant moment $= -P_1l_1 + P_2l_2 - P_3l_3 + P_4l_4$

Since the body is in rotational equilibrium, according to the principle of moments the resultant moment should be zero.

$$\therefore \quad -P_1l_1 + P_2l_2 - P_3l_3 + P_4l_4 = 0$$

or $$P_1l_1 + P_3l_3 = P_2l_2 + P_4l_4$$

or Sum of anti-clockwise moments = sum of clockwise moments

This verifies the principle of moments.

3.4. EQUILIBRIUM CONDITIONS FOR BODIES UNDER COPLANAR NON-CONCURRENT FORCES

When a body is under the action of a coplanar non-concurrent force system it may rotate due to resultant moment of the force system or it may set in a horizontal or vertical motion due to horizontal and vertical components of forces. The *body, thus can only be in equilibrium if the algebraic sum of all the external forces and their moments about any point in their plane is zero.*

Mathematically, the *conditions of equilibrium* may be expressed as follows :

1. $\Sigma H = 0$ (ΣH means sum of all the horizontal forces)
2. $\Sigma V = 0$ (ΣV means sum of all the vertical forces)
3. $\Sigma M = 0$ (ΣM means sum of all the moments)

When coplanar forces meet in a point the system is known as *coplanar concurrent force system.* This system will be in equilibrium if it satisfies the conditions of equilibrium *viz.* $\Sigma H = 0$ and $\Sigma V = 0$.

When coplanar forces do not meet in a point the system is known as **coplanar non-concurrent force system.** This system will be in equilibrium if it satisfies all the three conditions of equilibrium *viz.* $\Sigma H = 0$, $\Sigma V = 0$, $\Sigma M = 0$.

The conditions $\Sigma H = 0$ and $\Sigma V = 0$ *ensure that the system does not reduce to a single force and condition* $\Sigma M = 0$ *ensures that it does not reduce to a couple.*

In the case of coplanar non-concurrent force system ΣM may be equal to zero but the system may not still be called in equilibrium because the point where the moments are taken about may be lying on the line of action of the resultant. Hence in this case, all the three conditions of equilibrium have to be fulfilled.

3.5. VARIGNON'S THEOREM

It states as under :

"The algebraic sum of the moments of two forces about any point in their plane is equal to the moment of their resultant about that point."

Proof.

Case 1. *When the two forces meet at a point.*

Fig. 3.6 shows two forces P and Q acting at A. The magnitude of P is represented by AB and that of Q is represented by AD. Complete the parallelogram $ABCD$. AC represents the resultant R of P and Q.

Take any other point O in the plane of the forces P and Q and in the line CD produced as shown. Join OB and OA.

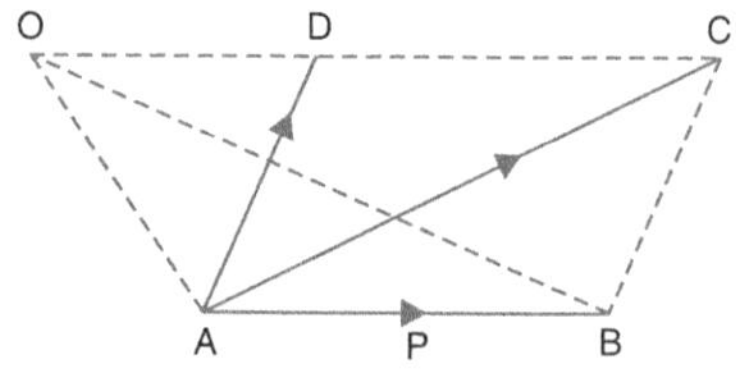

Fig. 3.6

Moment of P about $O = 2\ \Delta OAB$

Moment of Q about $O = 2\ \Delta OAD$

Moment of R about $O = 2\ \Delta OAC$

But area of ΔOAB = area of ΔABC = area of ΔACD

Adding algebraically the moments of P and Q

$= 2\ \Delta OAB + 2\ \Delta OAD$

$= 2\ \Delta ACD + 2\ \Delta OAD$

(Substituting $\Delta 2ACD$ for ΔOAB which are equal)

$= 2\ (\Delta ACD + 2\ \Delta OAD)$

$= 2\ \Delta OAC$

= Moment of R about O.

Note. Refer Fig. 3.7. Consider a force P which can be represented in magnitude and direction by the line AB. Let O be the point, about which the moment of this force is required. Draw OM perpendicular to AB and join OA and OB.

Now moment of force P about $O = P \times OM$

$= AB \times OM$

But $AB \times OM$ is equal to twice the area of triangle OAB because geometrically the area of this triangle is equal to $(AB \times OM)/2$.

$\therefore$ Moment of force P about O *i.e.*, $AB \times OM = 2\ \Delta AOB$.

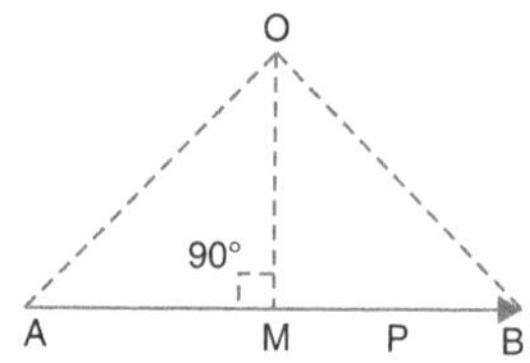

Fig. 3.7

Case 2. *When the two forces are parallel to each other :*

Let P and Q be the two parallel forces as shown in Fig. 3.8. Draw a line AB perpendicular to the forces to meet their lines of action in A and B. Locate any point O in the plane of the two forces on AB produced. The resultant of P and Q will be R which is equal to sum of forces P and Q. Let it act through a point C in AB so that

$$Q \times CB = P \times CA$$

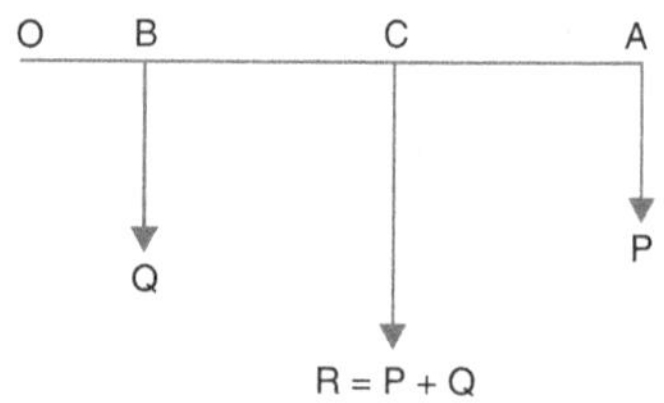

Fig. 3.8

The sum of moments of P and Q about O

$= P \times OA + Q \times OB$

$= P(OC + CA) + Q(OC - CB)$

$= (P + Q)\ OC + P \times CA - Q \times CB$

$= (P + Q)\ OC$ $\qquad (\because\ Q \times CB = P \times CA)$

= Moment of R about O.

Note. *Varignon's theorem is applicable to cases, where the two forces give out a single resultant and it does not apply when the forces form a couple since the resultant force of a couple is zero.*

The principle of moments (discussed earlier) is the extension of Varignon's theorem.

3.6. PARALLEL FORCES

Forces whose lines of action are parallel are called parallel forces. They are said to be *like* when they act in the *same sense*, they are said to be *unlike* when they act in *opposite sense*.

(*i*) **Resultant of two like parallel forces.** Refer Fig. 3.9. Let the like parallel forces P and Q act at the points A and B and let their resultant R cut AB at C. By resolving parallel to P or Q we find that $R = P + Q$ and that R is parallel to P and Q.

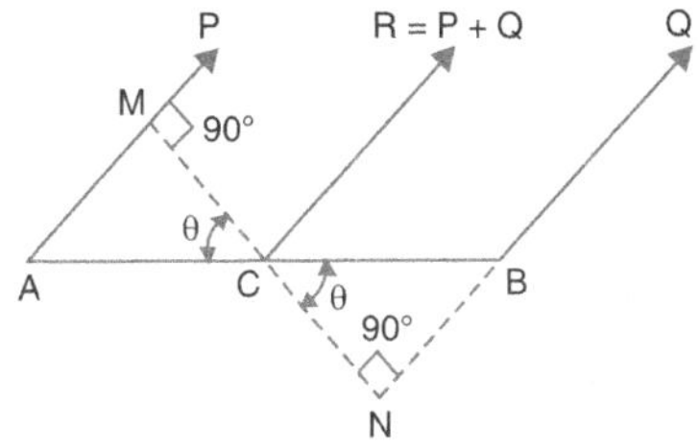

Fig. 3.9

The moment of R about C is zero, so that the algebraic sum of moments of P and Q about C must also be zero.

Through C draw MN perpendicular to P and Q

then
$$CM = AC \cos \theta$$
$$CN = BC \cos \theta$$

Taking moments about C,
$$P \times CM = Q \times CN$$
$$P \times AC \cos \theta = Q \times BC \cos \theta$$
$$P \times AC = Q \times BC$$
or
$$P/BC = Q/AC$$

$\therefore$ C divides AB internally in the inverse ratio of the forces.

(*ii*) **Resultant of two unlike parallel forces.** Refer Fig. 3.10. Let two unlike parallel forces P and Q act at A and B, and let their resultant R meet AB at C. Let P be greater than Q. By resolving parallel to P or Q, we get
$$R = P - Q$$
acting in the same sense as P.

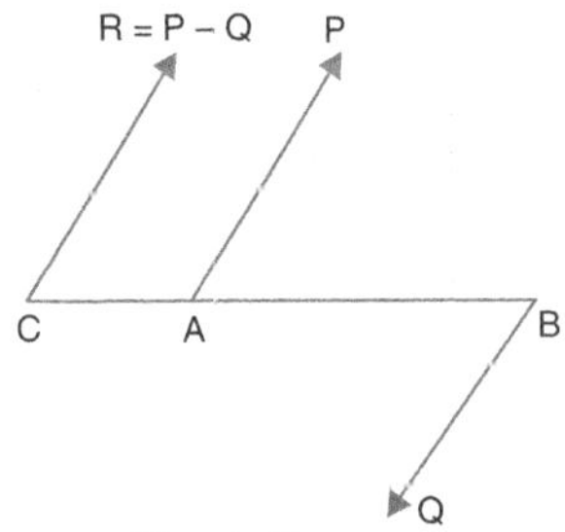

Fig. 3.10

The algebraic sum of the moments of P and Q about C must be zero so that these moments must be equal and opposite. Hence C must lie outside AB, and it must be nearer to A than to B.

Taking moments about C, we get
$$P \times AC = Q \times BC$$
or
$$P/BC = Q/AC$$

$\therefore$ C divides AB externally in the inverse ratio of the forces.

The point C is called the *centre of parallel forces*. It is clear that the *position of C is independent of the inclination of the forces to AB.*

3.7. GRAPHICAL METHOD FOR FINDING THE RESULTANT OF ANY NUMBER OF LIKE OR UNLIKE PARALLEL FORCES

Refer Fig. 3.11 (*a*). P, Q, R, S are the parallel forces P, Q and S in one direction and R in the opposite direction.

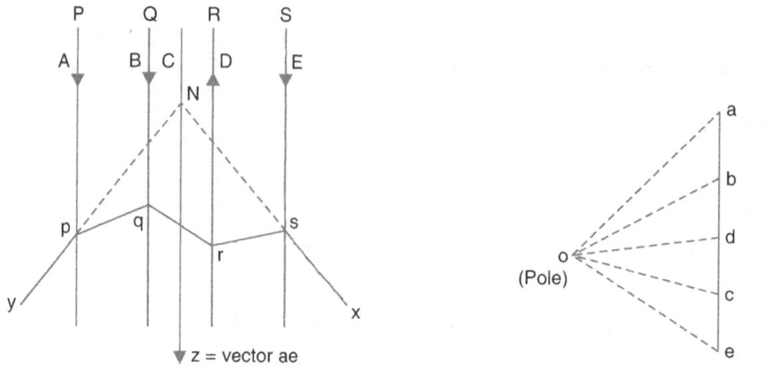

Space and funicular polygon diagram
(a)

Polar and vector diagram
(b)

Fig. 3.11

The various steps involved in finding the resultant by graphical method are given below :

(*i*) Put letters *A*, *B*, *C*, *D* and *E* in the space about *P*, *Q*, *R*, *S*.

(*ii*) To convenient scale mark off *ab* equal to force *P*, *bc* equal to force *Q*, *cd* equal to force *R* in the opposite direction and *de* equal to force *S*.

ae represents the resultant in magnitude and direction and its line of action is found as follows : Refer Fig. 3.11 (*b*).

(*i*) Take any point *O* (called pole) outside the vector diagram *abcde* and join *oa*, *ob*, *oc*, *od* and *oe*.

(*ii*) Take any points on force *S* and the point *s* in the space *E* draw a line parallel to *oe*.

(*iii*) Similarly, in the spaces *D*, *C*, *B* and *A* draw lines *sr, rq, qp* and *py* parallel to *od, oc, ob*, and *oa* respectively.

(*iv*) Produce lines *xs* and *yp* to meet each at *N*.

The resultant force *Z* which passes through point *N* is completely given by *ae* in magnitude and direction.

3.8. COUPLE

A couple is pair of two equal and opposite forces acting on a body in a such a way that the lines of action of the two forces are not in the same straight line.

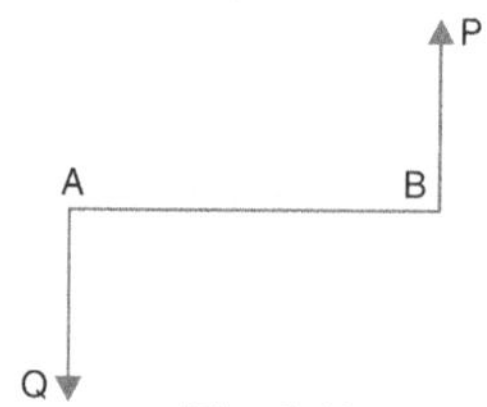

Fig. 3.12

The effect of a couple acting on a rigid body is to rotate it without moving it as a whole. The movement of the whole body is not possible because the resultant force is zero in the case of forces forming a couple. The perpendicular distance between the lines of action of two forces forming the couple is called the *arm* of *couple*. Thus, in Fig. 3.12 two equal forces of magnitude *P* and acting at points *A* and *B* in the opposite direction form a couple with *AB* as arm of the couple.

The moment of a couple is known as torque which is equal to one of the forces forming the couple multiplied by arm of the couple.

The following are the examples of couples in every day life.

1. Opening or closing a water tap. The two forces constitute a couple as shown in Fig. 3.13.
2. Turning the cap of a pen.
3. Unscrewing the cap of an ink bottle.
4. Twisting a screw driver.
5. Steering a motor-car (Fig. 3.14).
6. Winding a watch or clock with a key.

Fig. 3.13

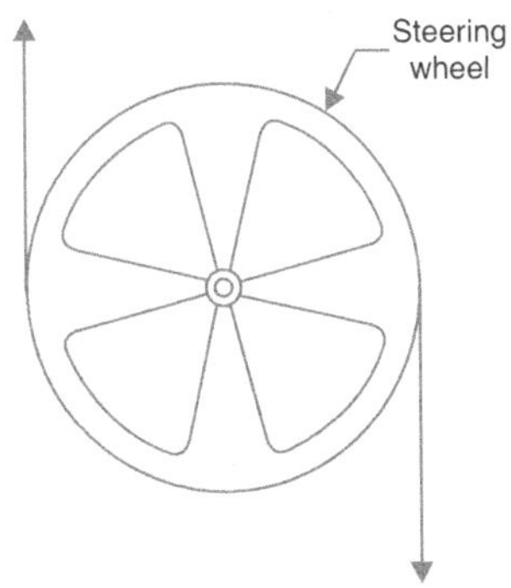

Fig. 3.14

3.9. PROPERTIES OF A COUPLE

1. The algebraic sum of the moments of the forces forming a couple about any point in their plane is constant. Let two parallel and unlike forces be of magnitude P each forming a couple $P \times AB$ where points A and B are the points where forces P and P act.

Refer Fig. 3.15 (a)

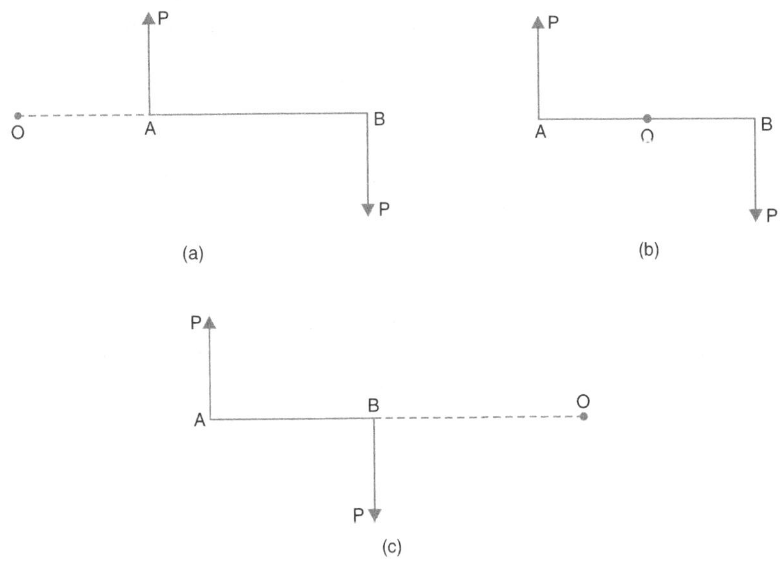

Fig. 3.15

Moments about O $= P \times OB - P \times OA$

$= P(OB - OA)$

$= P \times AB$

Refer Fig. 3.15 (*b*),

Moments about O $= P \times OB + P \times OA$

$= P(OB + OA)$

$= P \times AB$

Refer Fig. 3.15 (*c*),

Moments about O $= P \times OA - P \times OB$

$= P(OA - OB)$

$= P \times AB.$

In all the three cases, we find that the sum of the moments in each case is independent of the position of the point O, and depends only on the constant arm of the couple, so **the algebraic sum of moments of the forces forming a couple about any point in their plane is constant.**

2. Any two couples of equal moments and sense, in the same plane are equivalent in their effect. This result is quite useful as it clearly states that moment is the only important thing about a couple. Thus, in a couple we may change the magnitude or direction of the forces or the arm of the couple itself without changing its effect provided that the new couple with changed values has the same moment in the same sense.

3. Two couples acting in one place upon a rigid body whose moments are equal but opposite in sense, balance each other.

4. A force acting on a rigid body can be replaced by an equal like force acting at any other point and a couple whose moment equals the moment of the force about the point where the equal like force is acting.

5. Any number of coplanar couples are equivalent to a single couple whose moment is equal to the algebraic sum of the moments of the individual couples.

3.10. ENGINEERING APPLICATIONS OF MOMENTS

Some of the important engineering applications of moments are discussed below :

1. The levers (simple curved, bent or cranked and compound levers).
2. The balance.
3. The common steel yard.
4. Lever safety valve.

3.10.1. The Lever. The lever is defined as a *rigid bar, straight or curved which can turn about a fixed point* called the **'fulcrum'.** It works on the *principle of moments i.e.*, when the lever is in equilibrium the algebraic sum of the moments, about the fulcrum, of the forces acting on it is zero.

The principle of lever was first developed by Archimedes. Some common examples of the use of lever are :

(*i*) Crow bar ;

(*ii*) A pair of scissors ;

(*iii*) Fire tongs etc.

Refer to Fig. 3.16.

Power arm. The perpendicular distance between the fulcrum and the line of action of forces is known as *power arm*.

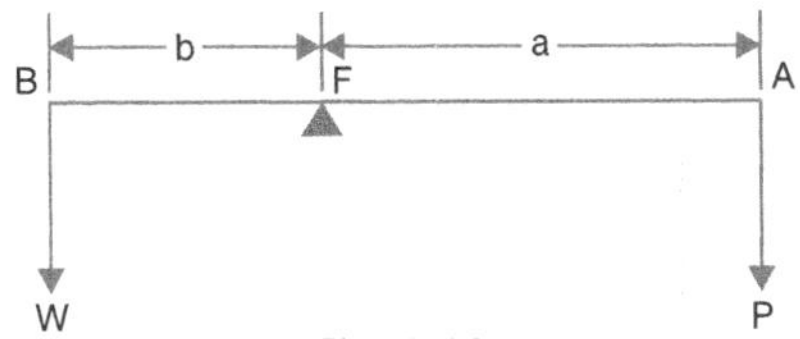

Fig. 3.16

Weight arm. The perpendicular distance between the fulcrum and the point where the load/weight acts is called *weight* (or load) arm.

The principle of moments is applicable when the lever is in equilibrium.

Taking moments about F

$$P \times a = W \times b$$

Mechanical advantage of the lever

$$= \frac{W}{P} = \frac{a}{b} = \frac{\text{power arm}}{\text{weight arm}}$$

or **Power × power arm = weight × weight arm.**

This is known as the *principle of the lever*. Fig. 3.17 shows a crow bar used to move a heavy log of wood with the help of a small effort applied at its end with fulcrum suitably placed.

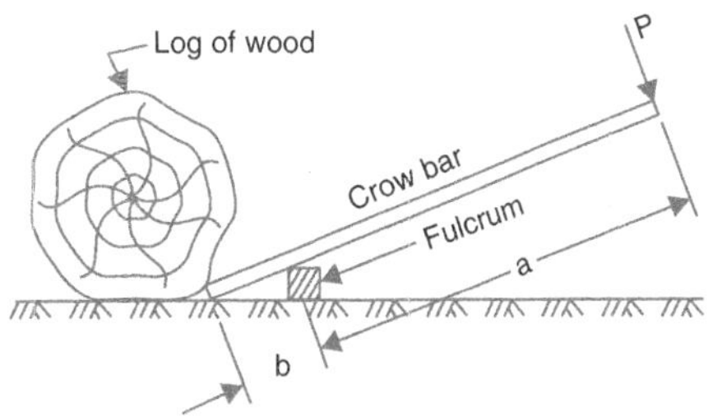

Fig. 3.17

The levers may be of the following types :

(*a*) Simple levers

(*b*) Curved bent or cranked levers

(*c*) Compound levers.

(*a*) **Simple levers.** *A simple lever is one which has only one fulcrum*. These levers consist of only one power arm and one weight arm as shown in Fig. 3.18.

Depending upon the position of their fulcrum, load and effort, simple levers are further divided into the following three types :

Type 1. *Fulcrum between W and P* [Fig. 3.18. (*a*)]

Reaction R on the fulcrum = $P + W$

Type 2. *Weight between F and P* [Fig. 3.18 (*b*)]

Reaction R on the fulcrum = $W - P$

Type 3. *Power between W and F* [Fig. 3.18 (*c*)]

Reaction R on the fulcrum = $P - W$

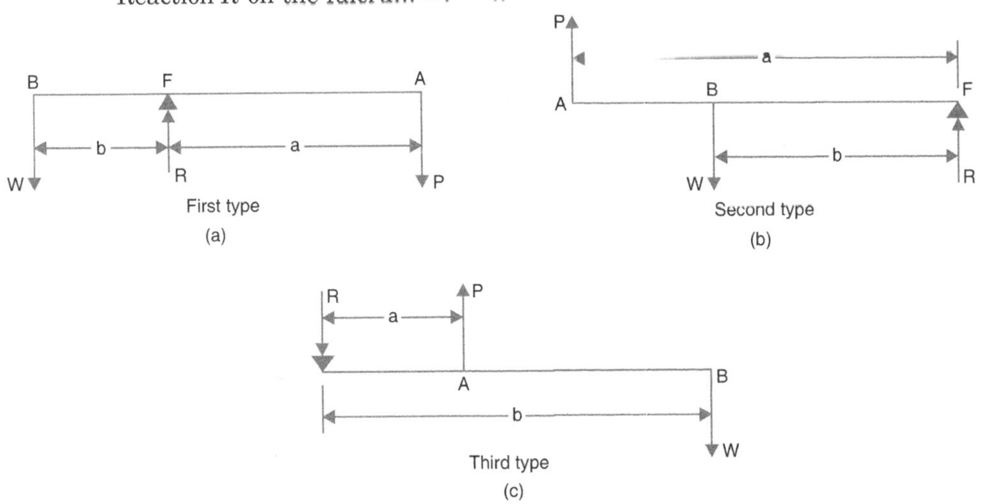

Fig. 3.18

In all the three types of lever, taking moments about fulcrum F, in each case, we have

$$P \times AF = W \times BF$$

∴ Mechanical advantage

$$\text{M.A.} = \frac{W}{P} = \frac{AF}{BF} = \frac{\text{power arm}}{\text{weight arm}}.$$

(*b*) **Curved bent or cranked levers.** In this category of levers, the two arms may be inclined to each other at any angle varying from 0° to 180°. The main plus point associated with these levers is that effort can be applied in convenient direction to overcome the resistance in the desired direction. Fig. 3.19 shows a cranked lever in which effort P is acting horizontally at a perpendicular distance 'y' from the fulcrum and weight W is acting vertically downwards at a perpendicular distance 'x' from the fulcrum F.

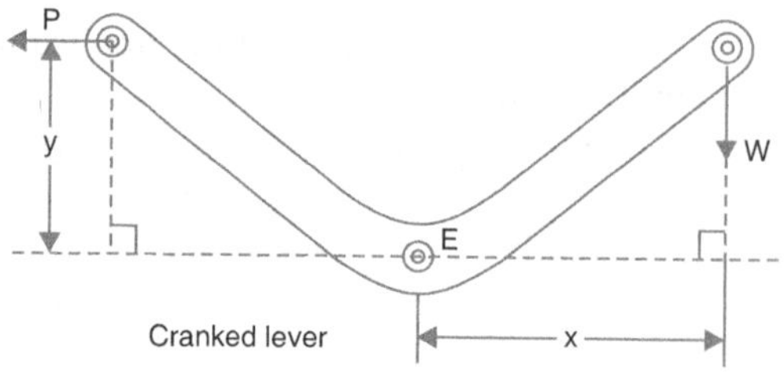

Fig. 3.19

By principle of moments,

$$P \times y = W \times x$$

Fig. 3.20 shows a *bell crank lever* which is an important type of cranked lever. In this case, the two arms of the lever are at 90°. Effort is applied horizontally and load is lifted vertically upwards.

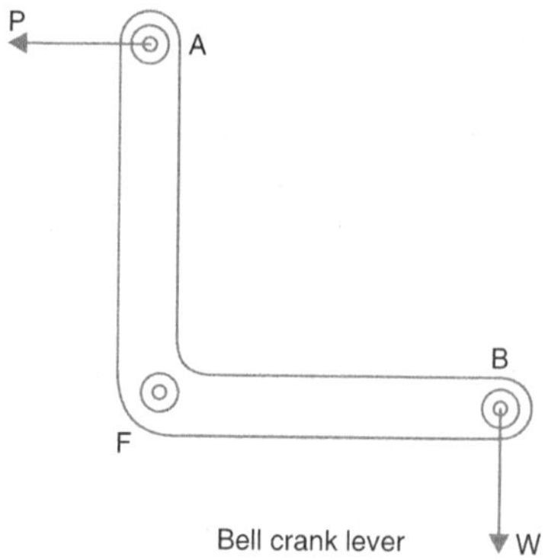

Fig. 3.20

By principle of moments,

$$P \times AF = W \times BF$$

(*c*) **Compound levers.** A *compound lever* is one which comprises of *several levers* to get an increased leverage to overcome greater resistance than that which can be handled by the force in a single lever. A compound lever and the free body diagrams for the solution are shown in Fig. 3.21. In this case $R_M = R_N$.

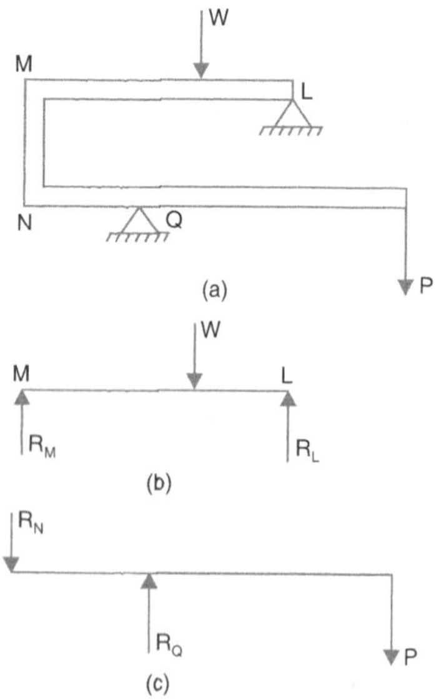

Fig. 3.21

3.10.2. The Balance. Fig. 3.22 shows a common balance which is just a lever of the first type. It is device employed for weighing goods. It consists of a rigid beam having two scale pans suspended from each end. The beam can turn freely about a fulcrum F which is outside the beam but rigidly connected with it. A pointer is attached to the beam at its middle point M. A scale is also provided as shown in Fig. 3.22.

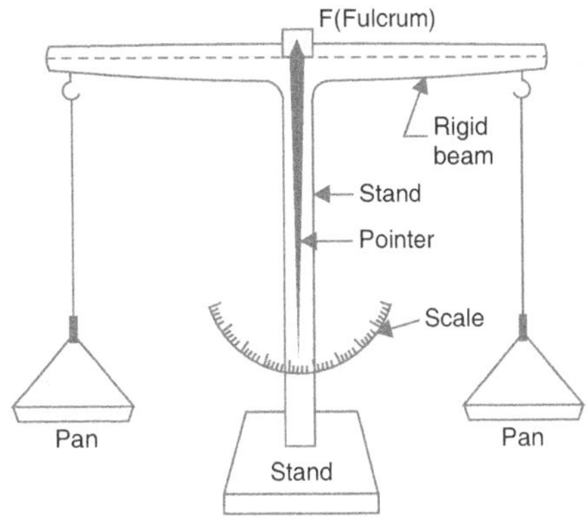

Fig. 3.22. The balance.

For true balance the following conditions must be satisfied :

1. The arms of the balance must be of equal size.
2. The weights of scale pans must be equal.
3. The fulcrum, the centre of rigid beam and *c.g.* of the beam including its connected parts must lie on the line perpendicular to the beam.

If the above conditions are not satisfied the problem of finding the unknown values from known values may be solved by taking moments about the fulcrum F.

3.10.3. The Common Steel Yard. The common steel yard is a device used for weighing goods based on type 1 lever. It consists of a rod AB which can move about a fixed fulcrum F which is kept near one of the ends. The rod is made heavier on the shorter side so that the *c.g.* of the whole rod and pan which is attached to the end of the shorter side lies on the shorter side. The rider (movable weight) is kept on the longer side as shown in Fig. 3.23. The longer arm on which the rider moves is graduated. The position of rider determines the weight of the body acting on shorter side end when the rod rests in a horizontal position.

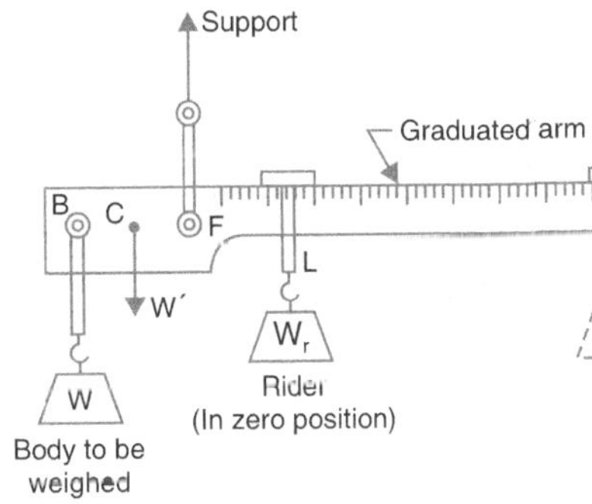

Fig. 3.23. The common steel yard.

Let W' be the weight of rod AB along with the pan attached to the pointer on point L will give zero graduation.

Taking moments about F in this case

$$W' \times GF = W_r \times LF$$

$$\therefore \qquad LF = \frac{W'}{W_r} \times GF \qquad ...(i)$$

W', W_r and GF are known, the value of LF can be calculated which gives zero reading from the fulcrum F.

Now, put a known weight W say 10 units in the pan and move the rider of weight W_r towards the end A on to point M so that the rod becomes horizontal.

Taking moments about F in this case

$$W \times FB + W' \times GF = W_r \times MF$$

$$\therefore \quad MF = \frac{W \times FB + W' \times GF}{W_r} \quad ...(ii)$$

The position of movable weight W_r on M indicates 10 units weight of the body. From the equation (*i*), we have already got the position L indicating zero reading. Thus, $MF - LF = LM$ ranges from zero to 10 units weight of the body. LM is divided into 10 equal parts and position of movable weight on the end of first part from L will give reading of the weight as 1 unit, placed on the end of second part from L will give reading of the weight as 2 units and so on.

It may be noted that the steel yard will give the value of the weight of the body in the same units in which its calibration is made.

3.10.4. Lever Safety Valve. A lever safety valve is *a boiler mounting the purpose of which is to keep the steam pressure in the boiler upto certain safe values* and to the release the same when the pressure increases the safe limits.

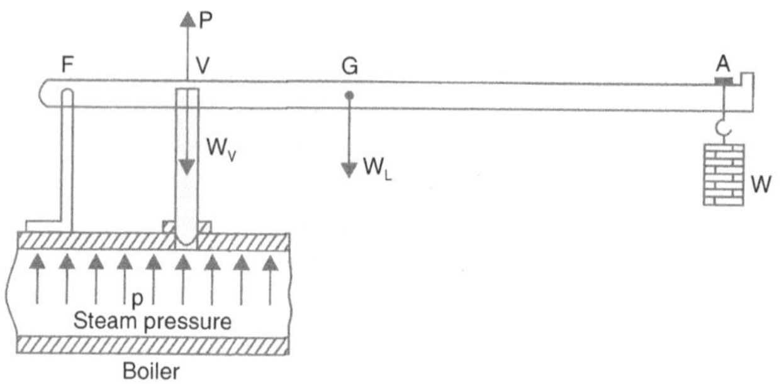

Fig. 3.24. Lever safety valve.

Refer to Fig. 3.24. It consists of a valve V rigidly connected with the lever FA, whose fulcrum is at F. At the end A, a weight W is hung which exerts moment on to the valve to keep it placed on its seat against the steam pressure from below, which exerts a counter moment about the fulcrum F. As soon as the moment due to steam pressure increases, the valve lifts up its seat and releases excessive pressure into the atmosphere. When the steam pressure inside the boiler falls down to safe value the valve automatically occupies its seat and stops further escape of steam.

Let W_L = weight of the lever (acting at its *c.g.* point G)

W_V = weight of the valve

W = weight attached at end A

p = safe intensity of pressure of steam

A = area of the valve (= $\pi/4\ d^2$ where d is the diameter of the valve).

In order to find out the value of load W which will first keep the valve on its seat against steam pressure, let us take moments about the fulcrum F,

$$W \times AF + W_L \times GF + W_V \times VF = p \times A \times VF$$

$$\therefore \quad W = \frac{VF\,(P \times A - W_V) - W_L \times GF}{AF}$$

Since all quantities except W and AF known, therefore W can be calculated.

and reaction at $F = p \times A - W_L - W_V - W$

This reaction will act downward when steam pressure is more and upwards when it is lesser than the other downward forces.

3.11. RESULTANT OF A COPLANAR, NON-CONCURRENT NON-PARALLEL FORCE SYSTEM

(*i*) The magnitude, direction and position of the resultant of a given coplanar, non-concurrent, non-parallel force system are found analytically as follows :

$$R = \sqrt{(\Sigma H)^2 + (\Sigma V)^2}$$

where ΣH = algebraic sum of the horizontal components of all the forces.

ΣV = algebraic sum of vertical components of all the forces.

(*ii*) The *direction* of the resultant is determined by using the relation,

$$\tan \alpha = \frac{\Sigma V}{\Sigma H}.$$

(*iii*) The *position* of the resultant is determined by taking moments of all the rectangular components of forces about a point in their plane and equating the algebraic sum of moments of all the forces to that of the resultant by using the relation,

Moment of resultant '*R*' about the point

= algebraic sum of rectangular components of all the forces.

WORKED EXAMPLES

Example 3.1. *A man and a boy carry a weight of 300 N between them by means of uniform pole 2 m long and weighing 100 N. Where must the weight be placed so that the man may bear twice as much of the weight as that of the boy ?*

Sol. Refer to Fig. 3.25. The weight of the pole acts at its centre *G*. Let boy bears weight *W* and the man bears 2*W*.

When the weight of 300 N acts at a distance *x* metres from the man.

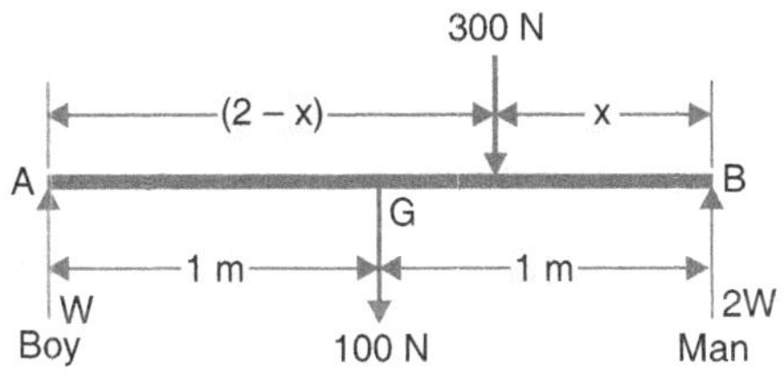

Fig. 3.25

$$\Sigma V = 0$$
$$W + 2W = 300 + 100 = 400$$
$$\therefore \quad W = 133.3 \text{ N}$$

Taking moments about *A* (boy)

$$2W \times 2 = 300\,(2 - x) + 100 \times 1$$
$$4W = 600 - 300x + 100$$
$$4 \times 133.3 = 700 - 300x$$
$$533.2 = 700 - 300x$$
$$300x = 166.8$$
$$\therefore \quad \boldsymbol{x = 0.556 \text{ m.}} \textbf{ (Ans.)}$$

Example 3.2. *A rod of length 5 metres and weight 15 N has its centre of gravity at a distance of 2 metres from A. It rests on two parallel smooth pegs at a distance of 3 metres apart in the same horizontal plane so that equal portions of the rod project beyond the peg. Find the reactions at the pegs.*

Sol. Refer to Fig. 3.26. Let *G* be the *c.g.* of the rod *AB*, *C* and *D* be the pegs. Let *R* and *S* be the reactions at the pegs *C* and *D* respectively.

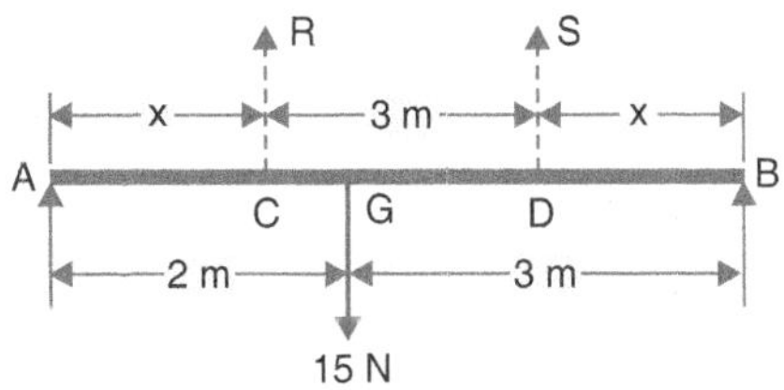

Fig. 3.26

Now $\quad R + S = 15$

$$S = 15 - R \quad \quad ...(i)$$

Let $\quad AC = x, DB = x$

Taking moments about C, we get

$$15 \times CG = S \times 3$$
$$15(2 - x) = S \times 3$$
$$10 - 5x = S$$
$$10 - 5x = 15 - R \quad \text{From } (i)$$
$$R - 5x = 5 \quad ...(ii)$$

Taking moments about D, we get

$$R \times CD = 15 \times GD$$
$$R \times 3 = 15 \times (3 - x)$$
$$R = 5(3 - x) = 15 - 5x \quad ...(iii)$$

On solving eqns. (*ii*) and (*iii*), we get

$$x = 1 \text{ m}, \mathbf{R = 10 \text{ N}, S = 5 \text{ N}}. \textbf{ (Ans.)}$$

Example 3.3. *Fig. 3.27 shows a bell crank lever. Determine the magnitude of the load lifted perpendicular to the weight arm FB, 5 cm long when a force of 100 N is applied at A perpendicular to power arm FA, 20 cm long.*

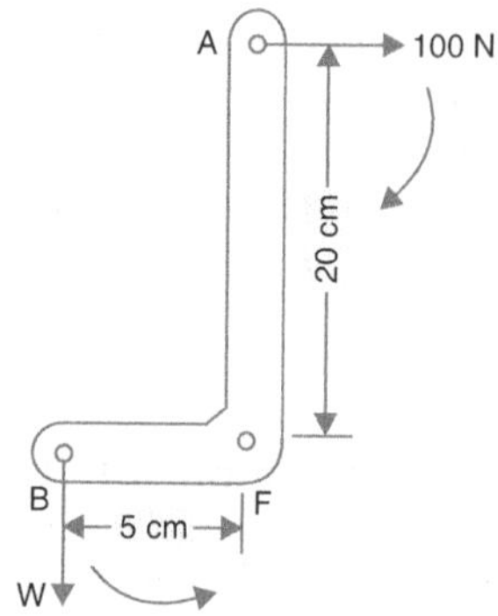

Fig. 3.27

Sol. Taking moments about F,

$$100 \times 20 = W \times 5$$

$$\therefore \quad W = \frac{100 \times 20}{5}$$
$$= \mathbf{400 \text{ N}}. \textbf{ (Ans.)}$$

Example 3.4. *The lever of a safety valve weighs 40 N and its c.g. lies at a distance 9 cm from the centre of the valve which is at 8 cm from the fulcrum. If the diameter and weight of the valve are respectively 7 cm and 60 N find the minimum weight that will be hung to the end of 80 cm long lever to conserve steam in the boiler at a pressure of 100 N/cm².*

Sol. Refer to Fig. 3.28. (Art. 3.10.4)

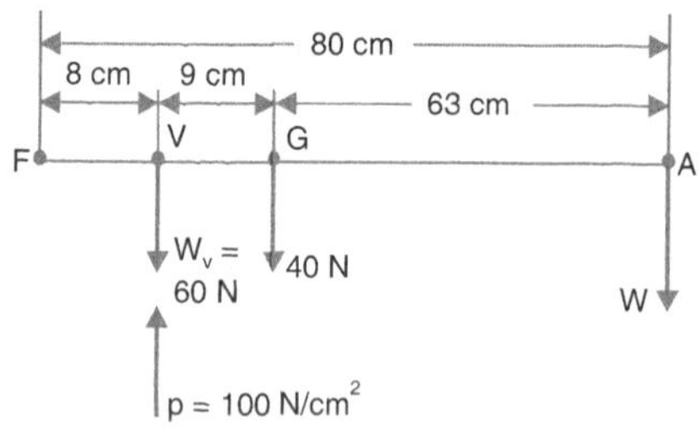

Fig. 3.28

Area of the valve, $A = \frac{\pi}{4} d^2$, where d is the diameter of the valve

$$= \frac{\pi}{4} \times 7^2 = 38.48 \text{ cm}^2$$

Let W be the weight attached to the end A.

Taking moments about the fulcrum F, we get

$$W \times 80 + 40 \times (9 + 8) + 60 \times 8 = p \times A \times 8 \quad (p = \text{pressure of steam})$$
$$W \times 80 + 680 + 480 = 100 \times 38.48 \times 8$$
$$\therefore \quad \mathbf{W = 370.3 \text{ N}}. \textbf{ (Ans.)}$$

Example 3.5. *The lever of a common steel yard without hook connections weighs 80 N and acts at 5 cm to the left of the fulcrum. The hook which carries the load, weighs 30 N and is also attached on the left hand side of the fulcrum at a distance of 12 cm. If the graduation mark 75 cm from the fulcrum reads 550 N, find*

(*i*) *The weight of the rider.*

(*ii*) *How far is the zero graduation from the fulcrum ?*

(*iii*) *What is the movement of rider per 10 N of graduation ?*

Sol. Refer to Fig. 3.29. (Art 3.10.3)

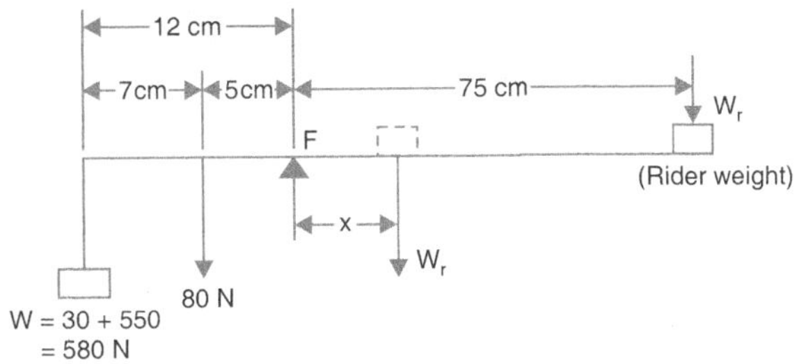

Fig. 3.29

(*i*) Let W_r be the weight of rider. With its position at a distance of 75 cm from the fulcrum the weight attached at the hook is 550 N (*i.e.*, total weight = 550 + 30 N = 580 N).

Taking moments about fulcrum F, we get

$$W_r \times 75 = 80 \times 5 + 580 \times 12$$

$$= 400 + 6960$$

$\therefore$ $\mathbf{W_r = 98.1\ N.}$ **(Ans.)**

(*ii*) Let the position of the rider be at a distance of x with no load on the hook.

Taking moments about the fulcrum, we have

$$W_r \times x - 30 \times 12 + 80 \times 5$$

$$98.1x = 360 + 400$$

$\therefore$ $\mathbf{x = 7.74\ cm.}$ **(Ans.)**

(*iii*) Distance from zero graduation to 550 N graduation

$$= 75 - 7.74 = 67.26 \text{ cm}$$

Distance per 10 N graduation

$$= \frac{67.26}{55} = \mathbf{1.22\ cm.\ (Ans.)}$$

Example 3.6. *Forces 1P, 3P, – 4P act in order round the sides of an equilateral triangle of side 20 mm drawn on a rigid lamina. Find the magnitude, direction and position of their resultant (line of action).*

Sol. Refer to Fig. 3.30.

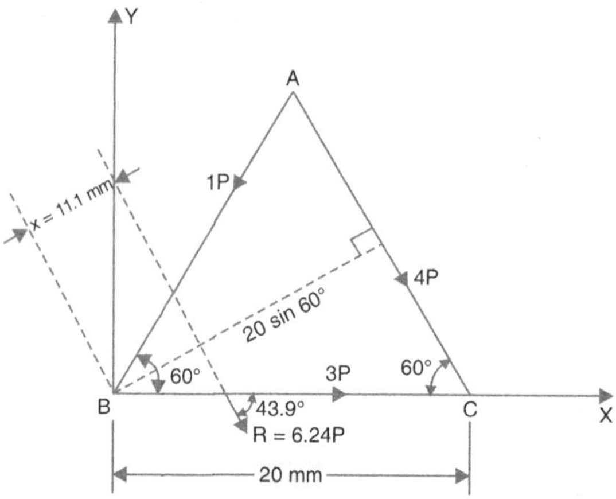

Fig. 3.30

Magnitude of the resultant force R = ?

Resolving the forces *horizontally* :

$$\Sigma H = -1P \cos 60° + 3P + 4P \cos 60°$$
$$= -0.5P + 3P + 2P = 4.5P$$

Resolving the forces *vertically* :

$$\Sigma V = -1P \sin 60° - 4P \sin 60° = -4.33P$$

Resultant force, $R = \sqrt{(\Sigma H)^2 + (\Sigma V)^2} = \sqrt{(4.5P)^2 + (-4.33P)^2}$

$$= \mathbf{6.24\ P.\ (Ans.)}$$

Direction of the resultant θ = ?

$$\tan \theta = \frac{\Sigma V}{\Sigma H} = \frac{-4.33P}{4.5P} = -0.962$$

∴ $\theta = -$ **43.9° (with horizontal). (Ans.)**

$$\left[\begin{array}{c}\Sigma H\ (+\text{ve})\text{ and }\Sigma V\ (-\text{ve})\text{ indicate that angle}\\ \theta\text{ lies in the 4th quadrant (Fig. 3.31)}\end{array}\right]$$

Position of the resultant force = ?

Let x = perpendicular distance between B and the line of the resultant force.

Now, taking moments about B, we get

$$\Sigma M_B : 6.24P \times x = P \times 0 + 3P \times 0 + 4P \times 20 \sin 60°$$

∴ $\mathbf{x = 11.1\ mm.\ (Ans.)}$

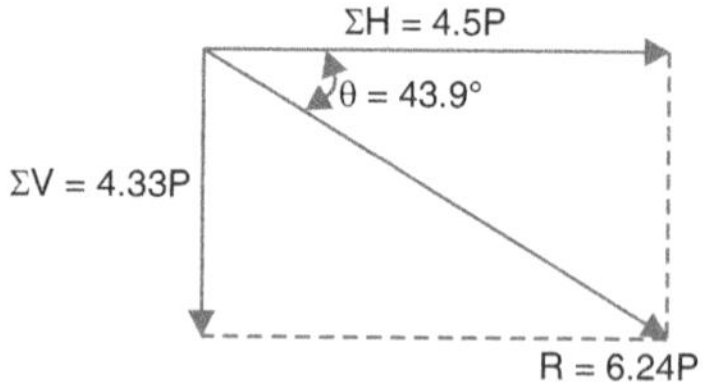

Fig. 3.31

Example 3.7. *Forces 1 kN, 2 kN and 3 kN act along the sides AB, BC and CA of a given equilateral triangle ABC. Find the magnitude and direction of their resultant, and find the point in which its line of action meets the side BC.*

Sol. Refer to Fig. 3.32.

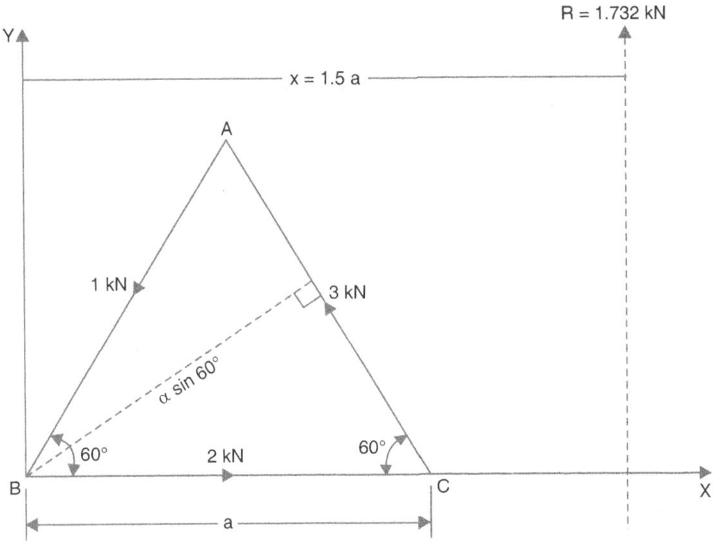

Fig. 3.32

Take B as the origin, BC as X-axis and a perpendicular through B as Y-axis. Let each side of the equilateral triangle, ABC be 'a'.

Magnitude of the resultant, R = ?

Resolving the forces horizontally :

$$\Sigma H = -1 \cos 60° + 2 - 3 \cos 60°$$
$$= -0.5 + 2 - 1.5 = 0$$

Resolving the forces *vertically* :

$$\Sigma V = -1 \sin 60° + 3 \sin 60°$$
$$= 1.732 \textbf{ kN}\uparrow$$

Resultant, $$R = \sqrt{(\Sigma H)^2 + (\Sigma V)^2}$$
$$= \sqrt{0 + (1.732)^2} = \textbf{1.732 kN} \uparrow \textbf{(Ans.)}$$

Direction of the resultant, θ = ?

$$\tan \theta = \frac{\Sigma V}{\Sigma H} = \frac{1.732}{0} = \infty$$

i.e., $\theta = \mathbf{90°}$.

Position of the resultant force = ?

Taking moments about B, we get

$$1.732 \times x = 3 \times a \sin 60°$$

$\therefore$ $\boldsymbol{x = 1.5\ a.}$ **(Ans.)**

Example 3.8. *Four forces equal to 10 N, 20 N, 30 N and 40 N are respectively acting along the four sides (1 m each) of a square ABCD, taken in order. Find the magnitude, direction and position of the resultant force.*

Sol. Fig. 3.33 shows the system of the given forces.

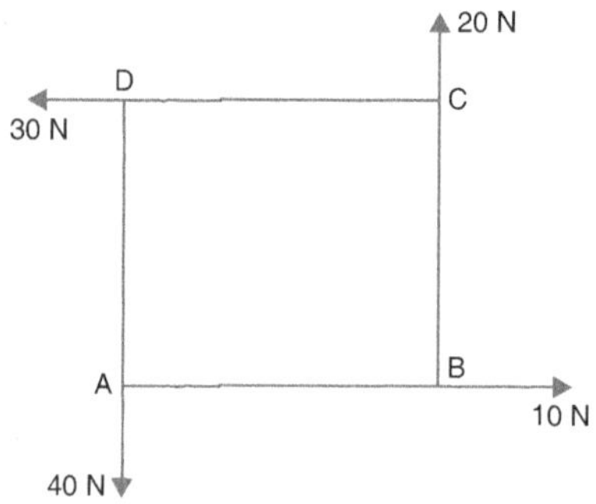

Fig. 3.33

Magnitude of the resultant force, R = ?

Resolving all the forces horizontally,

$$\Sigma H = 10 - 30 = -20 \text{ N}$$

and resolving all the forces vertically,

$$\Sigma V = 20 - 40 = -20 \text{ N}$$

Now, resultant force

$$R = \sqrt{(\Sigma H)^2 + (\Sigma V)^2}$$

$$= \sqrt{(-20)^2 + (-20)^2}$$

$$= 28.28 \text{ N}$$

i.e., $\boldsymbol{R = 28.28}$ **N. (Ans.)**

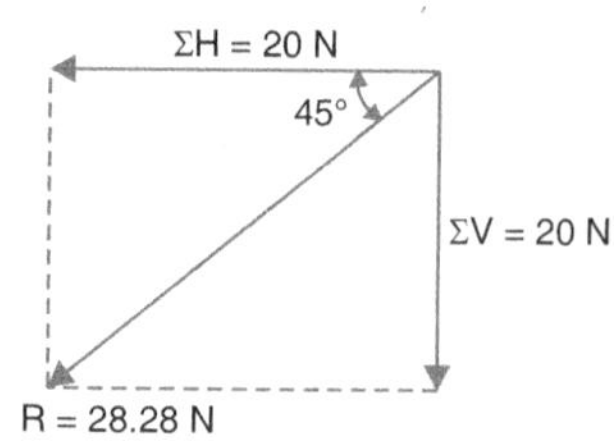

Fig. 3.34

Direction of the resultant force ?

Let θ = angle which the resultant makes with the horizontal

$$\therefore \quad \tan\theta = \frac{\Sigma V}{\Sigma H} = \frac{-20}{-20} = 1$$

or $\theta = 45°$

Since ΣH as well ΣV is –ve, therefore θ lies between 180° and 270°

∴ Actual θ = 180 + 45° = **225°. (Ans.)**

Position of the resultant force = ?

Let x = perpendicular distance between A and the line of the resultant force.

Now taking moments about A, we get

$$\Sigma M_A : 28.28 \times x = 10 \times 0 + 20 \times 1 + 30 \times 1 + 40 \times 0 = 50$$

$$\therefore \quad \boldsymbol{x = \frac{50}{28.28} = 1.768} \textbf{ m. (Ans.)}$$

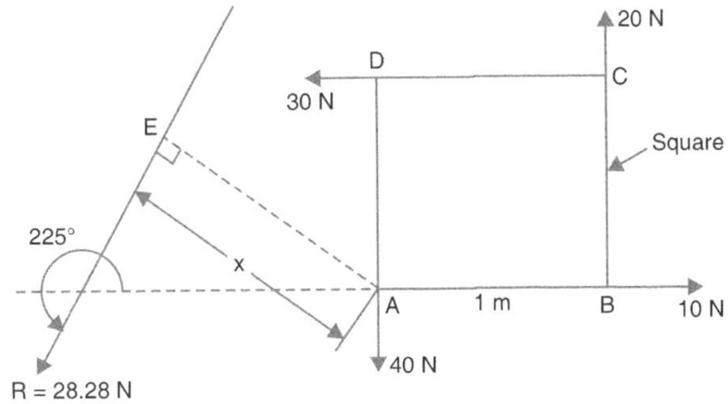

Fig. 3.35

Example 3.9. *A body is under the action of four coplanar forces as shown in Fig. 3.36. Find the magnitude, direction and position of resultant of the given force system.*

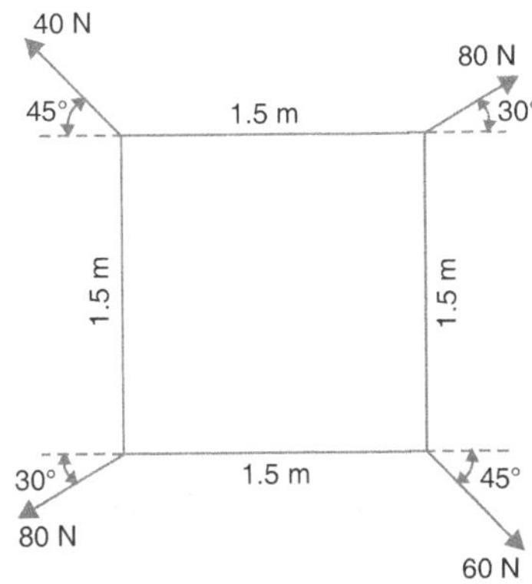

Fig. 3.36

Sol. Refer to Fig. 3.37.

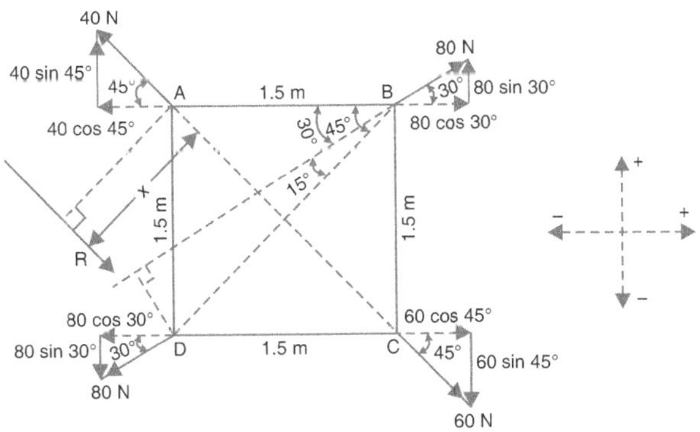

Fig. 3.37

Magnitude of resultant, *R* = ?

Resolving the forces *horizontally*,

$$\Sigma H = 80 \cos 30° + 60 \cos 45° - 80 \cos 30° - 40 \cos 45°$$

$$= 80 \times 0.866 + 60 \times 0.707 - 80 \times 0.866 - 40 \times 0.707 = 14.14 \text{ N}.$$

Resolving the forces *vertically* :

$$\Sigma V = 80 \sin 30° - 60 \sin 45° - 80 \sin 30° + 40 \sin 45° = -14.14 \text{ N}$$

Resultant, $R = \sqrt{(\Sigma H)^2 + (\Sigma V)^2}$

$$= \sqrt{(14.14)^2 + (-14.14)^2}$$

$$= 20 \text{ N}$$

i.e., **R = 20 N. (Ans.)**

Direction of resultant, α = ?

The resultant will act at an angle with the horizontal so that,

$$\tan \alpha = \frac{\Sigma V}{\Sigma H} = \frac{14.14}{14.14} = 1$$

or $\alpha = 45°$. (Fig. 3.38).

Position of the resultant = ?

The position of the resultant 'R' can be determined by using the relation :

Moment of resultant about A = algebraic sum of moments of the rectangular components of all forces about A.

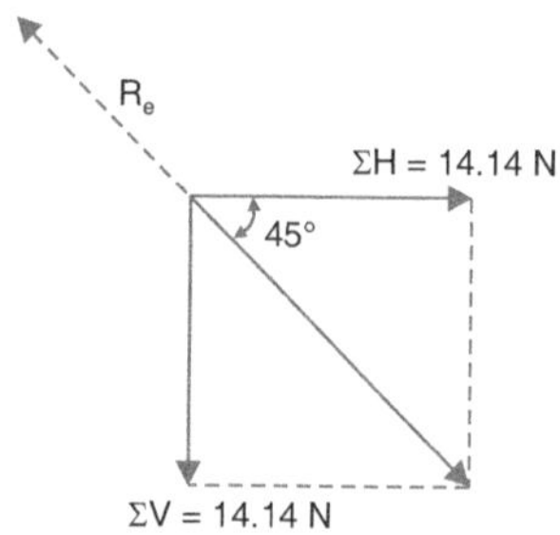

Fig. 3.38

$$-R \times x = 40 \sin 45° \times 0$$
$$+ 40 \cos 45° \times 0$$
$$+ 80 \sin 30° \times 1.5$$
$$+ 80 \cos 30° \times 0$$
$$+ 60 \cos 45° \times 1.5 - 60 \sin 45° \times 1.5 - 80 \cos 30° \times 1.5$$
$$+ 80 \sin 30° \times 0$$

$$-20 \times x = 0 + 0 + 60 + 0 + 63.63 - 63.63 - 103.92 + 0$$

or $-20\,x = -43.92$

or **x = 2.196 m from A (Fig. 3.37). (Ans.)**

Example 3.10. *The lever LMN of a component of a machine is hinged at M, and is subjected to a system of coplanar forces as shown in Fig. 3.39. Neglecting friction determine :*

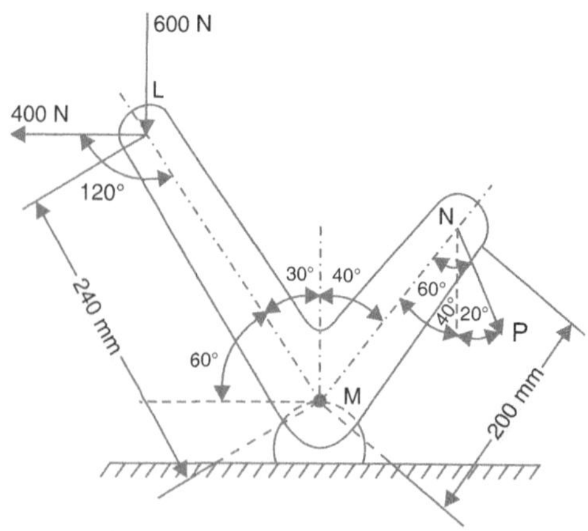

Fig. 3.39

(i) *The magnitude of the force P to keep the lever in equilibrium.*

(ii) *The magnitude and direction of the reaction at M.*

Sol. Magnitude of force P = ?

Taking moments about the hinge M, we get

$$\Sigma M : P \times 200 \sin 60° = 600 \times 240 \cos 60° + 400 \times 240 \sin 60°$$

or $$P \times 200 \times 0.866 = 600 \times 240 \times 0.5 + 400 \times 240 \times 0.866$$

$$173.2\, P = 155136$$

$\therefore$ $$\mathbf{P = 895\ N.\ (Ans.)}$$

Magnitude of reaction at M, R_M = ?

Resolving the forces horizontally,

$$\Sigma H = +\,400 - P \sin 20°$$
$$= +\,400 - 895 \times 0.342 = 93.9 \text{ N}$$

Resolving the forces vertically,

$$\Sigma V = 600 + P \cos 20°$$
$$= 600 + 895 \times 0.939 = 1440.4 \text{ N}$$

$\therefore$ $$R_M = \sqrt{(\Sigma H)^2 + (\Sigma V)^2} = \sqrt{(93.9)^2 + (1440.4)^2}$$
$$= \mathbf{1443.46\ N.\ (Ans.)}$$

Direction of the reaction at M = ?

Let θ = angle, which the reaction at M makes with the horizontal.

Then $$\tan \theta = \frac{\Sigma V}{\Sigma H} = \frac{93.9}{1440.4} = 0.065$$

or $$\theta = \mathbf{3.73° \text{ or } 3°44'.\ (Ans.)}$$

Example 3.11. *A square LMNS has forces acting along its sides as shown in Fig. 3.40. Find the values of F_1 and F_2, if the system reduces to a couple. Also find magnitude of the couple, if the side of the square is 2 m.*

Sol. Refer to Fig. 3.40.

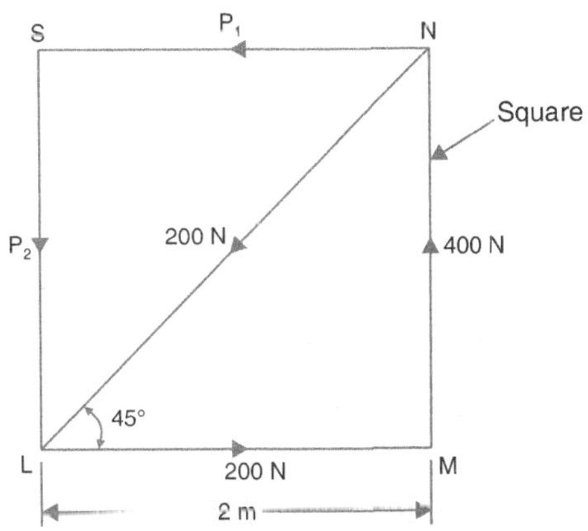

Fig. 3.40

Values of P_1 and P_2 = ?

We know that if *the system reduces to a couple, the resultant force in horizontal and vertical directions is zero.* Therefore, resolving the forces horizontally, we get

$$200 - 200 \cos 45° - P_1 = 0$$

$$\therefore \quad P_1 = 200 - 200 \times 0.707 = \mathbf{58.6\ N.\ (Ans.)}$$

Now resolving the forces vertically

$$400 - 200 \sin 45° - P_2 = 0$$

$$\therefore \quad P_2 = 400 - 200 \times 0.707 = \mathbf{258.6\ N.\ (Ans.)}$$

Magnitude of the couple = ?

We know that moment of the couple is equal to the algebraic sum of the moments about any corner. Therefore, moment of the couple (taking moments about L)

$$= -400 \times 2 - P_1 \times 2 = -800 - 58.6 \times 2$$

$$= \mathbf{-917.2\ Nm.\ (Ans.)}$$

(Minus sign taken due to anti-clockwise moments).

Example 3.12. *ABCDEF is a regular hexagon ; forces P, 2P, 3P, 2P, 5P, 6P act along AB, BC, DC, ED, EF, AF respectively. Show that the six forces are equivalent to a couple, and find the moment of this couple.*

Sol. Refer to Fig. 3.41.

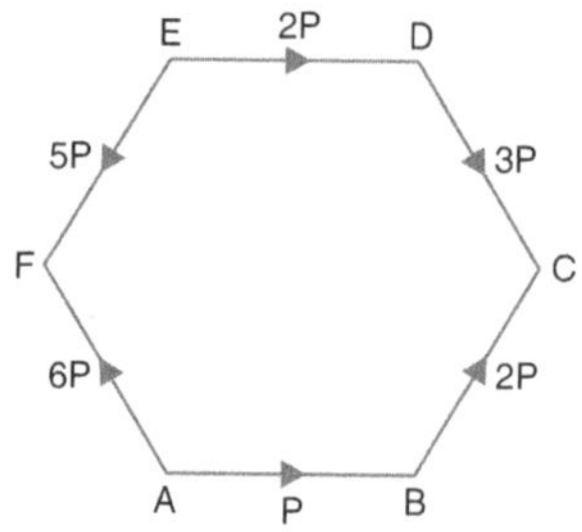

Fig. 3.41

Let A be the origin and let X and Y-axis be along AB and AE respectively.

Then,
$$\Sigma H = P + 2P \cos 60° + 3P \cos 60° + 2P - 5P \cos 60° - 6P \cos 60°$$

$$= P + P + \frac{3P}{2} + 2P - \frac{5P}{2} - 3P = 0$$

$$\Sigma V = 0 + 2P \sin 60° - 3P \sin 60° + 0 - 5P \sin 60° + 6P \sin 60°$$

$$= (2P - 3P - 5P + 6P) \frac{\sqrt{3}}{2} = 0$$

Hence, there is no resultant force.

If there be a resultant couple its moment

= the sum of moments of the forces about A. (Each side = 1)

$$= 2P\left(\frac{\sqrt{3}}{2}\right) - 3P(\sqrt{3}) - 2P(\sqrt{3}) + 5P\left(\frac{\sqrt{3}}{2}\right)$$

$$= \sqrt{3}P - 3\sqrt{3}P - 2\sqrt{3}P + \frac{5}{2}\sqrt{3}P$$

$$= \sqrt{3}P\ (1 - 3 - 2 + 5/2)$$

$$= \frac{-3\sqrt{3}\,P}{2}$$

Hence, **moment of couple** = $\mathbf{\dfrac{-3\sqrt{3}P}{2}}$. **(Ans.)**

Example 3.13. *A uniform rod of weight 250 N is hinged at its bottom end B to a fixed point. The rod rests in a position inclined at 60° to the horizontal by a string at A in a vertical position. Find the reaction at the hinge and tension in the string.*

Sol. Refer to Fig. 3.42.

Let l = length of the rod AB

T = tension in the string

Taking moments about point B,

$$T \times l \cos 60° = 250 \times l/2 \cos 60°$$

$$T \times l/2 = 250 \times l/2 \times \frac{1}{2}$$

$$T = \mathbf{125\ N.\ (Ans.)}$$

Also, $\Sigma V = 0$

$\therefore$ $R_B + T = 250$

or $R_B + 125 = 250$

or $\mathbf{R_B = 125\ N.\ (Ans.)}$

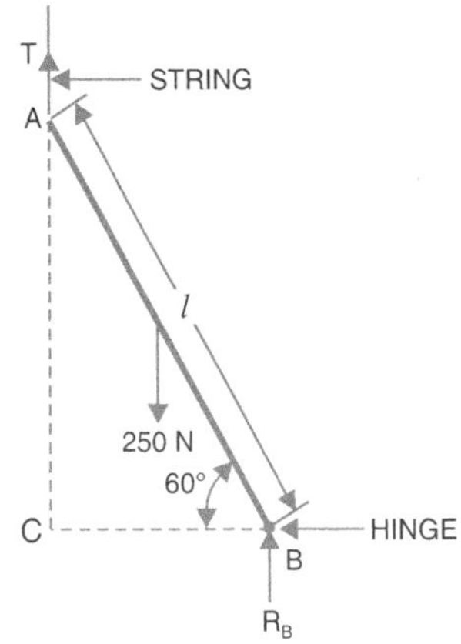

Fig. 3.42

Example 3.14. *A uniform rod 4 m long and of weight 200 N is pivoted at a point 1 m from A. A weight of 1000 N is suspended from A. What force applied at B, in a direction perpendicular to the rod, will keep it in equilibrium with A below B and AB inclined at 60° with the horizontal.*

Sol. Refer to Fig. 3.43.

Let O be the position of the pivot and G be the middle point of rod.

The rod will be in equilibrium when the sum of clockwise moments is equal to sum of anti-clockwise moments about the pivot point O.

$$P \times BO + 200 \times GO \cos 60° = 1000 \times AO \cos 60°$$

$$P \times 3 + 200 \times 1 \times \cos 60° = 1000 \times 1 \times \cos 60°$$

or $$3P + 200 \times \frac{1}{2} = 1000 \times \frac{1}{2}$$

or $$3P + 100 = 500$$

or $$3P = 400$$

or $$\mathbf{P = 133.3\ N.\ (Ans.)}$$

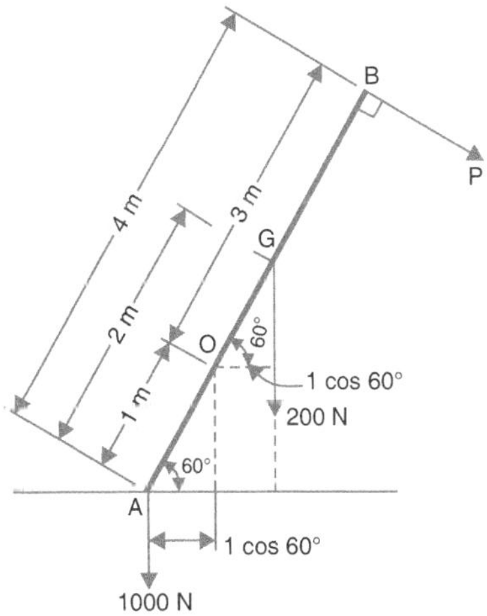

Fig. 3.43

Example 3.15. *A uniform ladder AB whose weight is 600 N and length 4 metres rests against a smooth vertical wall making with it an angle of 30°. The other end rests on the ground surface. Find the reactions given to the ladder by the wall and the floor and their inclination to the vertical, when a man weighing 700 N climbs the ladder by a distance 1 m along the length of the ladder.*

Sol. Refer to Fig. 3.44.

Since, the vertical wall is smooth, therefore, reaction at B will be perpendicular to the surface of the wall.

Taking moments about A, we get

$$700 \times 1 \cos 60° + 600 \times 2 \cos 60° = R_B \times 4 \sin 60°$$

or $$700 \times \frac{1}{2} + 600 \times 2 \times \frac{1}{2} = R_B \times 4 \times 0.866$$

or $$350 + 600 = R_B \times 3.464$$

or $$R_B = \frac{950}{3.464} = 274.2 \text{ N}$$

Again, $$\Sigma H = 0$$

$\therefore$ $$R_B = R_{AH} = 274.2 \text{ N}$$

and $$\Sigma V = 0$$

$$\uparrow R_{AV} = 700 \downarrow + 600 \downarrow = 1300 \text{ N}$$

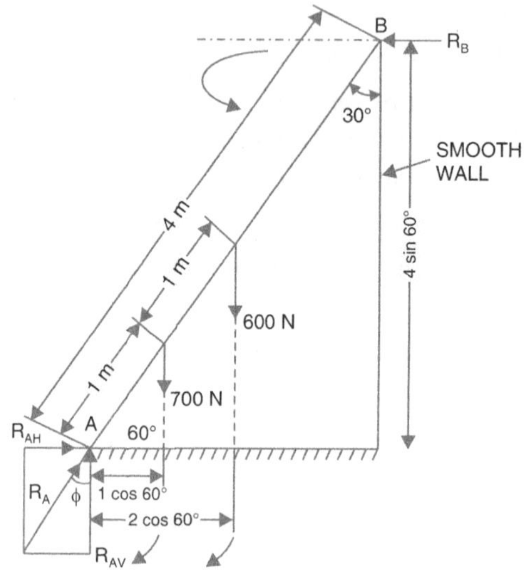

Fig. 3.44

$\therefore$ $$R_A = \sqrt{(R_{AH})^2 + (R_{AV})^2} = \sqrt{(274.2)^2 + (1300)^2}$$

$$= \sqrt{7518.5 + 169000} = 1328.6 \text{ N}$$

or $$\boldsymbol{R_A = 1328.6 \text{ N. (Ans.)}}$$

$$\tan \phi = \frac{R_{AH}}{R_{AV}} = \frac{274.2}{1300} = 0.2109$$

or $$\boldsymbol{\phi = 11.91° \text{ or } 11° 54'. \text{ (Ans.)}}$$

Example 3.16. *A uniform rod AB of weight 270 N and length equal to 3 metres is hinged at point A. At point B a weight of 120 N is hung. The rod is kept in horizontal position by a string BC which makes an angle of 30° with the rod. Find the tension in the string and the reaction at A.*

Sol. Refer to Fig. 3.45.

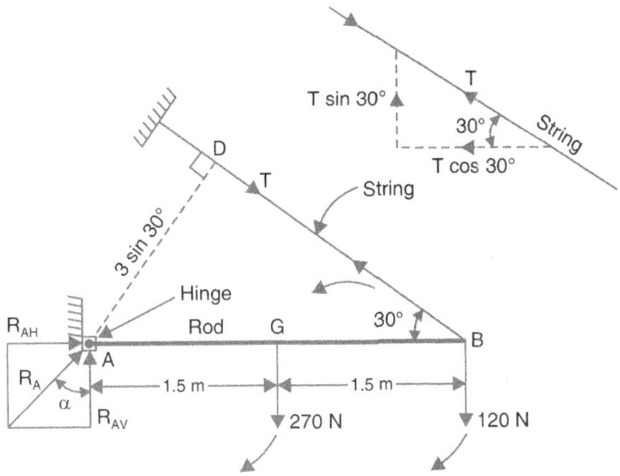

Fig. 3.45

Let T be the tension in the string and R_A be the reaction at A.

Tension in the string, T = ?

Since, the rod is in equilibrium, therefore, sum of clockwise moments of the forces acting on the rod about any point must be equal to anti-clockwise moments about the same point.

Taking moments about A :

$$T \times AD = 270 \times 1.5 + 120 \times 3$$

or $$T \times 3 \sin 30° = 405 + 360$$

or $$T \times 3 \times 0.5 = 765$$

or $$\mathbf{T = 510 \text{ N. (Ans.)}}$$

Reaction at A, R_A = ?

Again, $$\Sigma H = 0$$

$$\rightarrow$$

∴ $$R_{AH} = T \cos 30° = 51 \times 0.866$$
$$= 441.7 \text{ N}$$

and $$\Sigma V = 0$$

∴ $$R_{AV} + T \sin 30° = 270 + 120$$

or $$R_{AV} + 510 \times 0.5 = 270 + 120$$

or $$R_{AV} = 135 \text{ N.}$$

∴ Reaction at A,

$$R_A = \sqrt{(R_{AH})^2 + (R_{AV})^2} = \sqrt{(431.7)^2 + (135)^2}$$

or $$\mathbf{R_A = 461.8 \text{ N. (Ans.)}}$$

Example 3.17. *The wires passing round a telegraph pole are horizontal and two portions attached to it are inclined at 60° to each other. Pole is supported by a wire attached to the middle of the pole and inclined at 60° to the horizontal. Show that the tension in the supporting wire is $4\sqrt{3}$ times the telegraph wire.*

Sol. Refer to Fig. 3.46.

Let h = height of telegraph pole AB, having two portions BC and BD of telegraphic wire attached to it at B such that $\angle CBD = 60°$

T = tension in each portion of the telegraphic wire in N

T' = tension in the supporting wire LM.

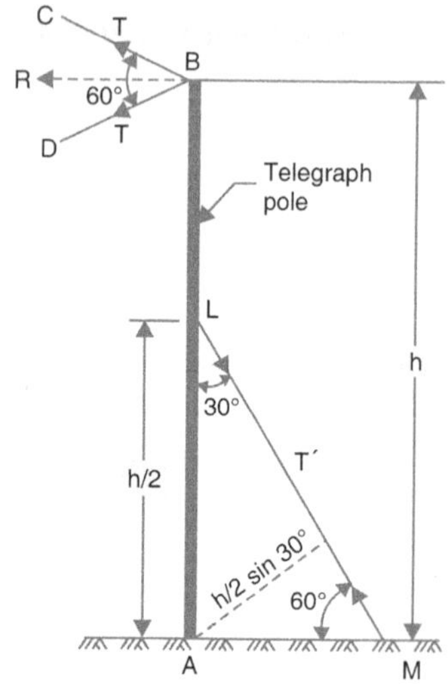

Fig. 3.46

Let us first find out the resultant of tensions T each in portions BC and BD of telegraphic wire by using the relation,

$$R = \sqrt{P^2 + Q^2 + 2PQ \cos \theta}$$

$$= \sqrt{T^2 + T^2 + 2 \times T \times T \cos 60°}$$

$$= \sqrt{2T^2 + 2T^2 \times 0.5}$$

$$= T\sqrt{3} \text{ N.}$$

The pole is in equilibrium under the action of the following forces :

(*i*) Resultant of tensions T each in portions BC and BD, R,

(*ii*) Tension in the supporting wire, T'.

Now taking moments about bottom A of the pole, we get

$$T' \times \frac{h}{2} \sin 30° = R \times h$$

$$T' \times \frac{h}{2} \times \frac{1}{2} = T\sqrt{3} \times h$$

$$\therefore \qquad \boldsymbol{T' = 4\sqrt{3}\,T} \text{ ......\textbf{Proved.}}$$

Hence, tension in the supporting wire is $4\sqrt{3}$ times the tension in the telegraph wire.

Example 3.18. *A ladder rests at an angle of 30° to the horizontal with its ends resting on a smooth floor and against a smooth vertical wall. The lower end being attached to the junction of the wall and floor by a string, find the tension in the string.*

Find also tension in the string when a man whose weight is one-half of the weight of ladder, has ascended the two-third of its length.

Sol. Case I. Refer to Fig. 3.47.

The ladder is in equilibrium under the action of following forces :

(*i*) Weight of the ladder, W ;

(*ii*) Reaction at the floor, R_A ;

(*iii*) Reaction at the wall, R_B ; and

(*iv*) Tension in the string,

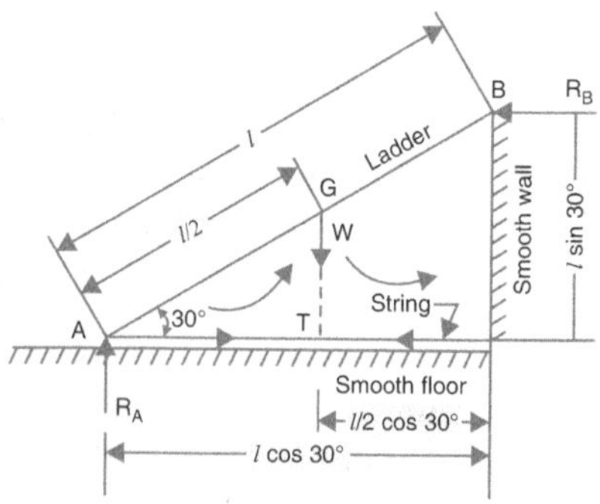

Fig. 3.47

Considering horizontal equilibrium of the ladder ($\Sigma H = 0$)

$$R_B = T \qquad ...(i)$$

Considering vertical equilibrium of the ladder ($\Sigma V = 0$)

$$R_A = W \qquad ...(ii)$$

Also $\Sigma M = 0$

Taking moments about B, we get

$$R_A \times l \cos 30° = W \times \frac{l}{2} \cos 30° + T \times l \sin 30°$$

or $$R_A \times l \times 0.866 = W \times l/2 \times 0.866 + T \times l \times 0.5$$

or $$R_A \times 0.866 = W \times 0.433 + 7 \times 0.5 \qquad [\because \; R_A = W \text{ from eqn. } (ii)]$$

or $$W \times 0.866 = W \times 0.433 + 0.5\, T$$

or $$\mathbf{T = 0.866\, W.\ (Ans.)}$$

Case II. Refer to Fig. 3.48. M shows the position of man having weight $\frac{W}{2}$ (one-half the weight of ladder W). The ladder is in equilibrium under the action of the following forces :

(*i*) Weight of the ladder, W ;

(*ii*) Weight of the man, $\frac{W}{2}$;

(*iii*) Reaction at the floor, R_A ;

(*iv*) Reaction of the wall R_B ; and

(*v*) Tension in the string T.

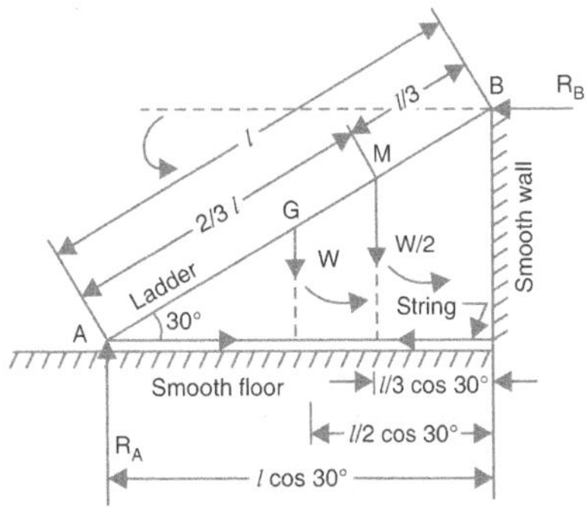

Fig. 3.48

Considering horizontal equilibrium of the ladder ($\Sigma H = 0$)

$$R_B = T \qquad ...(i)$$

Considering vertical equilibrium of the ladder ($\Sigma V = 0$),

$$R_A = W + \frac{W}{2} = \frac{3W}{2} \qquad ...(ii)$$

Also, $\Sigma M = 0$

Taking moments about B, we get

$$R_A \times l \cos 30° = W \times \frac{l}{2} \cos 30° + \frac{W}{2} \times \frac{l}{3} \cos 30° + T \times l \sin 30°$$

$$\frac{3W}{2} \times l \times 0.866 = W \times \frac{l}{2} \times 0.866 + \frac{W}{2} \times \frac{l}{3} \times 0.866 + T \times l \times 0.5$$

$$1.3\, W = 0.433\, W + 0.144\, W + 0.5T$$

$\therefore$ $$\mathbf{T = 1.446\, W.\ (Ans.)}$$

Example 3.19. *A beam simply supported at both the ends carries load system as shown in Fig. 3.49. Find the reactions at the two ends.*

Sol. Refer to Fig. 3.49. Since, all the loads acting on the beam are vertically downwards therefore, the reactions at the ends shall be vertically upwards. Let R_A and R_B be the reactions at the ends A and B respectively.

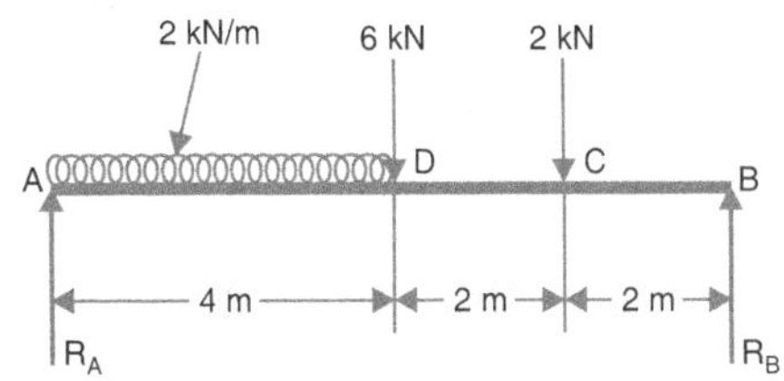

Fig. 3.49

Since, the beam AB is in equilibrium,

$\therefore$ $$\Sigma H = 0, \Sigma V = 0 \text{ and } \Sigma M = 0.$$

There is no horizontal force.

$\therefore$ Considering vertical equilibrium of the beam ($\Sigma V = 0$)

$$R_A + R_B = 2 \times 4 + 6 + 2$$

or $$R_A + R_B = 16 \qquad ...(i)$$

Taking moments about $A(\Sigma M = 0)$, we get

$$R_B \times 8 = 2 \times 4 \times \frac{4}{2} + 6 \times 4 + 2 \times 6$$

$$= 16 + 24 + 12 = 52$$

$\therefore$ $\boldsymbol{R_B = 6.5}$ **kN. (Ans.)**

Substituting this value of R_B in eqn. (*i*), we get

$$R_A + 6.5 = 16$$

or $\boldsymbol{R_A = 9.5}$ **kN. (Ans.)**

Example 3.20. *Find the reaction R_1, R_2 and R_3 in the case of two beams placed one over the other and loaded as shown in Fig. 3.50.*

Sol. Refer to Fig. 3.50.

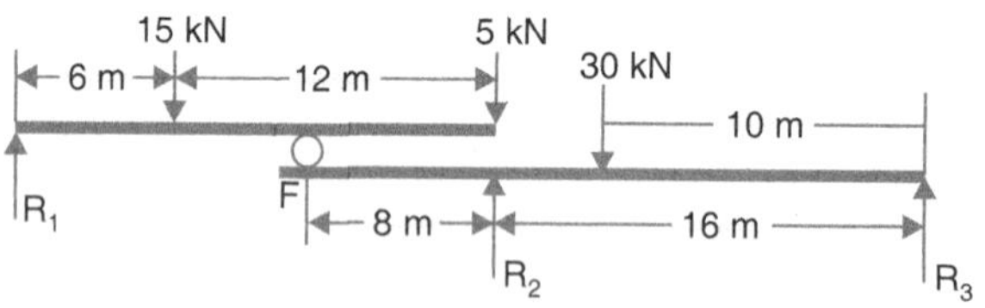

Fig. 3.50

Taking moments about F for the **top beam**

$$R_1 \times 10 + 5 \times 8 = 15 \times 4$$

$\therefore$ $\boldsymbol{R_1 = 2}$ **kN. (Ans.)**

Out of total load of (15 + 5 = 20 kN) 20 kN, rection R_1 takes up 2 kN and remaining (20 – 2 = 18 kN) 18 kN acts at the edge of the bottom beam as shown in Fig. 3.51.

Taking moments about R_3,

$$R_2 \times 16 = 18 \times 24 + 30 \times 10$$

$$= 432 + 300$$

$\therefore$ $\boldsymbol{R_3 = 45.765}$ **kN. (Ans.)**

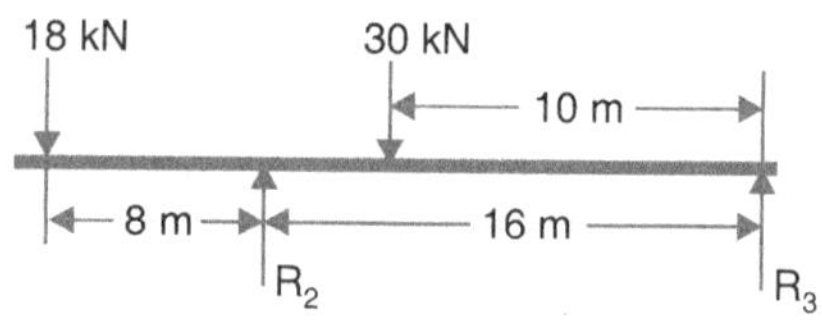

Fig. 3.51

Now, $\Sigma V = 0$

$\therefore$ $R_2 + R_3 = 18 + 30$

or $45.75 + R_3 = 48$

or $\boldsymbol{R_3 = 2.25}$ **kN. (Ans.)**

Example 3.21. *A simply supported overhanging beam 18 metres long carries a system of loads as shown in Fig. 3.52. Determine the reactions at the supports.*

Sol. Refer to Fig. 3.52.

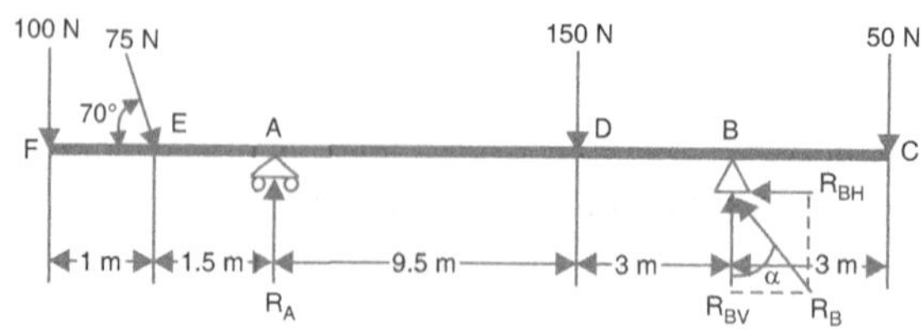

Fig. 3.52

Reaction at A : will be *vertical* since there is a roller *support* at A.

Reaction at B : will be *inclined* since there is a *hinged support* at B.

Considering horizontal equilibrium of the beam : $(\Sigma H = 0)$

$$R_{BH} = 75 \cos 70° = 25.65 \text{ N} \quad ...(i)$$

Considering vertical equilibrium of the beam $(\Sigma V = 0)$

$$R_A + R_{BV} = 100 + 75 \sin 70° + 150 + 50$$
$$= 100 + 75 \times 0.9396 + 150 + 50$$
$$= 370.47 \text{ N} \quad ...(ii)$$

Taking moments about B : $(\Sigma M = 0)$

$$R_A \times 12.5 + 50 \times 3 = 100 \times 15 + 75 \sin 70° \times 14 + 150 \times 3$$
$$R_A \times 12.5 + 150 = 1500 + 986.67 + 450$$

$\therefore$ $\mathbf{R_A = 222.93}$ **N. (Ans.)**

Substituting this value of R_A in eqn. (*i*), we get

$$222.93 + R_{BV} = 370.47$$

$\therefore$ $R_{BV} = 147.54$ N

$\therefore$ $R_B = \sqrt{(R_{BH})^2 + (R_{BV})^2}$

$$= \sqrt{(25.65)^2 + (147.54)^2} = 149.75 \text{ N}$$

or $\mathbf{R_B = 149.75}$ **N. (Ans.)**

$$\tan \alpha = \frac{R_{BH}}{R_{BV}} = \frac{25.65}{147.54} = 0.1738$$

or $\alpha =$ **9° 51′ with the vertical. (Ans.)**

Example 3.22. *Calculate the effort required on the end of the iron crow bar of weight 100 N and length 6 metres to lift a load of 1500 N on the other end. Fulcrum is kept at a distance of 4.5 m from the end where effort it applied.*

Sol. Refer to Fig. 3.53. This is a lever of type 1 as the fulcrum lies between W and P.

Taking moments about F (fulcrum) considering the weight of the crow bar to act at the centre of the bar.

$$P \times 4.5 + 100 \times 1.5 = 1500 \times 1.5$$
$$P \times 4.5 + 150 = 2250$$

or $\mathbf{P = 466.6}$ **N. (Ans.)**

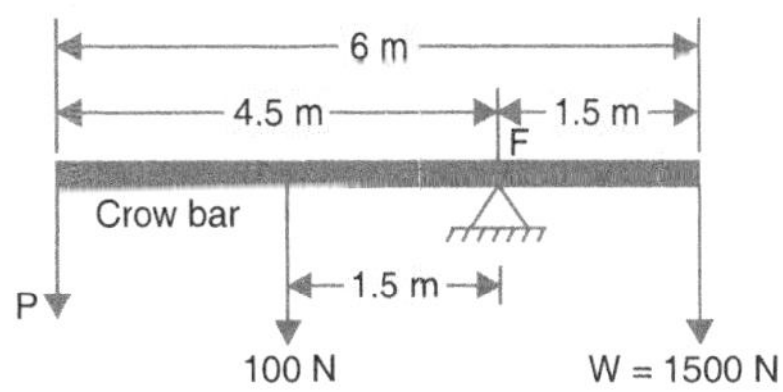

Fig. 3.53

SELECTED QUESTIONS FROM EXAMINATION PAPERS

Example 3.23. *Find the sum of moments of the forces shown in Fig. 3.54 about point A.*

Sol. Refer to Figs. 3.54 and 3.55.

$$\Sigma M_A = 200 \times 6 + 400 \cos 30° \times 0 - 400 \sin 30° \times 4$$
$$= 1200 - 800 = \mathbf{400 \text{ Nm. (Ans.)}}$$

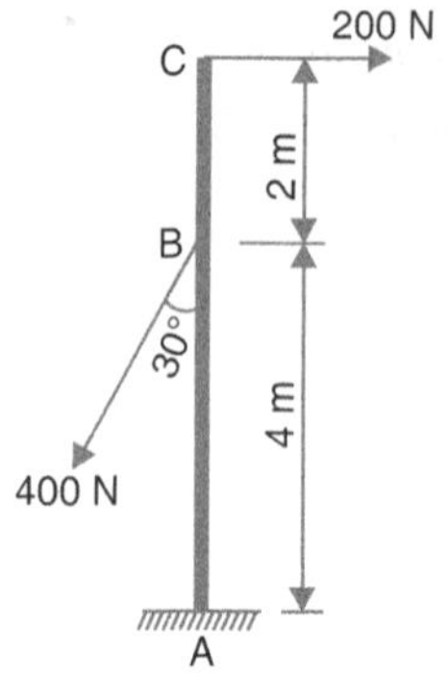

Fig. 3.54

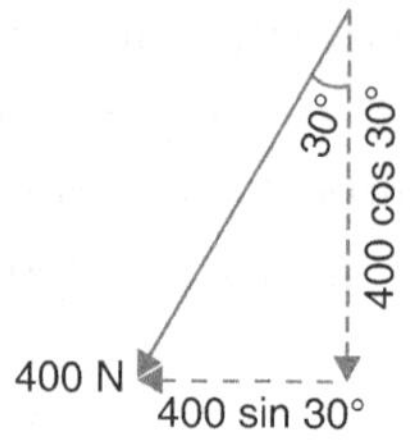

Fig. 3.55

Example 3.24. *Find the sum of the moments of the forces shown in Fig. 3.56 about point A.*

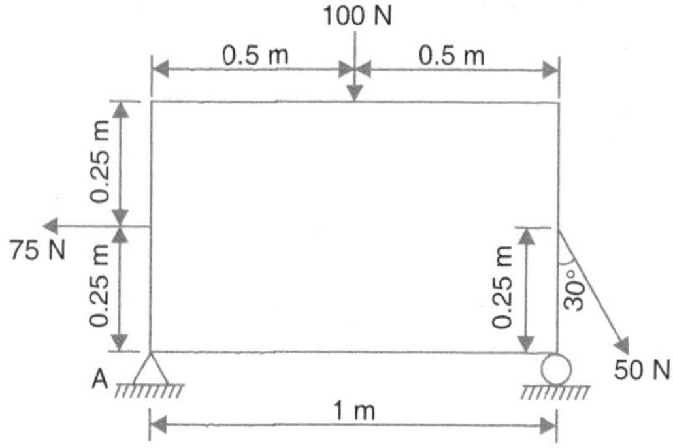

Fig. 3.56

Sol.

$$\Sigma M_A = 50 \sin 30° \times 0.25 + 50 \cos 30° \times 1 + 100 \times 0.5 - 75 \times 0.25$$
$$= 6.25 + 43.3 + 50 - 18.75$$
$$= \mathbf{80.8\ Nm.\ (Ans.)}$$

Example 3.25. *LMNS is a square, each side being 0.4 m and Q is the middle point of LM. Forces of 14, 16, 24, 10, 18 and 12 N act in the line of directions LM, QN, MN, MS, NL and SQ respectively. Find the magnitude, direction and position of the resultant force.*

Sol. Refer to Fig. 3.57.

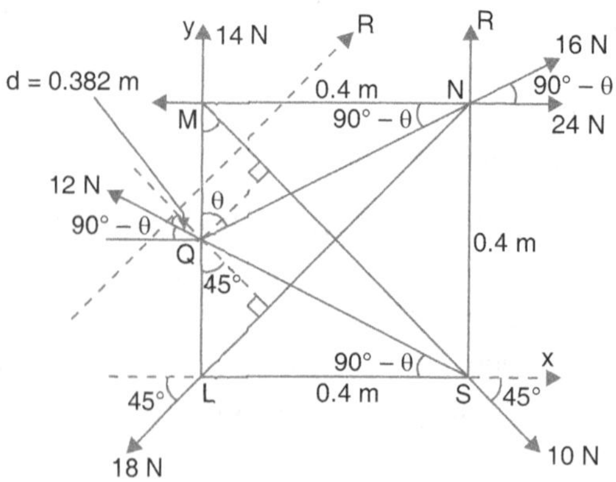

Fig. 3.57

Let $\angle MQN = \theta$

$\therefore \quad \tan \theta = \dfrac{0.4}{0.2} = 2$

$\therefore \quad \theta = \tan^{-1}(2) = 63.4°$

$\therefore \quad \sin \theta = \sin 63.4° = 0.894$

and $\quad \cos \theta = \cos 63.4° = 0.447$

Magnitude of resultant force, R :

Resolving all forces along *X-axis,* we get

$$\begin{aligned}\Sigma P_x &= 16 \cos(90° - \theta) + 24 + 10 \cos 45° - 18 \cos 45° - 12 \cos(90° - \theta)\\ &= 16 \times 0.894 + 24 + 10 \times 0.707 - 18 \times 0.707 - 12 \times 0.894\\ &= 14.304 + 24 + 7.07 - 12.726 - 10.728\\ &= 21.92 \text{ N}\end{aligned}$$

Resolving all the forces along *Y*-axis we get

$$\begin{aligned}\Sigma P_y &= 16 \sin(90° - \theta) - 10 \sin 45° - 18 \sin 45° + 12 \sin(90° - \theta) + 14\\ &= 16 \times 0.447 - 10 \times 0.707 - 18 \times 0.707 + 12 \times 0.447 + 14\\ &= 7.152 - 7.07 - 12.726 + 5.364 + 14\\ &= 6.72 \text{ N}\end{aligned}$$

Now, magnitude of resultant force (R) is given by

$$\begin{aligned}R &= \sqrt{(\Sigma P_x)^2 + (\Sigma P_y)^2}\\ &= \sqrt{(21.92)^2 + (6.72)^2} = 22.92 \text{ N}\end{aligned}$$

i.e., Resultant force, $\quad$ **R = 22.92 N. (Ans.)**

Direction of the resultant force, α :

Let $\quad \theta$ = angle, the resultant force makes with MN *i.e.,* with the horizontal.

$\therefore \quad \tan \alpha = \dfrac{\Sigma P_y}{\Sigma P_y} = \dfrac{6.72}{21.92} = 0.3066$

or $\quad \alpha = 17°3'$

Since ΣP_x and ΣP_y are +ve therefore θ lies between 0° and 90°

Actual $\quad \alpha$ = **17°3′. (Ans.)**

Position of the resultant force :

Let d = perpendicular distance between the point Q and the line of action of the resultant force.

Taking moments about Q, we get

$\Sigma M_Q : 22.92\, d = 14 \times 0 + 16 \times 0 + 24 \times 0.2 + 10 \times 0.2$
$\sin 45° + 18 \times 0.2 \sin 45° + 12 \times 0$

or $\quad 22.92\, d = 4.8 + 1.414 + 2.545$
$= 8.759$

$\therefore \quad$ ***d* = 0.382 m or 382 mm. (Ans.)**

HIGHLIGHTS

1. The rotational tendency of a force is called '*Moment*'. The force multiplied by the perpendicular distance from the point to the line of action of the force is called *moment about that point i.e.*, Moment of a force = force × perpendicular distance.
2. Conventionally, clockwise moments are taken as *negative* moments and *anti-clockwise moments* as positive moments.
3. Principle of moments states : "When a body acted upon by several forces is in rotational equilibrium, the sum of clockwise moments of the forces about any point is equal to sum of the anti-clockwise moments of the forces about the same point".
4. The body can be in equilibrium if the algebraic sum of all the external forces and their moments about any point in their plane is zero.

   Mathematically, the conditions of equilibrium may be expressed as follows :

   (*i*) $\Sigma H = 0$ (ΣH means sum of all the *horizontal* forces)

   (*ii*) $\Sigma V = 0$ (ΣV means sum of all the *vertical* forces)

   (*iii*) $\Sigma M = 0$ (ΣM means sum of all the *moments*).
5. Varignon's theorem states : "The algebraic sum of the moments of two forces about any point in their plane is equal to moment of their resultant about that point."
6. Forces whose lines of action are parallel are called *parallel forces*. They are said to be like when they act in the *same sense* ; they are said to be unlike when they act in *opposite sense*.
7. A couple is a pair of two equal and opposite forces acting on a body in such a way that the lines of action of the two forces are not in the same straight line.

   The moment of couple is known as *torque* which is equal to one of the forces forming the couple multiplied by the arm of the couple.
8. Examples of couples in every day life :

   (*i*) Opening or closing a water tap.

   (*ii*) Turning the cap of a pen.

   (*iii*) Unscrewing the cap of an ink bottle.

   (*iv*) Turning a screw driver.

   (*v*) Steering a motor car.

   (*vi*) Winding a watch or clock with a key.
9. Some important engineering applications of moments are

(*i*) The balance ; (*ii*) The steel yard ;

(*iii*) Bell crank lever ; (*iv*) Lever safety valve ; and

(*v*) Cranked lever.

OBJECTIVE TYPE QUESTIONS

Fill in the blanks :

(*i*) The ...... is a rotational tendency of a force.

(*ii*) Moment of a force = ... × perpendicular distance.

(*iii*) Conventionally, clockwise moments are taken as ...... moments.

(*iv*) Sum of clockwise moments = sum of ...... moments.

(*v*) When coplanar forces meet in a point the system is known as ...... force system.

(*vi*) The algebraic sum of the moments of two forces about any point in their plane is equal to the moment of their ...... about that point.

(*vii*) Forces whose lines of action are parallel are called ...... forces.

(*viii*) Parallel forces are said to be like when they act in the ...... sense.

(*ix*) A ...... is a pair of two equal and opposite forces acting on a body in such a way that the lines of action of two forces are not in the same straight line.

(*x*) The moment of couple is known as ...... which is equal to one of the forces forming the couple multiplied by arm of the couple.

Answers

(*i*) moment (*ii*) force (*iii*) negative (*iv*) anti-clockwise
(*v*) coplanar concurrent (*vi*) resultant (*vii*) parallel (*viii*) same
(*ix*) couple (*x*) torque.

EXERCISES

Theoretical Questions

1. (*a*) Define the terms 'Moment' and 'Moment of a force'.
   (*b*) State and explain the principle of moments.
2. State equilibrium conditions for bodies under coplanar forces.
3. State and prove Varignon's theorem.
4. Explain the graphical method for finding the resultant of any number of like or unlike parallel forces.
5. (*a*) What is a couple ?
   (*b*) Enumerate the properties of a couple.
6. State some engineering applications of moments.
7. Explain the method of finding resultant of a coplanar, non-concurrent, non-parallel force system.

Unsolved Examples

1. A man and a body carries a weight of 250 N between them by means of a uniform pole 1.7 metres long and weighing 90 N. Where must the weight be placed so that man may bear twice as much of the weight as that of the boy ? [**Ans.** 0.465 m]
2. A simply supported overhanging beam 15 metres long carries a system of loads as shown in Fig. 3.58. Determine the reactions at the supports. [**Ans.** 7.75 kN ; 9.27 kN acting at 26° 36′ with the vertical]

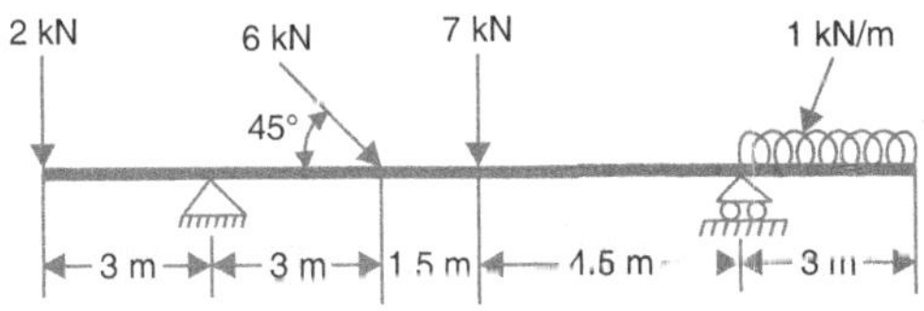

Fig. 3.58

3. Forces act along the sides *AB, AD, CB* and *CD* of a square *ABCD* equal to 5, 10, 15 and 20 N respectively. Also, a force of $8\sqrt{2}$ N acts along *BD*. Find the algebraic sum of moments about the point *A*. The sides of the square being 2 metres long. [**Ans.** 26 N]
4. A pair of nut crackers is 25 cm long and a nut is placed at a distance of 5 cm from the hinge. Find the force required to crack the nut if it requires a weight of 100 N for its cracking. [**Ans.** 20 N]
5. *ABCD* is a square whose side is 2 m. Along *AB, BC, CD* and *DA* act forces equal to 1, 2, 8 and 5 N and along *AC* and *DB* forces equal to $5\sqrt{2}$ N and $2\sqrt{2}$ N. Show that they are equivalent to a couple whose moment is equal to 16 Nm.
6. Find out the effort required on one end of the iron crow bar of weight 10 N and length 5 metres to lift a weight of 200 N on the other end. Fulcrum is kept at a distance of 4 metres from the end where effort is to be applied. [**Ans.** 46.25 N]

7. Find the equilibrant of the system of forces in magnitude and direction as shown in Fig. 3.59.
[**Ans.** R_e = 6.3 kN acting clockwise from point A at an angle 17° 48′ with the horizontal]

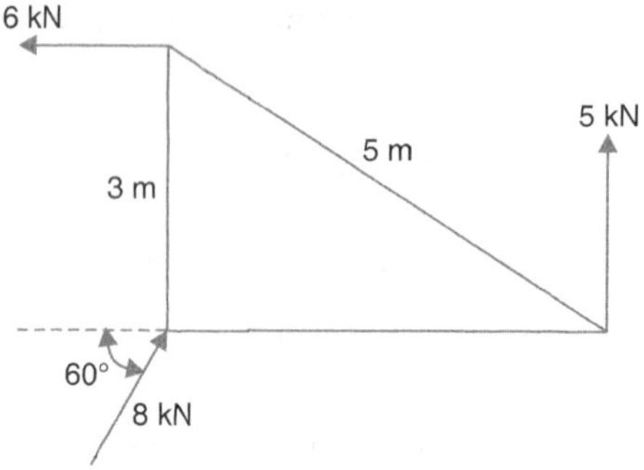

Fig. 3.59

8. The lever of common steel yard without hook connections weighs 70 N and acts at 5 cm to the left of the fulcrum. The hook which carries the load, weighs 40 N and is also attached on the left hand side of the fulcrum at a distance of 12 cm. If the graduation mark 72 cm from the fulcrum reads 500 N, find

(*i*) The weight of the rider.

(*ii*) How far is the zero graduation from the fulcrum ?

(*iii*) What is the movement of the rider per N of graduation ? [**Ans.** 94.8 N, 8.75 cm ; 1.265 cm]

9. The lever of a safety valve weighs 30 N and its *c.g.* lies at a distance of 8 cm from the centre of the valve which is at 7 cm from the fulcrum. If the diameter and weight of the valve are respectively 6 cm and 50 N, find the minimum weight that will be hung to the end of 75 cm long lever to conserve steam in the boiler at a pressure of 80 N/cm^2. [**Ans.** 205.8 N]

10. A lever safety valve 6 cm diameter is just on the point of blowing off steam when the weight of the valve is 10 N, weight on the lever is 300 N acting at 7 cm from the fulcrum and weight of the lever is 40 N and acting at 5 cm from the fulcrum towards the weight of 300 N. The valve is 4 cm from the fulcrum acting on the same side as weight. Find the intensity of pressure of the steam. [**Ans.** p = 20.8 N/cm^2]

11. A uniform ladder AB whose weight is 200 N and length 6 m rests against a smooth vertical wall making with it an angle of 30°. The other end rests on the ground surface. Find the reactions given to the ladder by the wall and the floor and their inclination to the vertical, when a man weighing 700 N climbs the ladder by a distance 1.5 m along the length of ladder. [**Ans.** 332 N ; 1536 N, 12° 27′]

12. Find the magnitude of two like parallel forces acting at a distance of 100 cm apart which is equivalent to a force of 30 N acting at a distance of 20 cm from one of the forces. [**Ans.** 24 N]

13. Find the magnitude and line of action of the resultant of parallel forces 70, 50 and 90 N in one direction and 120 N in the opposite direction acting at points A, B, C and D respectively in a straight line where AB = 12 cm, BC = 36 cm and CD = 50 cm. [**Ans.** Resultant = 90 N, acting at 76 cm from 70 N force]

14. Four coplanar forces equal to 20 N, 30 N, 50 N and 70 N are acting on a square body of each side equal to 1 m as shown in Fig. 3.60. Determine the magnitude, direction and position of the force which will keep the body in equilibrium. [**Ans.** R_e = 65.3 N acting anti-clockwise from point C at an angle 20° 30′ with the horizontal and at a distance of 0.42 m]

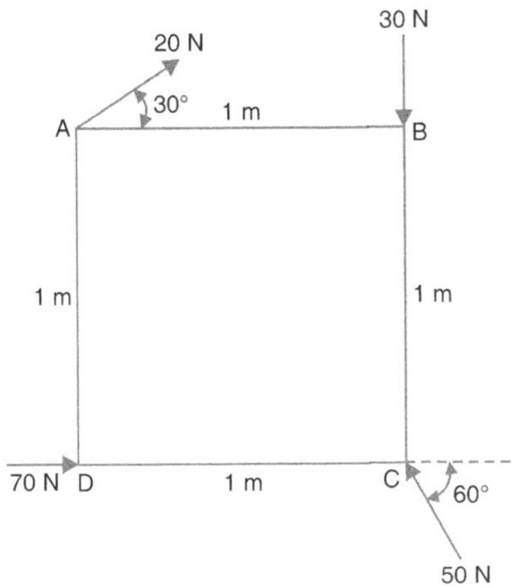

Fig. 3.60

15. The equilateral triangle shown in Fig. 3.61 has a side of 100 mm. Find:

(*i*) The magnitude and direction of the resultant force.

(*ii*) The point at which the resultant cuts *QR*.

[**Ans.** 10.6 N, 41° to *QR*, 150 mm to right of *R* or *QR* produced]

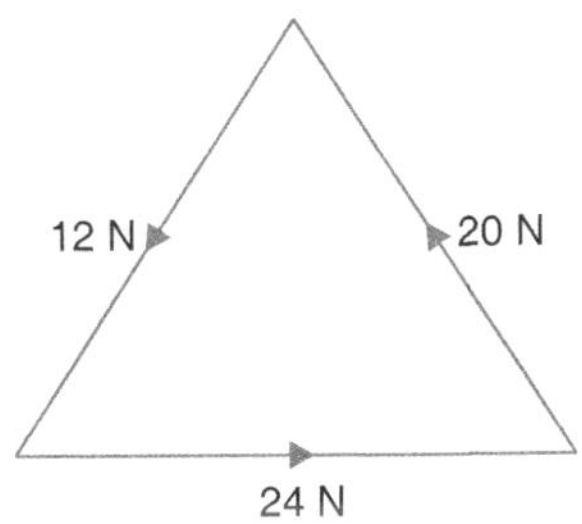

Fig. 3.61

16. The rectangle shown in Fig. 3.62 has *AB* = 200 mm and *BC* = 150 mm. Calculate the magnitude, direction, and position of the resultant force.

[**Ans.** 10.2 kN, 11° 19′ to *BA* ; cuts 40 mm from *A* on *DA*]

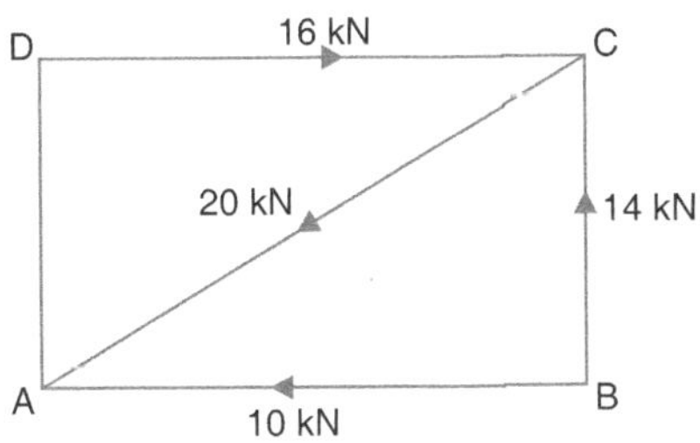

Fig. 3.62

17. *ABCD* is a square each side being 500 mm and *E* is the mid-point of *AP*. Forces of 3.18, 3.63, 5.44, 2.27, 4.10 and 2.72 N act on a body along the lines and directions of *AB*, *EC*, *BC*, *BD*, *CA* and *DE* respectively. Find the magnitude and direction and position with respect to *ABCD* of the force required to keep the body in equilibrium.

[**Ans.** *R* – 5.22 N, acts in a direction making an angle of 197.25° with *AD* and at a distance x = 274.5 mm from *E*, so as to cause an anti-clockwise couple]

18. A square *ABCD* has forces acting along its sides as shown in Fig. 3.63. Find the values of P_1 and P_2, if the system reduces to a couple. Also find magnitude of the couple, if the sides of the square is 1 m.

[**Ans.** P_1 = 58.6 N, P_2 = 258.6 N ; couple = 458.6 Nm (anti-clockwise)]

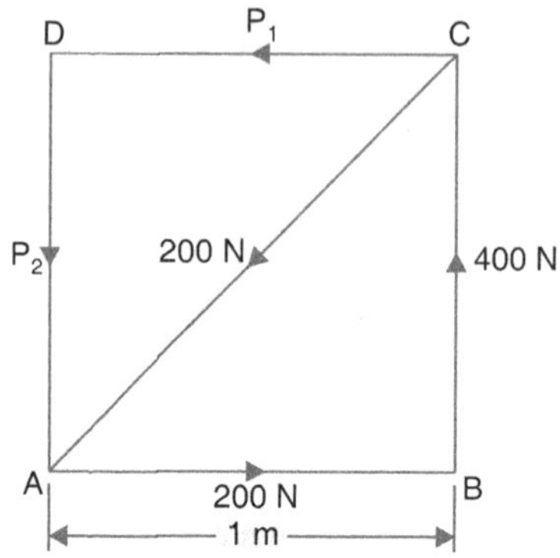

Fig. 3.63

4

Centre of Gravity and Centroid

4.1. CENTRE OF GRAVITY OF A BODY

A body comprises of several parts and its every part possesses weight. *Weight is the force of attraction between a body and the earth and is proportional to mass of the body.* The weights of all parts of a body can be considered as parallel forces directed towards the centre of the earth. Therefore, they may be combined into a resultant force whose magnitude is equal to their algebraic sum. If a supporting force, equal and opposite to the resultant, is applied to the body along the line of action of the resultant, the body will be in equilibrium. This line of action will pass through the centre of gravity of the body. Thus, centre of gravity of the body may be defined as the *point through which the whole weight of a body may be assumed to act.* The centre of gravity of a body or an object is usually denoted by c.g. or simply by *G*. The position of c.g. depends upon shape of the body and this may or may not necessarily be within the boundary of the body.

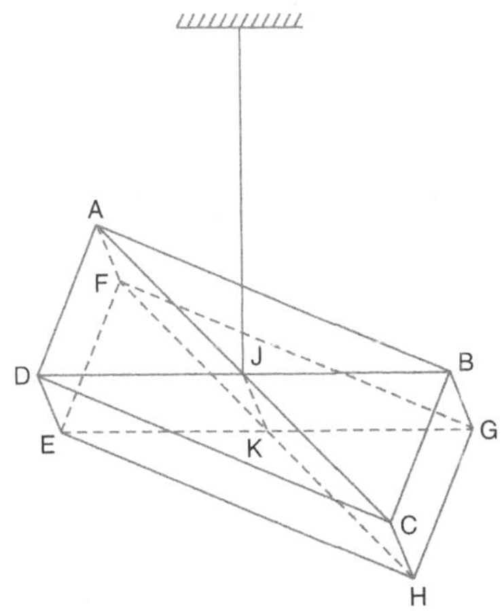

Fig. 4.1

4.2. DETERMINATION OF CENTRE OF GRAVITY

The centre of gravity of some objects may be found by *balancing the object on a point*. Take a thin plate of thickness *t*, shown in Fig. 4.1. Draw the diagonals of the upper and lower forces to intersect at *J*, and *K* respectively. It the plate is placed on point at *K*, the plate will not fall. That is, it is balanced. If suspended from *J*, the plate will hang horizontally. The centre of gravity of the plate is at the centre of the line *JK*.

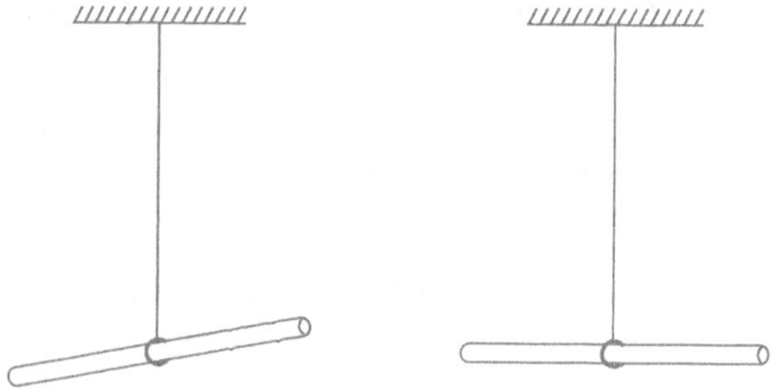

Fig. 4.2

Again, if we suspend a uniform rod by a string (Fig. 4.2) and move the position of the string until the rod hangs vertically, we can determine that the centre of gravity of the rod lies at its centre. Through the use of similar procedures it can be established that a body which has an axis, or line, symmetry has its centre of gravity located on that line, or axis. Of course, if a body has more than one axis of symmetry, the centre of gravity must lie at the intersection of the axes.

Another method for determining the centre of gravity is by *suspension*. Take an object, the section of which is shown in Fig. 4.3. Suspend it from point *L*. The body will not come to rest until its resultant weight is vertically downward from *A*. Through *L*, draw a vertical line *LN*. Then suspend the body from a point *M*, and let come to rest. Through *M*, draw a vertical line *MT*. The point of *LN* and *MT* is the position of the centre of intersection of gravity.

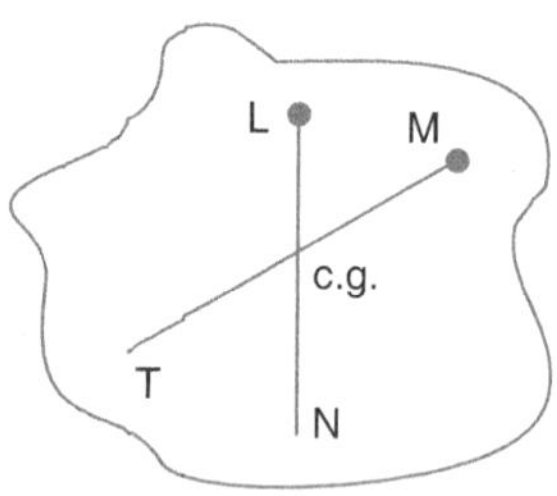

Fig. 4.3

4.3. CENTROID

The centroid or centre of area is defined as the point where the whole area of the figure is assumed to be concentrated. Thus, centroid can be taken as quite analogous to centre of gravity when bodies have area only and not weight.

4.4. POSITIONS OF CENTROIDS OF PLANE GEOMETRICAL FIGURES

Table 4.1 gives the positions of centroids of some plane geometrical figures.

Table 4.1 Centroids of Plane Geometrical Figures

Shape	*Area*	$\bar{x}$	$\bar{y}$	*Figures*
Rectangle	bh	$\frac{b}{2}$	$\frac{h}{2}$	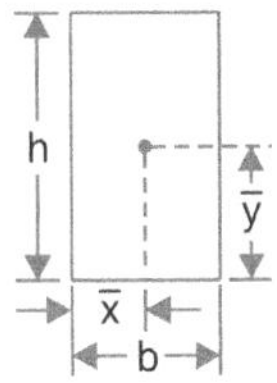 Fig. 4.4
Triangle	$\frac{bh}{2}$	$\frac{b}{3}$	$\frac{h}{3}$	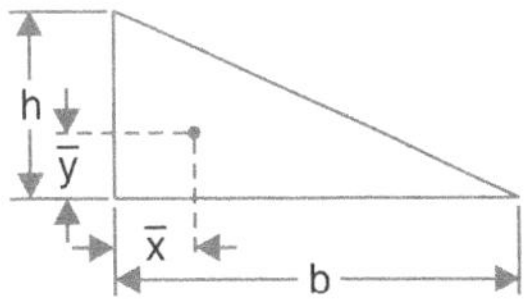 Fig. 4.5
Circle	$\frac{\pi}{4}d^2$	$\frac{d}{2}$	$\frac{d}{2}$	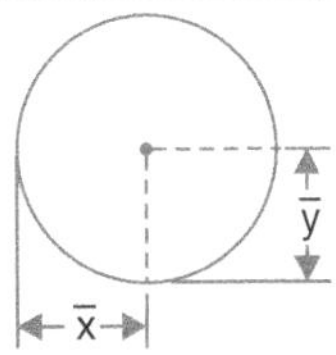 Fig. 4.6
Semicircle	$\frac{\pi}{8}d^2$	$\frac{d}{2}$	$\frac{4r}{3\pi}$ (= 0.424 r)	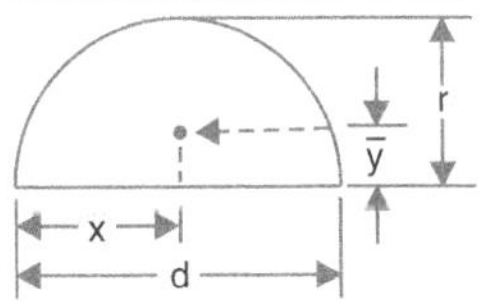 Fig. 4.7
Quadrant	$\frac{\pi}{16}d^2$	0.424 r	0.424 r	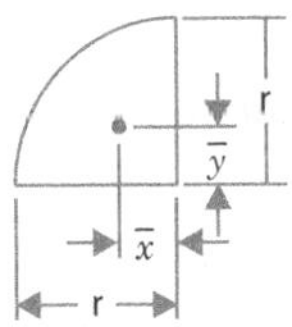 Fig. 4.8
Trapezium	$(a+b)\frac{h}{2}$	$\frac{a^2+b^2+ab}{3(a+b)}$	$\frac{(2a+b)}{(a+b)} \times \frac{h}{3}$	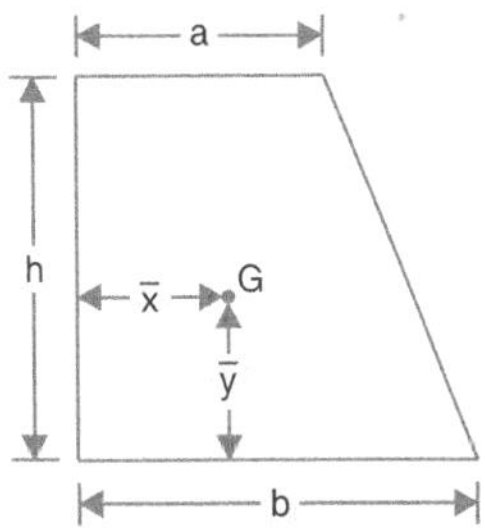 Fig. 4.9

4.5. POSITIONS OF CENTRE OF GRAVITY OF REGULAR SOLIDS

Table 4.2 gives the positions of centre of gravity of regular solids.

Table 4.2. C.G. of Regular Solids

Shape	*Volume*	*Regular solids*
Cylinder	$\pi r^2 h$	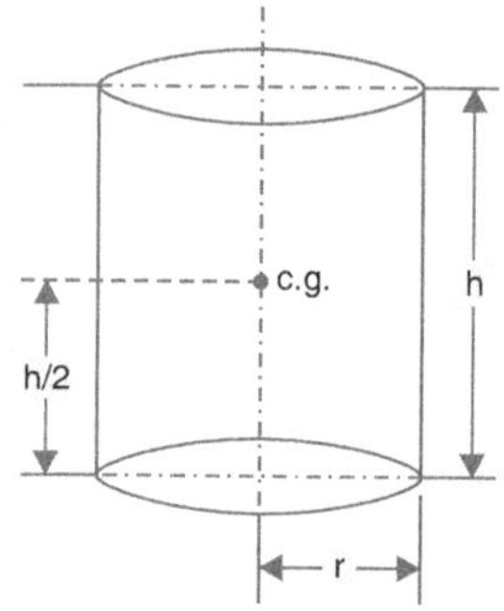 Fig. 4.10
Sphere	$\frac{4}{3}\pi r^3$	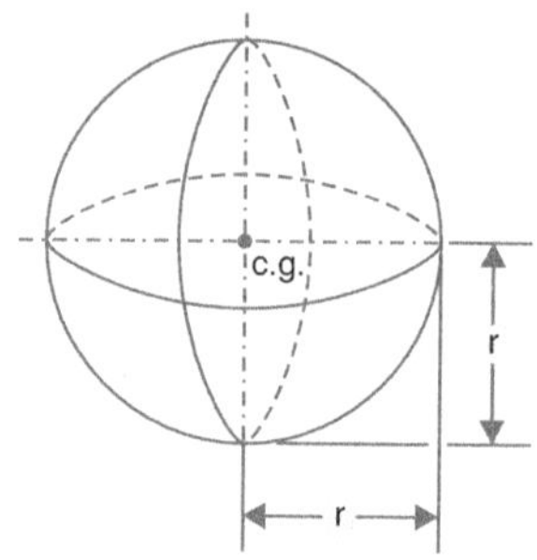 Fig. 4.11
Hemisphere	$\frac{2}{3}\pi r^3$	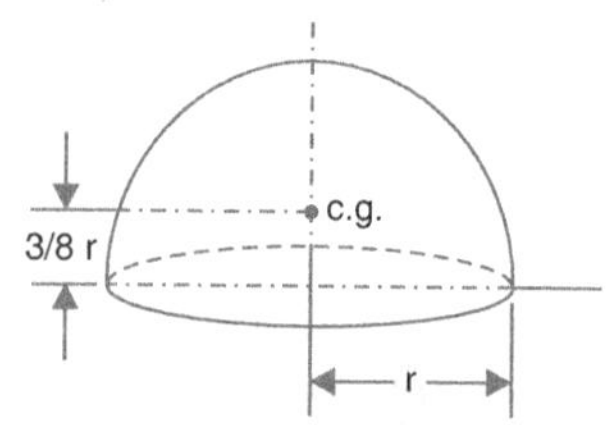 Fig. 4.12
Right circular cone	$\frac{1}{3}\pi r^2 h$	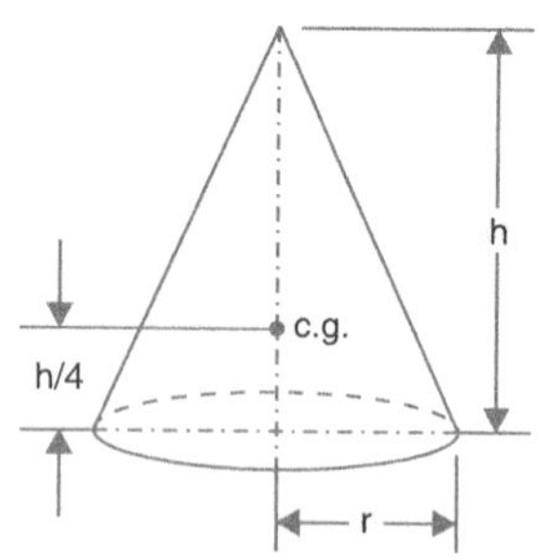 Fig. 4.13

4.6. (*a*) CENTROIDS OF COMPOSITE AREAS

The location of the centroid of a plane figure can be thought of as the average distance of the area to an axis. Usually the axes involved will be the X and Y-axes. In determining the location of the centroid it is found advantageous to place the X-axis through the lowest point and the Y-axis through the left edge of the figure. This places the plane area entirely within the first quadrant where x and y distances are positive (Fig. 4.14). Then divide the area into simple areas such as rectangles,

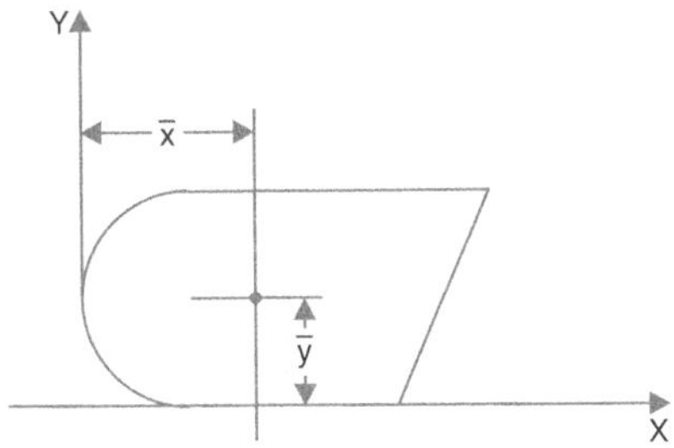

Fig. 4.14. Centroid of a composite area.

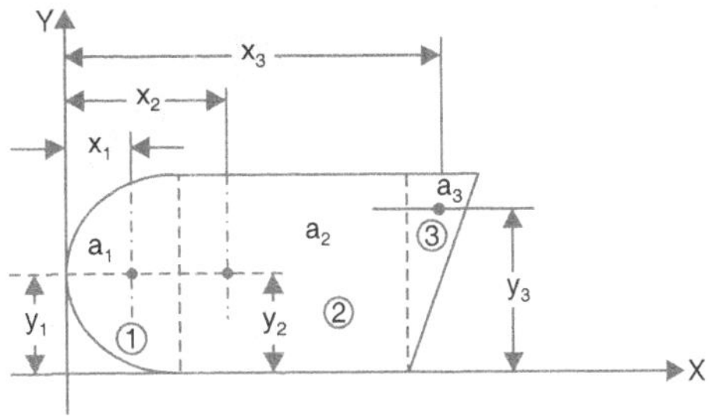

Fig. 4.15. Composite area divided into simple areas.

triangles, etc. (Fig. 4.15). Take the moment of each single area about the Y-axis. Sum up the moments about the Y-axis. Since the centroid of the composite figure is the point at which the entire area about the Y-axis must be equal to the moments of its component parts about the Y-axis, therefore

$$(a_1 + a_2 + ... + a_n)\,\bar{x} = a_1x_1 + a_2x_2 + ... + a_nx_n$$

or

$$\bar{x} = \frac{a_1x_1 + a_2x_2 + ... + a_nx_n}{a_1 + a_2 + ... + a_n}$$

or

$$\bar{x} = \frac{\Sigma ax}{\Sigma a} \qquad ...(4.1)$$

Following the same procedure for moments about the X-axis

$$y = \frac{a_1y_1 + a_2y_2 + ... + a_ny_n}{a_1 + a_2 + ... + a_n}$$

or

$$\bar{y} = \frac{\Sigma ay}{\Sigma a} \qquad ...(4.2)$$

Note. *If a hole exists in the plane figure, treat it as a negative area. The moment of a negative area will be negative provided that the entire figure lies in the first quadrant.*

4.6. (*b*) CENTRE OF GRAVITY OF SIMPLE SOLIDS

The weight of the body is a force acting at its own centre of gravity and directed towards the centre of the earth. The position of the centres of bodies weighing W_1, W_2, W_3 etc. is found in the same manner as the resultant of parallel forces.

$$\left.\begin{aligned} \bar{x} &= \frac{\Sigma Wx}{\Sigma W} \\ \bar{y} &= \frac{\Sigma Wy}{\Sigma W} \\ \bar{z} &= \frac{\Sigma Wz}{\Sigma W} \end{aligned}\right\} \qquad ...(4.3)$$

If all the bodies are of the same material and have the same density (ρ), then

$$W = \rho V_1,\ W_2 = \rho V_2,\ W_3 = \rho V_3 \text{ etc.}$$

Substituting in eqn. (4.3) above, we have

$$\left.\begin{aligned}\bar{x} &= \frac{\Sigma \rho V x}{\Sigma \rho V} = \frac{\Sigma V x}{\Sigma V}\\ \bar{y} &= \frac{\Sigma \rho V y}{\Sigma \rho V} = \frac{\Sigma V y}{\Sigma V}\\ \bar{z} &= \frac{\Sigma \rho V z}{\Sigma \rho V} = \frac{\Sigma V z}{\Sigma V}\end{aligned}\right\} \quad \text{...(4.4)}$$

That is, if the bodies are made of the same material and are of the same density throughout, the centre of gravity of the bodies is their centre of volume. If the bodies are of the same cross-section but perhaps of different lengths,

$$V = al_1,\ V_2 = al_2,\ V_3 = al_3 \text{ etc.}$$

Substituting in eqn. (4.4) above, we get

$$\left.\begin{aligned}\bar{x} &= \frac{\Sigma a l x}{\Sigma a l} = \frac{\Sigma l x}{\Sigma l}\\ \bar{y} &= \frac{\Sigma a l y}{\Sigma a l} = \frac{\Sigma l y}{\Sigma l}\\ \bar{z} &= \frac{\Sigma a l z}{\Sigma a l} = \frac{\Sigma l z}{\Sigma l}\end{aligned}\right\} \quad \text{...(4.5)}$$

If the bodies are parts of a wire, pipe or rod of constant cross-section, then their centre of gravity may be found from the centre of their lengths.

Positions of centre of gravity of some regular solids are given in Table 4.2.

4.7. AREAS AND VOLUMES—CENTROID METHOD

Since the centre of gravity of an area or a body is the point at which the area or mass of the body may be assumed to be concentrated, it can be said that the distance through which an area or a body moves is the same as the distance described by its centre of gravity. This relation is used in finding areas and volumes. Thus, a line moving parallel to its original position is said to generate an area that is equal to the *length of the line multiplied by the distance through which its centroid moves.* That is, area is equal to length times width. Also, an area moving parallel to its original position is said to develop the same volume of prism that is equal to the *area multiplied by the distance through which the centroid moves.* That is, a *volume is equal to the area of the base times the altitude.*

Similarly, a line rotating about one end will develop the area of a circle. A right triangle rotating about either leg will develop the volume of a cone. *In each case a line or area moves through a distance equal to length of a path described by the centroid of either the line or the area.* Many determinations of areas or volumes are simplified by the use of this method.

4.8. CENTRE OF GRAVITY IN A FEW SIMPLE CASES

1. C.G. of a solid right circular cone

Refer to Fig. 4.16. Let ABC be the cone and AD its axis. Consider an elementary circular plate PQ cut off by two planes parallel to the base BC at distance y and $y + dy$ from A, and having its centre at M.

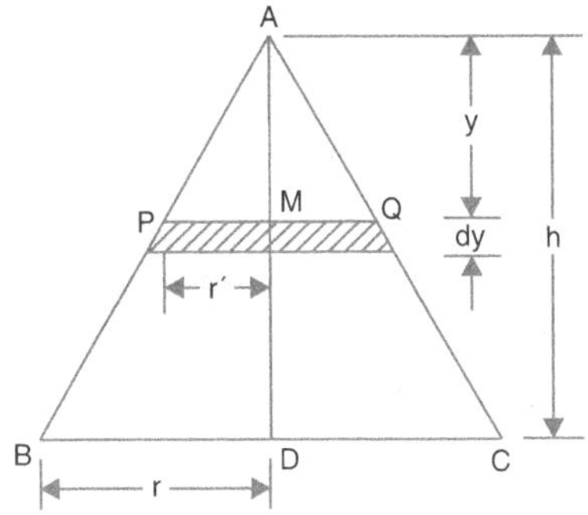

Fig. 4.16

Let $AD = h, BD = r, PM = r'$

Triangles APM, ABD are similar.

$$\therefore \quad \frac{AM}{MP} = \frac{AD}{BD}$$

$$\therefore \quad \frac{y}{r'} = \frac{h}{r} \text{ i.e., } r' = \frac{yr}{h}$$

If w be the density of the material, mass of PQ

$$= \pi r'^2 \, dy \, . \, w$$

$$= \frac{\pi r^2 y^2}{h^2} \, . \, dy \, . \, w$$

The c.g. of PQ is at M. Hence, the distance of c.g. of the cone from A

$$= \frac{\sum_{y=0}^{y=h} \frac{\pi^2 y^2}{h^2} dy \, . \, w \, . \, y}{\sum_{y=0}^{y=h} \frac{\pi r^2 y^2}{h^2} dy \, . \, w} = \frac{\int_0^h y^3 dy}{\int_0^h y^2 dy} \qquad \left[\begin{array}{l} w = \text{weight density}) \\ w = \rho g, \text{where } \rho \text{ is} \\ \quad \text{mass density} \end{array}\right]$$

$$= \frac{\frac{h^4}{4}}{\frac{h^3}{3}} = \frac{3h}{4}$$

$$= \left(h - \frac{3h}{4}\right) = \frac{h}{4} \text{ from the base.} \qquad ...(4.6)$$

Hence, the c.g. of a solid cone lies on the axis at a height one-fourth of the total height from the base.

2. C.G. of a thin hollow right circular cone

Refer to Fig. 4.17. Let ABC be the cone and AD its axis. Consider a circular ring cut off by planes PQ and $P'Q'$ to parallel to the base BC at distances y and $y + dy$ from A.

Let the radius of $PQ = r'$

$$BD = r, AD = h$$

Semi-vertical angle of the cone

$$= \angle BAD = \alpha$$

Clearly, $PP' = dy \sec \alpha$

Also $r' = \frac{ry}{h}$ ...(Refer previous case)

Area of elementary ring

$$= 2\pi r' PP' = 2\pi \cdot \frac{ry}{h} dy \sec \alpha$$

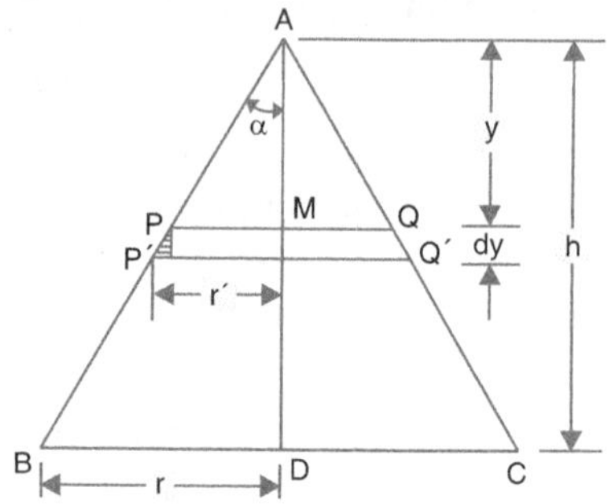

Fig. 4.17

If w be the weight per unit area of the material, weight of the ring

$$= 2\pi \frac{ry}{h} \cdot dy \sec \alpha \cdot w$$

The c.g. of the ring lies on AD at distance y from A.

Hence, the distance of the c.g. of the cone from A

$$= \frac{\sum_{y=0}^{y=h} 2\pi \frac{ry}{h} dy \sec \alpha w . y}{\sum_{y=0}^{y=h} 2\pi \frac{ry}{h} dy \sec \alpha . w} = \frac{\int_0^h y^2 dy}{\int_0^h y\, dy} = \frac{\frac{h^3}{3}}{\frac{h^2}{2}}$$

$$= \frac{2}{3} h$$

$$= \left(h - \frac{2}{3} y\right) = \frac{h}{3} \text{ from the base} \quad \text{...(4.7)}$$

Hence the c.g. of a thin hollow cone lies on the axis at a height one-third of the total height above the base.

3. C.G. of a solid hemisphere

Refer to Fig. 4.18. Let ACB be the hemisphere of radius r, and OC its central radius. Consider an elementary circular plate PQ cut off by planes parallel to AB at distances y and $y + dy$ from AB.

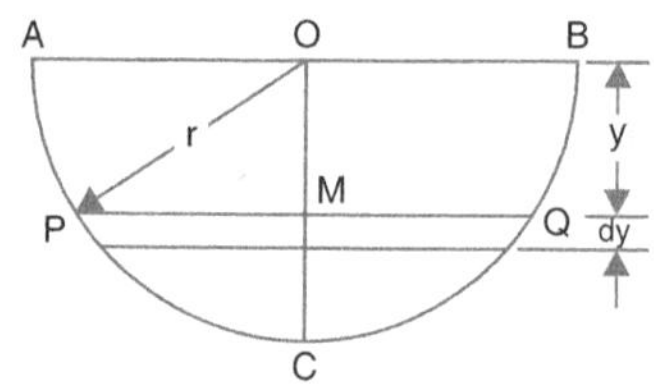

Fig. 4.18

$$PM^2 = OP^2 - OM^2$$

$$= r^2 - y^2$$

Weight of PQ $= \pi (r^2 - y^2) \times dy \cdot w$

where w is the weight of unit volume of the material.

The c.g. of PQ lies on OC at distance y from O.

Therefore, the distance of the c.g. of the hemisphere from O

$$= \frac{\sum_{y=0}^{y=r} \pi (r^2 - y^2)\, dy \cdot w \cdot y}{\sum_{y=0}^{y=r} \pi (r^2 - y^2)\, dy \cdot w}$$

$$= \frac{\int_0^r (r^2 - y^2)\, y\, dy}{\int_0^r (r^2 - y^2)\, dy} = \frac{\left| \frac{r^2 y^2}{2} - \frac{y^4}{4} \right|_0^r}{\left| \left(r^2 y - \frac{y^3}{3} \right) \right|_0^r}$$

$$= \frac{\frac{r^4}{4}}{\frac{2r^3}{3}} = \frac{3r}{8} \qquad ...(4.8)$$

Hence c.g. of a solid hemisphere lies on the central radius at distance $\frac{3r}{8}$ from the plane base, where r is the radius of the hemisphere.

4. C.G. of a thin hollow hemisphere

In Fig. 4.18, if the hemisphere is hollow of negligible thickness, then *PQ* is a ring whose area = $2\pi r\, dy$, by mensuration.

Weight of $PQ = 2\pi r\, dy \,.\, w$

where w is the weight per unit area of the material.

$\therefore$ Distance of the c.g. of the hemisphere from $O = \bar{y}$ (say)

$$= \frac{\int_0^r 2\pi r\, dy\, w \,.\, y}{\int_0^r 2\pi r\, dy \,.\, w} = \frac{\int_0^r y\, dy}{\int_0^r dy} = \frac{\frac{r^2}{2}}{r}$$

$$= \frac{r}{2} \qquad ...(4.9)$$

Hence the c.g. of a hollow hemisphere bisects the central radius.

5. C.G. of a semi-circular lamina

Let Fig. 4.18 represents a semi-circular plate of radius r.

The length of the elementary strip $PQ = 2 \times PM = 2\sqrt{(r^2 - y^2)}$

$$\text{Area of } PQ = 2\sqrt{r^2 - y^2}\; dy$$

If w be the weight per unit area of the material, weight of

$$PQ = 2\sqrt{r^2 - y^2}\; dy \,.\, w$$

The c.g. of the *PQ* lies on *OC* at distance y from *O*.

$$\frac{\int_0^r 2\sqrt{r^2 - y^2}\; dy \,.\, wy}{\int_0^r 2\sqrt{r^2 - y^2}\; dy \,.\, w} = \frac{\int_0^r y\sqrt{r^2 - y^2}\; dy}{\int_0^r \sqrt{r^2 - y^2}\; dy}$$

Now $\quad \int_0^r y\sqrt{r^2 - y^2}\; dy = -\frac{1}{2}\int_0^r (r^2 - y^2)(-2y)\, dy$

$$= -\frac{1}{2}\left|\frac{2}{3}(r^2 - y^2)^{\frac{3}{2}}\right|_0^r = \frac{r^3}{3}$$

and $$\int_0^r \sqrt{r^2 - y^2}\, dy = \left| y\frac{(r^2 - x^2)}{2} + \frac{r^2}{2}\sin^{-1}\frac{y}{r}\right|_0^r$$

$$= \frac{r^2}{2} \times \frac{\pi}{2} = \frac{\pi r^2}{4}$$

$\therefore$ $$\bar{y} = \frac{\frac{r^3}{3}}{\frac{\pi r^2}{4}} = \frac{4r}{3\pi} \quad ...(4.10)$$

Hence the c.g. of a semi-circular lamina lies on the central radius at distance $\frac{4r}{3\pi}$ *from the bounding diameter, where r is the radius of the plate.*

6. C.G. (or centroid) of semi-circular arc

Refer to Fig. 4.19. Let *OC* be the central radius and *P* an element of arc subtending angle $d\theta$ at *O*.

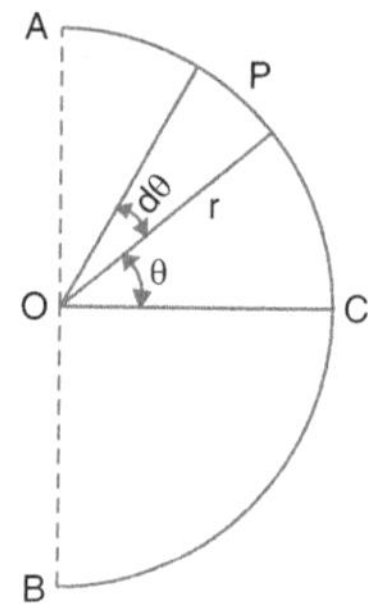

Fig. 4.19

By symmetry, the c.g. of the arc lies on *OC*.

Length of elemental arc $P = rd\theta$, where *r* is the radius of the arc.

Distance of *P* from $AB = r\cos\theta$

$\therefore$ The distance of c.g. of the whole arc from *AB*

$$= \frac{\int_{-\pi/2}^{\pi/2} rd\theta \,.\, r\cos\theta}{\int_{-\pi/2}^{\pi/2} r\, d\theta}$$

$$= \frac{r\left|\sin\theta\right|_{-\pi/2}^{\pi/2}}{\left|\theta\right|_{-\pi/2}^{\pi/2}}$$

$$= \frac{2r}{\pi} \quad ...(4.11)$$

WORKED EXAMPLES

Example 4.1. *Find out the position of the centroid of L section as shown in Fig. 4.20.*

Sol. Refer to Fig. 4.20.

Divide the composite figure into two simple areas :

(*i*) Rectangle (16 cm × 4 cm) ...(1)

(*ii*) Rectangle (8 cm × 4 cm) ...(2)

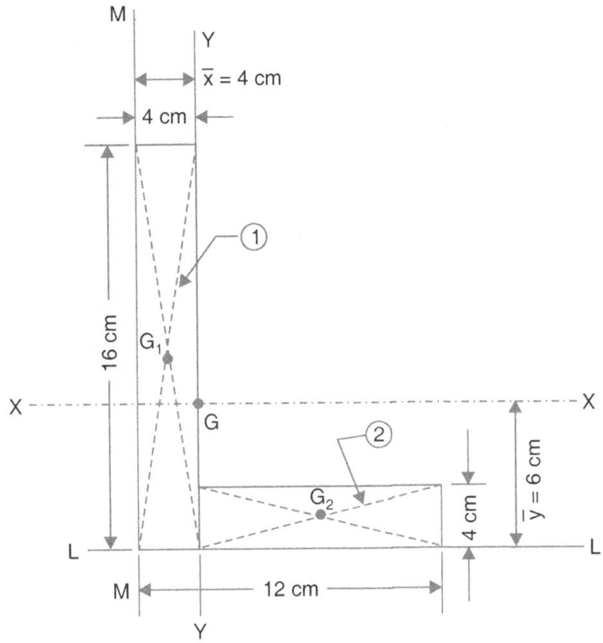

Fig. 4.20

To determine the location of the centroid of the plane figure we have the following table :

Components	*Area a (cm²)*	*Centroidal distance 'x' from MM (cm)*	*Centroidal distance 'y' from LL (cm)*	*ax (cm³)*	*ay (cm³)*
Rectangle (1)	16 × 4 = 64	2	8	128	512
Rectangle (2)	8 × 4 = 32	8	2	256	64
	96 (Σa)	—	—	384 ($\Sigma\, ax$)	576 ($\Sigma\, ay$)

$$\bar{x} = \frac{\Sigma ax}{\Sigma a} = \frac{384}{96} = \textbf{4 cm. (Ans.)}$$

$$\bar{y} = \frac{\Sigma ay}{\Sigma a} = \frac{576}{96} = \textbf{6 cm. (Ans.)}$$

Example 4.2. *Determine the position of the centroid of I-section as shown in Fig. 4.21.*

Sol. Refer to Fig. 4.21.

Divide the composite figure into there simple areas :

(*i*) Rectangle (10 cm × 2 cm) – top flange ...(1)

(*ii*) Rectangle (25 cm × 2 cm) – web ...(2)

(*iii*) Rectangle (15 cm × 3 cm) – bottom flange ...(3)

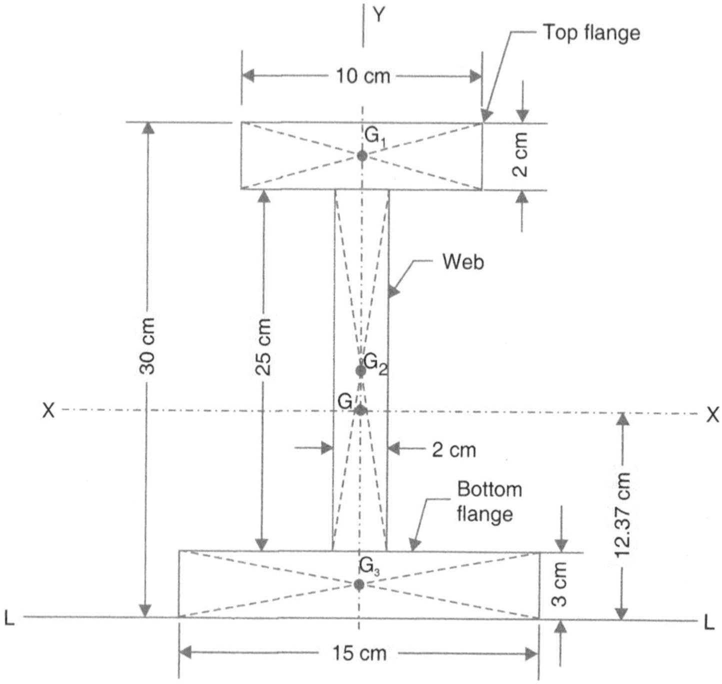

Fig. 4.21

To determine the location of the centroid of the plane figure we have the following table.

Components	*Area 'a'* *(cm²)*	*Centroidal distance 'y' from LL (cm)*	*ay* *(cm³)*
Rectangle (1)	10 × 2 = 20	29	580
Rectangle (2)	25 × 2 = 50	15.5	775
Rectangle (3)	15 × 3 = 45	1.5	67.5
	115 (Σa)	—	1422.5 (Σay)

$$\bar{y} = \frac{\Sigma ay}{\Sigma a} = \frac{1422.5}{115} = \textbf{12.37 cm. (Ans.)}$$

Example 4.3. *Using the analytical method, determine the centre of gravity of the plane uniform lamina shown in Fig. 4.22.*

Sol. Refer to Fig. 4.22.

The lamina may be divided into three parts :

(*i*) A triangle marked (1)

(*ii*) A semi-circle marked (2)

(*iii*) A rectangle marked (3)

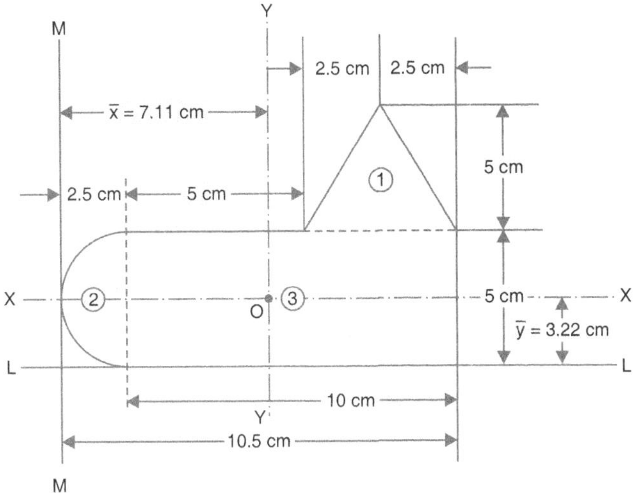

Fig. 4.22

The area of these components, their centroidal distances from the *LL*-axis and *MM*-axis and the moments of the areas of individual components about *LL*-axis and *MM*-axis are tabulated below :

Components	*Area (a)* *(cm^2)*	*Centroidal distance 'x' from MM (cm)*	*Centroidal distance 'y' from LL (cm)*	*ax* *(cm^3)*	*ay* *(cm^3)*
Triangle (1)	$\frac{5 \times 5}{2} = 12.50$	2.5 + 5 + 2.5 = 10	5 + 5/3 = 6.67	125	83.4
Semi-circle (2)	$\frac{\pi \times 2.5^2}{2} = 9.82$	$2.5 - \frac{4 \times 2.5}{3\pi}$ = 1.44	2.5	14.14	24.55
Rectangle (3)	10 × 5 = 50.00	2.5 + 5 = 7.5	2.5	375	125
Total	72.32 (Σ *a*)	—	—	514.14 (Σ *ax*)	232.95 (Σ *ay*)

Distance of the centroid from *MM*-axis,

$$\bar{x} = \frac{\Sigma ax}{\Sigma a} = \frac{514.13}{72.32} = \textbf{7.11 cm. (Ans.)}$$

Distance of the centroid from *LL*-axis

$$\bar{y} = \frac{\Sigma ay}{\Sigma a} = \frac{232.95}{72.32} = \textbf{3.22 cm. (Ans.)}$$

Example 4.4. *Determine the location of the centroid of the plane figure shown in Fig. 4.23.*

Sol. Refer to Fig. 4.23.

Divide the composite figure into three simple areas :

(*i*) a rectangle (7 cm × 5 cm) ...(1)

(*ii*) a quadrant (2 cm radius) ...(2)

(*iii*) a circle (3 cm dia.) ...(3)

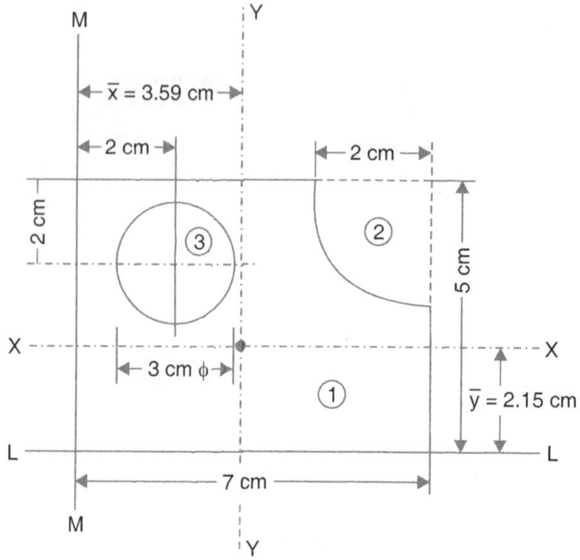

Fig. 4.23

The rectangle is a positive area. The quadrant and hole are treated as negative areas.

To determine the location of the centroid of the plane figure, we have the following table :

Components	*Area 'a'* *(cm^2)*	*Centroidal distance 'x' from M-M (cm)*	*Centroidal distance 'y' from L-L (cm)*	*ax* *(cm^3)*	*ay* *(cm^3)*
Rectangle (1)	$7 \times 5 = 35$ (+)	3.5	2.5	122.5(+)	87.5(+)
Quadrant (2)	$\frac{\pi}{16}d^2 = \frac{\pi}{16} \times 4^2$ = 3.14 (–)	6.15	4.15	19.31(–)	13.03(–)
Circle (3)	$\frac{\pi}{4}d^2 = \frac{\pi}{4} \times 3^2$ = 7.07 (–)	2	3	14.14(–)	21.21(–)
Total	24.79 (Σa)	—	—	89.05 ($\Sigma\, ax$)	53.26 ($\Sigma\, ay$)

Distance of centroid from *MM*-axis

$$\bar{x} = \frac{\Sigma ay}{\Sigma a} = \frac{89.05}{24.79} = \textbf{3.59 cm. (Ans.)}$$

Distance of centroid from *LL*-axis

$$\bar{y} = \frac{\Sigma ay}{\Sigma a} = \frac{53.26}{24.79} = \textbf{2.15 cm. (Ans.)}$$

Example 4.5. *Determine the centroid of the lamina shown in Fig. 4.24.*

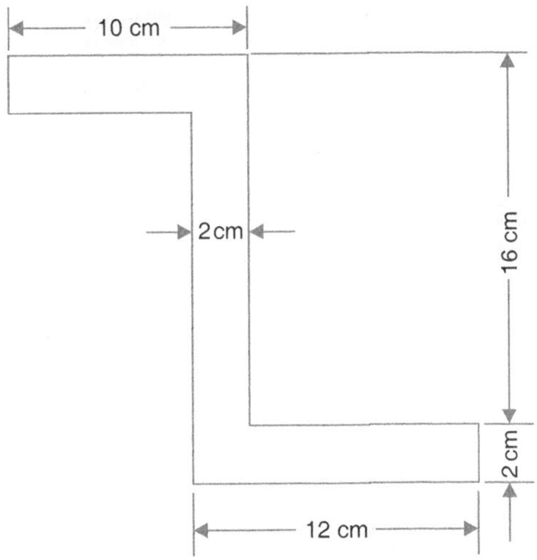

Fig. 4.24

Sol. (*a*) **Analytical method :**

Refer to Fig. 4.25.

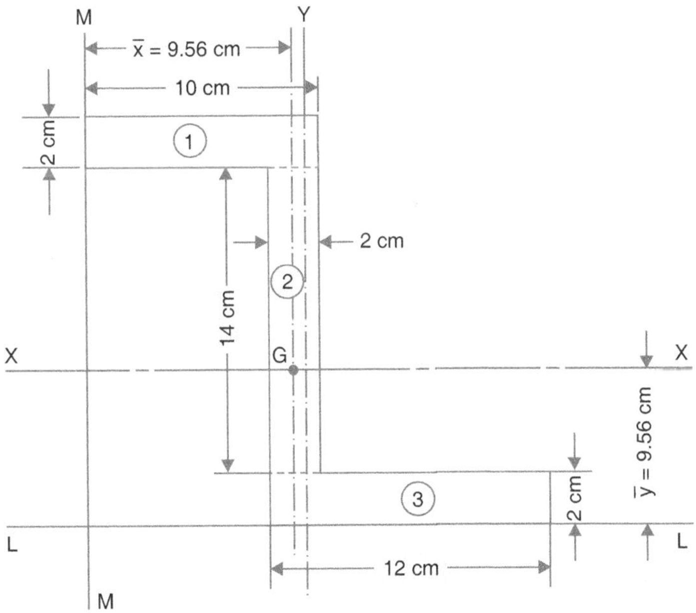

Fig. 4.25

Divide the figure into three areas :

(*i*) Rectangle (10 × 2) ...(1)

(*ii*) Rectangle (14 × 2) ...(2)

(*iii*) Rectangle (12 × 2) ...(3)

To determine the location of the centroid of the plane figure we have the following table :

Components	*Area 'a'* *(cm²)*	*Centroidal distance 'x' from MM* *(cm)*	*Centroidal distance 'y' from LL* *(cm)*	*ax* *(cm³)*	*ay* *(cm³)*
Rectangle (1)	10 × 2 = 20	5	17	100	340
Rectangle (2)	14 × 2 = 28	9	9	252	252
Rectangle (3)	12 × 2 = 24	14	1	336	24
Total	72 (Σ *a*)	—	—	688 (Σ *ax*)	616 (Σ *ay*)

$$\bar{x} = \frac{\Sigma ax}{\Sigma a} = \frac{688}{72} = \textbf{9.56 cm. (Ans.)}$$

$$\bar{y} = \frac{\Sigma ay}{\Sigma a} = \frac{616}{72} = \textbf{8.56 cm. (Ans.)}$$

(*b*) **Graphical method.** Refer to Fig. 4.26.

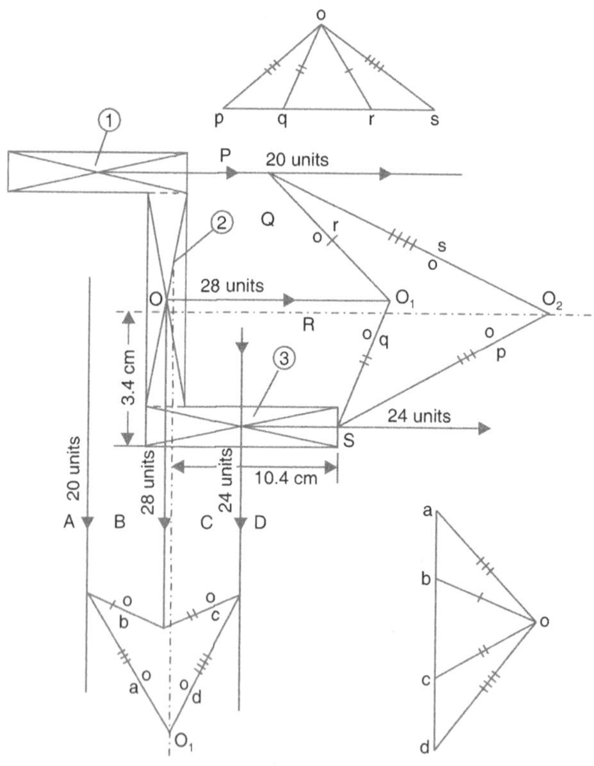

Fig. 4.26

To find $\overline{\mathbf{x}}$. The given section is split upto into three rectangles as shown. The areas of the three rectangles are respectively 20 cm², 28 cm², 24 cm².

Now forces of magnitudes 20 units, 28 units and 24 units are assumed acting vertically through the centroids of the respective rectangles. Let the forces be represented by *AB, BC* and *CD*. Adopting a convenient scale mark off *ab* = 20 units. *bc* = 28 units and *cd* = 24 units. A pole *O* is chosen and *ao, bo, co* and *do* are joined. Starting from any point on the load line *AB* the funicular polygon is drawn and the point O_1, is obtained. Through O_1, a vertical line is drawn. This line represents the vertical centroidal axis. This axis is at a distance of 9.56 cm from left end.

To find $\overline{\mathbf{y}}$. Similarly, taking three horizontal forces of 20 units, 28 units and 24 units through the centroids of the respective rectangles, a polar diagram and the corresponding furnicular polygon are drawn and the point O_2 is obtained. Now a horizontal line is drawn through O_2. This represents the horizontal centroids axis. This axis is found to be 8.56 cm above the bottom edge.

Example 4.6. *A square hole is punched out of a circular lamina of diameter 'a' with one diagonal of the square coinciding with a radius of the circle. Show that the centre of remainder is at a distance of* $\frac{a}{8\pi - 4}$ *from the centre of the circle.*

Sol. Refer to Fig. 4.27. Since, the remaining body is symmetrical about *XX*-axis.

$$\bar{y} = 0$$

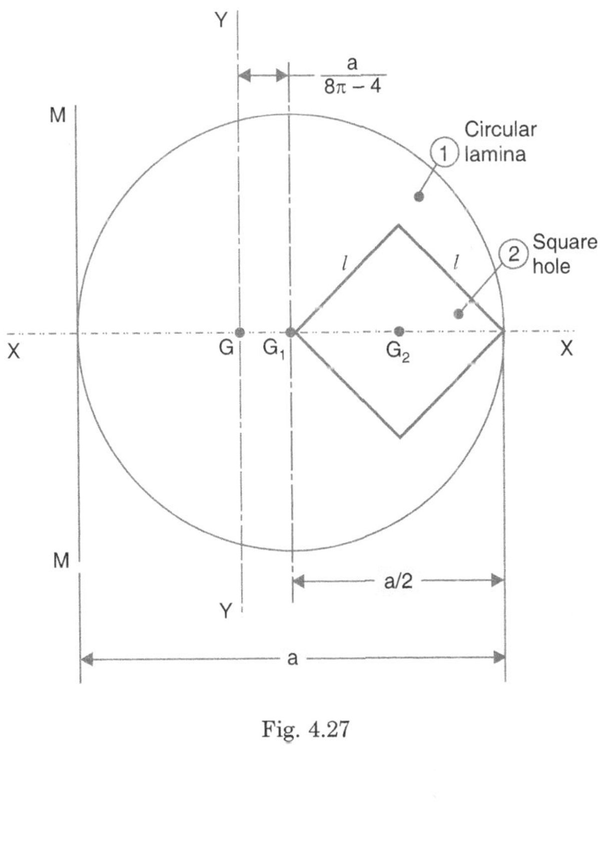

Fig. 4.27

For finding out location of centroid, we have the following table :

Components	*Area 'A'*	*Centroidal distance 'x' from MM*	*Ax*
Circular lamina (1)	$\frac{\pi a^2}{4}$	$\frac{a}{2}$	$\frac{\pi a^3}{8}$
Square hole (2)	$-\frac{a^2}{8}(=l^2)$ $\left[\because\ l^2+l^2=\frac{a^2}{4} \text{ or } l^2=\frac{a^2}{8}\right]$	$\frac{3a}{4}$	$\frac{-3a^3}{32}$
	$\Sigma a = \frac{\pi a^2}{4} - \frac{a^2}{8} = a^2\left(\frac{\pi}{4}-\frac{1}{8}\right)$	—	$\Sigma ax = \frac{\pi a^3}{8} - \frac{3a^3}{32}$

$$\bar{x} = \frac{\Sigma Ax}{\Sigma A} = \frac{\frac{\pi a^3}{8} - \frac{3a^3}{32}}{a^2\left(\frac{\pi}{4}-\frac{1}{8}\right)} \quad \text{from } MM$$

$$= \frac{\frac{a^3}{8}\left(\pi - \frac{3}{4}\right)}{a^2\left(\frac{\pi}{4}-\frac{1}{8}\right)} \quad \text{from } MM$$

$$= \frac{\frac{a^3}{8}\left(\frac{4\pi-3}{4}\right)}{a^2\left(\frac{2\pi-1}{8}\right)} \quad \text{from } MM$$

$$= \frac{a}{4}\left(\frac{4\pi-3}{2\pi-1}\right) \quad \text{from } MM$$

Centre of the remainder from the centre of the circle

$$= \frac{a}{2} - \frac{a}{4}\left(\frac{4\pi-3}{2\pi-1}\right) = \frac{a}{4}\left(2 - \frac{4\pi-3}{2\pi-1}\right)$$

$$= \frac{a}{4}\left(\frac{4\pi-2-4\pi+3}{2\pi-1}\right) = \frac{\mathbf{a}}{\mathbf{8\pi-4}}. \textbf{ (Ans.)}$$

Example 4.7. *From a uniform circular disc of radius 'a' is cut off a circular portion of a radius a/2. Find the distance between the centres of the disc and hole if the centre of the mass of the remainder is on the circumference of the hole.*

Sol. Refer to Fig. 4.28. Since, the remaining body is symmetrical about *X*-axis, therefore, $\bar{y} = 0$.

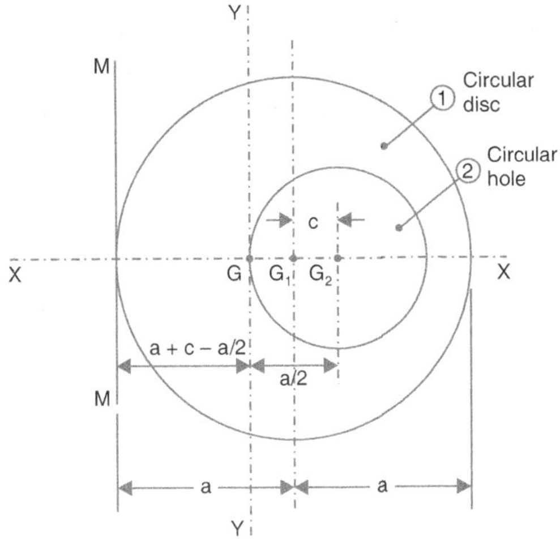

Fig. 4.28

Let 'c' be the distance between the centre of the disc and hole. For finding the value of 'c' we have the following table :

Components	*Area 'A'*	*Centroidal distance 'x' from MM*	*Ax*
Circular disc (1)	πa^2	a	πa^3
Circular hole (2)	$-\dfrac{\pi a^2}{4}$	$(a + c)$	$-\dfrac{\pi a^2}{4}(a + c)$
	$\Sigma A = \dfrac{3\pi a^2}{4}$	—	$Ax = \pi a^3 - \dfrac{\pi a^2}{4} \times (a + c)$

$$\bar{x} = \frac{\Sigma Ax}{\Sigma A} = \frac{\pi a^3 - \frac{\pi}{4} a^2 (a + c)}{\frac{3\pi a^2}{4}}$$

But
$$\bar{x} = \left(a + c - \frac{a}{2}\right)$$

$\therefore$
$$\left(a + c - \frac{a}{2}\right) = \frac{\pi a^2 \left(a - \frac{a + c}{4}\right)}{\frac{3\pi a^2}{4}}$$

or
$$\left(\frac{a}{2} + c\right) = \frac{\left(\frac{3a - c}{4}\right)}{\frac{3}{4}} = \frac{3a - c}{3}$$

$$\frac{3a}{2} + 3c = 3a - c$$

$$4c = \frac{3a}{2}$$

or
$$\mathbf{c = \frac{3a}{8}}. \textbf{ (Ans.)}$$

Example 4.8. *Where must a circular hole of 1 metre radius be punched out of a circular disc of 3 metres radius so that the centre of gravity of the remainder be 2 cm from the centre of the disc ?*

Sol. Refer to Fig. 4.29. Since, the remaining body is symmetrical about *XX*-axis, therefore $\bar{y} = 0$.

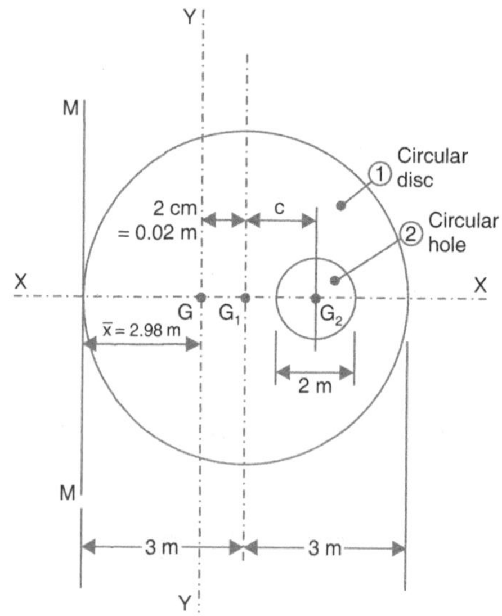

Fig. 4.29

Let '*c*' be the distance of the centre of the circular hole from the centre of the circular disc. For finding out the value of '*c*' we have the following table :

Components	*Area 'a'* *(m²)*	*Centroidal distance 'x' from MM (m)*	*ax* *(m³)*
Circular disc (1)	$\pi \times 3^2 = 9\pi$	3	27π
Circular hole (2)	$-\pi \times 1^2 = -\pi$	$(3 + c)$	$-\pi(3 + c)$
	$\Sigma a = 8\pi$	—	$\Sigma ax = 27\pi - \pi(3 + c)$

$$\bar{x} = \frac{\Sigma ax}{\Sigma a} = \frac{27\pi - \pi(3+c)}{8\pi} = \frac{24 - c}{8}$$

But $\bar{x} = 2.98 \text{ m} = \text{(given) from } MM$

$$\therefore \quad 2.98 = \frac{24 - c}{8}$$

or $24 - c = 23.84$

or $c =$ **0.16 m. (Ans.)**

Example 4.9. *A square hole is punched out of a circular lamina, a diagonal of such a square being along any radius of the circle with one vertex at the centre of the circular lamina. It is said that the length of the said diagonal is equal to the radius of circular lamina. Find the centre of gravity of the remainder, if r be the radius of the circle.*

Sol. Refer to Fig. 4.30. Since, the remaining body is symmetrical about X-axis, therefore, $\bar{y} = 0$.

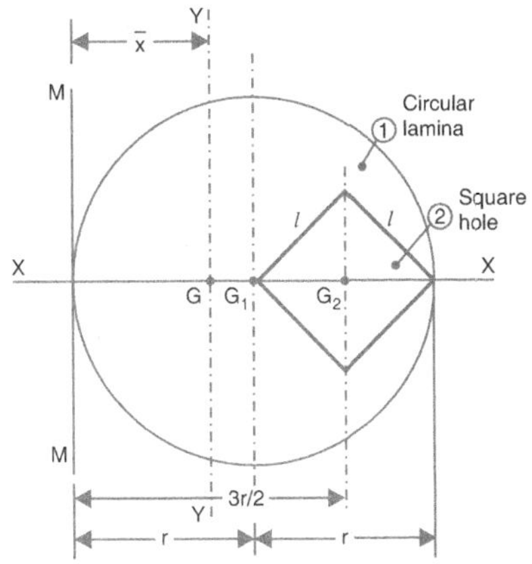

Fig. 4.30

To find out the centre of gravity of the remainder, we have the following table :

Components	*Area 'a'*	*Centroidal distance 'x' from MM*	*ax*
Circular lamina (1)	πr^2	r	πr^3
Square hole (2)	$-\frac{r^2}{2}$ $\left[\begin{array}{l} l^2 + l^2 = r^2 \\ \therefore \ l^2 = \frac{r^2}{2} \end{array}\right]$	$\frac{3r}{2}$	$-\frac{3r^3}{4}$
	$\Sigma a = \pi r^2 - \frac{r^2}{2}$	—	$\Sigma ax = \pi r^3 - \frac{3r^3}{4}$

$$\bar{x} = \frac{\Sigma ax}{\Sigma a} = \frac{\pi r^3 - \frac{3r^3}{4}}{\pi r^2 - \frac{r^2}{2}} = \frac{r^2\left(\pi - \frac{3}{4}\right)}{r^2\left(\pi - \frac{1}{2}\right)}$$

$$= \frac{r\,(4\pi - 3)}{2\,(2\pi - 1)} \text{ from } MM$$

$$= \frac{\mathbf{r\,(4\pi - 3)}}{\mathbf{4\pi - 2}} \textbf{ from MM. (Ans.)}$$

Example 4.10. *From the circular lamina of radius 'a' a smaller circular hole, one quarter the size of the given one is punched out so that its centre bisects a radius of the larger circle. Find the centre of gravity of the remainder.*

Sol. Refer to Fig. 4.31.

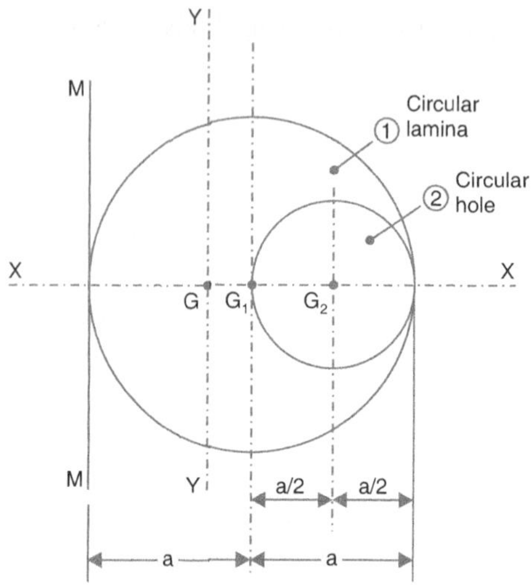

Fig. 4.31

Since, the remaining body is symmetrical about x-axis therefore, we shall find out $\bar{x}$ only ($\bar{y} = 0$).

For finding out $\bar{x}$, we have the following table :

Components	*Area 'a'*	*Centroidal distance 'x' from MM*	*ax*
Circular lamina (1)	πa^2	a	πa^3
Circular hole (2)	$-\pi \times \left(\frac{a}{2}\right)^2 = -\frac{\pi a^2}{4}$	$a + \frac{a}{2} = \frac{3a}{2}$	$-\frac{3\pi a^3}{8}$
	$\Sigma a = \pi a^2 - \frac{\pi a^2}{4}$	—	$\Sigma ax = \pi a^3 - \frac{3\pi a^3}{8}$

$$\bar{x} = \frac{\Sigma ax}{\Sigma a} = \frac{\pi a^3 - \frac{3\pi a^3}{8}}{\pi a^2 - \frac{\pi a^2}{4}} = \frac{\pi a^2\left(a - \frac{3a}{8}\right)}{\pi a^2\left(1 - \frac{1}{4}\right)} = \frac{\frac{5a}{8}}{\frac{3}{4}}$$

$$= \mathbf{\frac{5a}{6}} \textbf{ from MM. (Ans.)} \qquad \left[\text{or } \left(a - \frac{5a}{6}\right) = \frac{a}{6} \text{ from the centre.}\right]$$

Example 4.11. *Determine the position of the c.g. of the shaded area of spandrel OLM, the curve OM being a parabola with axis vertical.*

Sol. Refer to Fig. 4.32.

Let the equation of *OM* be $x^2 = ky$

The point *M*, whose co-ordinates are (a, b) lies on it.

$\therefore \qquad a^2 = kb$

or $\qquad k = a^2/b$

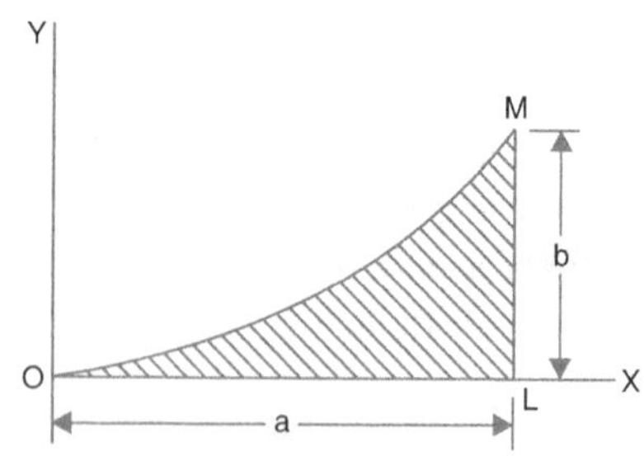

Fig. 4.32

$$x^2 = \frac{a^2}{b} y$$

$$\bar{x} = \frac{\int_0^a ydx \cdot x}{\int_0^a ydx} = \frac{\frac{b}{a^2}\int_0^a x^3\, dx}{\frac{b}{a^2}\int_0^a x^2 dx}$$

$$\frac{a^4/4}{a^3/3} = \mathbf{\frac{3a}{4}}. \textbf{ (Ans.)}$$

$$\bar{y} = \frac{\int_0^a ydx \cdot \frac{y}{2}}{\int_0^a ydx} = \frac{\frac{1}{2}\int_0^a \frac{b^2}{a^4} x^4\, dx}{\int_0^a \frac{b}{a^2} x^2\, dx}$$

$$\frac{b}{2a^2} \cdot \frac{\frac{a^5}{5}}{\frac{a^3}{3}} = \mathbf{\frac{3b}{10}}. \textbf{ (Ans.)}$$

Example 4.12. *Determine the c.g. of the area of a sector of angle α of a circle of radius r as shown in Fig. 4.33.*

Sol. Refer to Fig. 4.33.

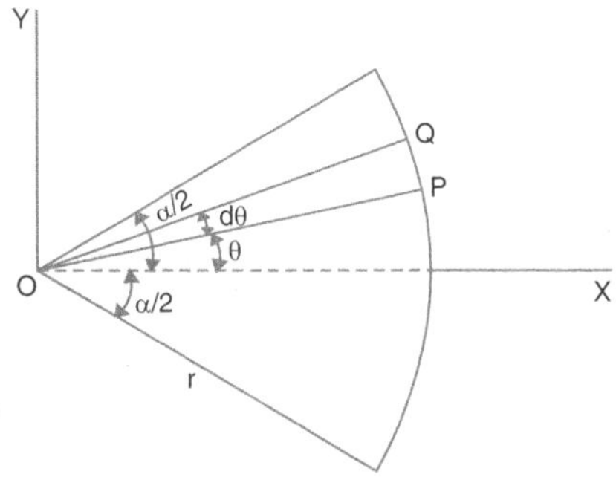

Fig. 4.33

Area $OPQ = \frac{1}{2} r^2\, d\theta$. Its c.g. will be on OP at a distance $\frac{2r}{3}$ from O, so that its distance from OY will be $\frac{2r}{3} \cos\theta$

$$\bar{x} = \frac{\int_0^{\alpha/2} \frac{2r}{3}\cos\theta \frac{r^2}{2}\, d\theta}{\int_0^{\alpha/2} \frac{r^2}{2}\, d\theta} = \frac{\frac{2r}{3}\int_0^{\alpha/2} \cos\theta\, d\theta}{\int_0^{\alpha/2} d\theta}$$

$$= \frac{2r}{3} \cdot \frac{\sin \alpha/2}{\alpha/2} = \mathbf{\frac{4r}{3\alpha} \cdot \sin\frac{\alpha}{2}}. \textbf{ (Ans.)}$$

$\bar{\mathbf{y}}$ **= 0. (Ans.)**

Example 4.13. *Find the position of centre of gravity of the plane lamina in the form of a quarter of an ellipse, shown in Fig. 4.34.*

Sol. Refer to Fig. 4.34.

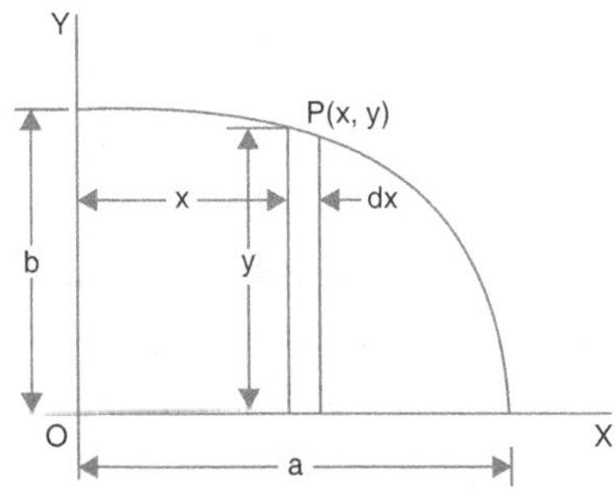

Fig. 4.34

$$\bar{x} = \frac{\int_0^a xy\, dx}{\int_0^a y\, dx}$$

$$\bar{y} = \frac{\int_0^a \frac{y}{2} y dx}{\int_0^a y\, dx}.$$

The equation of the ellipse is

$$\frac{x^2}{a^2} + \frac{y^2}{b^2} = 1$$

From which $$\frac{y^2}{b^2} = 1 - \frac{x^2}{a^2} = \frac{a^2 - x^2}{a^2}$$

$$y = b/a \sqrt{a^2 - x^2}$$

$\therefore$ $$\int_0^a xy\, dx = \int_0^a x . \frac{b}{a} \sqrt{a^2 - x^2}\, dx$$

$$= -\frac{b}{2a} \int_0^a (a^2 - x^2)^{1/2} (-2x)\, dx$$

$$= -\frac{b}{2a} . \frac{2}{3} \left| (a^2 - x^2)^{3/2} \right|_0^a = \frac{b}{3a} . a^3 = \frac{a^2 b}{3}$$

$$\int_0^a y\, dx = \int_0^a \frac{b}{a} \sqrt{a^2 - x^2}\, dx$$

$$= \frac{b}{a} \left| \frac{x\sqrt{a^2 - x^2}}{2} + \frac{a^2}{2} \sin^{-1} \frac{x}{a} \right|_0^a = \frac{b}{a} . \frac{a^2}{2} \quad \frac{\pi}{2} = ab \frac{\pi}{4}$$

$\therefore$ $$\bar{x} = \frac{a^2 b}{3} \times \frac{4}{ab\pi}$$

or $$\bar{x} = \mathbf{\frac{4a}{3\pi}}. \quad \textbf{(Ans.)}$$

Similarly, $$\bar{y} = \mathbf{\frac{4b}{3\pi}}. \quad \textbf{(Ans.)}$$

Example 4.14. *A hemisphere of diameter 60 mm is placed on the top of a cylinder, whose diameter is also 60 mm. The height of the cylinder is 75 mm. Find the common C.G. of the composite body.*

Sol. Refer to Fig. 4.35.

Since, the composite solid is symmetrical about Y-axis, so the c.g. will lie on Y-axis and we shall, therefore, find out $\bar{y}$ only.

Volume of hemisphere,

$$V_1 = \frac{2}{3} \pi r^3 = \frac{2}{3} \times \pi \times 30^3 = 18000\, \pi \text{ mm}^3$$

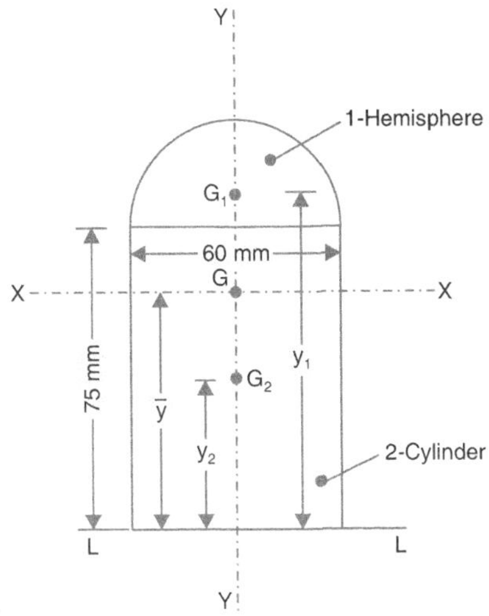

Fig. 4.35

Volume of cylinder,

$$V_2 = \pi r^2 h = \pi \times 30^2 \times 75 = 67500\ \pi\ \text{mm}^3.$$

To determine $\bar{y}$ of composite solid, we have the following table :

Components	*Volume (mm³)* *V*	*Centroidal distance 'y' from LL (mm)*	*V . y (mm⁴)*
Hemisphere (1)	18000π	$y_1 = 75 + \frac{3r}{8} = 75 + \frac{3 \times 30}{8} = 86.25$	1552500π
Cylinder (2)	67500π	$y_2 = \frac{75}{2} = 37.5$	2531250π
Total	85500π (ΣV)	—	4083750 π ΣVy

$$\bar{y} = \frac{\Sigma V \,.\, y}{\Sigma V} = \frac{4083750\pi}{85500\pi}$$

$$= \mathbf{47.76\ mm.\ (Ans.)}$$

Hence the c.g. of the composite body lies at a distance of 47.76 mm from *LL*, the axis of reference.

Note. In this case since the individual components (hemisphere and cylinder) are made of same material so we have considered volumes only (instead of weights) but in case the components are made of different materials then we should take weights instead of volumes (because densities are different).

Example 4.15. *A body consists of a right circular solid cone of height 18 cm and radius 15 cm placed on a solid hemisphere of radius 15 cm of the same material. Find the position of c.g.*

Sol. Refer to Fig. 4.36.

Since, the composite body is symmetrical about *Y*-axis, so the c.g. will lie on *Y*-axis and we shall, therefore, find out y only.

Volume of right circular cone,

$$V_1 = \frac{1}{3}\pi r^2 h = \frac{1}{3} \times \pi \times 15^2 \times 18 = 1350\,\pi \text{ cm}^3$$

Volume of hemisphere,

$$V_2 = \frac{2}{3}\pi r^3 = \frac{2}{3}\,\pi \times 15^3 = 2250\,\pi \text{ cm}^3$$

To determine $\bar{y}$ of composite body, we have the following table :

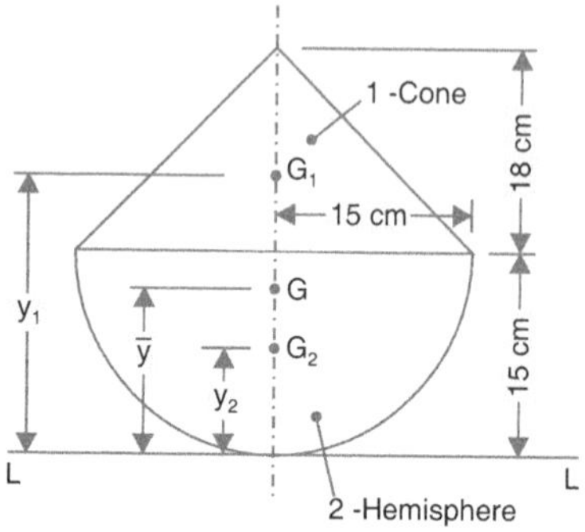

Fig. 4.36

Components	*Volume (cm³)* V	*Centroidal distance 'y' from LL*	*V.y (cm⁴)*
Cone (1)	1350π	$y_1 = 15 + \frac{h}{3} = 15 + \frac{18}{4} = 19.5$ cm	32265π
Hemisphere (2)	2250π	$y_2 = 15 - \frac{3r}{8}$ $= 15 - \frac{3 \times 15}{8} = 9.375$ cm	21093.75π
Total	3600π (ΣV)	—	47418.75π (ΣV.y)

$$\bar{y} = \frac{\Sigma V.y}{\Sigma V} = \frac{47418.75\pi}{3600\pi} = 13.17 \text{ cm.}$$

Hence, the c.g. of the composite body lies at a distance of **13.17 cm from LL**, the axis of reference. **(Ans.)**

Example 4.16. *The frustrum of a solid cone has base diameter 10 cm, the top diameter 5 cm and height 7.5 cm. Determine the c.g. of the solid.*

Sol. Refer to Fig. 4.37 (*a* and *b*)

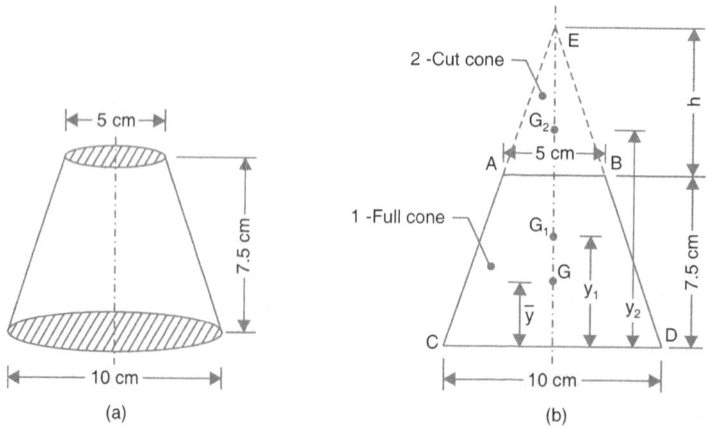

Fig. 4.37

To find h, from similar Δs EFB and EGD, we have

$$\frac{EF}{EH} = \frac{FB}{HD}$$

$$\frac{h}{h+7.5} = \frac{2.5}{5}$$

or $$5h = 2.5\,(h + 7.5) = 2.5h + 18.75$$

or $$2.5\,h = 18.75$$

$$h = 7.5 \text{ cm}$$

Volume of full cone CDE,

$$V_1 = \frac{1}{3} \times \pi \times 5^2 \times (7.5 + 7.5) = 125\pi$$

Volume of cut cone ABE,

$$V_2 = \frac{1}{3} \times \pi \times 2.5^2 \times 7.5 = 15.625\pi.$$

Since the solid is symmetrical about Y-axis, we shall find out $\bar{y}$ only as given below :

Components	*Volume (cm³)* V	*Centroidal distance 'y' from CD (cm)*	*Vy*
Full cone CDE (1)	125π	$y_1 = \frac{15}{4} = 3.75$	468.74π
Cut cone ABE (2)	-15.625π	$y_2 = 7.5 + \frac{7.5}{4} = 9.375$	-146.48π
Total	109.375π (ΣV)	—	322.26π (ΣVy)

$$\bar{y} = \frac{\Sigma Vy}{\Sigma V} = \frac{322.26\pi}{109.375\pi} = \mathbf{2.946\ cm.} \quad \textbf{(Ans.)}$$

Example 4.17. *A hemisphere and a cone have their bases jointed together, the two bases being of the same size. Find the ratio of the height of the cone to the radius of the base, so that their common c.g. may be at the centre of the common base.*

Sol. Refer to Fig. 4.38.

Let O be the centre of the common base, r its radius, h the height of the cone, G_1 the c.g. of the cone and G_2 the c.g. of the hemisphere.

Then $$OG_1 = 1/4\ h$$

and $$OG_2 = 3/8\ r$$

Now masses of the cone and hemisphere are as

$$\frac{1}{3}\pi r^2 h : 2/3\ \pi r^3$$

or $$h : 2r$$

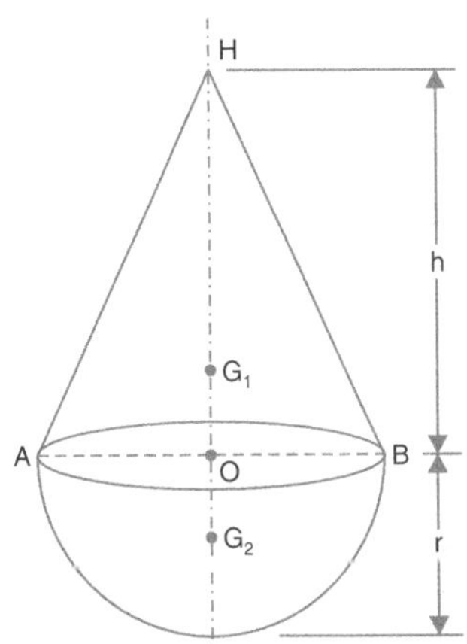

Fig. 4.38

Let the mass of the cone = kh and the mass of the hemisphere,

$$k \times 2r = 2rk$$

∴ Distance of the c.g. of the whole from O, taking OH as positive

$$= \frac{kh \times h/4 - 2rk \times 3/8r}{kh + 2rk} = 0 \qquad (\because \; O \text{ is the c.g.})$$

$$\therefore \qquad \frac{kh^2}{4} - \frac{3r^2k}{4} = 0$$

$$h^2 - 3r^2 = 0$$

$$h^2 = 3r^2 \quad \text{or} \quad \frac{h^2}{r^2} = 3$$

$$\frac{h}{r} = \sqrt{3}$$ **which is the required ratio. (Ans.)**

Example 4.18. *A solid right circular cylinder has its base scooped out so that the hollow is a right circular cone on the same base and having the same height as the cylinder. Find the c.g. of the remainder.*

Sol. Refer to Fig. 4.39.

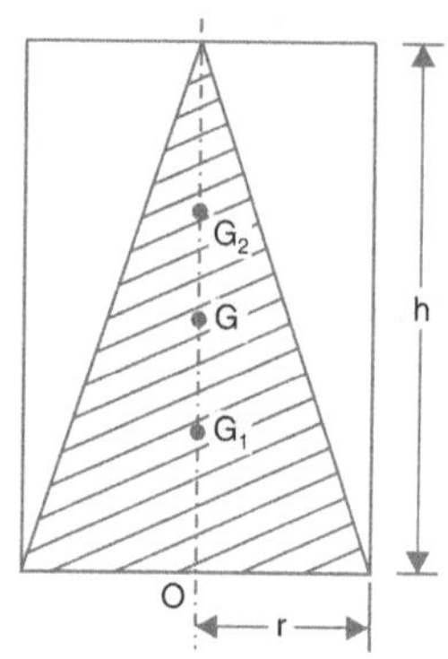

Fig. 4.39

The portion scooped out is shown shaded in the figure. O is the centre of the base, G is the c.g. of the cylinder, G_1 is the c.g. of the portion removed, and G_2 is the c.g. of the remainder. Let r be the radius of the base and h the height of the cylinder.

Weight of the cylinder = $\pi r^2 hw$

Here w is the weight of unit volume of the material.

Weight of the portion scooped out

$$= \frac{1}{3} \pi r^2 hw \text{ acting at } G_1$$

Weight of the remainder

$$= \pi r^2 h.w - \frac{1}{3} \pi r^2 h.w$$

$$= \frac{2}{3} \pi r^2 h.w \text{ acting at } G_2$$

$$OG_1 = \frac{1}{4} h, \; OG = \frac{1}{2} h$$

$$GG_1 = \frac{1}{2} h - \frac{1}{4} h = \frac{1}{4} h$$

G is the c.g. of a weight $\frac{1}{3} \pi r^2 hw$ acting at G_1 and a weight $\frac{2}{3} \pi r^2 hw$ acting at G_2.

$$\therefore \qquad \frac{1}{3} \pi r^2 hw \times GG_1 = \frac{2}{3} \pi r^2 hw \times GG_2$$

$$GG_2 = \frac{1}{2} GG_1 = \frac{1}{8} h$$

$$OG_2 = OG + GG_2$$

$$= \frac{1}{2} h + \frac{1}{8} h = \mathbf{\frac{5}{8} h.} \quad \textbf{(Ans.)}$$

☞ **Example 4.19.** *Determine the centre of gravity of a homogeneous solid body of revolution as shown in Fig. 4.40.*

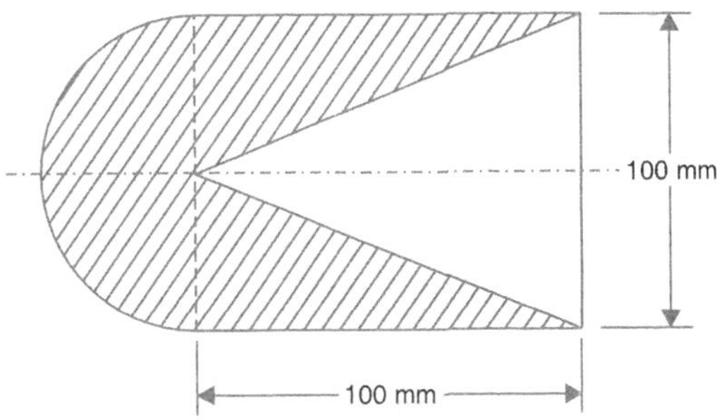

Fig. 4.40

Sol. Refer to Fig. 4.41.

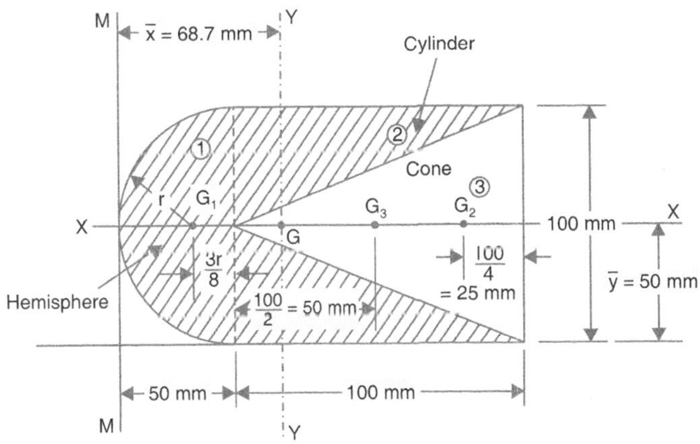

Fig. 4.41

The solid body consists of the following three parts :

(*i*) Hemisphere ...(1)

(*ii*) Cylinder ...(2)

(*iii*) Cone ...(3)

The solid body is obtained by *removing* the mass corresponding to the cone (3) from the composite body consisting of hemisphere and the cylinder.

Since, the body is symmetrical about X-axis, therefore we shall find out $\bar{y}$ only with the following table.

Components	*Volume (mm³)* *V*	*Centroidal distance 'x' from MM (mm)*	*Vx (mm⁴)*
Hemispherical (1)	$\frac{2}{3}\pi r^3 = \frac{2}{3}\pi \times 50^3 = 8.3333\pi(+)$	$50 - \frac{3r}{8} = 50 - \frac{3 \times 50}{8} = 31.25$	$2604156\pi(+)$
Cylinder (2)	$\pi r^2 h = \pi \times 50^2 \times 100 = 250000\pi(+)$	$50 + \frac{100}{2} = 100$	$25000000\pi(+)$
Cone (3)	$\frac{1}{3}\pi r^2 h = \frac{1}{3} \times \pi \times 50^2 \times 100 = 83333\pi(-)$	$50 + 100 - \frac{100}{4} = 125$	$10416625\pi(-)$
Total	$\Sigma V = 250000\pi$	—	$\Sigma Vx = 17187531\pi$

$$\bar{x} = \frac{\Sigma Vx}{\Sigma V} = \frac{17187531\pi}{250000\pi} = \textbf{68.7 mm.} \quad \textbf{(Ans.)}$$

and
$$\bar{y} = \frac{100}{2} = \textbf{50 mm. (Ans.)}$$

Example 4.20. *A frustrum of a solid right circular cone having base circle diameter 100 mm, height 100 mm and top diameter 50 mm has an axial hole of diameter 25 mm. Find the c.g. of the solid.*

Sol. Refer to Fig. 4.42. The frustrum of cone can be obtained by removing cone (2) of base diameter 50 mm and height 100 mm from another cone (1) of base diameter 100 mm and height 200 mm. The given solid will be obtained by removing a cylinder of (3) of diameter 25 mm and height 100 mm from this frustrum.

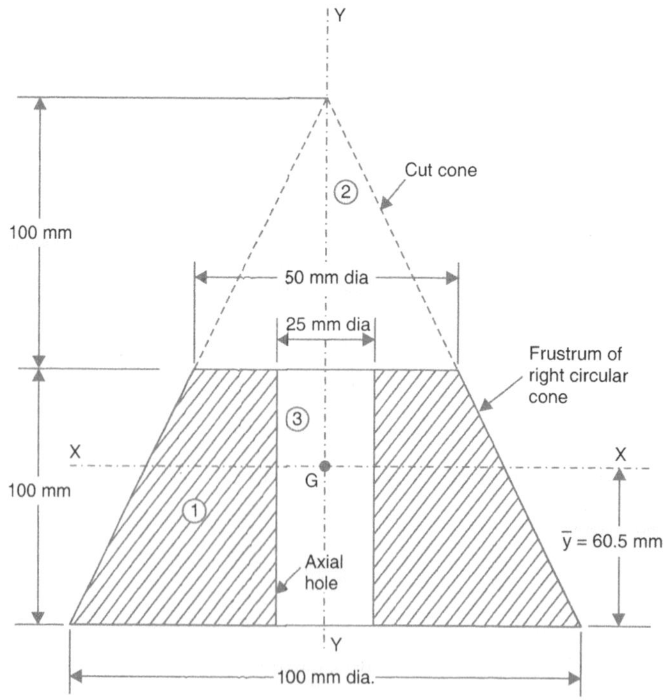

Fig. 4.42

Since, the solid is symmetrical about y-axis, we shall out $\bar{y}$ only with the following table :

Components	*Volume (mm³)* V	*Centroidal distance 'y' from base (mm)*	*Vy (mm⁴)*
Cone (1)	$\frac{1}{3} \times \pi \times 50^2 \times 200 = 166666\,\pi(+)$	$\frac{200}{4} = 50$	8333300π (+)
Cone (2)	$\frac{1}{3} \times \pi \times 25^2 \times 100 = 20833\,\pi(-)$	$\left(\frac{100}{4} + 100\right) = 125$	2604125π (–)
Cylinder (3)	$\pi \times (12.5)^2 \times 100 = 15625\pi$ (–)	$\frac{100}{2} = 50$	781250π (–)
Total	$\Sigma V = 130208\pi$	—	$\Sigma Vy = 4947925\pi$

$$\therefore \quad \bar{y} = \frac{\Sigma Vy}{\Sigma V} = \frac{4947925\pi}{130208\pi} = \textbf{38 mm. (Ans.)}$$

HIGHLIGHTS

1. Weight is force of attraction between a body and the earth and is proportional to mass of the body.
2. Centre of gravity of the body is the point through which the whole weight of the body may be assumed to act.
3. The centre of gravity of some objects may be found by balancing the object on a point.
4. The *centroid* or *centre of area* is defined as the point where the whole area of the figure is assumed to be concentrated.
5. Centroid of a triangle lies at intersection of radius ($h/3$ from base) of the triangle.
6. Centroid of semi-circle lies at $\frac{4r}{3\pi} = 0.424r$ from base, where r is radius of the semi-circle.
7. C.G. of a cylinder lies on the axis at a height one-half of the total height from the base.
8. C.G. of a solid cone lies on the axis at a height one-fourth of the total height from the base.
9. C.G. of thin hollow cone lies on the axis at a height one-third of the total height above the base.
10. C.G. of a solid hemisphere lies on the central radius at distance $3r/8$ from the plane base, where r is radius of hemisphere.
11. C.G. of a hollow hemisphere bisects the central radius.

OBJECTIVE TYPE QUESTIONS

Fill in the blanks :

(*i*) ...... is the force of attraction between a body and the earth and is proportional to mass of the body.

(*ii*) ...... is the point through which the whole weight of a body may be assumed to act.

(*iii*) ...... is the point where the whole area of the figure is assumed to be concentrated.

(*iv*) Centroid of a triangle lies at .... from the base, where h is the height of triangle.

(*v*) Centroid of a semi-circle lies at a distance of ...... from the base, where r is the radius of semi-circle.

(*vi*) Area of a quadrant is given by ......, where r is the radius.

(*vii*) C.G. of a cylinder lies on the axis at a height ...... of the total height from the base.

(*viii*) C.G. of a solid cone lies on the axis at a height ...... of the total height from the base.

(*ix*) C.G. of a hollow cone lies on the axis at a height ...... of the total height above the base.

(*x*) C.G. of a solid hemisphere lies on the central radius at a distance ...... from the plane base, where r is the radius of hemisphere.

Answers

(*i*) Weight	(*ii*) Centre of gravity	(*iii*) Centroid	(*iv*) $h/3$
(*v*) $0.424r$	(*vi*) $\pi r^2/4$	(*vii*) $h/2$	(*viii*) one-fourth
(*ix*) $h/3$	(*x*) $\frac{3r}{8}$.		

EXERCISES

Theoretical Questions

1. Define the terms 'Centre of gravity' and 'Centroid'.
2. Explain briefly how centre of gravity of a body can be determined.
3. Write down the positions of centroids of the following plane geometrical figures : Rectangle, circle, semi-circle, quadrant and trapezium.
4. Where do the C.Gs of the following regular solids lie ? Cylinder, sphere, hemisphere, right circular cone.
5. Derive expressions for centre of gravity in the following cases :
   (*i*) Solid right circular cone
   (*ii*) Thin hollow right circular cone
   (*iii*) Semi-circular lamina.

Unsolved Examples

1. Locate the centroid of the piece of sheet metal shown in Fig. 4.43. [**Ans.** $\bar{x}$ = 4.59 cm, $\bar{y}$ = 3.19 cm]

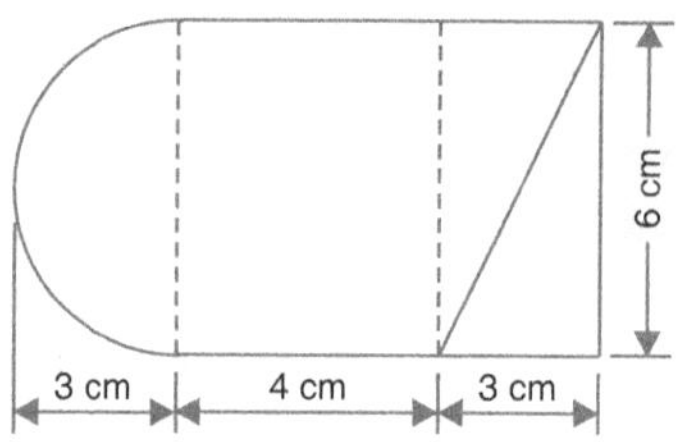

Fig. 4.43

2. Find the centroid of the lamina shown in Fig. 4.44. [**Ans.** $\bar{x}$ = 2.50 cm, $\bar{y}$ = 3.50 cm]

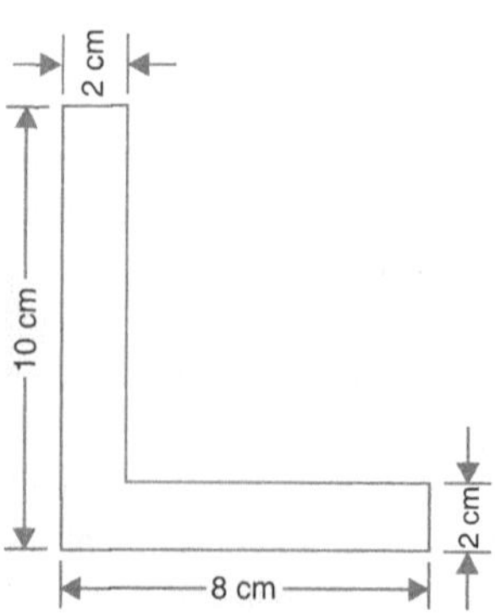

Fig. 4.44

3. Find the centroid of the lamina shown in Fig. 4.45. [**Ans.** $\bar{y}$ = 1.864 cm from the bottom]

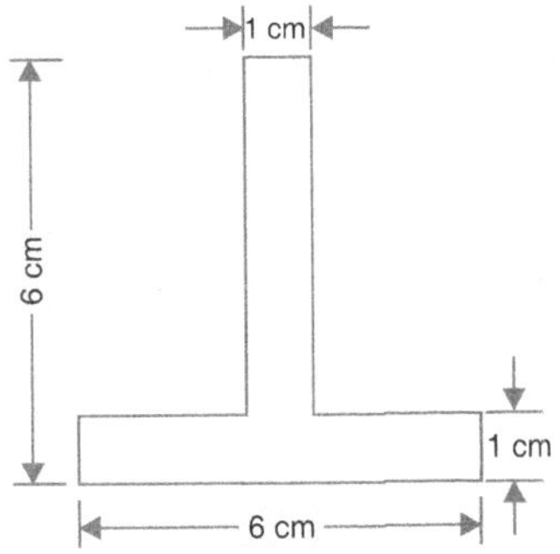

Fig. 4.45

4. Fig. 4.46 shows an unsymmetrical I section. The upper flange is 6 cm × 3/4 cm, the lower flange is 12 cm × 1 cm, the overall depth is 16 cm and the thickness of the metal at the web is $\frac{1}{2}$ cm. Find the distance of c.g. from the bottom of the lower flange. [**Ans.** 5.68 cm]

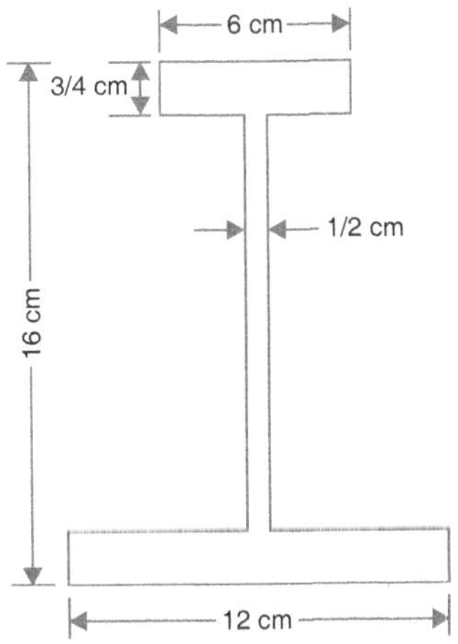

Fig. 4.46

5. A square sheet of metal has a square of one quarter of the original area cut from one corner as shown in Fig. 4.47. Calculate the position of the c.g. of the remaining portion of the sheet.
[**Ans.** Distance of c.g. from *AB* or *AD* = 5/12 *AB*]

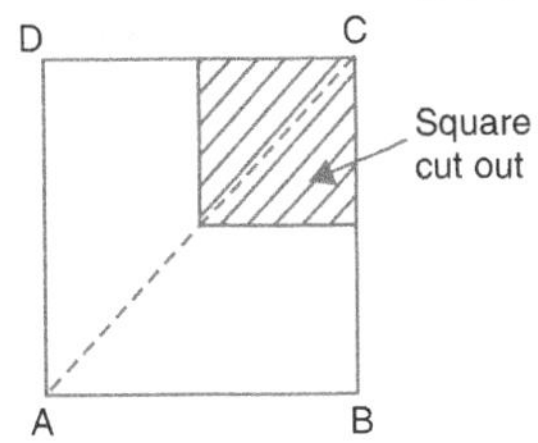

Fig. 4.47

6. From a circular plate of diameter 6 cm is cut out a circle whose diameter is a radius of the plate. Find the c.g. of the remainder. [**Ans.** 0.5 cm from the centre]

7. A hollow cost iron column has 30 cm external diameter and 22.5 cm inside diameter. In casting, the bore got eccentric such that the thickness varies from 2.5 cm on one side and 5 cm on the other side. Find the position of c.g. of the section. [**Ans.** 16.67 cm]

8. From a rectangular sheet of metal *ABCD,* in which $AB = 40$ cm and $BC = 60$ cm, a triangular piece *ABX* is removed, such that $AX = BX = 25$ cm. Calculate the distance of centre of gravity of the remainder. [**Ans.** 33.57 cm from the base]

9. Find the c.g. of a semi-circular section having the outer and inner radii 20 cm and 16 cm respectively. [**Ans.** $\bar{y} = 11.2$ cm]

10. Where must a hole of radius 0.25 cm be punched out of a circular disc of radius 1 m, if c.g. of the remainder is 5 cm from centre of disc. [**Ans.** 0.625 m]

11. *ABCD* is a square piece of paper of side 10 cm and *E* and *F* are mid-points of *AB* and *AD*. Find the c.g. of the portion left when ΔAEF is cut off as shown in Fig. 4.48. [**Ans.** $\bar{x} = 5.476$ cm ; $\bar{y} = 5.476$ cm]

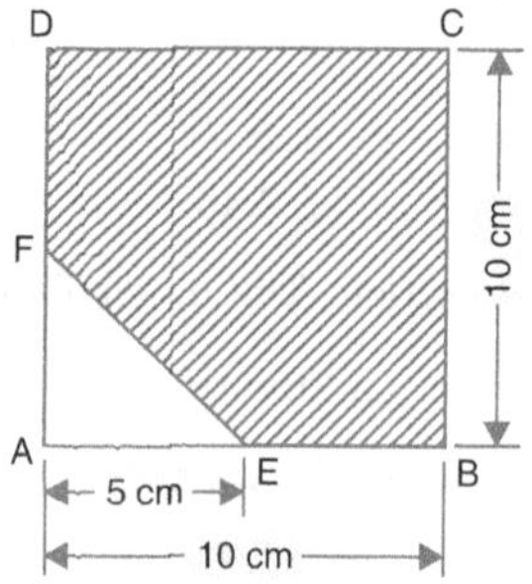

Fig. 4.48

12. Determine the position of centroid of an area enclosed by two semi-circles and their common diameter as shown in Fig. 4.49. Take $r_1 = 180$ mm and $r_2 = 135$ mm. [**Ans.** $\bar{x} = 0$, $\bar{y} = 103$ mm]

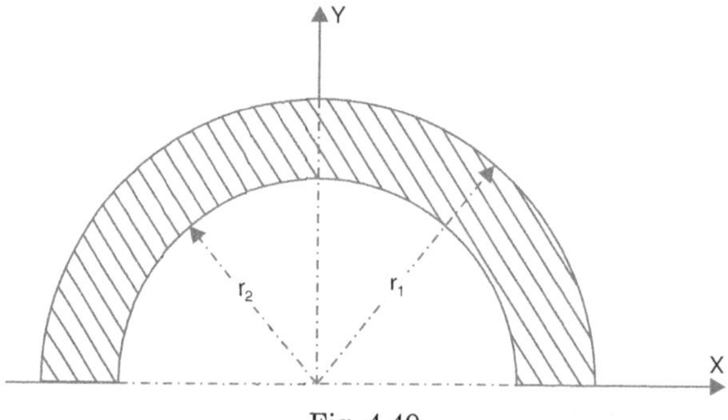

Fig. 4.49

13. A semi-circle of diameter 100 mm is drawn on a paper sheet. A triangle with diameter of the semi-circle as base is cut out of the semi-circle as shown in Fig. 4.50. [**Ans.** $\bar{x} = 0$, $\bar{y} = 29.2$ mm]

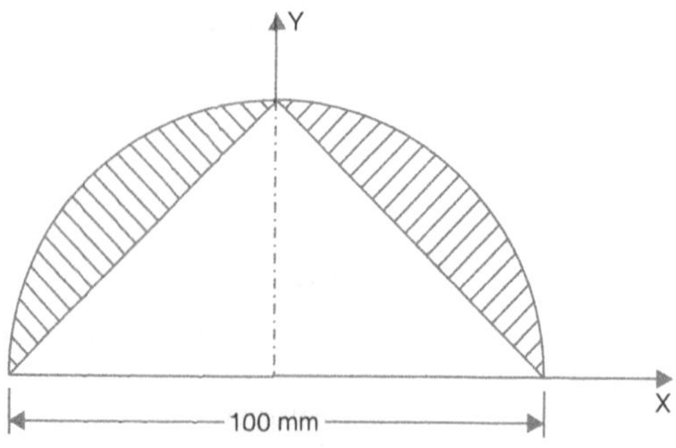

Fig. 4.50

14. Find the centroid of the shaded area shown in Fig. 4.51. [**Ans.** $\bar{x}$ = 47 mm, $\bar{y}$ = 64.44 m]

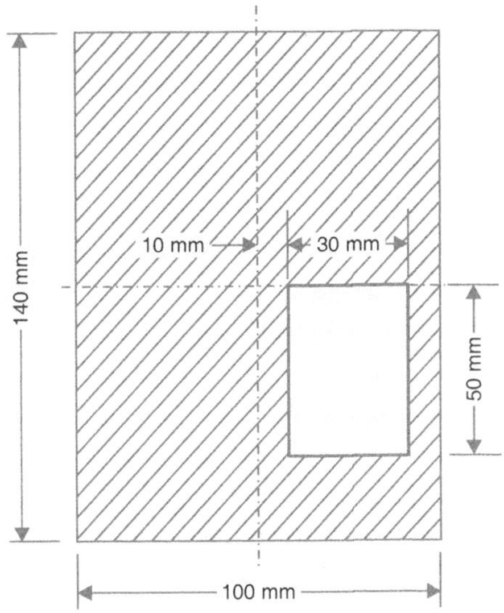

Fig. 4.51

15. Determine by calculation the position of the c.g. of the section shown in Fig. 4.52.

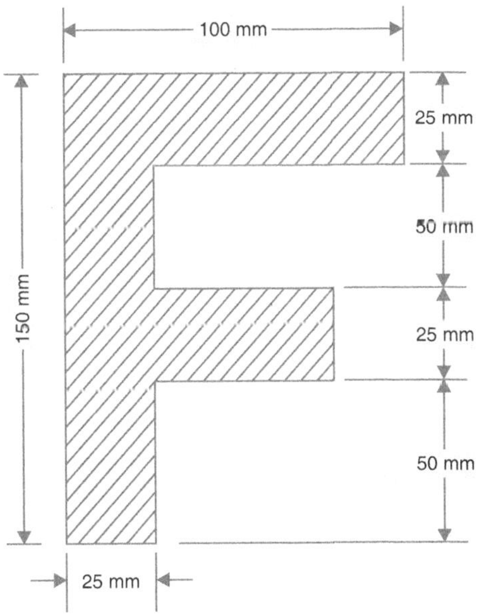

Fig. 4.52

16. Find the co-ordinates of the centroid of the area obtained after removing a semi-circular of radius 100 mm from a quadrant of a circle of radius 200 mm. (See Fig. 4.53). [**Ans.** 69.8 mm ; 127.2 mm]

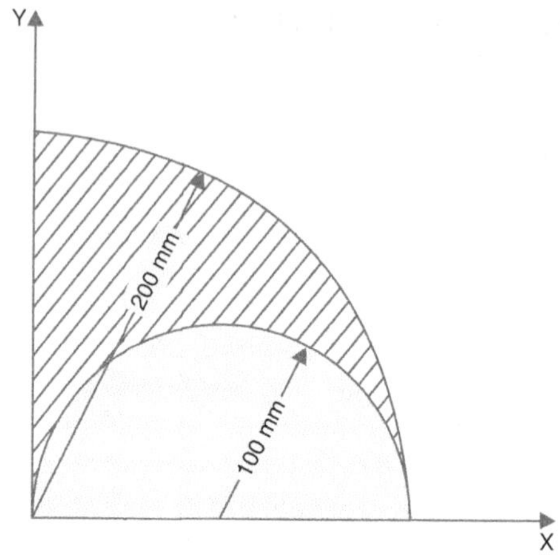

Fig. 4.53

17. A solid right circular cylinder has its base scooped out (Fig. 4.54) so that the hollow is a right circular cone on the same base and having the same height as the cylinder. Find the c.g. of the remainder.

[**Ans.** 5/8h]

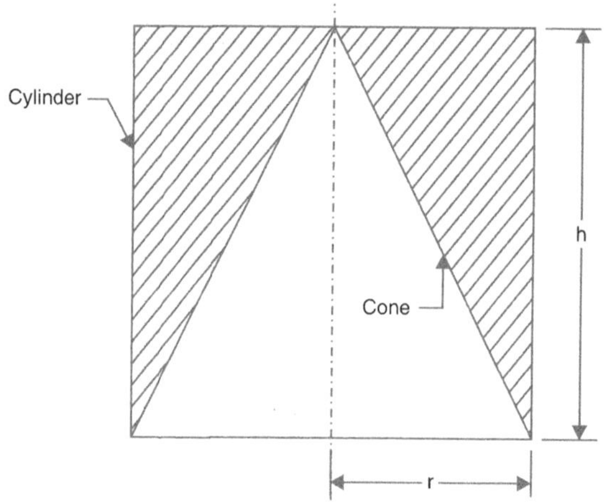

Fig. 4.54

18. A solid steel cone is made lighter by removing a part of the material from it in the form of a cylinder as shown in Fig. 4.55. Determine the position of the c.g. of the remaining metal. [**Ans.** 37.6 mm]

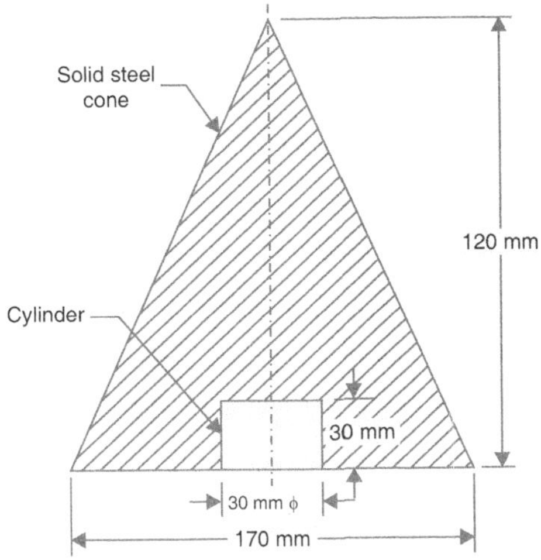

Fig. 4.55

19. A plate of uniform thickness is in the form of an isosceles triangle with base 24 cm and height 30 cm. A hole of 25 cm^2 is cut from it, with its centre 8 cm above the base and 2 cm to the left of vertex. Determine the centroid of the remaining plate. [**Ans.** $\bar{x}$ = 12.15 cm ; $\bar{y}$ = 10.15 cm]

20. A body consists of a solid hemisphere of radius 10 cm and a right circular solid cone of height 12 cm. The hemisphere and cone have a common base and are made of the same material. Find the position of the c.g. of the compound body. [**Ans.** 13.22 cm]

21. A solid right circular cone has its base scooped out, so that the hollow is right cone on the same base. Find the height upto which the cone should be scooped out so that the c.g. of the remainder coincide with the vertex of the hollow. [**Ans.** $h/3$]

22. A right circular cylinder and a hemisphere, having the same radius of 6 cm are joined together face to face. The density of the material of the hemisphere is twice that of the material of the cylinder. Find the greatest height of the cylinder so that the compound body may rest with any point of the hemisphere in contact with a horizontal plane. [**Ans.** 6 cm]

5

Moment of Inertia

5.1. MOMENT OF INERTIA

The cumulative product of area and square of its distance from an axis is called the moment of inertia of a section about that axis. It can be expressed as

$$I_x = \int^a y^2 \, da \qquad ...(5.1)$$

where I_x = moment of inertia (M.O.I.) of the section about the x-axis, and

y = the distance of infinitesimal area da from the x-axis as shown in Fig. 5.1.

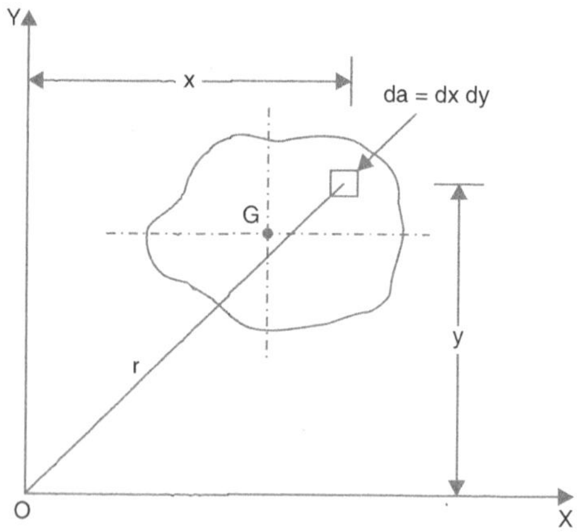

Fig. 5.1

Similarly, the moment of inertia of a section about the y-axis is given by

$$I_y = \int^a x^2 \, da \qquad ...(5.2)$$

The cumulative product of area and square of its distance from a point is known as the polar moment of inertia. It is given by

$$I_p = \int^a r^2 \, da \qquad ...(5.3)$$

$$= \int (x^2 + y^2) \; da$$

$$= I_x + I_y \qquad \text{...(5.4)}$$

Table 5.1 gives the centroidal moments of inertia for simple areas.

Table 5.1. Moments of Inertia for Simple Areas

Shape	*Moment of inertia*	*Simple areas*
Rectangle	$I_{xx} = \dfrac{bd^3}{12}$	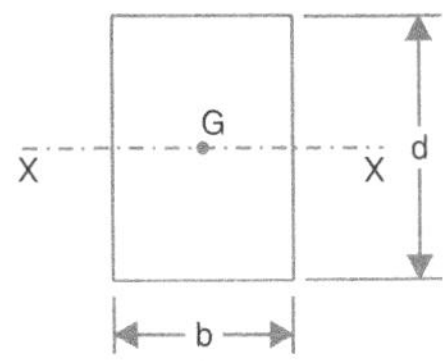 Fig. 5.2
Triangle	$I_{xx} = \dfrac{bh^3}{36}$	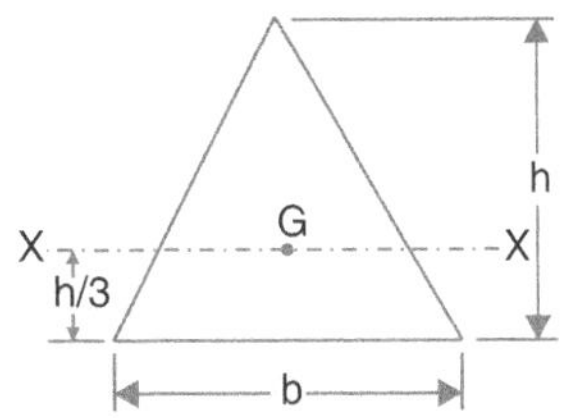 Fig. 5.3
Circle	$I_{xx} = \dfrac{\pi d^4}{64}$	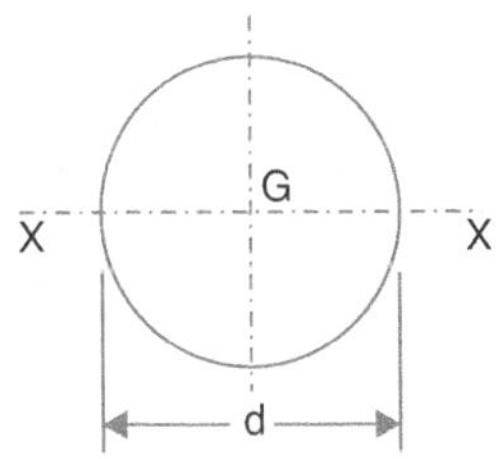 Fig. 5.4
Semi-circle	$I_{xx} = 0.11r^4$ $I_{yy} = \dfrac{\pi d^4}{128}$	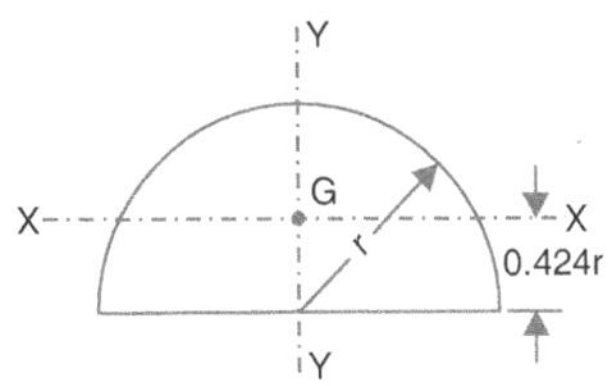 Fig. 5.5
Quadrant	$I_{xx} = 0.055r^4$	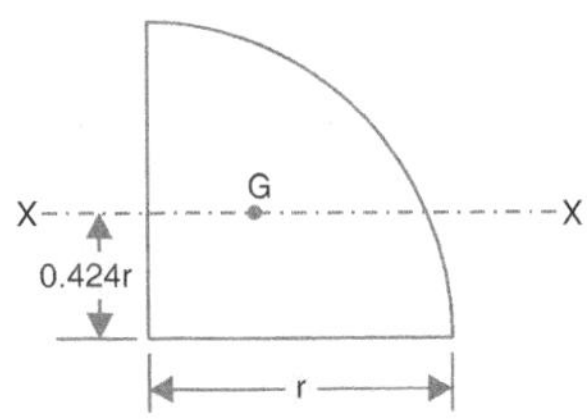 Fig. 5.6

5.2. THEOREM OF PARALLEL AXES

The theorem of parallel axes states as follows :

"The moment of inertia of a lamina about any axis in the plane of the lamina equals the sum of moment of inertia about a parallel centroidal axis in the plane of lamina and the product of the area of the lamina and square of the distance between the two axes."

In Fig. 5.7 is shown a lamina of area *A*. Let *LM* be the axis in the plane of lamina about which the moment of inertia of the lamina is required to be found out. Let *XX* be the centroidal axis in the plane of the lamina and parallel to the axis *LM*. Let '*h*' be the distance between the two *XX* and *LM*.

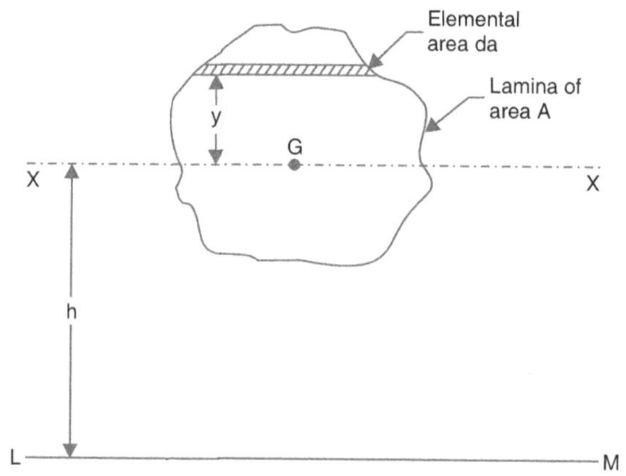

Fig. 5.7

It may be assumed that the lamina consists of an infinite number of small elemental components parallel to the axis *XX*. Consider one such elemental component at a distance *y* from the axis *XX*. The distance of the elemental component from the axis *LM* will be $(h + y)$ according as the elemental component and the axis *LM* are on opposite sides of *XX* or on the same side of *XX*.

Moment of inertia of the elemental component about the axis

$$LM = da\,(h \pm y)^2$$

$\therefore$ Moment of inertia of the whole lamina about the axis *LM*

$$= I_{lm} = \Sigma da\,(h \pm y)^2$$
$$= \Sigma dah^2 + \Sigma day^2 \pm 2\Sigma dahy$$
$$= h^2\Sigma da + \Sigma da\,y^2 \pm 2h\Sigma da\,y$$

But $\Sigma da = A,\; h^2\,\Sigma da = Ah^2$

Σday^2 = moment of inertia of the lamina about the axis *XX*

$\Sigma day = 0$ since *XX* is a centroidal axis

$$\therefore \quad I_{lm} = I_{xx} + Ah^2 \qquad \text{...(5.5)}$$

5.3. THEOREM OF PERPENDICULAR AXES

The theorem of perpendicular axes states as follows :

"If I_{ox} and I_{oy} be the moments of inertia of a lamina about mutually perpendicular axes OX and OY in the plane of the lamina and I_{oz} be the moment of inertia of the lamina about an axis (oz) normal to the lamina and passing through the point of intersection of the axes OX and OY,

then $$I_{oz} = I_{ox} + I_{oy}$$

Refer to Fig. 5.8. Let OX and OY be the two mutually perpendicular axes lying is the plane of the lamina. Let OZ be axis normal to the plane of the lamina and passing through O.

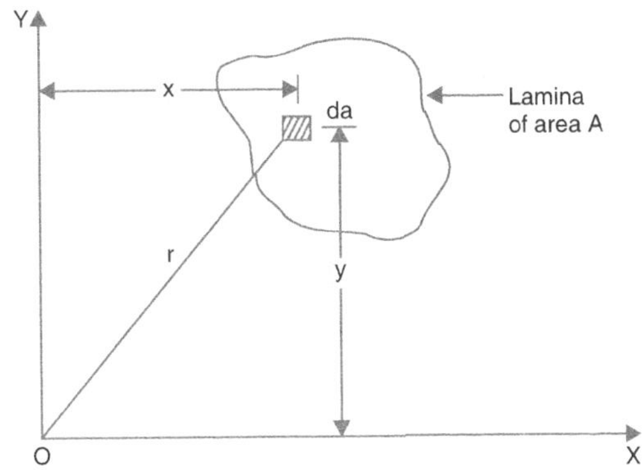

Fig. 5.8

Consider an elemental component of area da of the lamina. Let the distance of this elemental component from the axis OZ, *i.e.* from O be r.

∴ Moment of inertia of the elemental component OZ,

$$= da \times r^2$$

If the co-ordinates of the elemental component be (x, y) referred to the axes OX and OY, we have

$$r^2 = x^2 + y^2$$

Moment of inertia of the elemental component about the axis OZ

$$= da(x^2 + y^2)$$

$$= da\, x^2 + da\, y^2$$

∴ Total moment of inertia of the lamina about the axis OZ

$$= I_{oz} = \Sigma(da\, x^2 + da\, y^2)$$

$$= \Sigma da\, x^2 + \Sigma da\, y^2$$

But Σdax^2 = moment of inertia of the lamina about the axis $OY = I_{oy}$ ...(5.6)

and Σday^2 = moment of inertia of the lamina about the axis $OX = I_{ox}$

Hence, $I_{oz} = I_{ox} + I_{oy}$.

5.4. RADIUS OF GYRATION OF THE SECTION

One of the properties of cross-section which influence the structural behaviour of the members is *radius of gyration*.

$$k_i = \sqrt{\frac{I_i}{A}} \quad \text{...(5.7)} \quad (\because \; I_i = Aki^2)$$

where I_i = moment of inertia about ith axis ; and

k_i = radius of gyration of area about ith axis.

Members, when subjected to axial forces tend to buckle. The load at which members will buckle is proportional to the square of the radius of the gyration. The radius of gyration is usually referred to *with respect to centroidal axes system of the reaction.*

5.5. MOMENT OF INERTIA OF LAMINAE OF DIFFERENT SHAPES

1. Rectangular lamina. Refer to Fig. 5.9. Let a rectangular lamina be b units wide and d units deep.

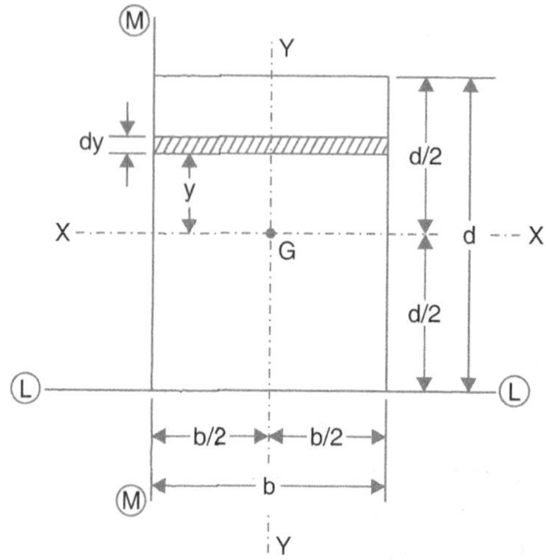

Fig. 5.9

(*a*) Moment of inertia about centroidal axis XX parallel to the width. Consider an elemental component of lamina at a distance y from the axis XX and having a depth dy.

Area of the elemental component

$$= da = b \times dy.$$

∴ Moment of inertia of the elemental component about the axis

$$XX = da \times y^2 = b \times dy \times y^2 = by^2 \, dy$$

∴ Total moment of inertia of the lamina about the axis XX,

$$I_{xx} = 2 \int_0^{d/2} by^2 \, dy = 2b \times \frac{1}{3} \times \frac{d^3}{8}$$

$$= \frac{bd^3}{12} \quad \text{...(5.8)}$$

Similarly, moment of inertia about the centroidal axis YY parallel to the depth,

$$I_{yy} = \frac{db^3}{12} \quad \text{...(5.9)}$$

(*b*) Moment of inertia about an axis LL passing through the bottom edge or top edge. By the theorem of parallel axis, the moment of inertia about the axis LL is given by

$$I_{ll} = L_{xx} + Ah^2$$

In this case $A = bd$

$$h = \frac{d}{2}$$

$$I_{xx} = \frac{bd^3}{12}$$

$$\therefore \qquad I_{ll} = \frac{bd^3}{12} + bd \times \left(\frac{d}{2}\right)^2$$

$$= \frac{bd^3}{12} + \frac{bd^3}{4} = \frac{bd^3}{3} \qquad \text{...(5.10)}$$

Similarly, the moment of inertia about the axis MM is given by,

$$I_{mm} = \frac{db^3}{3} \qquad \text{...(5.11)}$$

If G be the centroid of the lamina, the axis through the centroid and normal to the plane of the lamina is called the *polar axis.* Let I_p be the moment of inertia about the polar axis. I_p is called the polar moment of inertia.

By the theorem of perpendicular axes,

$$I_p = I_{xx} + I_{yy} = \frac{bd^3}{12} + \frac{db^3}{12} \qquad \text{...(5.12)}$$

2. Rectangular lamina with a centrally situated rectangular hole. Refer to Fig. 5.10. Let in a rectangular lamina $B \times D$, a rectangular hole $b \times d$ be made centrally.

Moment of inertia of the lamina about any axis = moment of inertia of bigger rectangle – moment of inertia of the smaller rectangle.

For example,

$$I_{xx} = \frac{BD^3}{12} - \frac{bd^3}{12} \qquad \text{...(5.13)}$$

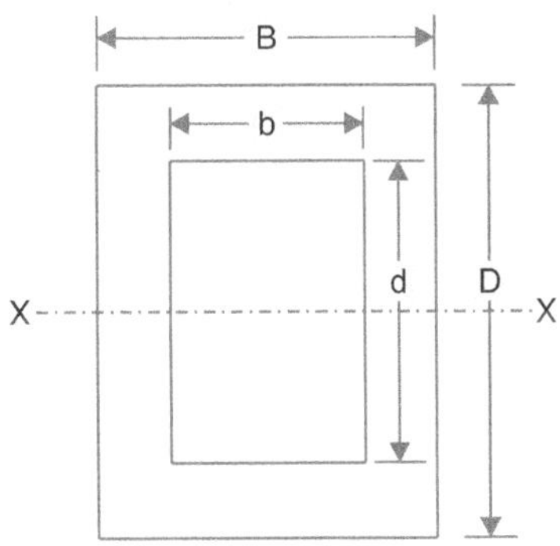

Fig. 5.10

3. Moment of inertia of a triangular lamina. Refer to Fig. 5.11. Let ABC be a triangular lamina of base b and altitude h.

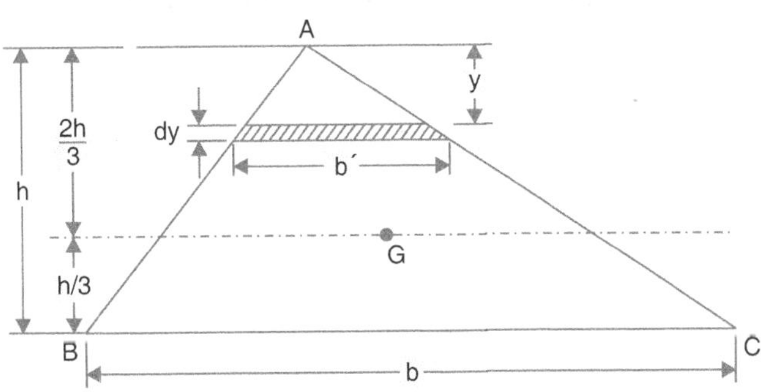

Fig. 5.11

(*a*) **Moment of inertia of a triangle about an axis LL through the vertex and parallel to the axis.** The triangle may be considered to consist of a number of infinitely small elemental components parallel to the base. Consider one such elemental component at a distance y from the vertex and of thickness dy. Width of elemental component

$$= b' = \frac{b}{h} y$$

∴ Area of elemental component

$$= b'\, dy = \frac{b}{h} y\, dy$$

∴ Moment of inertia of the elemental component about the axis LL

$$= \frac{b}{h} y\, dy\, y^2 = \frac{b}{h} y^3\, dy$$

∴ Moment of inertia of the lamina about the axis LL,

$$I_{ll} = \frac{b}{h} \int_0^h y^3 dy = \frac{bh^3}{4} \qquad \text{...(5.14)}$$

(*b*) **Moment of inertia of a triangle about the centroidal axis parallel to the base.** Let XX be the centroidal axis. This axis is at a distance of $\frac{2}{3}h$ from the vertex.

Applying the theorem of parallel axes, we have

$$I_{ll} = I_{xx} + A\left(\frac{2}{3}h\right)^2$$

∴
$$\frac{bh^3}{4} = I_{xx} + \frac{bh}{2} \times \frac{4}{9}h^2$$

∴
$$I_{xx} = \frac{bh^3}{4} - \frac{2}{9}bh^3$$

or
$$I_{xx} = \frac{bh^3}{36} \qquad \text{...(5.15)}$$

(*c*) **Moment of inertia of a triangle about the base.** Applying the theorem of parallel axis again, we have

$$I_{bc} = I_{xx} + A\left(\frac{h}{3}\right)^2$$

$$= \frac{bh^3}{36} + \frac{bh}{2} \times \frac{h^2}{9} = \frac{bh^3}{36} + \frac{bh^3}{18}$$

$$= \frac{bh^3}{12} \qquad \text{...(5.16)}$$

4. Moment of inertia of a circular lamina. Fig. 5.12 shows a circular lamina of radius R. The lamina may be considered as consisting of infinite number of elemental concentric rings. Consider one such elemental ring at a radius r and having a thickness dr.

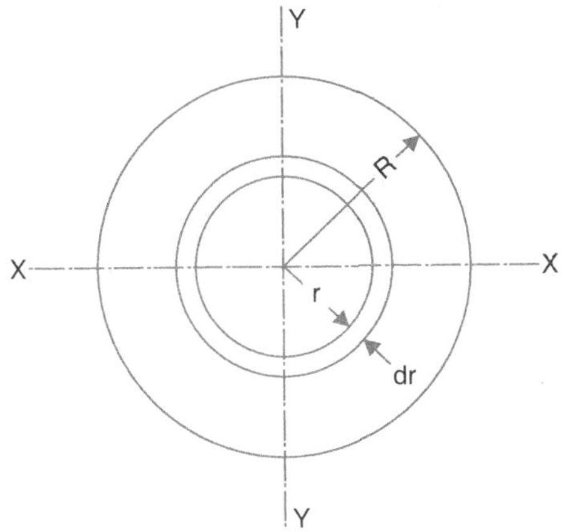

Fig. 5.12

The moment of inertia of elemental ring about the polar axis

$$= \text{area of ring} \times (\text{radius})^2$$

$$= 2\pi r\, dr \times r^2 = 2\pi r^3 \times dr$$

$\therefore$ Polar moment of inertia of the whole lamina,

$$I_p = \int_0^R 2\pi r^3\, dr$$

$$\therefore \quad I_p = \frac{2\pi R^4}{4} = \frac{\pi R^4}{2}$$

If D be the diameter of the lamina

$$D = 2R$$

$$\therefore \quad I_p = \frac{\pi}{2}\left(\frac{D}{2}\right)^4 = \frac{\pi}{32} D^4 \qquad ...(5.17)$$

But

$$I_{xx} = I_{yy}$$

and

$$I_{xx} + I_{yy} = I_p = \frac{\pi D^4}{32}$$

$$\therefore \quad I_{xx} = I_{yy} = \frac{\pi D^4}{64} \qquad ...(5.18)$$

5. Moment of inertia of a circular lamina with a centrally situated circular hole. Refer to Fig. 5.13. Let D and d be the external and internal diameters of the lamina respectively.

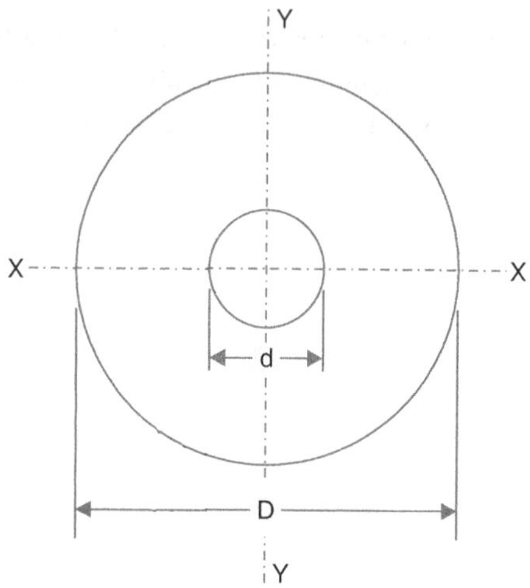

Fig. 5.13

Polar moment of inertia,

I_p = polar moment of inertia of the bigger circle-polar moment of inertia of the smaller circle

$$\therefore \qquad I_p = \frac{\pi D^4}{32} - \frac{\pi d^4}{32}$$

$$I_{xx} = I_{yy} = \frac{I_p}{2} = \frac{\pi D^4}{64} - \frac{\pi d^4}{64}.$$

6. Moment of inertia of a semi-circular lamina. In Fig. 5.14 is shown a semi-circle of radius R. Let LM be the base of the semi-circle.

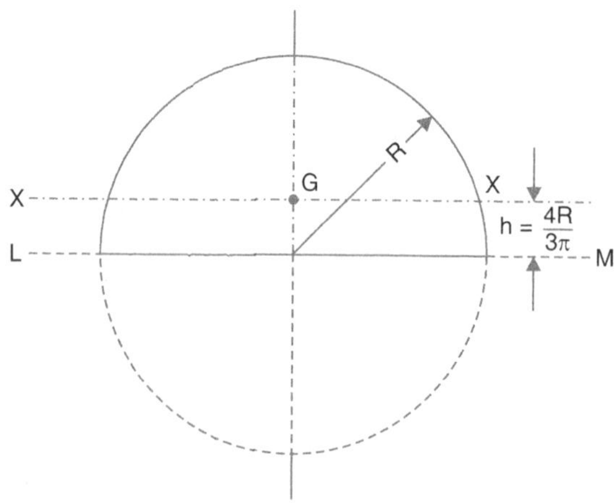

Fig. 5.14

The moment of inertia of a circular lamina about a diameter

$$LM = \frac{\pi R^4}{4} = \frac{\pi D^4}{64}$$

∴ Moment of inertia of the semi-circle about LM

$$I_{lm} = \frac{\pi R^4}{8} = \frac{\pi D^4}{128} \qquad ...(5.19)$$

Let XX be centroidal axis parallel to the base LM. Let h be the distance between the axis XX and LM. We have

$$h = \frac{4R}{3\pi} = \frac{4}{3\pi} \times \frac{D}{2} = \frac{2D}{3\pi}$$

By theorem of parallel axis, we have

$$I_{lm} = I_{xx} + Ah^2$$

∴ $$\frac{\pi R^4}{8} = I_{xx} + \frac{\pi R^2}{2} \times \left(\frac{4R}{3\pi}\right)^2$$

From which, $$I_{xx} = \frac{\pi R^4}{8} - \frac{8\pi R^4}{9\pi^2}$$

or, $$I_{xx} = \frac{\pi R^4}{8} - \frac{8R^4}{9\pi}$$

or, $$I_{xx} = 0.11R^4 \qquad ...(5.20)$$

$$I_{yy} = \frac{\pi R^4}{8} = \frac{\pi}{128} D^4 \qquad ...(5.21)$$

7. Moment of inertia of a quadrant. Refer to Fig. 5.15. Let LON be a quadrant of a circular plate of radius R.

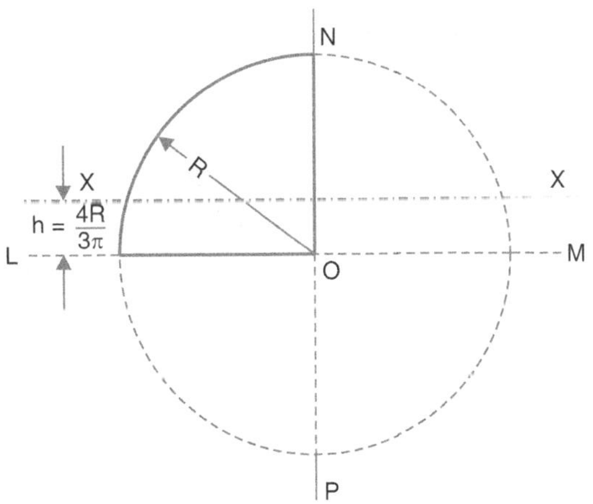

Fig. 5.15

Moment of inertia of the area LON = one-fourth of the moment of inertia of the a circular area about axis LM.

∴ $$I_{lm} = \frac{1}{4} \times \frac{\pi R^4}{4} = \frac{\pi R^4}{16}$$

Consider the semi-circle LNM. Distance of its centroid from LM

$$= \frac{4R}{3\pi}$$

The distance of the centroid of the quadrant *LON* from *LM* is also equal to $\frac{4R}{3\pi}$, *i.e.*, the axis *XX* is centroidal axis of the quadrant *LON* as well as the semi-circle *LNB*.

∴ Moment of inertia of the quadrant about axis *XX*,

$$I_{xx} = \text{one-half of the moment of inertia of the semi-circle about } XX.$$

$$\therefore \quad I_{xx} = \frac{1}{2}(0.11R^4)$$

$$= 0.055R^4 \qquad \text{...(5.22)}$$

8. Moment of inertia of a thin ring. Fig. 5.16 shows a ring of mean radius r and of thickness t. Consider an elemental component da of the lamina. The moment of inertia of this elemental component about the polar axis of the lamina = dar^2. Since all the elemental components are at the same distance r from the polar axis, moment of inertia of the ring about the polar axis,

$$I_p = \Sigma da\ r^2 = r^2\ \Sigma da$$

$$= r^2 \times \text{area of the whole ring}$$

$$= r^2 \times 2\pi\ rt \qquad \text{...(5.23)}$$

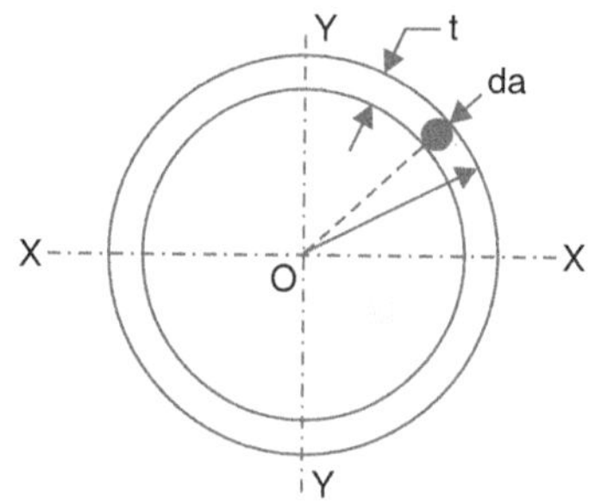

Fig. 5.16

Moment of inertia about the axis *XX*

= moment of inertia about the *YY*

$$= \frac{I_p}{2}.$$

WORKED EXAMPLES

Example 5.1. *Find the moment of inertia about the centroidal axes XX and YY of the section shown in Fig. 5.17.*

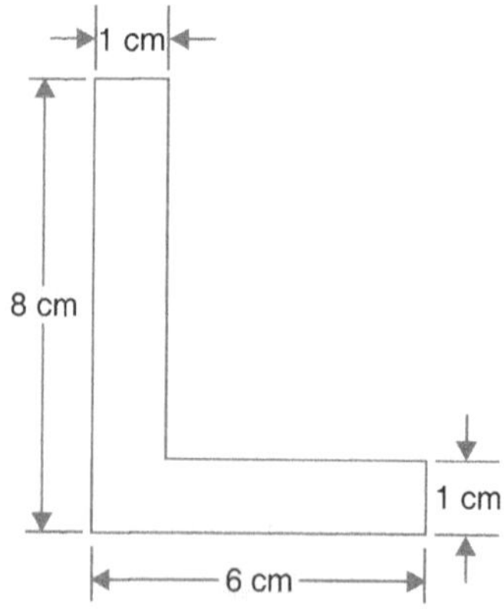

Fig. 5.17

Sol. Refer to Fig. 5.18.

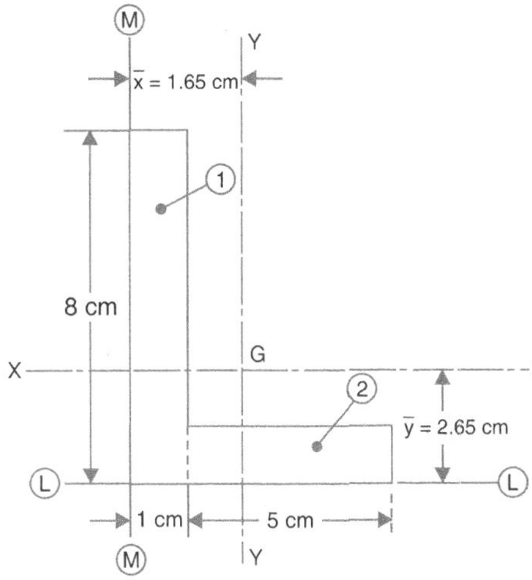

Fig. 5.18

To determine the location of centroid of the section we have the following table :

Components	*Area 'a' cm²*	*Centroidal distance 'x' from MM (cm)*	*Centroidal distance 'y' from LL (cm)*	*ax (cm³)*	*ay (cm³)*
Rectangle (1)	8 × 1 = 8	0.5	4	4	32
Rectangle (2)	5 × 1 = 5	3.5	0.5	17.5	2.5
Total	Σa = 13	—	—	Σax = 21.5	Σay = 34.5

∴ $$\bar{x} = \frac{\Sigma ax}{\Sigma a} = \frac{21.5}{13} = 1.65 \text{ cm}$$

and $$\bar{y} = \frac{\Sigma ay}{\Sigma a} = \frac{34.5}{13} = 2.65 \text{ cm.}$$

To find I_{xx} and I_{yy} using theorem of parallel axes, we have

$$I_{xx} = \left[\frac{1 \times 8^3}{12} + 8 \times 1 \times (4 - 2.65)^2\right] + \left[\frac{5 \times 1^3}{12} + 5 \times 1 \times (2.65 - 0.5)^2\right]$$

$$= (42.66 + 14.58) + (0.416 + 23.112)$$

$$= 57.24 + 23.53 = \mathbf{80.77\ cm^4.\ (Ans.)}$$

$$I_{yy} = \left[\frac{8 \times 1^3}{12} + 8 \times 1 \times (1.65 - 0.5)^2\right] + \left[\frac{1 \times 5^3}{12} + 5 \times 1 \times (3.5 - 1.65)^2\right]$$

$$= (0.66 + 10.58) + (10.42 + 17.11)$$

$$= 11.24 + 27.53 = \mathbf{38.77\ cm^4.\ \ (Ans.)}$$

Example 5.2. *Calculate the moment of inertia about horizontal and vertical gravity axes (I_{xx} and I_{yy}) of the section shown in Fig. 5.19.*

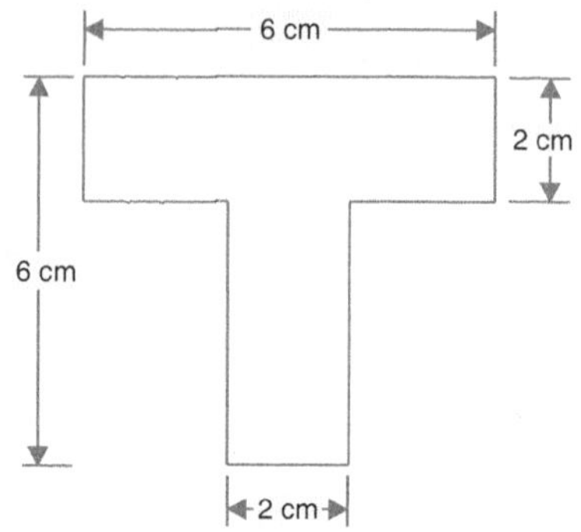

Fig. 5.19

Sol. Refer to Fig. 5.20.

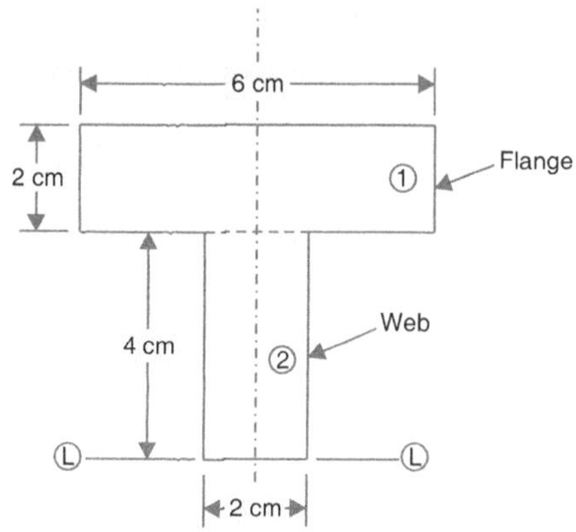

Fig. 5.20

Since, the section is symmetrical about *Y*-axis, therefore we shall find and $\bar{y}$ only, for which we have the following table :

Components	*Area 'a'* *(cm²)*	*Centroidal distance 'y' from LL (cm)*	*ay* *(cm³)*
Rectangle (1)	$6 \times 2 = 12$	$4 + \frac{2}{2} = 5$	60
Rectangle (2)	$4 \times 2 = 8$	$\frac{4}{2} = 2$	16
Total	$\Sigma a = 20$	—	$\Sigma ay = 76$

$$\therefore \quad \bar{y} = \frac{\Sigma ay}{\Sigma a} = \frac{76}{20} = 3.8 \text{ cm from } LL.$$

To find I_{xx} and I_{yy} using the theorem of parallel axes, we get

$$I_{xx} = \left[\frac{6 \times 2^3}{12} + 6 \times 2(5 - 3.8)^2\right] + \left[\frac{2 \times 4^3}{12} + 4 \times 2 \times (3.8 - 2)^2\right]$$

$$= (4 + 17.28) + (10.66 + 25.92)$$

$$= 21.28 + 36.58 = \mathbf{57.86\ cm^4}. \quad \textbf{(Ans.)}$$

$$I_{yy} = \left[\frac{2 \times 6^3}{12} + \frac{4 \times 2^3}{12} = 36 + 2.67\right]$$

$$= \mathbf{38.67\ cm^4}. \quad \textbf{(Ans.)}$$

Example 5.3. *A T-beam is made up of two plates and two angles as shown in Fig. 5.21. Determine the moment of inertia of T-section about an axis passing through the centroid of the section and parallel to the top plate.*

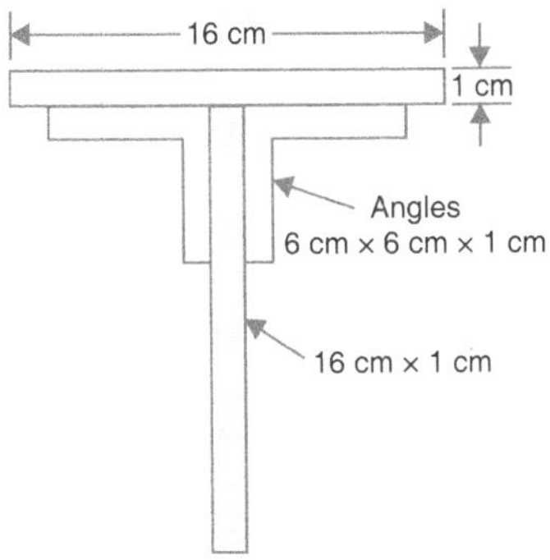

Fig. 5.21

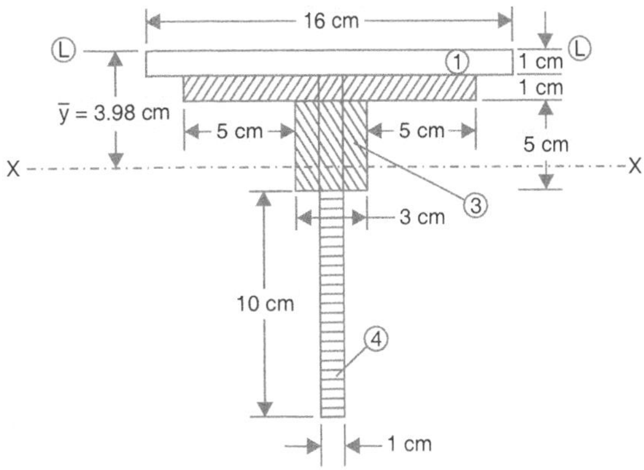

Fig. 5.22

Sol. Refer to Fig. 5.22.

The given section is divided into four rectangles as shown. Since the section is symmetrical about y-axis we shall find out $\bar{y}$ for which we have the follows table :

Components	*Area 'a' (cm^2)*	*Centroidal distance 'y' from LL (cm)*	*ay (cm^3)*
Rectangle (1)	16 × 1 = 16	$\frac{1}{2} = 0.5$	8
Rectangle (2)	13 × 1 = 13	$1 + \frac{1}{2} = 1.5$	19.5
Rectangle (3)	3 × 5 = 15	$\frac{5}{2} + 1 + 1 = 4.5$	67.5
Rectangle (4)	1 × 10 = 10	$\frac{10}{2} + 5 + 1 + 1 = 12$	120
Total	$\Sigma a = 54$	—	$\Sigma ay = 215$

$$\bar{y} = \frac{\Sigma ay}{\Sigma a} = \frac{215}{54} = 3.98 \text{ cm from } LL$$

I_{xx} = ?

$$I_{xx} = \left[\frac{16 \times 1^3}{12} + 16 \times 1 \times (3.98 - 0.5)^2\right] + \left[\frac{13 \times 1^3}{12} + 13 \times 1 \times (3.98 - 1.5)^2\right]$$

$$+ \left[\frac{3 \times 5^3}{12} + 3 \times 5 \times (4.5 - 3.98)^2\right] + \left[\frac{1 \times 10^3}{12} + 10 \times 1 \times (12 - 3.98)^2\right]$$

$$= (1.33 + 193.76) + (1.08 + 79.95) + (31.25 + 4.056) + (83.33 + 643.2)$$

$$= 195.09 + 81.03 + 35.31 + 726.53$$

$$= \mathbf{1037.96\ cm^4.}\quad \textbf{(Ans.)}$$

Example 5.4. *A compound section is formed by riveting 200 mm × 10 mm flat plates, one on each flange of (ISL 300) 300 mm × 150 mm I-section girder. Find for the compound section :*

(a) I_{xx}, *(b)* I_{yy} *and*

(c) The least radius of gyration.

The properties of I-section are as follows :

$I_{xx} = 7332.9\ cm^4$, $I_{yy} = 376.2\ cm^4$, *area* $= 48.08\ cm^2$.

Sol. Refer to Fig. 5.23.

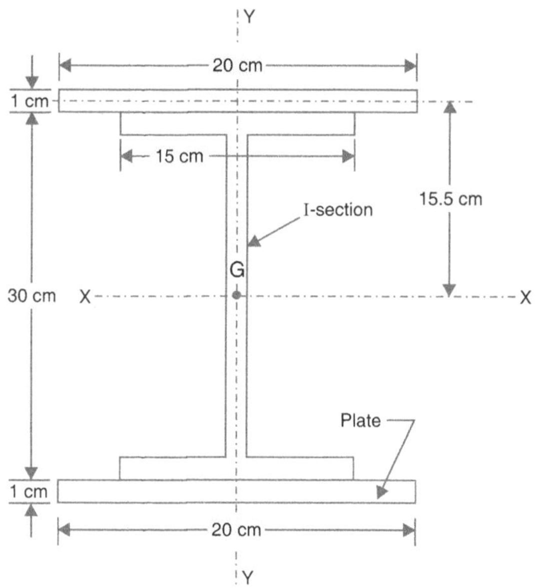

Fig. 5.23

Since, the section is symmetrical about both *xx* and *yy* axes, therefore the c.g. of the whole section will lie at *G* as shown in Fig. 5.23.

(*a*) Moment of inertia about *xx* axis,

$$I_{xx} = I_{xx} \text{ for } I\text{-section} + I_{xx} \text{ for two plates}$$

$$= 7332.9 + 2\left[\frac{20 \times 1^3}{12} + 20 \times 1 \times \left(15 + \frac{1}{2}\right)^2\right]$$

$$= 7332.9 + 2(1.67 + 4805)$$

$$= \mathbf{16946.24\ cm^4.}\quad \textbf{(Ans.)}$$

(*b*) Moment of inertia about *yy* axis,

$$I_{yy} = I_{yy} \text{ for } I\text{-section} + I_{yy} \text{ for two plates}$$

$$= 376.2 + 2 \times \left(\frac{1 \times 20^3}{12}\right)$$

$$= \mathbf{1709.53\ cm^4.} \quad \textbf{(Ans.)}$$

(*c*) Least value of radius of gyration,

$$k_{least} = \sqrt{\frac{\text{Least value of moment of inertia}}{\text{Total area of the section}}}$$

$$= \sqrt{\frac{1709.53}{48.08 + 2 \times 20 \times 1}} = \sqrt{\frac{1709.53}{88.08}}$$

$$= \mathbf{4.4\ cm.} \quad \textbf{(Ans.)}$$

Example 5.5. *A fabricated girder system consists of two plates 30 cm × 1.5 cm and two ISLC 300 channels 30 cm × 10 cm set 10 cm apart as shown in Fig. 5.24. The properties of a 30 cm × 10 cm channel section are as follows :*

$I_{xx} = 6047.9\ cm^4$, $I_{yy} = 346.0\ cm^4$, *area* $= 42.11\ cm^2$, *axis yy from back of channel* $= 2.55$ *cm.*

Calculate the value of moment of inertia and radii of gyration about the horizontal and vertical axes through the centroid of the combined section. Neglect the effect of rivets etc.

Sol. Refer to Fig. 5.24.

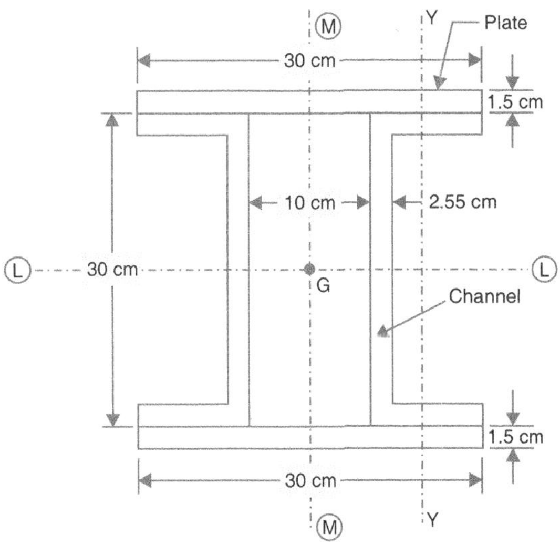

Fig. 5.24

Since, the section is symmetrical the centroid will lie on the intersection of two axes of symmetry, *LL* and *MM* as shown.

$$I_{ll} = I_{ll} \text{ or } I_{xx} \text{ for channels} + I_{ll} \text{ for plates}$$

$$= 2 \times 6047.9 + 2\left[\frac{30 \times (1.5)^3}{2} + 30 \times 1.5 \times \left(15 + \frac{1.5}{2}\right)^2\right]$$

$$= 12095.8 + 22342.5$$

$$= \mathbf{34438.3\ cm^4.} \quad \textbf{(Ans.)}$$

$$I_{mm} = I_{mm} \text{ for channels} + I_{mm} \text{ for plates}$$
$$= 2(I_{yy} \text{ for channels} + Ah^2) + I_{mm} \text{ for plates}$$
$$= 2\left[346 + 42.11\left(\frac{10}{2} + 2.55\right)^2\right] + 2 \times \frac{1.5 \times 30^3}{12}$$
$$= 5492.75 + 6750$$
$$= \mathbf{12242.75\ cm^4}. \quad \textbf{(Ans.)}$$

Radius of gyration about *LL* axis,

$$k_{ll} = \sqrt{\frac{I_{ll}}{\text{area}}} = \sqrt{\frac{34438.3}{2 \times 42.11 + 2 \times 30 \times 1.5}}$$
$$= \mathbf{14.06\ cm}. \quad \textbf{(Ans.)}$$

Radius of gyration about *MM* axis,

$$k_{mm} = \sqrt{\frac{I_{mm}}{\text{area}}} = \sqrt{\frac{12242.75}{2 \times 42.11 + 2 \times 30 \times 1.5}}$$
$$= \mathbf{8.38\ cm}. \quad \textbf{(Ans.)}$$

Example 5.6. *A stanchion is built up of three ISMB 225 as shown in Fig. 5.25. Calculate second moment of area of the built-up-section about the principal axes LL and MM, and the radius of gyration.*

Sol. Refer to Fig. 5.25.

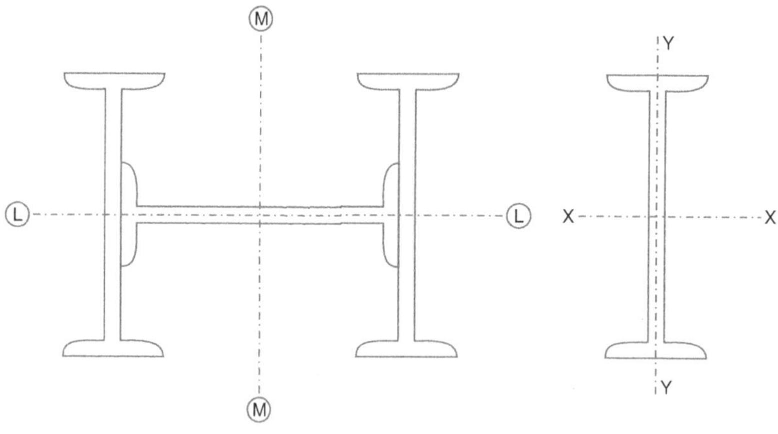

Fig. 5.25

From steel tables for one *ISMB 225*

Area, $a = 39.72 \text{ cm}^2$

Web thickness $= 6.5 \text{ mm}$

$I_{xx} = 3441.8 \text{ cm}^4$

$I_{yy} = 218.3 \text{ cm}^4$

Since, axis *LL* passes through the centroids of all the three beams,

$$I_{ll} = I_{xx} \text{ for two extreme beams} + I_{yy} \text{ for central beam}$$
$$= 2 \times 3441.8 + 218.3$$
$$= \mathbf{7101.9\ cm^4}. \quad \textbf{(Ans.)}$$

Axis *MM* passes through the centroids of central beam but is parallel to axis *YY* of two extreme beams.

$\therefore \quad I_{mm} = I_{xx}$ for central beam + 2[I_{yy} + area × (distance between axis *MM* and *YY* of extreme beam)^2]

$$= 3441.8 + 2\left[218.3 + 39.72\left(\frac{22.5}{2} + \frac{0.65}{2}\right)^2\right]$$

$= \mathbf{14521.82\ cm^4}$. **(Ans.)**

Least radius of gyration

$$= \sqrt{\frac{\text{least of } I_{ll} \text{ or } I_{mm}}{\text{area of compound beam}}} = \sqrt{\frac{7101.9}{3 \times 39.72}}$$

$= \mathbf{7.72\ cm}$. **(Ans.)**

SELECTED QUESTIONS FROM EXAMINATION PAPERS

Example 5.7. *Find the moment of inertia and radius of gyration of the area shown in Fig. 5.26 about the axis AB.*

Sol.
$$a_1 = \frac{1}{2} \times 50 \times 15 = 375 \text{ mm}^2$$
$$a_2 = 50 \times 20 = 1000 \text{ mm}^2$$
$$a^3 = \frac{\pi \times 15^2}{2} = 353.43 \text{ mm}^2$$

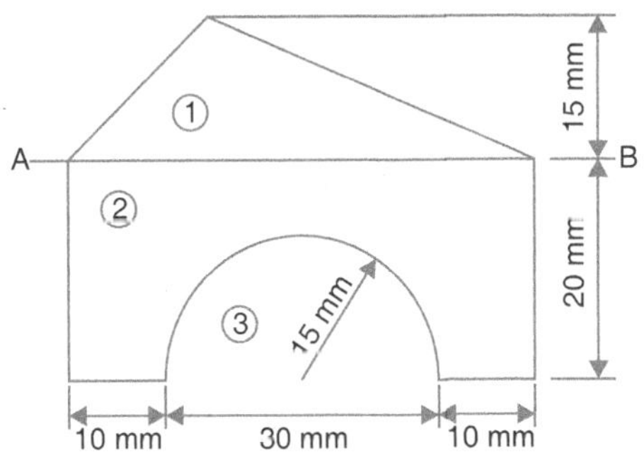

Fig. 5.26

M.O.I. about the axis AB :

$$I_{AB} = \left[\frac{50 \times 15^3}{36} + 375 \times \left(\frac{15}{3}\right)^2\right] + \left[\frac{50 \times 20^3}{12} + 1000 \times 10^2\right]$$
$$- \left[0.11 \times 15^4 \times 353.43 \times \left(20 - \frac{4 \times 15}{3\pi}\right)^2\right]$$
$$= 10^4\ [(0.469 + 0.937) + (3.333 + 10) - (0.557 + 6.569)]$$
$$= 7.613 \times 10^4 \text{ mm}^4$$

Hence $\quad I_{AB} = \mathbf{7.613 \times 10^4\ mm^4}$. **(Ans.)**

Radius of gyration about AB, k_{AB} :

Example 5.8. *Find the centroidal moment of inertia of the shaded area shown in Fig. 5.27.*

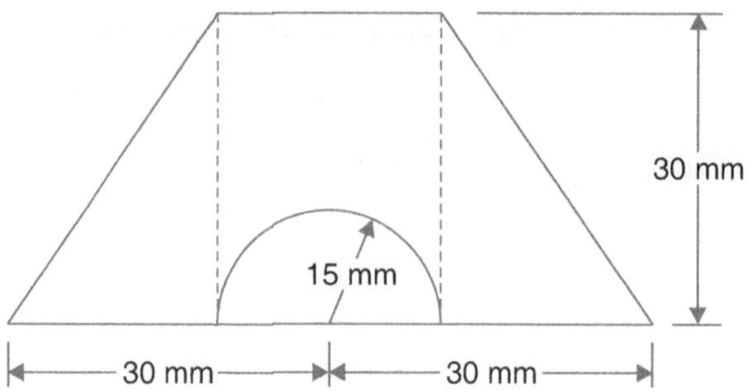

Fig. 5.27

Sol. Refer to Fig. 5.28.

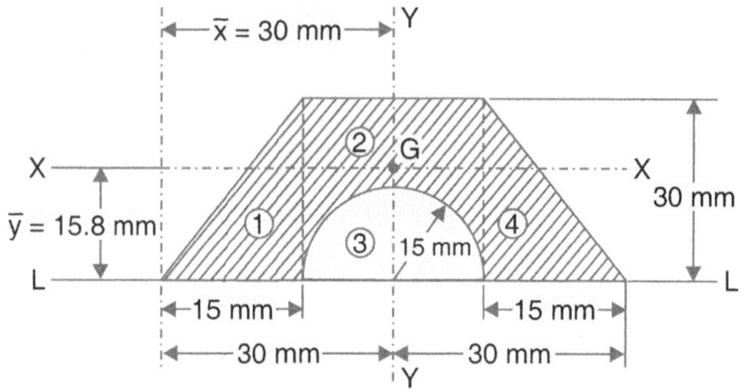

Fig. 5.28

Since the figure is symmetrical about *YY*-axis, therefore we shall find $\bar{y}$ for which we have the table given on the previous page.

$\mathbf{I_{xx}}$ **:**

$$I_{xx} = I_{xx_1} + I_{xx_2} - I_{xx_3} + I_{xx_4}$$

$$I_{xx} = \left[\frac{15 \times (30)^3}{36} + \frac{1}{2} \times 15 \times 30\left(15.8 - \frac{30}{3}\right)^2\right] + \left[\frac{30 \times 30^3}{12} + 30 \times 30 \times (15.8 - 15)^2\right]$$

$$- \left[0.11 \times 15^4 + \frac{\pi \times 15^2}{2} \times \left(15.8 - \frac{4 \times 15}{3\pi}\right)^2\right]$$

$$+ \left[\frac{15 \times 30^3}{36} + \frac{1}{2} \times 15 \times 30\left(15.8 - \frac{30}{3}\right)^2\right]$$

$$= 18819 + 68076 - 37023 + 18819$$

$$= 68691 \text{ mm}^4$$

i.e., $\mathbf{I_{xx} = 68691\ mm^4}$**. (Ans.)**

$\mathbf{I_{yy} = ?}$

$$I_{yy} = I_{yy_1} + I_{yy_2} - I_{yy_3} + I_{yy_4}$$

$$I_{yy} = \left[\frac{30 \times 15^3}{36} + \frac{1}{2} \times 30 \times 15 \times \left(30 - \frac{2}{3} \times 15\right)^2\right]$$

$$+ \left[\frac{30 \times 30^3}{12}\right] - \left[\frac{\pi \times 30^4}{64 \times 2}\right]$$

$$+ \left[\frac{30 \times 15^3}{36} + \frac{1}{2} \times 30 \times 15 \times \left(30 - \frac{2}{3} \times 15\right)^2\right]$$

$$= 92812.5 + 67500 - 19880.4 + 92812.5$$

$$= 233244.6 \text{ mm}^4$$

i.e., $\mathbf{I_{yy} = 233244.6 \text{ mm}^4}$. **(Ans.)**

Example 5.9. *For the shaded area shown in Fig. 5.29, find the following :*

(i) The position of the centroid

(ii) The second moment of area about the base

(iii) The radius of gyration about the base.

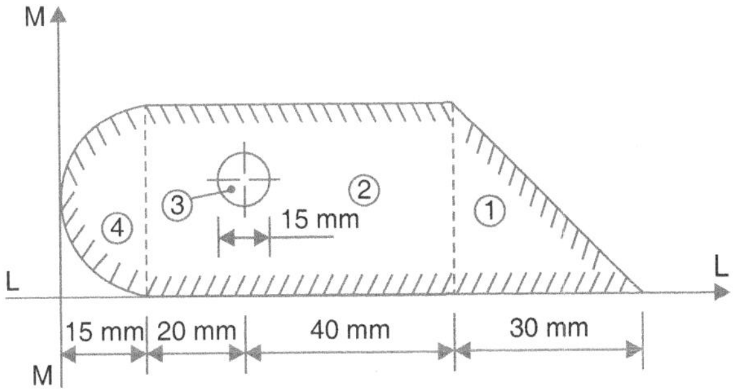

Fig. 5.29

Sol.

(i) **The position of the centroid :**

To determine the location of the centroid of the shaded area we have the table given below :

(ii) **The second moment of area (M.O.I.) about the base, I_{LL} :**

Component	*Area 'a' (mm²)*	*Centroidal distance 'x' from MM (mm)*	*Centroidal distance 'y' from MM (mm)*	*ax (mm²)*	*ay (mm²)*
Triangle (1)	$\frac{1}{2} \times 30 \times 30 = 450$	$15 + 20 + 40 + \frac{30}{3} = 85$	$\frac{30}{3} = 10$	38250	4500
Rectangle (2)	$60 \times 30 = 1800$	$15 + \frac{60}{2} = 45$	$\frac{30}{2} = 15$	81000	27000
Circle (3)	$\pi \times 7.5^2 = 176.71$ (–)	$15 + 20 = 35$	$\frac{30}{2} = 15$	6184.8 (–)	2650.6 (–)
Semicircle (4)	$\frac{\pi \times 15^2}{2} = 353.43$	$15 - \frac{4 \times 15}{3\pi} = 8.63$	$\frac{30}{2} = 15$	3050	5301.4
	$\Sigma a = 2426.72$			$\Sigma ax = 116115.2$	$\Sigma ay = 34150.8$

$$\bar{x} = \frac{\Sigma ax}{\Sigma a} = \frac{116115.2}{2426.72} = \mathbf{47.84\ mm.}\quad \textbf{(Ans.)}$$

$$\bar{y} = \frac{\Sigma ay}{\Sigma a} = \frac{34150.8}{2426.72} = \mathbf{14.07\ mm.}\quad \textbf{(Ans.)}$$

Applying parallel axes theorem :

$$I_{LL} = I_G + Ah^2$$

where I_G = M.O.I. about C.G.,

A = area of the lamina, and

h = distance between the C.G. and the reference line (*i.e., LL* in this case)

∴ M.O.I. about base *LL*,

$\mathbf{I_{LL}}$ = M.O.I. of triangle (1) + M.O.I. of *Rectangle* (2) – M.O.I. of *Circle* (3) + M.O.I. of *Semicircle* (4)

$$= \left(\frac{30 \times 30^3}{2}\right) + \left[\frac{60 \times 30^3}{12} + 60 \times 30 \times 15^2\right] - \left[\frac{\pi \times 15^4}{64} + \frac{\pi \times 15^2}{4} \times 15^2\right]$$

$$+ \left[\frac{\pi \times 30^4}{64 \times 2} + \frac{\pi \times 30^2}{8} \times 15^2\right]$$

$$= 67500 \times 540000 - 42245.8 + 99401.9$$

$$= 669656.1\ \text{mm}^4$$

$$= \mathbf{664656.1\ mm^4.}\quad \textbf{(Ans.)}$$

(*iii*) **The radius of gyration about the base, $\mathbf{k_{LL}}$:**

$$I_{LL} = A\,k_{LL}^2$$

$$\therefore \quad k_{LL} = \sqrt{\frac{I_{LL}}{A}}$$

$$= \sqrt{\frac{664656.1}{2426.72}} = \mathbf{16.55\ mm.}\quad \textbf{(Ans.)}$$

Example 5.10. *Determine the moment of inertia of the shaded area about the edge LM as shown in Fig. 5.30*

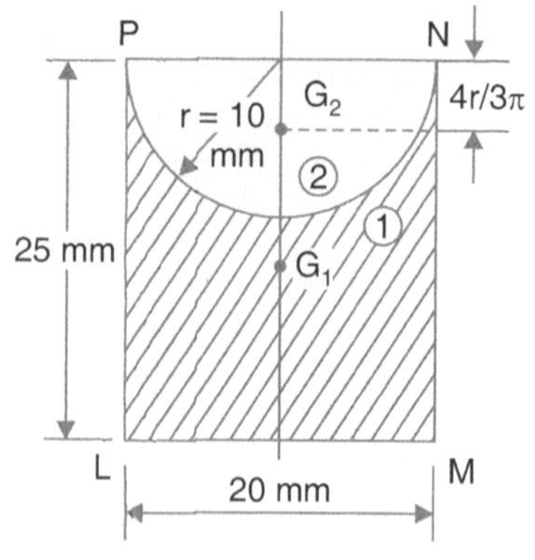

Fig. 5.30

Sol. I_{LM} :

I_{LM} = M.O.I. of *Rectangle* (1) – M.O.I. of *Semicircle* (2)

$$= \left[\frac{20 \times 25^3}{12} + 20 \times 25 \times \left(\frac{25}{2}\right)^2\right] - \left[0.11 \times 10^4 + \frac{\pi \times 10^2}{2} \times \left(25 - \frac{4 \times 10}{3\pi}\right)^2\right]$$

$= 104166 - 68771$

$=$ **35395 mm^4. (Ans.)**

Example 5.11. *(i) With reference to the shaded area of Fig. 5.31 locate the centroid w.r.t. OL and OM.*

(ii) Determine the moment of inertia of the area about the horizontal axis passing through the centroid.

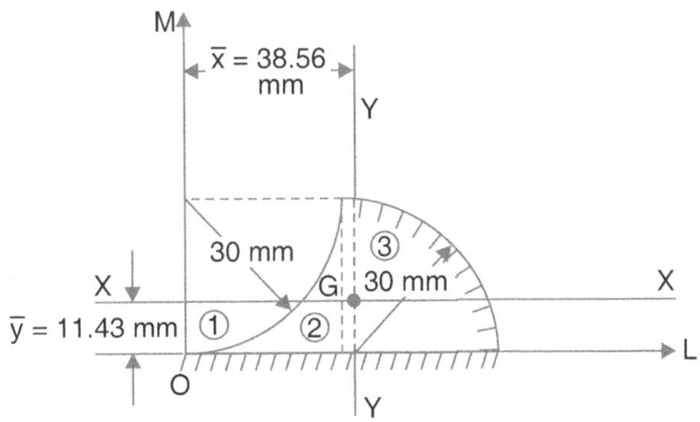

Fig. 5.31

Sol. *(i)* **Centroid of the area :**

For location of the centroid of the shaded area we have the following table.

Component	*Area 'a' (m^2)*	*Centroidal distance 'x' from OM (mm)*	*Centroidal distance 'y' from OL (mm)*	*ax (mm^2)*	*ay (mm^2)*
Quarter Circle (1)	$\frac{\pi \times 30^2}{4} = 706.85(-)$	$\frac{4 \times 30}{3\pi} = 12.73$	$30 - \frac{4 \times 30}{3\pi} = 17.27$	8998.2 (–)	12207.3 (–)
Square (2)	30 × 30 = 900 (+)	$\frac{30}{2} = 15$	$\frac{30}{2} = 15$	13500	13500 (+)
Quarter Circle (3)	$\frac{\pi \times 30^2}{4} = 706.85\ (+)$	$30 + \frac{4 \times 30}{3\pi} = 42.73$	$\frac{4 \times 30}{3\pi} = 12.73$	30203.7	8998.2 (+)
	$\Sigma a = 900$	—	—	$\Sigma ax = 34705.5$	$\Sigma ay = 10290.9$

$\therefore$ $$\bar{x} = \frac{\Sigma ax}{\Sigma a} = \frac{34705.5}{900} = \mathbf{38.56\ mm.}\quad \textbf{(Ans.)}$$

$$\bar{y} = \frac{\Sigma ay}{\Sigma a} = \frac{10290.9}{900} = \mathbf{11.43\ mm.}\quad \textbf{(Ans.)}$$

(*ii*) **M.O.I. about XX axis :**

$$I_{xx} = -I_{xx_1} + I_{xx_2} + I_{xx_3}$$

$$= -\left[0.055 \times 30^4 + \frac{\pi \times 30^2}{4} \times \left(18.57 - \frac{4 \times 30}{3\pi}\right)^2\right]$$

$$+ \left[\frac{30 \times (30)^3}{12} + 30 \times 30 \times (15 - 11.43)^2\right]$$

$$+ \left[0.055 \times 30^4 + \frac{\pi \times 30^2}{4} \times \left(\frac{4 \times 30}{3\pi} - 11.43\right)^2\right]$$

$$- [44550 + 24091] + [67500 + 11470] + [44550 + 1198]$$

$$= -68641 + 78970 + 45748 = 56077 \text{ mm}^4$$

Hence $\mathbf{I_{xx} = 56077 \text{ mm}^4}$. **(Ans.)**

HIGHLIGHTS

1. The cumulative product of area and square of its distance from an axis is called the *moment of inertia* of a section about that axis.
2. *Moment of inertia* of simple areas :

Rectangle : $I_{xx} = \dfrac{bd^3}{12}$

$I_{yy} = \dfrac{db^3}{12}$

Triangle : $I_{xx} = \dfrac{bh^3}{36}$

About the base : $= \dfrac{bh^3}{12}$

Circle : $I_{yy} = \dfrac{\pi d^4}{64}$

Semi-circle : $I_{xx} = 0.11r^4$

$I_{yy} = \dfrac{\pi d^4}{128}$

Quadrant : $I_{xx} = 0.055r^4$.

3. *Theorem of parallel axis :* It states that *moment of inertia of a lamina about any axis in the plane of lamina equals the sum of moment of inertia about a parallel centroidal axis in the plane of lamina and the product of area of the lamina and square of the distance between the two axes.*
4. *Theorem of perpendicular axes :* It states that *if I_{ox} and I_{oy} be the moment of inertia of a lamina about mutually perpendicular axes OX and OY in the plane of the lamina and I_{oz} be the moment of inertia of the lamina about an axis (OZ) normal to the lamina and passing through the point of intersection of the axes OX and OX, then*

$$I_{oz} = I_{ox} + I_{oy}.$$

5. *Radius of gyration of the section (k_i) :*

$$k_i = \sqrt{\frac{I_i}{A}}$$

where I_i = moment of inertia about its axis, and

A = area of cross-section.

OBJECTIVE TYPE QUESTIONS

Fill in the blanks :

1. The cumulative product of area and square of its distance from an axis is called the ...... of a section about that axis.
2. M.O.I. of a circle = ... where d is the diameter of the circle.
3. M.O.I. of a triangle about an axis passing through its c.g. = ... where b and h are the base and height of the triangle respectively.
4. Moment of inertia (M.O.I.) of a quadrant about axis XX (passing through c.g.) = ... where R is radius of the quadrant.

Answers

1. M.O.I. **2.** $\frac{\pi d^4}{64}$ **3.** $\frac{bh^3}{36}$ **4.** $0.055R^4$.

EXERCISES

Theoretical Questions

1. Define the term "Moment of inertia".
2. State and explain "Theorem of parallel axes".
3. Enunciate "Theorem of perpendicular axes".
4. Derive expressions for M.O.I. of the following :
   (*i*) Rectangular lamina (*ii*) Circular lamina
   (*iii*) Semi-circular lamina (*iv*) Thin ring.
5. What is "Radius of gyration" ?

Unsolved Examples

1. Find I_{xx} and I_{yy} for the unequal angle section 12.5 × 9.5 × 1 cm shown in Fig. 5.32.
[Ans. I_{xx} = 330.8 cm^4, I_{yy} = 161.3 cm^4]

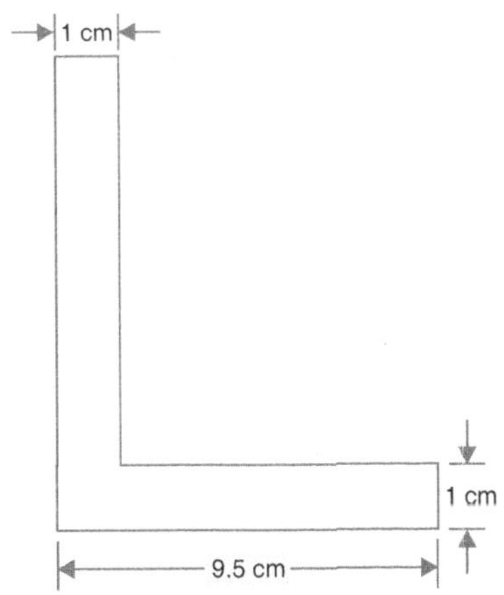

Fig. 5.32

2. Find the moment of inertia about the horizontal axis through the c.g. of the section shown in Fig. 5.33. [**Ans.** I_{xx} = 590.4 cm^4]

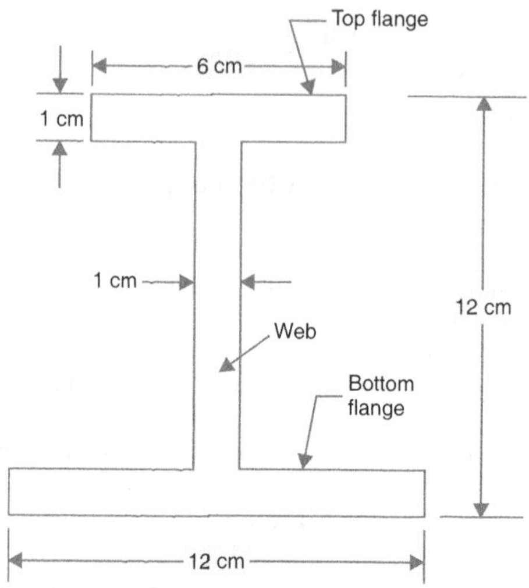

Fig. 5.33

3. Determine the moment of inertia of a *T*-section (as shown in Fig. 5.34) 10.16 × 10.16 × 1.27 cm about an axis passing through the centre of the section and perpendicular to the stem or vertical leg. [**Ans.** I_{xx} = 231.4 cm^4]

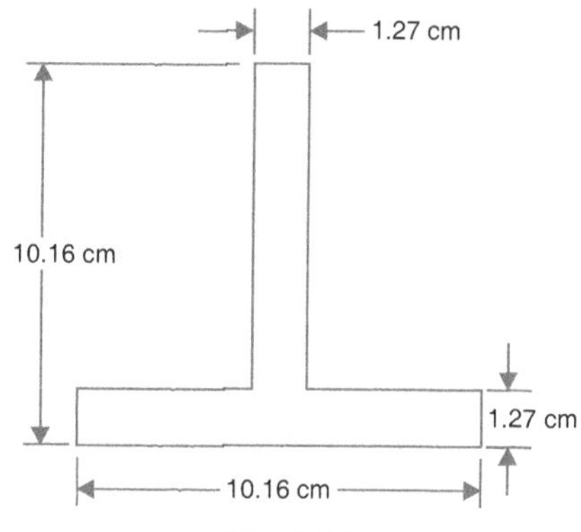

Fig. 5.34

4. Calculate the second moment of area of the built up section shown in Fig. 5.35 about *XX* and *YY*, *XX* and *YY* are the axes passing through the centroid of the section. [**Ans.** I_{xx} = 1152.35 cm^4 ; I_{yy} = 6463 cm^4]

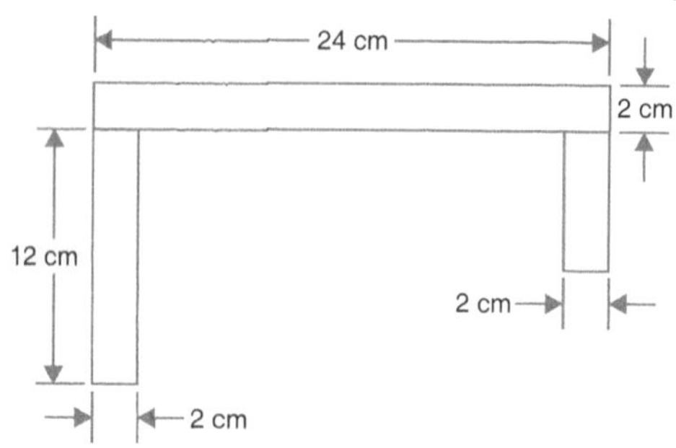

Fig. 5.35

5. Determine the moment of inertia of the Z-section in Fig. 5.36 about *XX* and *YY* axes, where *XX* and *YY* are the axes passing through the centroid of the figure. [**Ans.** I_{xx} = 1548 cm^4 ; I_{yy} = 2668 cm^4]

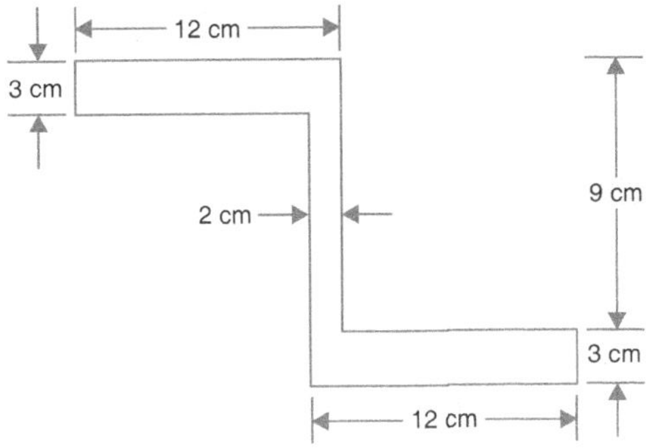

Fig. 5.36

6. Find the moment of inertia about the centroidal axis *XX* for the lamina shown in Fig. 5.37. [**Ans.** I_{xx} = 4015.38 cm^4]

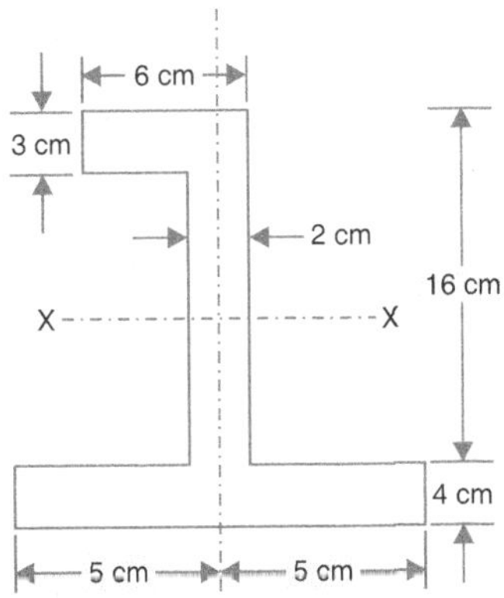

Fig. 5.37

7. Find I_{xx} of the channel section shown in Fig. 5.38. [**Ans.** I_{xx} = 6300 cm^4]

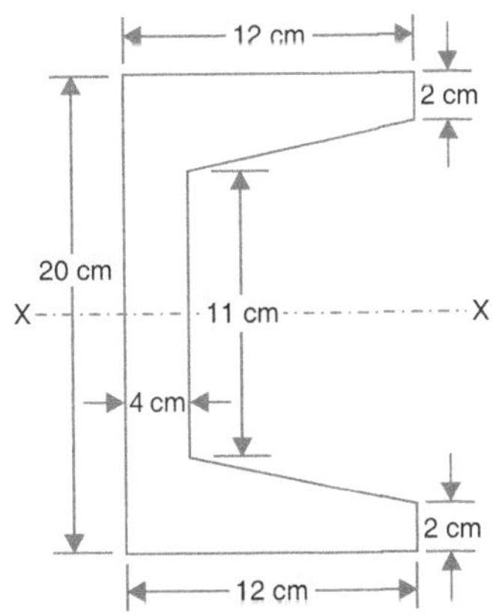

Fig. 5.38

8. Find the centroidal axis of the lamina shown in Fig. 5.39 parallel to the base. Find also the moment of inertia about this centroidal axis. **[Ans.** 4.08 cm ; 385.33 cm^4]

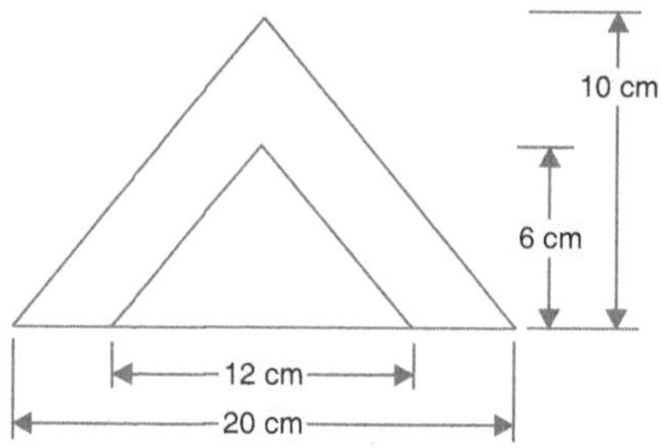

Fig. 5.39

9. A channel section 16 × 8 × 2 cm stands with flanges horizontal. Determine I_{xx} and I_{yy} passing through the centroid of the section. **[Ans.** I_{xx} = 1863.7 cm^4 ; I_{yy} = 302 cm^4]

10. A short column is built by riveting flanges of *T*-sections of size 30 mm × 60 mm × 5 mm to form across of 60 mm × 60 mm. Determine I_{xx}, I_{yy} and the least radius of gyration square. Neglect the effect of rivets.

[Ans. I_{xx} = 9.458 cm^4 ; I_{yy} = 18.052 cm^4 ; k^2_{least} = 1.113 cm^2]

11. A steel stanchion is built up of two rolled steel joists *ISJB* 150 placed 10 cm apart from centre to centre of webs, united by plates 15 cm wide and 1 cm thick, one on each side of the top and bottom flanges. If for *ISJB* 150 *RSJ* I_{xx} = 322.1 cm^4, I_{yy} = 9.2 cm^4 and area A = 9.01 cm^2, find the moment of inertia about (*a*) horizontal axis *LL* passing through the centroid ; and (*b*) vertical axis *MM* which also passes through the centroid of the section. **[Ans.** I_{LL} = 2566.7 cm^4 ; I_{mm} = 1031 cm^4]

12. Calculate the moment of inertia of the built up section shown in Fig. 5.40 about the centroidal axis *XX*. **[Ans.** I_{xx} = 336995.12 cm^4]

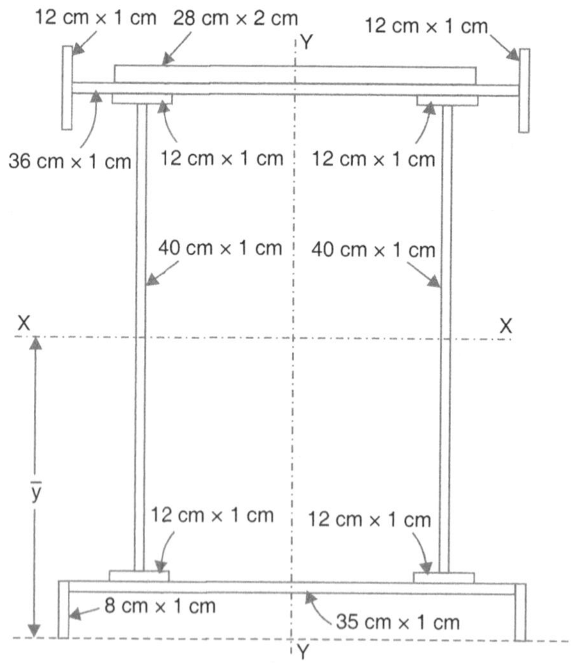

Fig. 5.40

6

Friction

6.1. Concept of friction. 6.2. Characteristics of frictional force. 6.3. Types of friction. 6.4. Static and Dynamic friction. 6.5. Limiting friction. 6.6. Laws of friction—Law of static friction—Laws of dynamic or kinetic friction. 6.7. Angle of friction. 6.8. Co-efficient of friction. 6.9. Angle of repose. 6.10. Cone of friction. 6.11. Motion of body on horizontal plane. 6.12. Motion up an inclined plane. 6.13. Motion down an inclined plane. 6.14. Screw friction. 6.15. Screw jack—Highlights—Objective Type Questions—Exercises—Theoretical Questions—Unsolved Examples.

6.1. CONCEPT OF FRICTION

It has been observed that surfaces of bodies, however smooth they may be, are not perfect and possess some irregularities and roughness. Therefore, if a block of one substance is placed over the level surface of another, a certain degree of interlocking of minutely projecting particles takes place. This interlocking properties of projecting particles oppose any tendency of the body to move. The resisting force acts in the direction opposite to that of the motion of the upper block and is called *friction*. Thus, *wherever there is a relative motion between two parts, a force of friction comes into play, and hence to overcome friction some energy is wasted.*

Hence, *force of friction or frictional force* may be defined as the *opposing force which is called into play in between the surfaces of contact of two bodies, when one body moves over the* surface of another body. (Figs. 6.1 and 6.2).

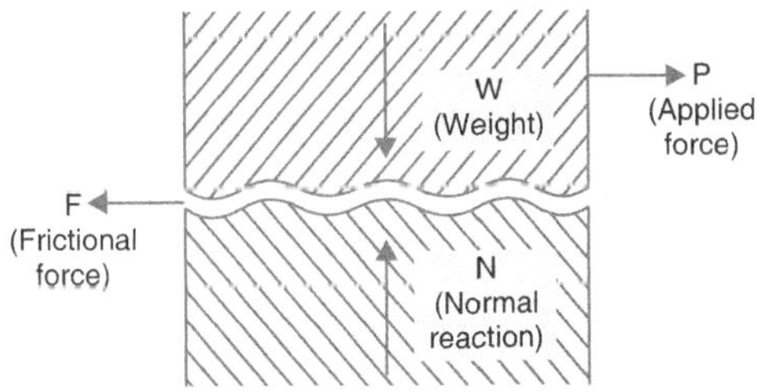

Fig. 6.1

In engineering applications friction is both desirable and undesirable.

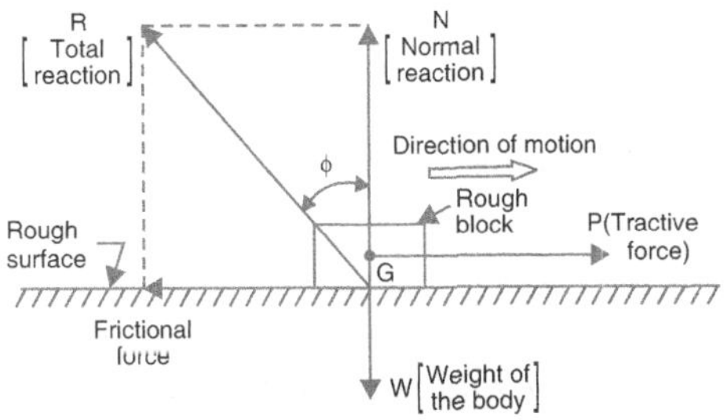

Fig. 6.2

There are appliances and devices known as *friction* devices such as belts and ropes, friction clutches, jib and cotter joints, brakes, nuts and bolts, in which friction is *desirable* and efforts are made to maximise it. On the contrary, the friction is very *undesirable* in *moving* parts of *machines.* It causes the loss of energy which manifests itself in the forms of heat energy. Due to friction a more force is required to cause motion of the parts. To improve the efficiency of the machines the friction force is reduced to the minimum possible by *lubrication.*

6.2. CHARACTERISTICS OF FRICTIONAL FORCE

The force of friction or frictional force entails the following characteristics :

(*i*) It is *self-adjusting.* As tractive force P increases, the frictional force F also increases, and at any instant only as much frictional force comes into play as is necessary to prevent the motion.

(*ii*) It always acts in a direction opposite to the motion (*i.e.*, always opposes the tractive force).

(*iii*) It is a *passive force* (since it exists only if the tractive force P exists).

6.3. TYPES OF FRICTION

Friction may be classified as follows :

1. Friction in unlubricated surfaces
2. Friction in lubricated surfaces.

Friction in unlubricated surfaces. The *friction that exists between two unlubricated surfaces is called solid friction or dry friction.*

It may be of the following two types :

(*i*) Sliding friction

(*ii*) Rolling friction.

The friction that exists when one surface *slides* over the other is called *sliding friction.*

The friction that exists between two surfaces separated by *balls* or rollers, is called the *rolling friction.*

It may be remembered that *rolling friction* is always *less than the sliding friction.*

Friction in lubricated surfaces. It may further be subdivided as follows :

(*i*) Boundry (or greasy or non-viscous) friction

(*ii*) Viscous friction.

If in between two rubbing surface there exists a thin film or layer of an oil or a lubricant, the oil gets absorbed in the surfaces and such a film is known as absorbed film. Instead of metal to metal contact of the surfaces, there is a contact between thin layer of the oil and obviously the frictional force is reduced. In such a case the frictional force is known as *boundary friction.*

In this chapter we shall only deal with friction between unlubricated parts.

6.4. STATIC AND DYNAMIC FRICTION

Static friction. *The static friction is the friction offered by the surfaces subjected to external forces until there is no motion between them.*

Dynamic friction. The dynamic friction is the *friction experienced by a body* when it is in motion. It is also known *as kinetic friction and is always less than static friction* (the kinetic friction is about 40 to 75 per cent of the limiting static friction).

6.5. LIMITING FRICTION

Fig. 6.3 shows a graph between the applied force and the friction. During static condition as the applied force is increased from zero value the frictional force increases in direct proportion to the applied force. A certain stage is reached when the applied force is just sufficient to overcome friction and motion of the body takes place. After this the friction *suddenly decreases* to a magnitude which remains constant throughout the motion period as shown in Fig. 6.3.

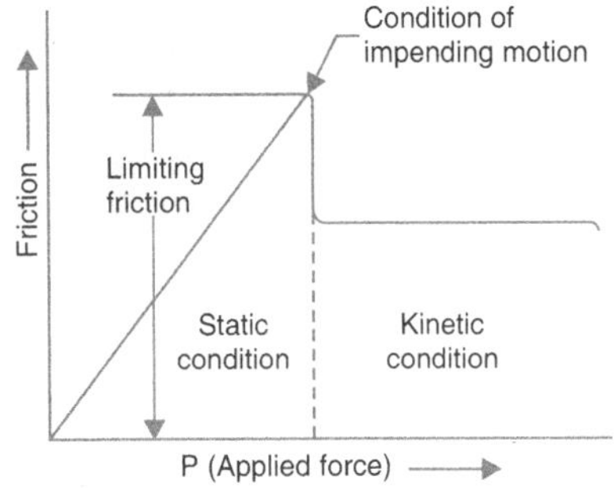

Fig. 6.3

When the motion is just to commence, maximum friction is encountered. This condition is known as limiting *equilibrium. The friction acting at this stage is termed as limiting friction.*

Hence, *limiting force of friction may be defined as the maximum value of friction force which exists when a body just begins to slide over the surface of the other body.* When the applied force or tractive force P is less than the limiting friction, the body remains at rest, and the friction is called *static friction*, which may have any value between zero and limiting friction.

6.6. LAWS OF FRICTION

6.6.1. Law of static friction

The laws of static friction are as follows :

1. The frictional force always acts in a direction *opposite* to that in which the body tends to move.
2. The frictional force is *directly proportional* to the normal reaction between the surfaces.
3. The frictional force *depends upon the nature of surfaces in contact.*
4. The frictional force is *independent of the area and shape of the contacting surfaces.*

6.6.2. Laws of dynamic or kinetic friction

1. The frictional force always acts in a direction opposite to that in which the body moves.
2. The frictional force is directly proportional to the normal reaction between the two contacting surfaces.
3. The magnitude of force of dynamic friction bears a constant ratio to the normal reaction between two surfaces but the ratio is slightly *less* than that in case of limiting friction.
4. The frictional force remains *constant for moderate speeds but it decreases slightly with the increase of speed.*

It may be noted that :

(*i*) For extremely low pressure and for very high pressures sufficient to produce excessive deformation, the co-efficient of static friction, *somewhat increases.*

(*ii*) For extremely low relative velocities, the co-efficient of kinetic friction increases and apparently becomes equal to the co-efficient of static friction.

(*iii*) For very high velocities co-efficient of kinetic friction decreases appreciably.

(*iv*) Ordinary changes in temperatures do not materially affect co-efficient of friction.

6.7. ANGLE OF FRICTION

Refer to Fig. 6.2. *It is defined as the angle which the resultant (R) of normal reaction (N) and limiting force of friction (F) makes with the normal (N).* It is denoted by ϕ.

$$\tan \phi = \frac{F}{N} \quad ...[6.1\ (a)]$$

$$\therefore \quad = \tan^{-1} \frac{F}{N} \quad ...[6.1\ (b)]$$

6.8. CO-EFFICIENT OF FRICTION

It is *defined as the ratio of limiting force of friction to the normal reaction between the two bodies.* It is denoted by μ.

Refer to Fig. 6.2.

$$\mu = \tan \phi = \frac{F}{N} \quad ...[6.2\ (a)]$$

$$\therefore \quad F = \mu N \quad ...[6.2\ (b)]$$

6.9. ANGLE OF REPOSE

Refer to Fig. 6.4. Consider a body of weight W acting on a rough horizontal plane inclined at angle α. The body is in equilibrium under the action of the following forces :

(*i*) Weight, W (which may be resolved into two components $W \sin \alpha$ and $W \cos \alpha$ as shown in Fig. 6.4)

(*ii*) Normal reaction, N and

(*iii*) Frictional force, F (= μN).

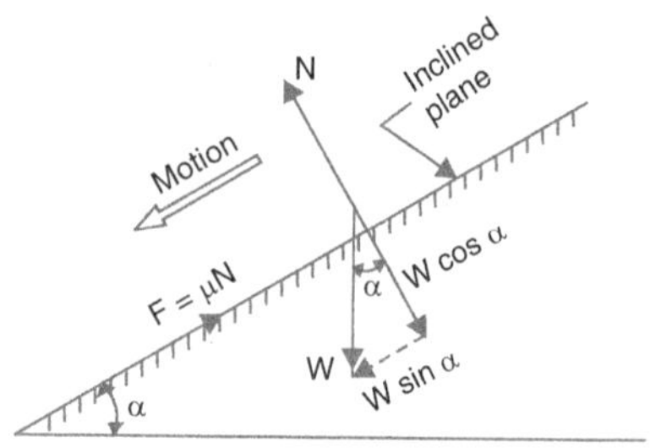

Fig. 6.4

In the limiting condition when the block is about to slide down the inclined plane, the frictional force must act up the plane and for equilibrium ; considering the forces along and perpendicular to the plane.

$$F = W \sin \alpha \quad ...(6.3)$$

and

$$N = W \cos \alpha \quad ...(6.4)$$

From eqns. (6.3) and (6.4), we get

$$\frac{F}{N} = \frac{W \sin \alpha}{W \cos \alpha} = \tan \alpha \quad ...(6.5)$$

But

$$\frac{F}{N} = \mu = \tan \phi$$

where ϕ is the angle of friction.

The angle α is called *angle of repose* and is *equal to the angle of friction when the body is in the condition of limiting equilibrium on an inclined plane.*

6.10. CONE OF FRICTION

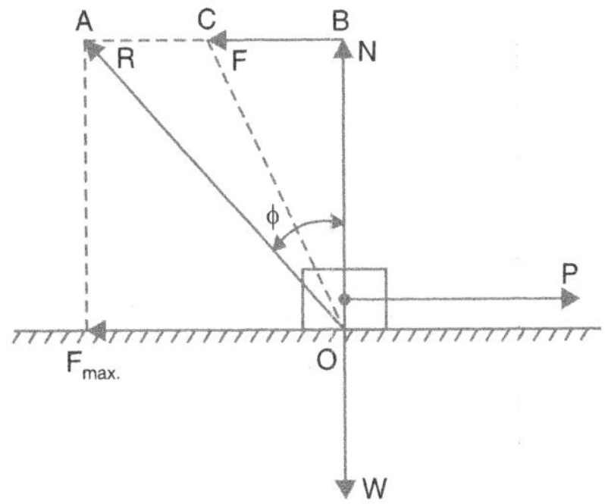

Fig. 6.5

If the line *OA* of Fig. 6.5 making the maximum angle of friction ϕ with the normal is revolved about *OB* as an axis, the cone generated is called the cone of *friction*. If the resultant *R* of the normal reaction and the friction falls within the cone of friction, the forces acting on the body are not great enough to cause motion. This principle is used in *self-locking* mechanisms and also in *taper pins*. For example, if the angle of taper pin is less than the angle of friction, no force at right angles in the axis of the pin could cause it to move in the direction of its axis.

6.11. MOTION OF BODY ON HORIZONTAL PLANE

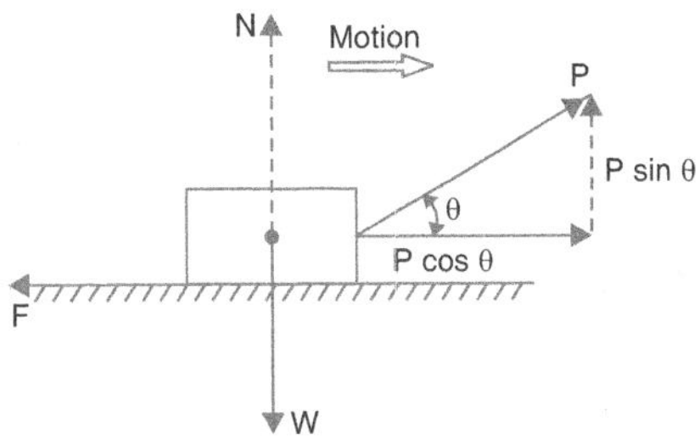

Fig. 6.6

Fig. 6.6 shows a body lying on a horizontal plane under the influence of force *P* which is inclined at angle θ to the surface (or horizontal plane). The value of force *P* can be determined by considering the limiting equilibrium.

Resolving the forces parallel to the plane (*i.e.*, horizontally), we get

$$F = P\cos\theta$$

or
$$\mu N = P\cos\theta \qquad ...(i)$$

Resolving the forces perpendicular to the plane (*i.e.*, vertically), we get

$$N + P\sin\theta = W$$

$$N = W - P\sin\theta \qquad ...(ii)$$

Substituting the value of *N* in (*i*), we get

$$\mu(W - P\sin\theta) = P\cos\theta$$

$$\mu W - \mu P\sin\theta = P\cos\theta$$

$$P(\cos\theta + \mu\sin\theta) = \mu W$$

$$P\left(\cos\theta + \frac{\sin\phi}{\cos\phi}.\sin\theta\right) = \frac{\sin\phi}{\cos\phi}.W \qquad \left(\because\ \mu = \tan\phi = \frac{\sin\phi}{\cos\phi}\right)$$

$$P(\cos\theta.\cos\phi + \sin\theta\sin\phi) = W\sin\phi$$

$$P\cos(\theta - \phi) = W\sin\phi$$

$$\therefore \qquad P = \frac{W\sin\phi}{\cos(\theta - \phi)} \qquad ...(6.6)$$

For *P* to be minimum, $\cos(\theta - \phi)$ should be maximum

or
$$\cos(\theta - \phi) = 1$$

$$0 - \phi = 0$$

or
$$\theta = \phi$$

or angle of inclination of force *P* should be equal to angle of friction.

6.12. MOTION UP AN INCLINED PLANE

Fig. 6.7 shows a body lying on an inclined plane under the influence of a force P. Let W be weight of the body, α be the inclination of the plane to the horizontal and μ be the co-efficient of friction.

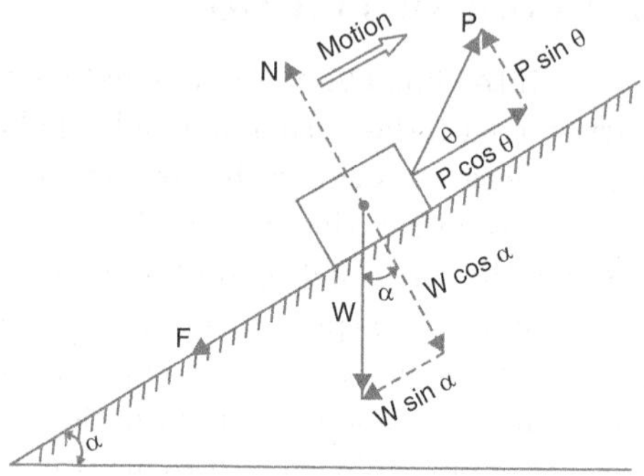

Fig. 6.7

The value of P can be determined by considering the limiting equilibrium as follows :

Case I. *When angle of inclination of the force to the plane is* θ :

Resolving the forces parallel to the plane, we have

$$F + W \sin \alpha = P \cos \theta$$

or $$F = P \cos \theta - W \sin \alpha$$

or $$\mu N = P \cos \theta - W \sin \alpha \qquad (\because \; F = \mu N) \quad ...(i)$$

Resolving the forces perpendicular to the plane.

$$N + P \sin \theta = W \cos \alpha$$

or $$N = W \cos \alpha - P \sin \theta \quad ...(ii)$$

Substituting the value of N from (*ii*) in (*i*), we get

$$\mu (W \cos \alpha - P \sin \theta) = P \cos \theta - W \sin \alpha$$

$$\mu W \cos \alpha - \mu P \sin \theta = P \cos \theta - W \sin \alpha$$

$$P (\cos \theta + \mu \sin \theta) = W (\mu \cos \alpha + \sin \alpha)$$

or $$P\left(\cos \theta + \frac{\sin \phi}{\cos \phi} . \sin \theta\right) = W\left(\frac{\sin \phi}{\cos \phi} \cos \alpha + \sin \alpha\right) \qquad \left(\because \; \mu = \tan \phi = \frac{\sin \phi}{\cos \phi}\right)$$

or $$P (\cos \theta \cos \phi + \sin \theta \sin \phi) = W (\sin \phi \cos \alpha + \sin \alpha \cos \phi)$$

or $$P \cos (\theta - \phi) = W \sin (\alpha + \phi)$$

or $$P = \frac{W \sin (\alpha + \phi)}{\cos (\theta - \phi)} \quad ...(6.7)$$

P is least when $\cos (\theta - \phi)$ is maximum

i.e., $$\cos (\theta - \phi) = 1$$

or $$\theta - \phi = 0$$

$$\theta = \phi.$$

Thus, the force required to move a body up the inclined plane is least when the force is inclined to the plane at an angle equal to angle of friction.

Case II. *When the force is parallel to the plane :*

$$\theta = 0$$

From $$P = \frac{W \sin (\alpha + \phi)}{\cos (\theta - \phi)}$$

$$P = \frac{W \sin (\alpha + \phi)}{\cos \phi} \quad ...(6.8)$$

Case III. *When there is no force of friction :*

$$\mu = 0$$

$\therefore$ $$\tan \phi = 0$$

or $$\phi = 0$$

From $$P = \frac{W \sin(\alpha + \phi)}{\cos(\theta - \phi)}$$

$$P = \frac{W \sin \alpha}{\cos \theta} \quad \text{...(6.9)}$$

6.13. MOTION DOWN AN INCLINED PLANE

Fig. 6.8 shows a body of weight W lying on an inclined plane inclined at an angle α, α is greater than angle of friction ϕ. The force P acts up the inclined plane. Let F be the limiting force of friction called into play as shown.

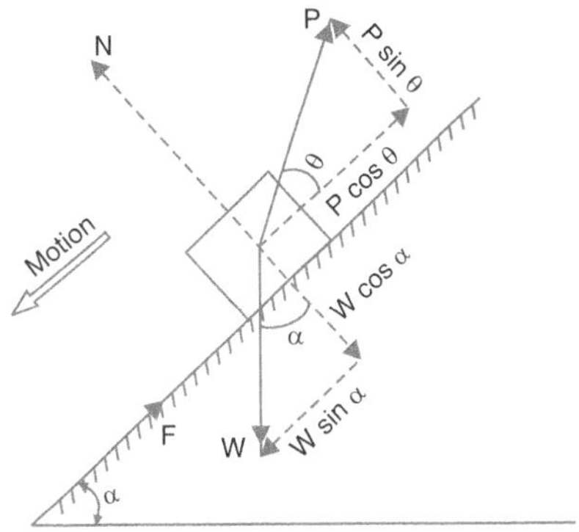

Fig. 6.8

Let us consider the limiting equilibrium for calculating the value of P.

Case I. *When P is inclined to the plane at angle θ :*

Resolving the forces parallel to the plane

$$F + P \cos \theta = W \sin \alpha$$

or $$\mu N + P \cos \theta = W \sin \alpha \quad \text{...}(i)$$

Resolving the forces perpendicular to the plane

$$N + P \sin \theta = W \cos \alpha \quad \text{...}(ii)$$

Substituting the value of N from (ii) in (i), we get

$$\mu (W \cos \alpha - P \sin \theta) + P \cos \theta = W \sin \alpha$$

$$\mu W \cos \alpha - \mu P \sin \theta + P \cos \theta = W \sin \alpha$$

or $$P (\cos \theta - \mu \sin \theta) = W (\sin \alpha - \mu \cos \alpha)$$

or $$P \left(\cos \theta - \frac{\sin \phi}{\cos \phi} \sin \theta \right) = W \left(\sin \alpha - \frac{\sin \phi}{\cos \phi} . \cos \alpha \right)$$

or $$P (\cos \theta \cos \phi - \sin \theta \sin \phi) = W (\sin \alpha \cos \phi - \cos \alpha \sin \phi)$$

or $$P \cos (\theta + \phi) = W \sin (\alpha - \phi)$$

or $$P = \frac{W \sin (\alpha - \phi)}{\cos (\theta + \phi)} \quad \text{...(6.10)}$$

Case II. *When the force P acts parallel to the plane :*

$$\theta = 0$$

From $$P = \frac{W \sin (\alpha - \phi)}{\cos (\theta + \phi)}$$

$$= \frac{W \sin (\alpha - \phi)}{\cos \phi}.$$

Case III. *When the plane is perfectly smooth and no frictional force acts on the plane :*

$$F = 0$$

i.e., $$\mu = 0$$

or $$\tan \phi = 0$$

or $$\phi = 0$$

From $$P = \frac{W \sin (\alpha - \phi)}{\cos (\theta + \phi)}$$

$$P = \frac{W \sin \alpha}{\cos \theta} \quad \text{...(6.11)}$$

Example 6.1. *A pull of 25 N at 30° to the horizontal is necessary to move a block of wood on a horizontal table. If the co-efficient of friction between the bodies in contact is 0.2, what is the weight of the block ?*

Sol. Let W = weight of the body

P = effort applied (= 25 N)

N = normal reaction

μ = co-efficient of friction (= 0.2)

Refer to Fig.6.9.

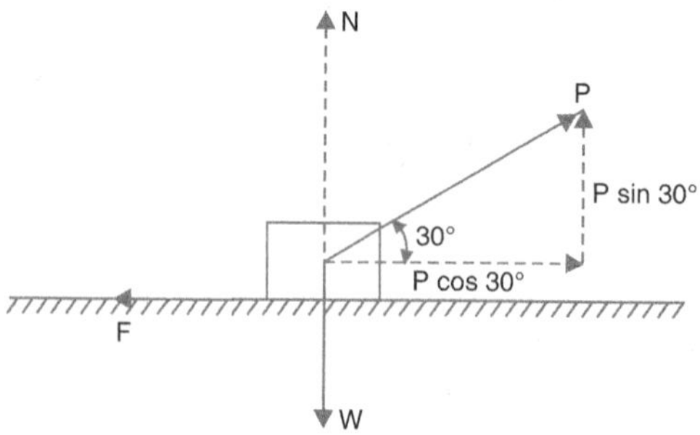

Fig. 6.9

Resolving forces **parallel** to the plane,

$$F = P \cos 30°$$

or $$\mu N = P \cos 30° \quad ...(i) \; (\because \; F = \mu N)$$

and resolving forces **perpendicular** to the plane,

$$N + P \sin 30° = W$$

or $$N = W - P \sin 30° \quad ...(ii)$$

Substituting the value of N in eqn. (i), we get

$$\mu (W - P \sin 30°) = P \cos 30°$$

or $$0.2 (W - 25 \times 0.5) = 25 \times 0.866$$

or $$0.2 (W - 12.5) = 21.65$$

or $$W = \frac{21.65}{0.5} + 12.5$$

or $$\mathbf{W = 120.75 \text{ N (newtons). (Ans.)}}$$

☞ **Example 6.2.** *A body resting on a rough horizontal plane required a pull of 18 N inclined at 30° to the plane just to move it. It was found that a push of 20 N inclined at 30° to the plane just moved the body. Determine the weight of the body and co-efficient of friction.*

Sol. Let W = weight of the body

P = effort applied

N = normal reaction

μ = co-efficient of friction.

Refer to Fig. 6.10.

Case I. Resolving forces **parallel** to the plane,

$$F = P \cos 30°$$

or $$\mu N = 18 \cos 30° \quad ...(i)$$

and resolving forces **perpendicular** to the plane,

$$N + P \sin 30° = W \quad ...(ii)$$

or $$N + 18 \sin 30° = W$$

or $$N = W - 18 \sin 30° \text{ or } N = W - 9$$

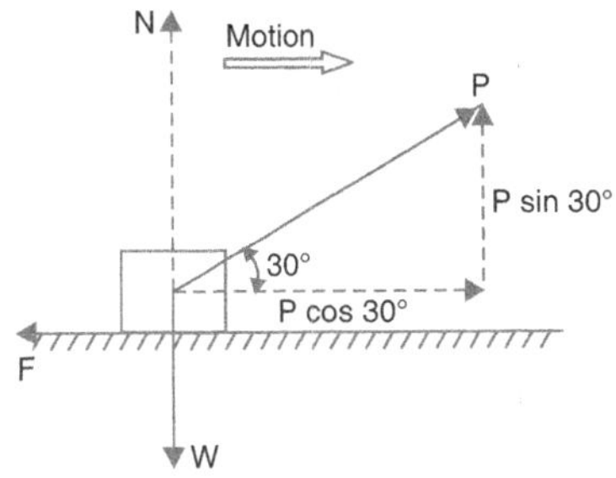

Fig. 6.10

Substituting the value of N in eqn. (i), we get

$$\mu(W - 9) = 18 \cos 30° \quad ...(1)$$

Refer to Fig. 6.11.

Case II. Resolving forces **parallel** to the plane,

$$F = P \cos 30°$$

or $$\mu N = 22 \cos 30° \quad ...(i)$$

and resolving forces **perpendicular** to the plane,

$$N = W + P \sin 30°$$

or $$N = W + 22 \sin 30°$$

or $$N = W + 11 \quad ...(ii)$$

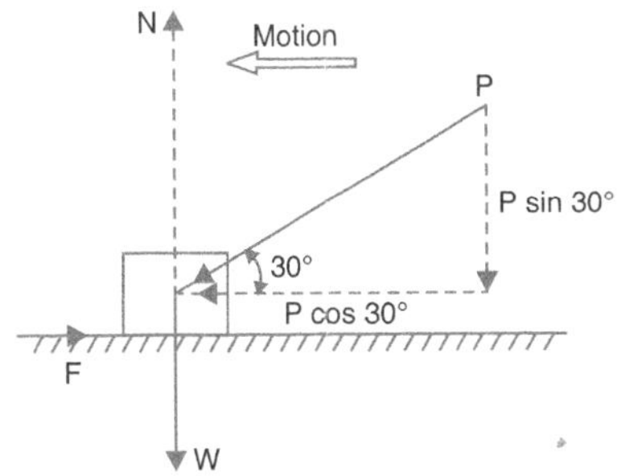

Fig. 6.11

Substituting the value of N in eqn. (i), we get

$$\mu (W + 11) = 22 \cos 30° \quad ...(2)$$

From (1) and (2), we get

$$\frac{\mu (W - 9)}{\mu (W + 11)} = \frac{18 \cos 30°}{22 \cos 30°}$$

$$22(W - 9) = 18 (W + 11)$$

$$22 W - 198 = 18 W + 198$$

$$4 W = 396$$

$$\mathbf{W = 99 \text{ N.}} \quad \textbf{(Ans.)}$$

Inserting this value of W in eqn. (1), we get

$$\mu (99 - 9) = 18 \cos 30°$$

$$\mu = \frac{18 \cos 30°}{90}$$

or $$\mu = \mathbf{0.1732.} \quad \textbf{(Ans.)}$$

Example 6.3. *Find the force required to move a load of 300 N up a rough inclined plane, the force being applied parallel to the plane. The inclination of the plane is such that when the same body is kept on a perfectly smooth plane inclined at that angle, a force of 60 N applied at an inclination of 30° to the plane keeps the same in equilibrium. Assume that co-efficient of friction between the rough plane and load is equal to 0.3.*

Sol. Refer to Fig. 6.12.

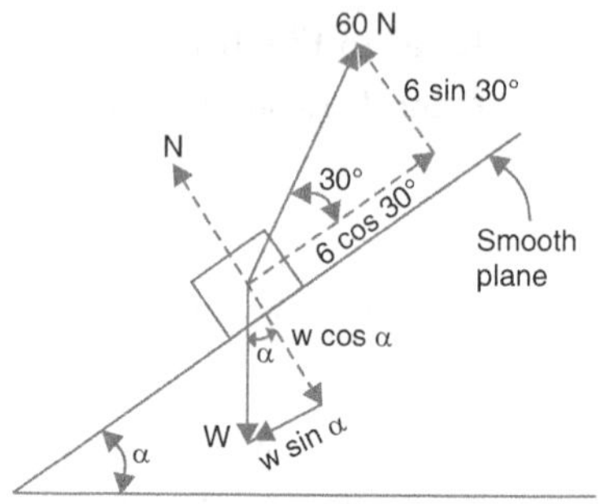

Fig. 6.12

Resolving forces **parallel** to the plane,

$$W \sin \alpha = 60 \cos 30°$$

or $$30 \sin \alpha = 60 \cos 30°$$

or $$\sin \alpha = \frac{60 \cos 30°}{300} = 0.1732$$

or $$\alpha = 9.97° \text{ or } 9° \ 58'.$$

When the force is applied parallel to rough plane.

Refer to Fig. 6.13.

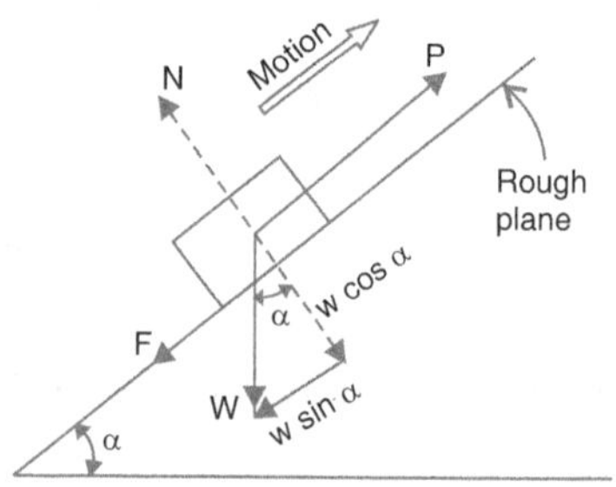

Fig. 6.13

Resolving forces **parallel** to the plane,

$$P = F + W \sin \alpha$$

or $$P = \mu N + W \sin \alpha \quad ...(i)$$

and resolving forces **perpendicular** to the plane,

$$N = W \cos \alpha \quad ...(ii)$$

Substituting the value of N in (i), we get

$$P = \mu W \cos \alpha + W \sin \alpha$$

or $$P = W (\mu \cos \alpha + \sin \alpha)$$

or $$P = W \left(\frac{\sin \phi}{\cos \phi} . \cos \alpha + \sin \alpha \right) \qquad \left(\because \ \mu = \tan \phi = \frac{\sin \phi}{\cos \phi} \right)$$

or, $$P = \frac{W \sin (\alpha + \phi)}{\cos \phi}$$

Given $W = 300$ N

$\mu = 0.3$

or, $\tan \phi = 0.3$

or, $\phi = 16.7°$

$\therefore$ $$P = \frac{300 \sin (9.97° + 16.7°)}{\cos 16.7°}$$

or, $$P = \frac{300 \times 0.5023}{0.9578}$$

or, **P = 157.3 N. (Ans.)**

Example 6.4. *An effort of 2000 N is required just to move a certain body up an inclined plane of 15°, the force acting parallel to the plane. If the angle of inclination of the plane is made 20°, the effort required, again applied parallel to the plane, is found to be 2300 N. Find the weight of the body and co-efficient of friction.*

Sol. Let W = weight of the body

N = normal reaction

F = force of friction

μ = co-efficient of friction.

Refer to Fig.6.14.

Case I. Resolving forces **parallel** to the plane,

$$F + W \sin 15° = 2000$$

or, $$\mu N + W \sin 15° = 2000 \quad ...(i)$$

and resolving forces **perpendicular** to the plane

$$N = W \cos 15° \quad ...(ii)$$

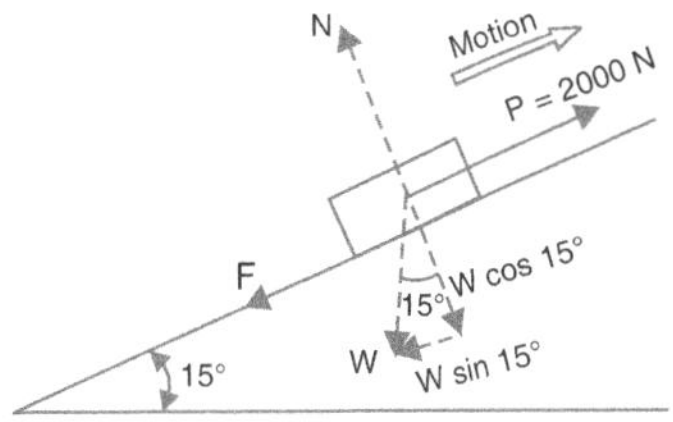

Fig. 6.14

Substituting the value of N in eqn. (i), we get

$$\mu W \cos 15° + W \sin 15° = 2000 \text{ N}$$

or, $$W (\mu \cos 15° + \sin 15°) = 2000 \text{ N} \quad ...(1)$$

Refer to Fig. 6.15

Resolving forces **parallel** to the plane,

$$F + W \sin 20° = 2300$$

or, $$\mu N + W \sin 20° = 2300 \quad ...(i)$$

and resolving forces perpendicular to the plane,

$$N = W \cos 20° \quad ...(ii)$$

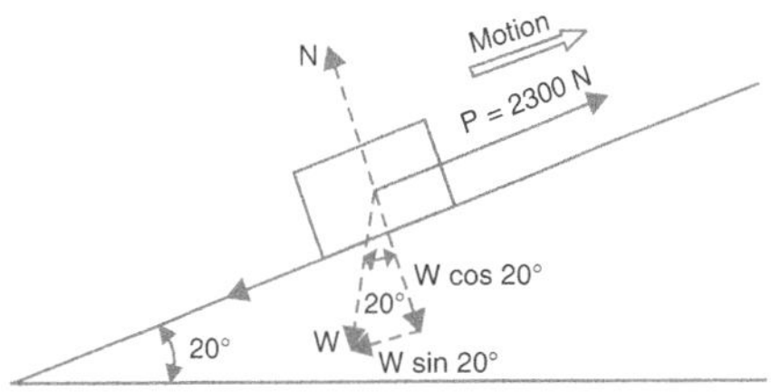

Fig. 6.15

Substituting this value of N in eqn. (i), we get

$$\mu W \cos 20° + W \sin 20° = 2300$$

or, $$W (\mu \cos 20° + \sin 20°) = 2300 \quad ...(2)$$

From (1) and (2), we get

$$\frac{W (\mu \cos 15° + \sin 15°)}{W (\mu \cos 20° + \sin 20°)} = \frac{2000}{2300}$$

$$\therefore \quad 2300 (\mu \cos 15° + \sin 15°) = 2000 (\mu \cos 20° + \sin 20°)$$

$$2300 \mu \cos 15° + 2300 \sin 15° = 2000 \mu \cos 20° + 2000 \sin 20°$$

$$2220 \mu + 595.3 = 1879 \mu + 684$$

$$341\mu = 88.7$$

$$\mu = \mathbf{0.26.} \quad \textbf{(Ans.)}$$

Substituting this value of μ in eqn. (1), we get

$$W (0.26 \cos 15° + \sin 15°) = 2000$$

$$W = \frac{2000}{0.26 \cos 15° + \sin 15°}$$

$$\therefore \quad \mathbf{W = 3922 \text{ N.}} \quad \textbf{(Ans.)}$$

Example 6.5. *Two blocks of equal weights and connected by a string passing over a frictionless pulley, rest on surfaces ; one on a horizontal surface and the other on an inclined surface as shown in Fig. 6.16. For both the surfaces the co-efficient of static friction* $\mu = 0.2$. *For what value of angle* α *the motion of the two blocks will impend ?*

Sol. Fig. 6.16 shows the various forces acting on the two blocks.

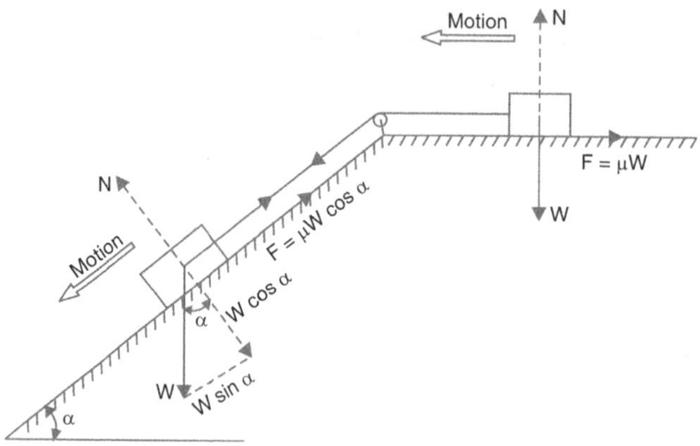

Fig. 6.16

Considering the equilibrium of the two blocks,

$$W \sin \alpha = \mu W \cos \alpha + \mu W$$

$$\sin \alpha = \mu \cos \alpha + \mu = \mu(1 + \cos \alpha)$$

i.e.,
$$\mu = \frac{\sin \alpha}{1 + \cos \alpha}$$

$$0.2 = \frac{2 \sin \alpha/2 \cos \alpha/2}{2 \cos^2 \alpha/2}$$

or
$$0.2 = \tan \frac{\alpha}{2}$$

∴
$$\frac{\alpha}{2} = 11.30° \text{ or } 11° \; 18'$$

and
$$\alpha = \mathbf{22° \; 36'}. \quad \textbf{(Ans.)}$$

Example 6.6. *A rectangular prism block weighing 150 N is resting on a rough inclined plane as shown in Fig. 6.17. The block is tied up by a horizontal string, which has a tension of 50 N. Find*

(*i*) *The frictional force on the block.*

(*ii*) *The normal reaction of the inclined plane.*

(*iii*) *The co-efficient of friction between the surfaces of the contact.*

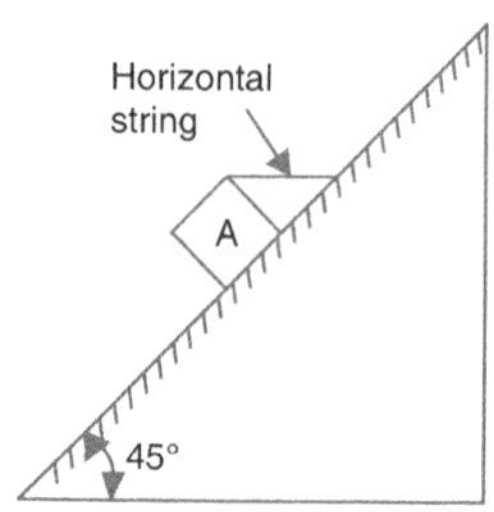

Fig. 6.17

Sol. Refer to Fig. 6.18.

(*i*) **Frictional force on the block, *F* = ?**

Resolving forces **parallel** to the plane,

$$F + 50 \cos 45° = W \sin 45°$$

$$F + 50 \cos 45° = 150 \sin 45°$$

$$\mathbf{F = 70.7 \; N.} \quad \textbf{(Ans.)}$$

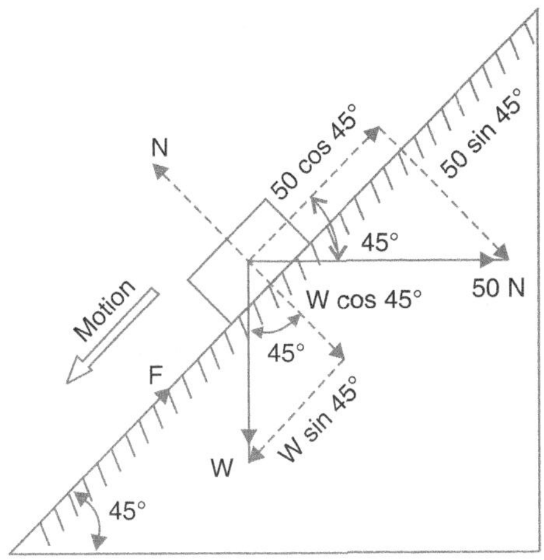

Fig. 6.18

(*ii*) **Normal reaction of the inclined plane, N = ?**

Resolving forces **perpendicular** to the plane,

$$N = W \cos 45° + 50 \sin 45°$$

$$= 150 \cos 45° + 50 \sin 45°$$

$$\mathbf{N = 141.4\ N.} \quad \textbf{(Ans.)}$$

(*iii*) **Co-efficient of friction, μ = ?**

Using the relation :

$$\mu = \frac{F}{N} = \frac{70.7}{141.4}$$

$$\mu = \mathbf{0.5.} \quad \textbf{(Ans.)}$$

Example 6.7. *An object of weight 100 N is kept in position on a plane inclined at 30° to the horizontal by the applied force P. The co-efficient of friction of the surface of the inclined plane in 0.25. Determine the minimum magnitude of the force P.*

Sol. Fig. 6.19 shows the various force acting on the object.

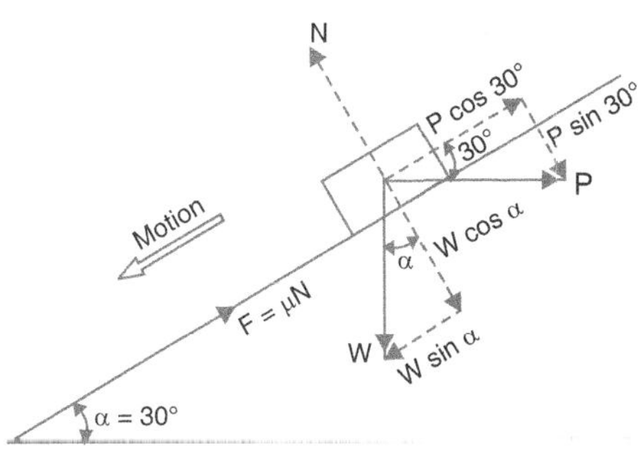

Fig. 6.19

Presuming that P is sufficient to hold the block W on the inclined plane, we have

Resolving all the forces **parallel** and **perpendicular** to the plane,

$$\mu N + P \cos 30° = W \sin \alpha$$

$$0.25\, N + P \cos 30° = W \sin 30° \quad ...(i)$$

and $$N = W \cos \alpha + P \sin 30°$$

or $$N = W \cos 30° + P \sin 30° \quad ...(ii)$$

From (*i*) and (*ii*), we get

$$0.25\,(W \cos 30° + P \sin 30°) + P \cos 30° = W \sin 30°$$

$$0.25\,(100 \cos 30° + P \sin 30°) + P \cos 30° = 100 \sin 30°$$

$$0.25\,(86.67 + 0.5\,P) + 0.867\,P = 50$$

$$21.67 + 0.125\,P + 0.867\,P = 50$$

$$0.992\,P = 28.33 \text{ N (newton)}$$

$$\mathbf{P = 28.56\ N\ (newtons).} \quad \textbf{(Ans.)}$$

Example 6.8. *Two blocks which are connected by a horizontal link AB are supported on two rough planes as shown in Fig. 6.20. The co-efficient of friction of block A is 0.4. The angle of friction for the block B on the inclined plane is* $\phi = 20°$. *Find the smallest weight W of the block A for which equilibrium can exist.*

Sol. Fig. 6.20 shows the two blocks connected by the link AB.

Let μ_A = co-efficient of friction between block A and rough plane = 0.4.

and μ_B for block $B = \tan 20° = 0.364$

Refer to Fig. 6.21.

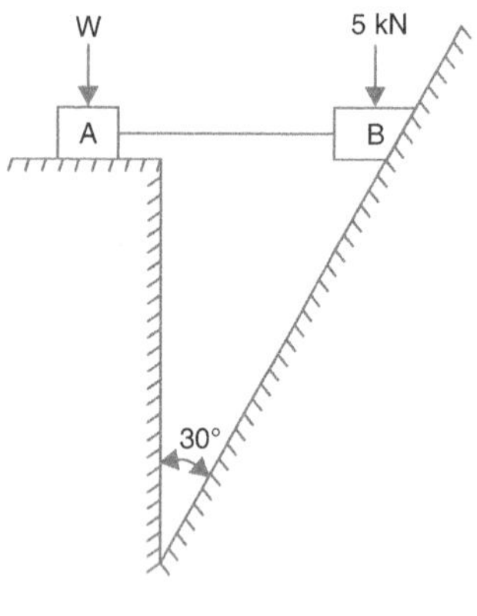

Fig. 6.20

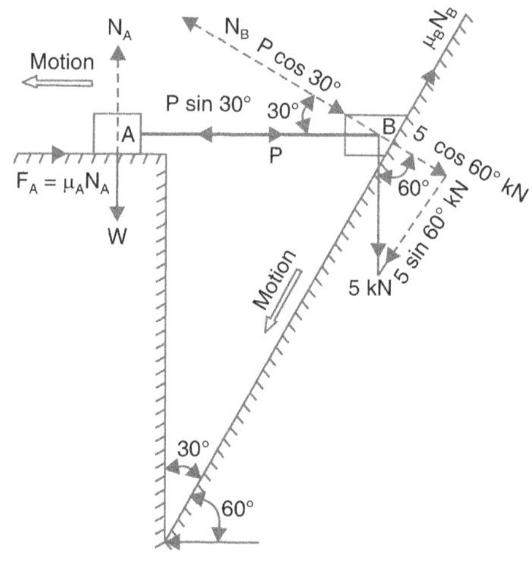

Fig. 6.21

The **block *B*** is in equilibrium under the action of forces :

(*i*) Its weight (5 kN) ;

(*ii*) Reaction of inclined plane, N_B (say) ;

(*iii*) Frictional force, $F_B = \mu_B\, N_B$; and

(*iv*) Force in the link AB , say P.

Resolving all forces **parallel** to the plane,

$$P \sin 30° + \mu_B N_B = 5 \sin 60° \quad ...(i)$$

and resolving all forces **perpendicular** to the plane,

$$N_B = P \cos 30° + 5 \cos 60°$$

or $$N_B = P \cos 30° + 5 \times 0.5$$

or $$N_B = P \cos 30° + 2.5 \quad ...(ii)$$

Substituting this value of N_B in eqn. (*i*), we get

$$P \sin 30° + 0.364\,(P \cos 30° + 2.5) = 5 \sin 60°$$

$$0.5\,P + 0.315\,P + 0.91 = 433$$

$$\therefore \quad P = \frac{0.342}{0.815} = 4.196 \text{ kN.}$$

Now considering the **block *A*.**

It is in equilibrium under the action of the forces :

(*i*) Its weight, W ;

(*ii*) Force in the line *AB*, P ;

(*iii*) Reaction of the plane, N_A ; and

(*iv*) Frictional force, $\mu_A N_A$.

Resolving the forces **parallel** to the plane,

$$F_A = P, \mu_A N_A = P$$

or $$0.4\,N_A = 4.196 \text{ kN}$$

or $$N_A = \frac{4.196}{0.4} = 10.49 \text{ kN.}$$

and resolving the forces **perpendicular** to the plane,

$$N_A = W$$

W = 10.49 kN. (Ans.)

Example 6.9. *A flywheel of radius 40 cm supported on a bracket and rotating about an axis O, is braked by means of a lever as shown in Fig. 6.22. The force applied at A is 15 N. If the co-efficient of kinetic friction between the flywheel and the brake shoe is 0.3, what braking moment will be applied on the wheel with respect to axis ?*

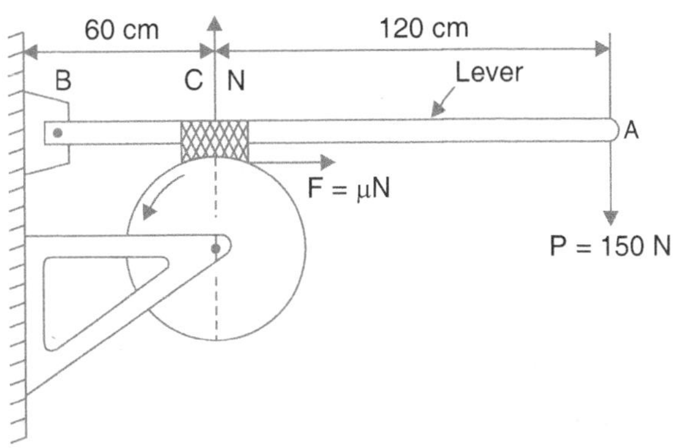

Fig. 6.22

Sol. Radius of wheel, $r = 40$ cm

Effort applied, $P = 150$ N

Co-efficient of friction, $\mu = 0.3$

Suppose the pressure exerted on the flywheel at C through the brake shoe is N.

Taking moments about B, we get

$$N \times 60 = 150 \times 180$$

$$N = 450 \text{ N}$$

Frictional force, $F = \mu N$

$$= 0.3 \times 45 = 135 \text{ N}$$

$\therefore$ Braking moments to be applied

$$= 135 \times r$$

$$= 135 \times 40 = 5400 \text{ N cm}$$

$$= \mathbf{54 \text{ Nm. (Ans.)}}$$

LADDER FRICTION

Example 6.10. *A ladder 5 m long and of 250 N weight is placed against a vertical wall in a position where its inclination to the vertical is 30°. A man weighing 800 N climbs the ladder. At what position will he induce slipping ? The co-efficient of friction for both the contact surfaces of the ladder viz., with the wall and the floor is 0.2.*

Sol. Refer to Fig. 6.23.

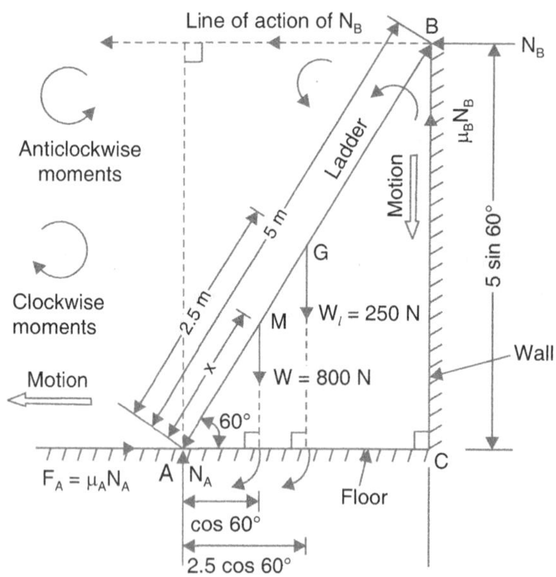

Fig. 6.23

$$\mu_A = \mu_B = 0.2$$

Weight of ladder, $W_l = 250$ N

Weight of man, $W = 800$ N

Let M be the position of the man when the ladder is at the point of sliding/slipping.

Let $AM = x$

Considering equilibrium of the ladder.

Resolving the forces *horizontally* and *vertically*, we get

$$N_B = \mu_A N_A \quad ...(i)$$

and

$$N_A + \mu_B N_B = W + W_l$$

$$N_A + \mu_B N_B = 800 + 250 = 1050 \text{ N} \quad ...(ii)$$

where N_A and N_B are the normal reactions at A and B respectively. Substituting the value of N_B from (*i*) in (*ii*), we get

$$N_A + \mu_B \mu_A N_A = 1050$$

or

$$N_A (1 + \mu_A \mu_B) = 1050$$

or

$$N_A (1 + 0.2 \times 0.2) = 1050$$

$$\therefore \quad N_A = \frac{1050}{1.04} = 1009 \text{ N}$$

and

$$N_B = 0.2 \times 100.9 = 201.8 \text{ N}$$

Taking moment about A,

$$800 \times x \cos 60° + 250 \times 2.5 \cos 60°$$

$$= N_B \times 5 \sin 60° + \mu_B N_B \times 5 \cos 60°$$

$$400\,x + 3125 = 201.8 \times 5 \sin 60° + 0.2 \times 201.8 \times 5 \cos 60°$$

$$= 974.7$$

$$\therefore \quad \boldsymbol{x = 1.65 \text{ m.}} \textbf{ (Ans.)}$$

Example 6.11. *A 13 m ladder weighing 250 N is placed against a smooth vertical wall with its lower end 5 m from the wall. The co-efficient of friction between the ladder and the floor is 0.3. Show that the ladder will remain in equilibrium in this position. What is the frictional force acting on the ladder at the point of contact between the ladder and the floor ? The centre of gravity of the ladder is at the mid-point of its length.*

Sol. Refer to Fig. 6.24.

Let N_A = normal reaction at A

N_B = normal reaction at B

Considering equilibrium of the ladder.

Resolving forces **horizontally** and **vertically,** we get

$$N_B = \mu_A N_A = 0.3\, N_A \quad ...(i) \quad (\because \mu_A = 0.3)$$

and

$$N_A = W_l = 250 \text{ N} \quad ...(ii)$$

$$\therefore \quad N_B = 0.3 \times 250 = 75 \text{ N}$$

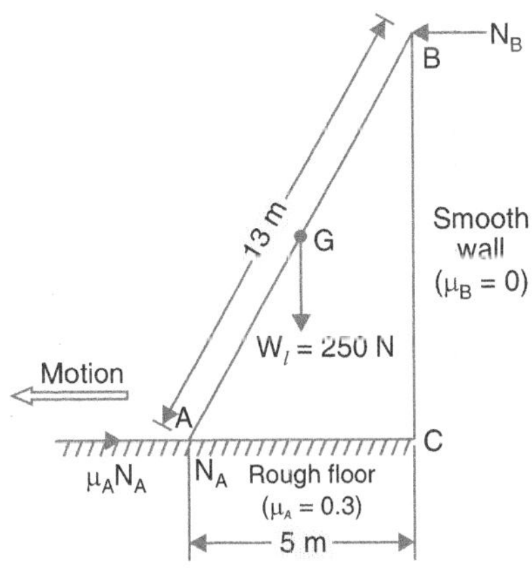

Fig. 6.24

Now, taking moments about A, we get

$$250 \times 6.5 \cos \alpha = N_B \times 13 \sin \alpha$$

But

$$\cos \alpha = \frac{5}{13}$$

and

$$\sin \alpha = \frac{\sqrt{13^2 - 5^2}}{13} = \frac{12}{13}$$

$$\therefore \quad 250 \times 6.5 \times \frac{5}{13} = N_B \times 13 \times \frac{12}{13}$$

$$N_B = 52.1 \text{ N}$$

The maximum amount of frictional force

$$= \mu_A N_A$$
$$= 0.3 \times 250 = 75 \text{ N}$$

which is greater than required *i.e.,* 52.1 N

Hence the ladder will remain in equilibrium. (Ans.)

The frictional force at the point of contact is **52.1 N. (Ans.)**

Example 6.12. *A uniform ladder rests with one end against a smooth vertical wall and the other on the ground, the co-efficient of friction being 0.75. If the inclination of the ladder to the ground be 45° show that a man whose weight is equal to that of the ladder, can just ascend to the top of the ladder without its slipping.*

Sol. Refer to Fig. 6.25

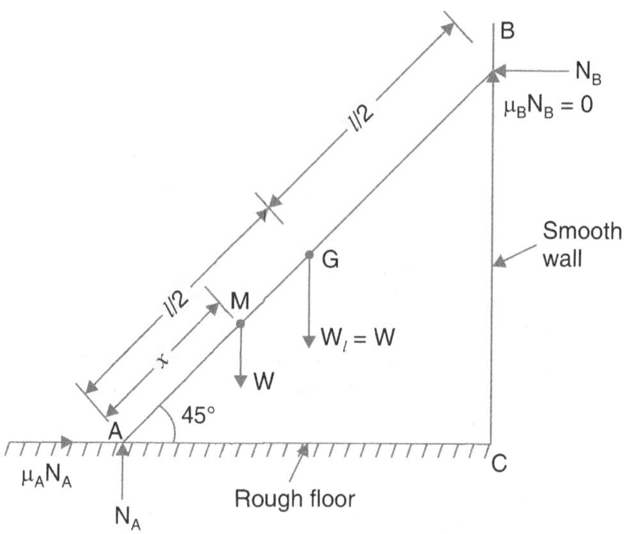

Fig. 6.25

Let *AB* be the ladder, of length *l*, resting against the rough floor and the smooth wall.

Let N_A = normal reaction of the floor (ground).

Let N_B = normal reaction of the wall

$F_A = \mu_A N_A$ = frictional force of the ground.

The weight 'W_l' of the ladder acts at its c.g., *G*. Let the man of weight '*W*' ascend on the ladder upto a distance '*x*' metre along the ladder as shown in Fig. 6.25.

Resolving all the forces horizontally and putting $\Sigma H = 0$

$$N_B = \mu_A N_A$$
$$= 0.75 N_A \quad ...(i)$$

Also, resolving all the forces **vertically** and putting $\Sigma V = 0$

$$N_A + \mu_B N_B = W_l + W$$
$$N_A = W + W = 2W \quad ...(ii)$$

$$\therefore \quad N_B = 0.75 \times 2W = 1.5 W \quad [\because \; W_l = W \text{ and } \mu_B = 0]$$

Taking moments of all the forces about A, we get

$$W \times x \cos 45° + W \times \frac{l}{2} \cos 45° = N_B \times l \sin 45°$$

$$W \, . \, x \cos 45° + W \times \frac{l}{2} \cos 45° = 1.5 \, W \times l \sin 45°$$

$$x \times \frac{1}{\sqrt{2}} + \frac{1}{2} \cdot \frac{1}{\sqrt{2}} = 1.51 \times \frac{1}{\sqrt{2}}$$

$$\therefore \quad \boldsymbol{x = 1.} \quad \textbf{(Ans.)}$$

Hence, the man can ascend to the top of the ladder without slipping.

WEDGE FRICTION

Example 6.13. *A block overlying a 10° wedge on a horizontal floor and leaning against a vertical wall and weighing 15 kN is to be raised by applying a horizontal force to the wedge. Assuming the co-efficient of friction between all the surfaces in contact to be 0.3, determine the minimum horizontal force, to be applied to raise the block.*

Sol. Fig. 6.26 shows the block, wedge and unknown force P.

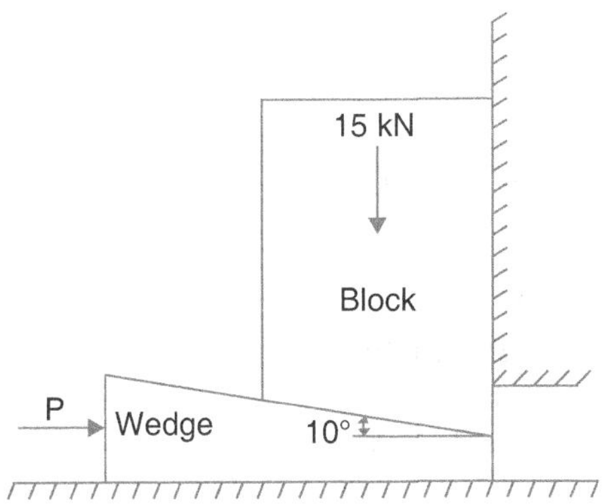

Fig. 6.26

Refer to Fig. 6.27 (a)

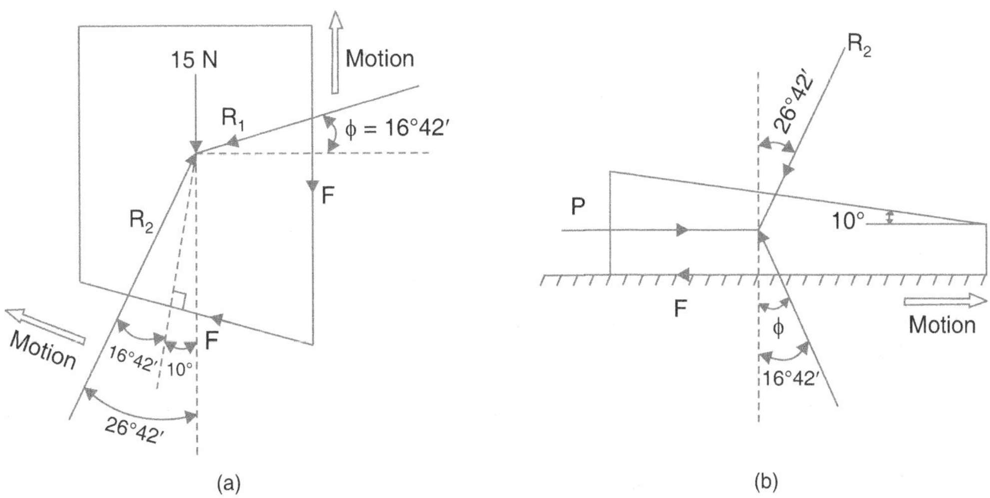

Fig. 6.27

The block is in equilibrium under the forces 15 kN weight, R_1 and R_2 (reactions).

$$\begin{bmatrix} \mu = \tan \phi = 0.3 \\ \therefore \quad \phi = 16°42' \end{bmatrix}$$

Using Lami's theorem, we get

$$\frac{15}{\sin (16° 42' + 90° + 26° 42')} = \frac{R_2}{\sin (90° - 16° 42')}$$

or
$$\frac{15}{\sin 133° 24'} = \frac{R_2}{\sin 73° 18'}$$

$\therefore$
$$R_2 = \frac{15 \times \sin 73° 18'}{\sin 133° 24'} = 19.77 \text{ kN}$$

Refer to Fig. 6.27 (*b*).

The wedge is in equilibrium under the forces P, R_2 and R_3.

Using Lami's theorem, we get

$$\frac{P}{\sin [180 - (26° 42' + 16° 42')]} = \frac{R_2}{\sin (90° + 16° 42')}$$

or
$$\frac{P}{\sin 136° 36'} = \frac{19.77}{\sin 106° 42}$$

$$\frac{P}{0.687} = \frac{19.77}{0.958}$$

$\therefore$
$$\mathbf{P = 14.18 \text{ kN. (Ans.)}}$$

Example 6.14. *Two wedge blocks A and B are employed to raise a load of 20 kN resting on another block C by application of force P as shown in Fig. 6.28. Neglecting weights of the blocks and assuming co-efficient of friction μ = 0.25 for all the surfaces, determine the value of P for impending upward motion of block C.*

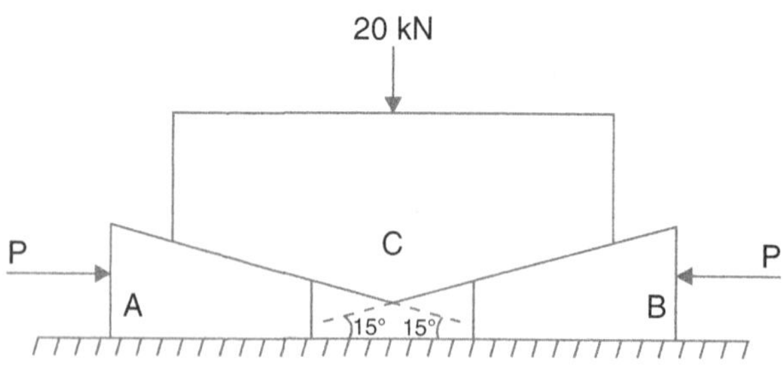

Fig. 6.28

Sol. The block *C*, under the action of forces *P* on blocks *A* and *B*, tends to move upward. Hence the frictional forces will act downward. What holds good for block *A*, the same will hold good for block *B*.

$$\tan \phi = \mu = 0.25 \text{ (given)}$$

where ϕ is the angle of friction.

$\therefore$
$$\phi = 14°$$

Refer to Fig. 6.29 (*a* and *b*).

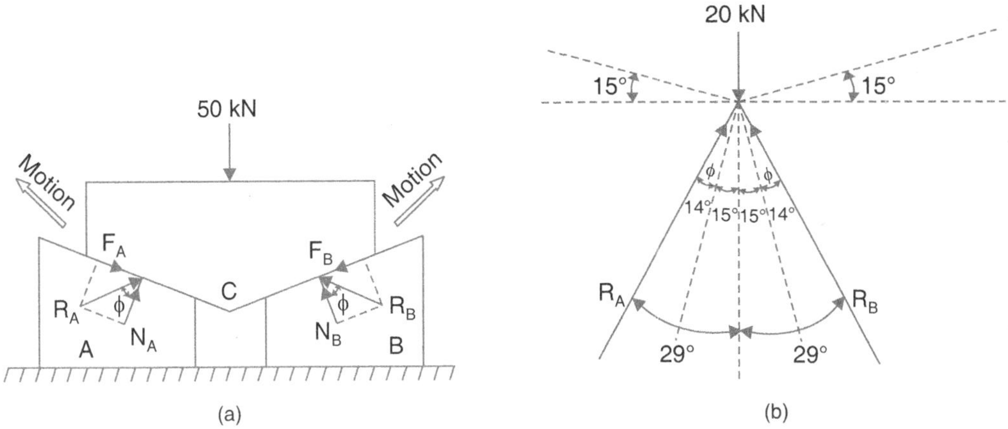

Fig. 6.29

Considering equilibrium of block C. It is acted upon by the forces

(*i*) Load 20 kN,

(*ii*) Total reaction R_A offered by wedge block A ; and

(*iii*) Total reaction R_B offered by wedge block B.

Using Lami's theorem, we get

$$\frac{20}{\sin 58^\circ} = \frac{R_A}{\sin (180^\circ - 29^\circ)} = \frac{R_B}{\sin (180^\circ - 29^\circ)}$$

or
$$\frac{20}{\sin 58^\circ} = \frac{R_A}{\sin 29^\circ} = \frac{R_B}{\sin 29^\circ}$$

$\therefore$
$$R_A = R_B = \frac{20 \times \sin 29^\circ}{\sin 58^\circ} = 11.43 \text{ kN}$$

Refer to Fig. 6.30 (*a* and *b*).

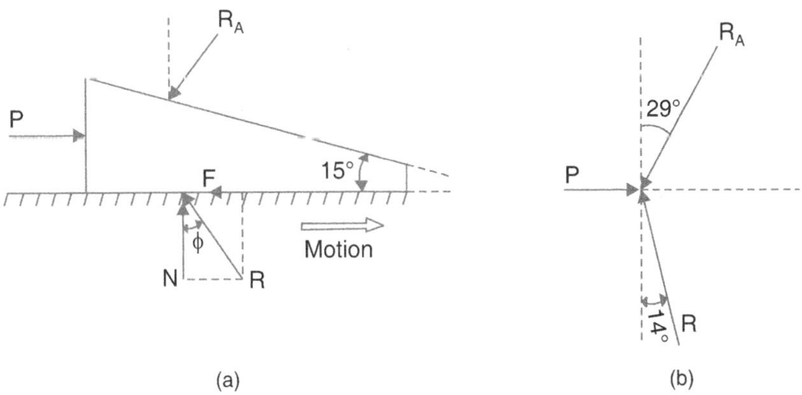

Fig. 6.30

Considering equilibrium of the block *A*. It is acted upon by the forces :

(*i*) Force *P* ;

(*ii*) R_A (from block *C*) ; and

(*iii*) Total reaction *R* offered by horizontal surface.

Using Lami's theorem, we get

$$\frac{P}{\sin [180° - (29° + 14°)]} = \frac{R_A}{\sin (90° + 14°)}$$

or
$$\frac{P}{\sin 137°} = \frac{R_A}{\sin 104°}$$

$\therefore$
$$P = \frac{11.43 \times \sin 137°}{\sin 104°} \qquad (\because \; R_A = 11.43 \text{ kN})$$

$$= \frac{11.43 \times 0.682}{0.97} = 8.03 \text{ kN}$$

Hence **P = 8.03 kN. (Ans.)**

6.14. SCREW FRICTION

The screw friction is the friction experienced by screw threads made by cutting a continuous helical groove on a cylindrical surfaces such as screws, bolts, nuts, studs etc. These parts are widely used in various machines and structures for fastening. The screw threads may be of (*i*) *V* shape and (*ii*) square shape. The *V* threads are stronger and offer more *frictional force to motion than square threads. V-threads are used for lighter load and square threads are used for heavier loads.*

Square threads are used in screw jacks, vice screws etc. *Screw jack* is a device used for raising/ lifting heavy loads by applying a small effort at its handle. *It works on the principle of an inclined plane.*

6.15. SCREW JACK

(*a*) **With square threads.** Fig. 6.31 shows a common form of a screw jack which consists of a threaded rod *A*, called screw rod or simply screw. The screw has *square threads*, in its outer surface, which fit into the inner threads of jack *B*. The load to be raised or lowered, is placed on the head of the screw, which is rotated by application of an effort at the end of the lever/handle for lifting or lowering the load.

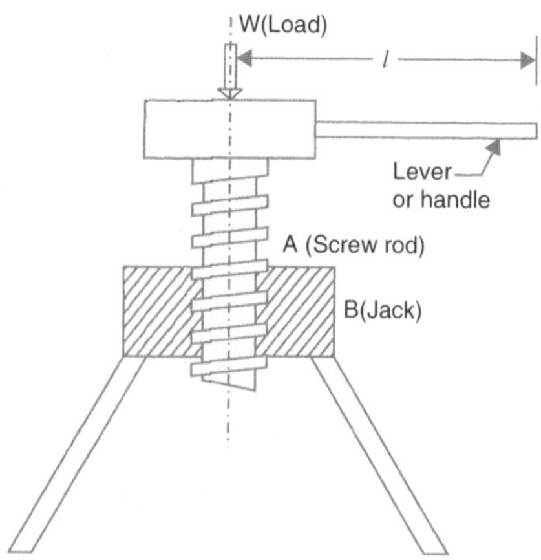

Fig. 6.31

With one rotation of the lever/handle the weight is lifted through a distance of equal to pitch. In case of multithreaded screws the actual pitch is np,

where p = apparent pitch and is equal to the distance between the two consecutive threads,

and n = number of starts.

Refer to Fig. 6.32. When a screw is given one revolution, it will move up by pitch p axially, and horizontally by πd_m (d_m is the mean diameter of the threads). Therefore, inclination of the threads can be calculated by,

$$\tan \alpha = \frac{p}{\pi d_m} \quad ...(6.12)$$

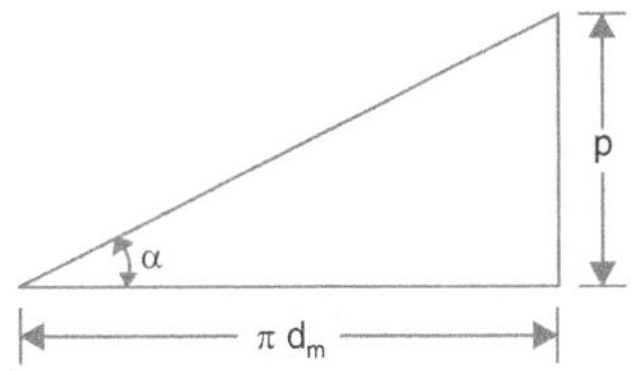

Fig. 6.32

where α is the helix angle.

(*i*) **Effort required to lift the load**

Let P = effort required to lift the load when applied at mean radius ;

W = weight of the body to the lifted ; and

μ = co-efficient of friction between the screw and the nut.

As the principle on which screw jack works, is similar to that of an inclined plane, therefore, the force applied on the lever of a screw jack is considered to be horizontal as shown in Fig. 6.33. Moreover, as the weight is being lifted, therefore, the force of friction F will act *downwards.* All the forces acting on the body are shown in Fig. 6.33.

Resolving the forces **parallel** to the plane,

$$P \cos \alpha = W \sin \alpha + F$$

or

$$P \cos \alpha = W \sin \alpha + \mu N \quad ...(6.13)$$

and resolving the forces **perpendicular** to the plane,

$$N = P \sin \alpha + W \cos \alpha \quad ...(6.14)$$

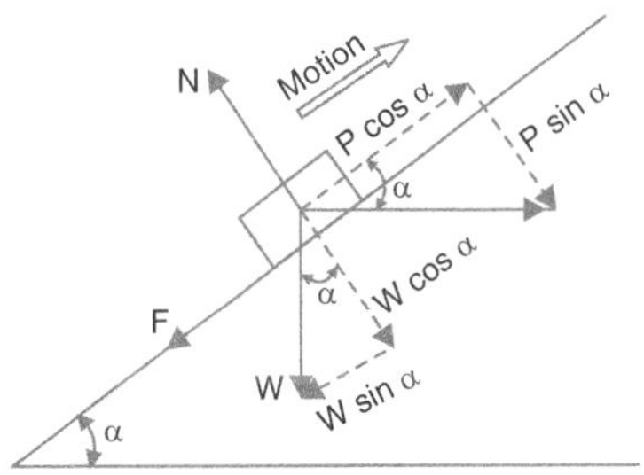

Fig. 6.33

Substituting the value of N in eqn. (6.13)

$$P \cos \alpha = W \sin \alpha + \mu (P \sin \alpha + W \cos \alpha)$$

$$= W \sin \alpha + \mu P \sin \alpha + \mu W \cos \alpha$$

$$P \cos \alpha - \mu P \sin \alpha = W \sin \alpha + \mu W \cos \alpha$$

$$P (\cos \alpha - \mu \sin \alpha) = W (\sin \alpha + \mu \cos \alpha).$$

Substituting $\mu = \tan \phi = \frac{\sin \phi}{\cos \phi}$, in the above expression, we get

$$P\left(\cos \alpha - \frac{\sin \phi}{\cos \phi} \sin \alpha\right) = W\left(\sin \alpha + \frac{\sin \phi}{\cos \phi} . \cos \alpha\right)$$

$$P (\cos \alpha \cos \phi - \sin \alpha \sin \phi) = W (\sin \alpha \cos \phi + \cos \alpha \sin \phi)$$

$$P \cos (\alpha + \phi) = W \sin (\alpha + \phi)$$

$$\therefore \quad P = W \frac{\sin (\alpha + \phi)}{\cos (\alpha + \phi)}$$

or

$$P = W \tan (\alpha + \phi) \quad ...(6.15)$$

Let P_h – horizontal force applied at the handle

L = length of the handle

Taking moments of P_h and P about the axis of the screw jack, we get

$$P_h \times L = P \frac{d_m}{2}$$

$$\therefore \quad P_h = P \frac{d_m}{2L}$$

or $$P_h = \frac{d_m}{2L} W \tan(\alpha + \phi)$$

or $$P_h = W \frac{d_m}{2L} \tan(\alpha + \phi) \quad ...(6.16)$$

(*ii*) **Effort required to lower the load.** Since the weight is being lowered, the force of friction F will act upwards. All the forces acting on the body are shown in Fig. 6.34.

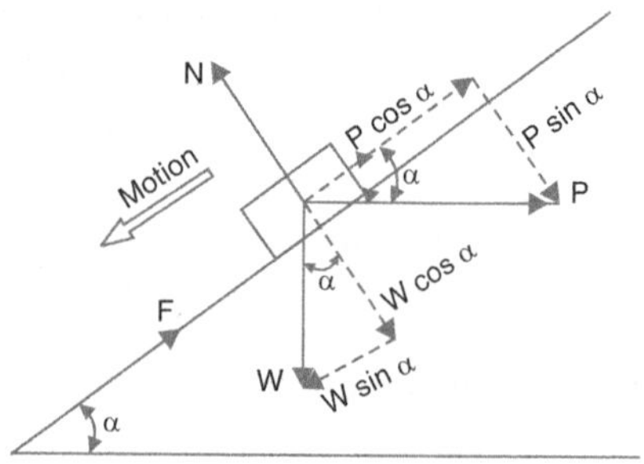

Fig. 6.34

Resolving all the forces **parallel** to the plane,

$$P \cos \alpha + F = W \sin \alpha$$

or $$P \cos \alpha = W \sin \alpha - F$$

or $$P \cos \alpha = W \sin \alpha - \mu N \quad ...(6.17)$$

and resolving the forces **perpendicular** to the plane,

$$N = W \cos \alpha + P \sin \alpha \quad ...(6.18)$$

Substituting the value of N in eqn. (6.17), we get

$$P \cos \alpha = W \sin \alpha - \mu (W \cos \alpha + P \sin \alpha)$$
$$= W \sin \alpha - W \cos \alpha - \mu P \sin \alpha$$

or $$P \cos \alpha + \mu P \sin \alpha = W \sin \alpha - \mu W \cos \alpha$$
$$P (\cos \alpha + \mu \sin \alpha) = W (\sin \alpha - \mu \cos \alpha)$$

Substituting $\mu = \tan \phi = \frac{\sin \phi}{\cos \phi}$ in the above equation, we get

$$P\left(\cos \alpha + \frac{\sin \phi}{\cos \phi} \sin \alpha\right) = W\left(\sin \alpha - \frac{\sin \phi}{\cos \phi} . \cos \alpha\right)$$

$$P (\cos \alpha \cos \phi + \sin \alpha \sin \phi) = W (\sin \alpha \cos \phi - \cos \alpha \sin \phi)$$
$$P \cos (\alpha - \phi) = W \sin (\alpha - \phi)$$

or $$P = W \frac{\sin (\alpha - \phi)}{\cos (\alpha - \phi)}$$

or $$P = W \tan (\alpha - \phi) \quad ...(6.19)$$

If $\alpha > \phi$, the nut or the weight placed on it will start moving downward by the effort component of the weight and the force P shall have to be applied to hold it on. To guard against this undesirable effect in *screws* α *is always kept less than* ϕ and in that case P will become negative which simply means that force shall have to be applied in opposite direction to cause the load to move downwards. Thus, the force required to move the load down,

$$P = W \tan (\phi - \alpha) \quad ...(6.20)$$

Efficiency of a screw jack. We know that effort P required at the mean radius of a screw jack to lift the load W is given by,

$$P = W \tan(\alpha + \phi)$$

If there would have been no friction between the screw and the nut, then ϕ will be zero. The value of effort P_0 necessary to raise the load, will then be given by the equation,

$$P_0 = W \tan \alpha$$

∴ Efficiency,
$$\eta_{up} = \frac{\text{ideal effort}}{\text{actual effort}} = \frac{P_0}{P}$$

$$= \frac{W \tan \alpha}{W \tan(\alpha + \phi)} = \frac{\tan \alpha}{\tan(\alpha + \phi)} \quad \text{...(6.21)}$$

It shows that the efficiency of a screw jack is independent of the weight lifted or effort applied.

Condition for *maximum efficiency :*

For determining condition for maximum efficiency

$$\frac{d\eta_{up}}{d\alpha} = 0 \quad \text{...(6.22)}$$

or
$$\frac{d}{d\alpha}\left[\frac{\tan \alpha}{\tan(\alpha + \phi)}\right] = 0$$

or
$$\frac{\sec^2 \alpha \tan(\alpha + \phi) - \sec^2(\alpha + \phi) \tan \alpha}{\tan^2(\alpha + \phi)} = 0$$

$$\therefore \quad \sec^2 \alpha \tan(\alpha + \phi) = \sec^2(\alpha + \phi) \tan \alpha$$

$$\sin(\alpha + \phi) \cos(\alpha + \phi) = \sin \alpha \times \cos \alpha$$

$$\sin 2(\alpha + \phi) = \sin 2\alpha$$

$$2(\alpha + \phi) = \pi - 2\alpha$$

or
$$\alpha = \frac{\pi}{4} - \frac{\phi}{2}$$

$$\therefore \quad \alpha = 45^\circ \quad \frac{\phi}{2} \quad \text{...(6.23)}$$

Maximum efficiency, $\eta_{up(max)}$:

Substituting the value of α in eqn. (6.21), we get

$$\eta_{up\,(max)} = \frac{\tan\left(45^\circ - \frac{\phi}{2}\right)}{\tan\left(45^\circ + \frac{\phi}{2}\right)} = \frac{\dfrac{1 - \tan\frac{\phi}{2}}{1 + \tan\frac{\phi}{2}}}{\dfrac{1 + \tan\frac{\phi}{2}}{1 - \tan\frac{\phi}{2}}}$$

$$= \frac{\left(1 - \tan\frac{\phi}{2}\right)^2}{\left(1 + \tan\frac{\phi}{2}\right)^2} = \frac{\left(\cos\frac{\phi}{2} - \sin\frac{\phi}{2}\right)^2}{\left(\cos\frac{\phi}{2} + \sin\frac{\phi}{2}\right)^2}$$

$$= \frac{1 - 2\sin\frac{\phi}{2}\cos\frac{\phi}{2}}{1 + 2\sin\frac{\phi}{2}\cos\frac{\phi}{2}}$$

$$= \frac{1 - \sin\phi}{1 + \sin\phi}$$

$\therefore$ Maximum efficiency, $\eta_{up\ (max)}$

$$= \frac{1 - \sin\phi}{1 + \sin\phi} \qquad ...(6.24)$$

(*b*) **With V-threads.** The square threads (as per configuration) take the load *W* perpendicular to them, whereas in *V*-threads the axial load does not act perpendicular to the surface of the threads as shown in Fig. 6.35. The normal reaction *N* between the threads and the screw should be such that its axial component is equal and opposite to the axial load *W*.

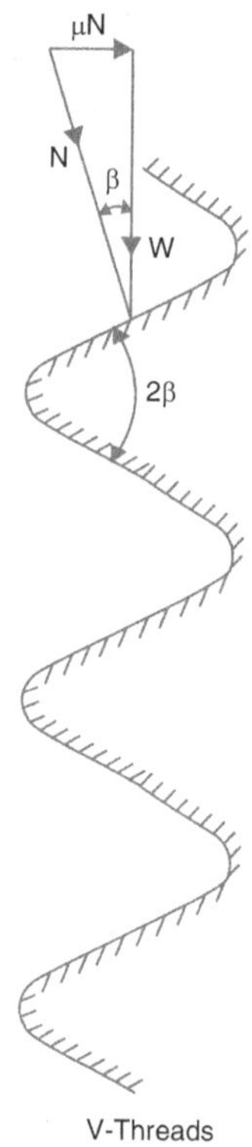

Fig. 6.35

Let 2β = angle inclined between the two sides of the thread.

Then, $N \cos\beta = W$

$$N = \frac{W}{\cos\beta} \qquad ...(6.25)$$

Frictional force (which acts tangential to the surface of the threads)

$$F = \mu N$$

or $$F = \frac{\mu W}{\cos\beta}$$

or $$F = \mu' W \qquad ...(6.26)$$

where μ' may be regarded as virtual co-efficient of friction

$$= \frac{\mu}{\cos\beta}$$

While solving problems on *V*-threads, for finding out effort *P* or η etc. μ may be substituted by μ' in all the relevant equations meant for the square threads.

It may be observed that the *force required to lift a given load with V-threads will be more than the square threads.*

Note. *Square threads are not only used for raising or lowering the load but are also used for transmission of power such as in lathes (lead screw), milling machines etc. They can transmit power without any side thrust but are difficult to cut when lead is long on account of the difficulty in cleaning the tool from the groove. Further, they cannot be used conveniently with split or half nuts on account of difficulty of disengagement. The Acme threads, though not as efficient as square threads are easier to cut or mill, are stronger than square threads and permit the use of split nut which can be used to take up the wear.*

SCREW FRICTION

Example 6.15. *A screw jack carries a load of 4 kN. It has a square threaded single start screw of 20 mm pitch and 50 mm mean diameter. The co-efficient of friction between the screw and its nut is 0.29. Calculate the torque required to raise the load and efficiency of the screw. What is the torque required to lower the load ?*

Sol. Load, $W = 4$ kN

Mean diameter, $d_m = 50$ mm $= 5$ cm

Pitch, $p = 20$ mm $= 2$ cm

Co-efficient of friction, $\mu = 0.29$

Torque required to raise the load,

$$T_{up} = W \tan(\alpha + \phi) \cdot \frac{d_m}{2}$$

But, $$\tan\alpha = \frac{p}{\pi d_m} = \frac{2}{\pi \times 5} = 0.1273$$

$\therefore$ $$\alpha = 7.25° \text{ or } 7° \, 15'$$

and $$\tan\phi = \mu = 0.29$$

$\therefore$ $$\phi = 16.17° \text{ or } 16° \, 10'$$

$\therefore$ $$T_{up} = 4 \tan(7.25 + 16.17) \times \frac{5}{2}$$

$$= \mathbf{4.3 \text{ kN cm.}} \quad \textbf{(Ans.)}$$

Efficiency, $$\eta = \frac{\tan\alpha}{\tan(\alpha + \phi)}$$

$$= \frac{\tan 7.25°}{\tan(7.25° + 16.17°)} = \frac{0.1273}{0.4331}$$

$$= \mathbf{0.294 \text{ or } 29.4\%.} \quad \textbf{(Ans.)}$$

Torque required to lower the load,

$$T_{lower} = W \tan(\phi - \alpha) \cdot \frac{d_m}{2}$$

$$= 4 \tan(16° \, 10' - 7° \, 15') \times \frac{5}{2}$$

$$= \mathbf{1.569 \text{ kN cm.}} \quad \textbf{(Ans.)}$$

Example 6.16. *A screw jack square threads has a mean diameter of 62.5 mm. The pitch of the threads is 12.5 mm. The load on the jack revolves with the screw. Co-efficient of friction at the screw thread is 0.05.*

(a) Find the tangential force which is applied at it at 300 mm radius required to lift a load of 540 N.

(b) State whether the jack is self locking. If it is; find the torque necessary to lower the load. If it is not find the torque which must be applied to keep the load from descending.

Sol. Mean diameter of screw thread,

$$d_m = 62.5 \text{ mm}$$

Pitch of the screw thread,

$$p = 12.5 \text{ mm}$$

Co-efficient of friction,

$$\mu = 0.05$$

$\therefore$ $\mu = \tan \phi = 0.05$

or $\phi = 2.86°$ or $2° 52'$

Length of lever, $l = 300$ mm.

(a) Tangential force = ?

Let P = tangential force applied at the mean radius of the screw.

P' = tangential force applied at a radius of 300 mm (lever)

α = helix angle.

We know that

$$\tan \alpha = \frac{p}{\pi d_m} = \frac{12.5}{\pi \times 62.5} = 0.064$$

$\therefore$ $\alpha = 3.66°$ or $3° 40'$

Using the relation,

$$P = W \tan (\alpha + \phi)$$
$$= 540 \tan (3.66° + 2.86)$$
$$= 61.7 \text{ N}$$

Now effort applied at the end of 300 mm lever may be found out by using the relation,

$$P' \times l = P \times \frac{d_m}{2}$$

$$P' \times 300 = 61.7 \times \frac{62.5}{2}$$

$\therefore$ $\boldsymbol{P' = 6.43}$ **N. (Ans.)**

(b) Is the screw self locking ?

First of all let us find efficiency of the screw jack.

Using the relation,

$$\eta = \frac{\tan \alpha}{\tan (\alpha + \phi)} = \frac{0.064}{\tan (3.66 + 2.86)}$$
$$= 0.56 \text{ or } 56\%.$$

Since, the efficiency of the screw jack is more than 50% therefore, it is **not self locking. Ans.**

Torque which must be applied to keep the load from descending ?

Let P_1 = tangential force required at the mean radius of the screw to keep the load from descending (*i.e.*, to prevent the load from descending).

Now, using the relation,

$$P_1 = W \tan (\alpha - \phi)$$
$$= 540 \tan (3.66 - 2.86)$$
$$= 7.54 \text{ N}$$

$\therefore$ Torque to be applied,

$$T = P_1 \times \frac{d_m}{2} = 7.54 \times \frac{62.5}{2} = 235.6 \text{ N mm}$$

Hence, $\boldsymbol{T = 0.2356}$ **Nm. (Ans.)**

Example 6.17. *A turn buckle is used to tighten a wire rope. The threads are right and left handed and square in section. The pitch is 10 mm and the mean diameter of the screw is 40 mm. Assuming that the co-efficient of friction between the screw and nut is 0.15, determine the turning moment necessary (i) to tighten, (ii) to slacken the wires when the pull is 1500 N. The wire rope is to be assumed not to twist.*

Sol. Assuming the threads to be single start,

$$\tan \alpha = \frac{p}{\pi d_m} = \frac{10}{\pi \times 40} = 0.0796$$

$\therefore$ $\alpha = 4.55°$ or $4° 33'$

Also, $\mu = \tan \phi = 0.15$

$\therefore$ $\phi = 8.53°$ or $8° 32'$

(*i*) *Now tightening the wires is just to move the axial load up the inclined plane.*

$\therefore$ Torque required for tightening,

$$T_{tight} = W \tan (\alpha + \phi) . \frac{d_m}{2}$$

$$= 1500 \tan (4.55° + 8.53°) \times \frac{40}{2}$$

$= 6970$ Ncm or **6.97 Nm. (Ans.)**

(*ii*) *Slackening the wire is as moving the load down the plane.*

$\therefore$ **Torque for slackening**

$$T_{slack} = W \tan (\phi - \alpha) . \frac{d_m}{2}$$

$$= 1500 \times \tan (8.53° - 4.55°) \times \frac{40}{2}$$

$= 2087$ Nmm or **2.087 Nm. (Ans.)**

SELECTED QUESTIONS FROM EXAMINATIONS

Example 6.18. *A system consisting of two blocks connected by a cable is as shown in Fig. 6.36. The masses of the block A and block B are 7.5 kg and 25 kg respectively. Determine the magnitude of minimum force and its inclination with reference to the horizontal, to be applied on block B. The block have impending motion towards the right. Take the coefficient of friction at all contact surfaces to be 0.28.*

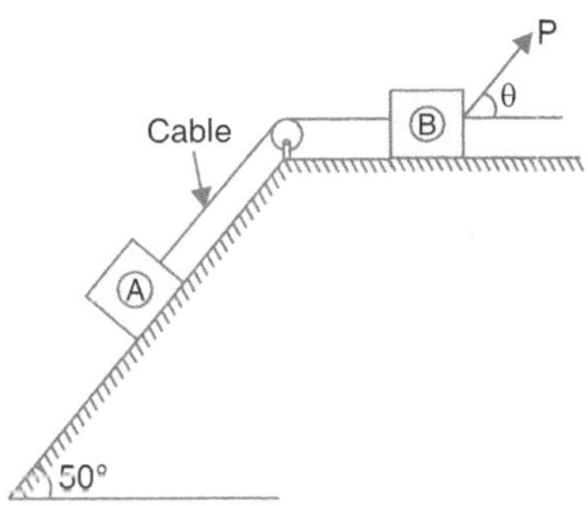

Fig. 6.36

Sol. Refer to Fig. 6.37

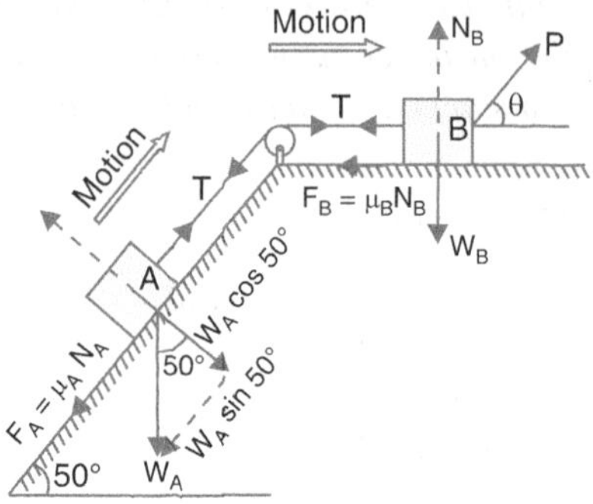

Fig. 6.37

Block A : The cable is subjected to tension T due to the forces acting on the blocks. The friction force acts down the plane on block A, since it has impending motion up the plane.

Resolving forces *parallel to the plane,* we get

$$T = W_A \sin 50° + \mu_A N_A$$

or $$T = (7.5 \times 9.81) \sin 50° + 0.28 N_A \quad ...(1)$$

Resolving forces *perpendicular to the plane,* we have

$$N_A = W_A \cos 50° = (7.5 \times 9.81) \cos 50° = 47.3 \text{ N} \quad ...(2)$$

Substituting the value of N_A from eqn. (2) into eqn (1), we get

$$T = (7.5 \times 9.81) \sin 50° + 0.28 \times 47.3 = 69.6 \text{ N}$$

Block B : The friction force on block B acts in the leftward direction, since the impending motion of block B is towards the right as shown in Fig. 6.37.

Resolving forces *parallel to the plane,* we get

$$T + \mu_B N_B = P \cos \theta$$

or $$T + 0.28 N_B = P \cos \theta \quad ...(3)$$

Resolving forces *perpendicular to the plane,* we get

$$N_B + P \sin \theta = W_B = 25 \times 9.81 \quad ...(4)$$

Substituting the value of N_B from eqn. (4) and $T = 69.6$ N into eqn. (3), we get

$$69.6 + 0.28 (25 \times 9.81 - P \sin \theta) = P \cos \theta$$

or $$P \cos \theta + 0.28 P \sin \theta = 138.27 \quad ...(5)$$

Differentiating w.r.t. θ and equating to zero, we have

$$-P \sin \theta + 0.28 P \cos \theta = 0$$

or $$\tan \theta = 0.28 \text{ } i.e., \theta = \tan^{-1} (0.28)$$

$$= \mathbf{15.64°.} \text{ } \mathbf{(Ans.)}$$

Substituting the value of θ in eqn. (5), we get

$$P \cos 15.64° + 0.28 P \sin 15.64° = 138.27$$

$$P (\cos 15.64° + 0.28 \sin 15.64°) = 138.27$$

$$\therefore \quad P = 133.15 \text{ N}$$

Hence the *magnitude of the minimum force* = **133.15 N. (Ans.)**

Example 6.19. *A block A weighing W newtons is placed on a rough inclined plane having $\mu = 0.2$ and is held in position by means of a horizontal rod hinged to the block B which presses against a rough vertical wall having $\mu = 0.4$ as shown in Fig. 6.38. If the block B weighs 500 N, find the minimum value of W consistent with equilibrium.*

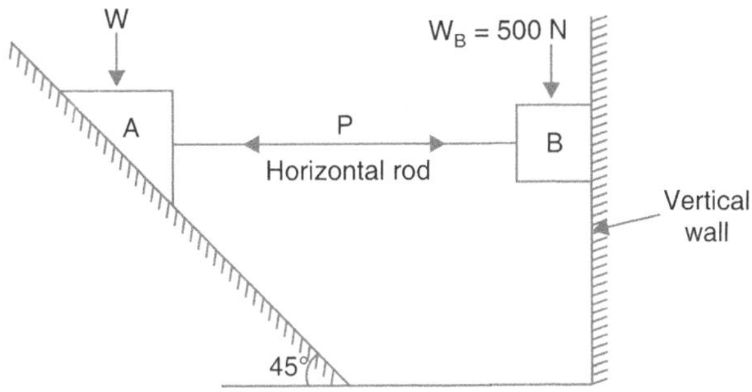

Fig. 6.38

Sol. Weight of block B, $W_B = 500$ N

$$\mu_A = 0.2$$

or $$\tan \phi_A \ (= \mu_A) = 0.2$$

or $$\phi_A = 11.3°$$

$$\mu_B = 0.4$$

or $$\tan \phi_B \ (= \mu_B) = 0.4$$

or $$\phi_B = 21.8°$$

Minimum value of weight of block A, W : We are to find out minimum value of W so that *block B* shall *just slide down.* Refer to Fig. 6.39. When the block B just slides down, the block B is in equilibrium under the action of the following:

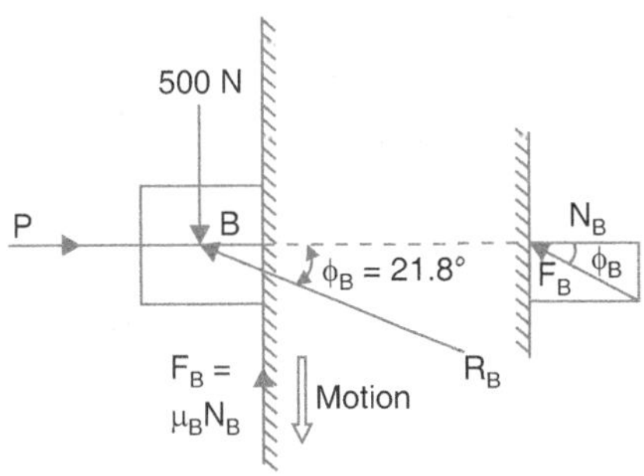

Fig. 6.39

(*i*) P—horizontal force (compressive) in the rod.

(*ii*) R_B—reaction of the vertical wall (resultant of N_B and F_B).

(*iii*) 500 N—weight of the block B.

Using *Lami's theorem,* we get

$$\frac{500}{\sin(180° - 21.8°)} = \frac{P}{\sin(90° + 21.8°)}$$

$$\frac{500}{\sin 158.2°} = \frac{P}{\sin 111.8°}$$

$$P = \frac{500 \times \sin 111.8°}{\sin 158.2°} = 1250 \text{ N}$$

Consider equilibrium of block A :

Fig. 6.40 shows the various forces under the action of which the block A is in equilibrium.

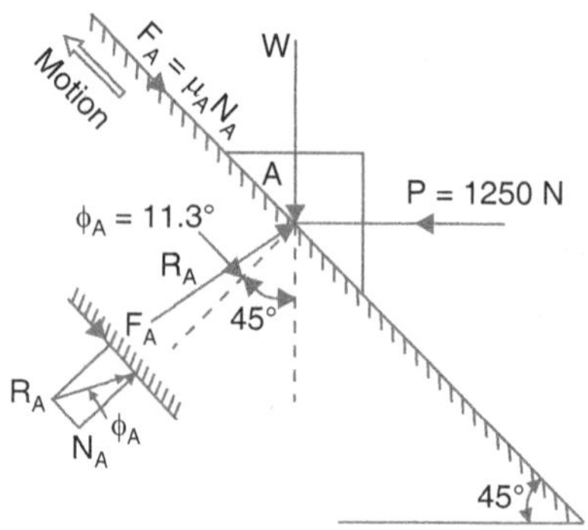

Fig. 6.40

Using *Lami's theorem,* we get

$$\frac{W}{\sin(90° + 45° + 11.3°)} = \frac{P}{\sin[180° - (45° + 11.3°)]}$$

$$= \frac{W}{\sin 146.3°} = \frac{1250}{\sin 123.7°}$$

$$\therefore \qquad W = 1250 \times \frac{\sin 146.3°}{\sin 123.7°} = 833.6 \text{ N}$$

Hence *minimum value of*

W = 833.6 N. (Ans.)

Example 6.20. *Find the maximum tension in the cord shown in Fig. 6.41 if the bodies have developed full friction.*

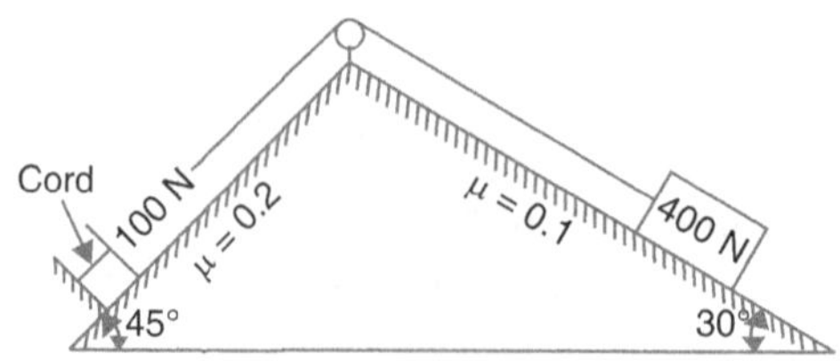

Fig. 6.41

Sol. Fig. 6.41 shows the two blocks connected through a cord.

Maximum tension in the cord, T :

Refer to Fig. 6.42

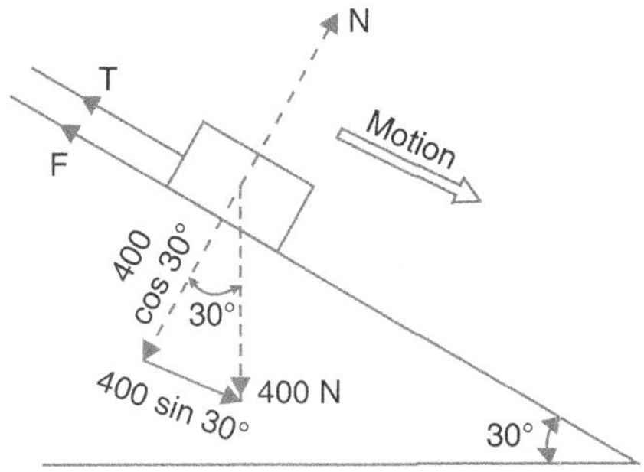

Fig. 6.42

Consider the equilibrium of the block resting on the plane of inclination 30°.

Resolving the forces *parallel* to the inclined plane

$$T + F = 400 \sin 30°$$

or $$T + \mu N = 200 \quad ...(i)$$

and resolving the forces *perpendicular* to the inclined plane

$$N = 400 \cos 30° = 346.4 \text{ N} \quad ...(ii)$$

Substituting this value of N in eqn. (i), we get

$$T + 0.1 \times 346.4 = 200$$

$$\therefore \quad T = 165.36 \text{ N}$$

Refer to Fig. 6.43.

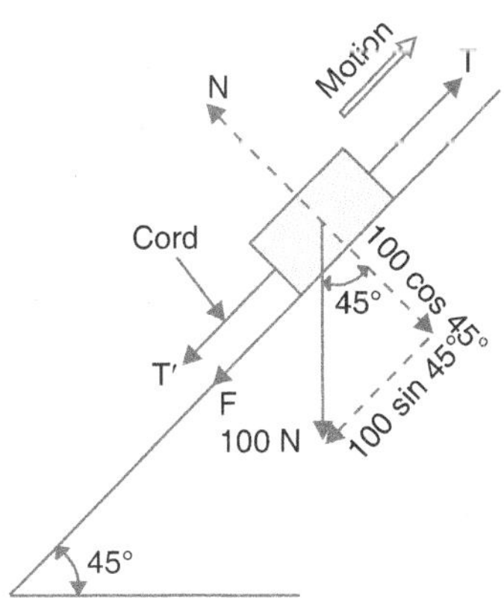

Fig. 6.43

Consider the equilibrium of the block resting on the plane of inclination 45°.

Let T' = tension in the cord.

Resolving the forces *parallel* to the inclined plane,

$$T = T' + F = 100 \sin 45°$$

or $$165.36 = T' + \mu N + 100 \sin 45°$$

or $$165.36 = T' + 0.2\, N + 100 \sin 45° \quad ...(i)$$

and resolving the forces *perpendicular* to the inclined plane,

$$N = 100 \cos 45° \quad ...(ii)$$

From (*i*) and (*ii*), we get

$$165.36 = T' + 0.2 \times 100 \cos 45° + 100 \sin 45°$$

$$T' = 165.36 - 0.2 \times 100 \cos 45° - 100 \sin 45°$$

or **T′ = 80.52 N. (Ans.)**

HIGHLIGHTS

1. *Frictional force* is the opposing force which is called into play in between the surfaces of contact of two bodies, when one body moves over the surface of another body.
2. Frictional force is self-adjusting, always acts in a direction opposite to the motion and a passive force.
3. The friction that exists between the two unlubricated surfaces is called *solid friction* or *dry friction*.
4. The friction that exists when one surface slides over the other is called *sliding friction*.
5. The friction that exists between two surfaces separated by *balls* or *rollers,* is called the rolling friction. Rolling friction is *always less* than the sliding friction.
6. The *static friction* is the friction offered by the surfaces subjected to external forces until there is no motion between them.
7. The *dynamic friction* is the friction experienced by a body when it is in motion. It is also known as kinetic friction and is *always less* than static friction.
8. *Limiting force of friction* may be defined as the maximum value of frictional force which exists when a body just begins to slide over the surface of the other body.
9. *Angle of friction* is the angle which the resultant (R) of normal reaction (N) and limiting force of friction (F) makes with the normal (N). It is denoted by ϕ.

$$\tan \phi = \frac{F}{N}, \quad \therefore \quad \phi = \tan^{-1} \frac{F}{N}.$$

10. Co-efficient of friction is defined as the ratio of limiting force of friction to the normal reaction between the two bodies. It is denoted by μ.

$$\mu = \tan \phi = \frac{F}{N}$$

$$\therefore \quad F = \mu N.$$

11. *Angle of repose is equal to angle of friction* when the body is in the condition of limiting equilibrium on an inclined plane.
12. Efficiency of screw jack, $\eta = \dfrac{\tan \alpha}{\tan (\alpha + \phi)}$.

Conditions for maximum efficiency,

$$\alpha = \frac{\pi}{4} - \frac{\phi}{2}$$

Maximum efficiency,

$$\eta_{max} = \frac{1 - \sin \phi}{1 + \sin \phi}.$$

OBJECTIVE TYPE QUESTIONS

Fill in the blanks

(*i*) ......is the opposing force which is called into play in between the surfaces of contact of two bodies, when one body moves over the surface of another body.

(*ii*) Friction is both......and......in engineering applications.

(*iii*) Frictional force is a......force.

(*iv*) Frictional force is self...... .

(*v*) ......friction is the friction that exists between two unlubricated surfaces.

(*vi*) The friction that exists when one surface slides over the other is called......friction.

(*vii*) Rolling friction is always......than the sliding friction.

(*viii*) The friction that exists between two surfaces separated by balls or rollers, is called......friction.

(*ix*) ......friction is the friction offered by the surfaces subjected to external forces until there is no motion between them.

(*x*) ......friction is the friction experienced by a body when it is in motion.

(*xi*) Dynamic friction is always......than static friction.

(*xii*) The maximum value of frictional force which exists when a body just begins to slide over the surface of the other body is known as......

(*xiii*) The frictional force is ......proportional to the normal reaction.

(*xiv*) Angle of repose is equal to angle of......when the body is in the condition of limiting equilibrium on an inclined plane.

(*xv*) the angle which the resultant (R) of normal reaction (N) and limiting force of friction (F) makes with the normal (N) is known as......

Answers

(*i*) Frictional force (*ii*) desirable and undesirable

(*iii*) passive (*iv*) adjusting (*v*) Solid or dry friction

(*vi*) sliding (*vii*) less (*viii*) rolling

(*ix*) Static (*x*) Dynamic (*xi*) less

(*xii*) limiting force of friction (*xiii*) directly (*xiv*) friction

(*xv*) angle of friction.

EXERCISES

Theoretical Questions

1. Define 'frictional force'. Explain how is friction both desirable and undesirable in engineering applications ?

2. State the characteristics of frictional force.

3. Write short note on :

(*i*) Friction in unlubricated surfaces

(*ii*) Friction in lubricated surfaces.

4. Differentiate between static and dynamic friction.

5. Explain the term 'Limiting friction'.

6. State the laws of static and dynamic friction.

7. Define the following :

(*i*) Angle of friction

(*ii*) Co-efficient of friction

(*iii*) Angle of repose

(*iv*) Cone of friction.

Unsolved Examples

1. A body which weights 1000 N rests on a horizontal plane, the co-efficient of friction between the body and the plane being 0.1. Find the force, which acting at 30° to the horizontal will just move the body. **[Ans.** 109.2 N]
2. A body resting on a rough horizontal plane required a pull of 82 N inclined at 30° to the plane just to move it. It was found that a push of 100 N inclined at 30° to the plane just moved the body. Determine the weight of the body and the co-efficient of friction. **[Ans.** 455 N, $\mu = 0.1714$**]**
3. Two equal bodies *A* and *B* of weight *W* each are placed on a rough inclined plane. The bodies are connected by a light string. If $\mu_A = 1/2$ and $\mu_B = 1/3$, show that bodies will be on the point of motion when the plane is inclined at $\tan^{-1}\left(\frac{5}{12}\right)$.
4. A 100 N mass is dragged by a rope up 10° slope, the rope being paralleled to the line of the sleepest slope. The co-efficient of friction is $\frac{1}{2}$. Calculate the tension in the rope. **[Ans.** 66.9 N]
5. A body is in equilibrium on an inclined plane. Show clearly by means of a diagram, the forces acting on the body, describing the forces *e.g.*, weight etc. The inclination of the plane is increased and it is found that uniform motion, once started, continues when the plane is inclined at 15° to the horizontal. What is the co-efficient of friction between the body and the plane ? What force parallel to the plane, will be required to keep the body, whose weight is 100 N at rest when the plane is inclined at 30° to the horizontal? **[Ans.** 0.268, 26.8 N]
6. The sliding face of a steam engine slide valve measures 25 cm by 40 cm and the steam pressure on the back of the valve is 50 N/cm^2. If the co-efficient of friction is 0.12, calculate the force required to move the valve. **[Ans.** 6 kN]
7. A uniform ladder rests with one end on a horizontal floor and the other end against a wall, the co-efficients of friction for the floor and wall are 4/9 and 1/4 respectively. Find the inclination of the ladder when it is about to slip. **[Ans.** 45°]
8. A ladder 5 m long and of 250 N weight is placed against a vertical wall in a position where its inclination to the vertical is 30°. A man weighing 800 N climbs the ladder. At what position will be induce slipping, the co-efficient of friction for both the contact surfaces of the ladder *viz* with the wall and floor is 0.2. **[Ans.** 1.79 m]
9. A uniform ladder of weight *P* kgf is inclined at 45° to the horizontal. If μ and μ' are the co-efficients of friction at horizontal and vertical surfaces, then show that the least horizontal for which will move the lower end towards the wall is just greater than
$$\frac{P}{2} \times \frac{1 + 2\mu + \mu\mu'}{1 - \mu'}.$$
10. A uniform ladder 3 m long weighs 200 N. It is placed against a wall making an angle of 60° with the floor. The co-efficient of friction between the wall and the ladder is 0.25 and that between the floor and ladder is 0.5. The ladder in addition to its own weight has to support a man weighing 1 kN its top. Calculate.
    (*i*) The horizontal force *P* to be applied to the ladder at the floor level to prevent slipping.
    (*ii*) If the force *P* is not applied, what should be the maximum inclination of the ladder with the horizontal so that there is no slipping of it with the man at top. **[Ans.** 183.6 N, 68° 57′]
11. A block overlying at 10° wedge on a horizontal floor and leaning against a vertical wall and weighing 1500 N is to be raised by applying a horizontal force to the wedge. Assuming the co-efficient of friction between all the surfaces in contact to be 0.3, determine the minimum horizontal force, to be applied to raise the block. **[Ans.** 1420 N]

12. A block weighing 10 kN is to be raised by means of a 20° wedge. Determine what minimum horizontal force *P*, should be applied to raise the block, the angles of friction at the contact surfaces *AC*, *AB* and *DE* are 11°, 14° and 10° respectively. (See Fig. 6.44) [**Ans.** 10.77 kN]

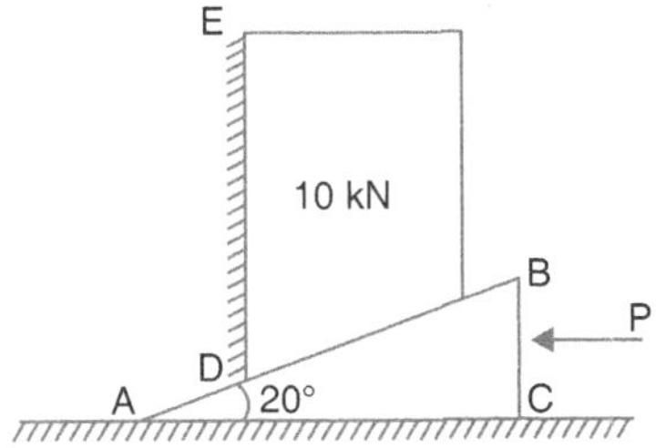

Fig. 6.44

13. Outside diameter of a square threaded spindle of a screw jack is 4 cm. The screw pitch is 1 cm. If the co-efficient of friction between the screw and nut is 0.15, neglecting friction between the nut and the collar, determine : (*i*) Force required to be applied at the screw to raise a load of 20 kN, (*ii*) Efficiency of screw jack, (*iii*) Force required to be applied at pitch radius to lower the same load of 20 kN and (*iv*) Efficiency while lowering the load, (*v*) What should be the pitch for maximum efficiency of the screw and what should be the value of the maximum efficiency ?

[**Ans.** 4.886 kN, 37.25%, 11.384 kN, 62.55%, 9.47 mm, 74.15%]

14. A screw jack has mean diameter of 80 mm and pitch 15 mm. The co-efficient of friction between its screw and nut is 0.475. Find the effort required to be normally applied at the end of its operating lever 800 mm long (*i*) to raise a load of 2 kN and (*ii*) to lower the same load. Find the efficiency under these loads.

[**Ans.** 73.52 N, 1.52 N, 44.2%]

7

Rectilinear Motion

7.1. CONCEPT OF MOTION

A body is said to be in motion if it changes its position with respect to its surroundings. The nature of path of displacement of various particles of a body determines the type of motion. The motion may be of the following types :

1. Rectilinear translation
2. Curvilinear translation
3. Rotary or circular motion.

Rectilinear translation is also known as *straight line motion*. Here particles of a body move in straight parallel paths. *Rectilinear means forming straight lines and translation means behaviour.* Rectilinear translation will mean behaviour by which straight lines are formed. Thus, when a body moves such that its particles form parallel straight paths the body is said to have rectilinear translation.

In a *curvilinear translation the particles of* a body move *along circular arcs or curved paths.*

Rotary or *circular motion* is a special case of curvilinear motion where particles of a body move along *concentric circles* and the *displacement* is *measured in terms of angle in radians or revolutions.*

7.2. DEFINITIONS

1. Displacement. If a particle has rectilinear motion with respect to some point which is assumed to be fixed, its displacement is its total change of position during any interval of time. The point of reference usually assumed is one which is at rest with respect to the surfaces of the earth.

The unit of displacement is same as that of distance or length. In M.K.S. or S.I. system it is *one metre*.

2. Rest and motion. A body is said to be at *rest* at an instant (means a small interval of time) if its position with respect to the surrounding objects remains unchanged during that instant.

A body is said to be in *motion* at an instant if it changes its position with respect to its surrounding objects during that instant.

Actually, nothing is absolutely at rest or absolutely in motion : *all rest or all motion is relative only*.

3. Speed. The speed of body is defined as *its rate of change of its position with respect to its surroundings irrespective of direction*. It is a *scalar quantity*. It is measured by distance covered per unit time.

Mathematically, speed

$$= \frac{\text{Distance covered}}{\text{Time taken}} = \frac{S}{t}$$

Its units are m/sec or km/hour.

4. Velocity. The velocity of a body is *its rate of change of its position with respect to its surroundings in a particular direction.* It is a *vector quantity.* It is measured by the distance covered in a *particular direction* per unit time.

i.e., $$\text{Velocity} = \frac{\text{Distance covered (in a particular direction)}}{\text{Time taken}}$$

$$v = \frac{S}{t}.$$

Its units are same as that of speed *i.e.,* m/sec or km/hour.

5. Uniform velocity. *If a body travels equal distances in equal intervals of time in the same direction it is said to be moving with a uniform or constant velocity.* If a car moves 50 metres with a constant velocity in 5 seconds, its velocity will be equal to,

$$\frac{50}{5} = 10 \text{ m/s.}$$

6. Variable velocity. *If a body travels unequal distances in equal intervals of time, in the same direction, then it is said to be moving with a variable velocity* or if it is changes either its speed or its direction or both shall again be said to be moving with a variable velocity.

7. Average velocity. *The average or mean velocity of a body is the velocity with which the distance travelled by the body in the same interval of time, is the same as that with the variable velocity.*

If u = initial velocity of the body

v = final velocity of the body

t = time taken

S = distance covered by the body

Then average velocity $$= \frac{u + v}{2} \quad \text{...(7.1)}$$

and $$S = \left(\frac{u + v}{2}\right) \times t \quad \text{...(7.2)}$$

8. Acceleration. The *rate of change of velocity of a body is called its acceleration.* When the velocity is increasing the acceleration is reckoned as *positive*, when decreasing as *negative.* It is represented by a or f.

If u = initial velocity of a body in m/sec

v = final velocity of the body in m/sec

t = time interval in seconds, during which the change has occurred,

Then acceleration, $$a = \frac{v - u}{t} \frac{\text{m/sec}}{\text{sec}}$$

or $$a = \frac{v - u}{t} \text{ m/sec}^2 \quad \text{...(7.3)}$$

From above, it is obvious that if *velocity of the body remains constant, its acceleration will be zero.*

9. Uniform acceleration. *If the velocity of a body changes by equal amounts in equal intervals of time, the body is said to move with uniform acceleration.*

10. Variable acceleration. *If the velocity of a body changes by unequal amount in equal intervals of time, the body is said to move with variable acceleration.*

7.3. DISPLACEMENT-TIME GRAPHS

Refer to Fig. 7.1 (*a*). The graph is parallel to the time-axis indicating that the *displacement is not changing with time*. The *slope of the graph is zero*. The body has no velocity and is at rest.

Refer to Fig. 7.1 (*b*). The displacement increases linearly with time. The displacement increases by equal amounts in equal intervals of time. *The slope of the graph is constant*. In other words, the body is moving with a *uniform velocity*.

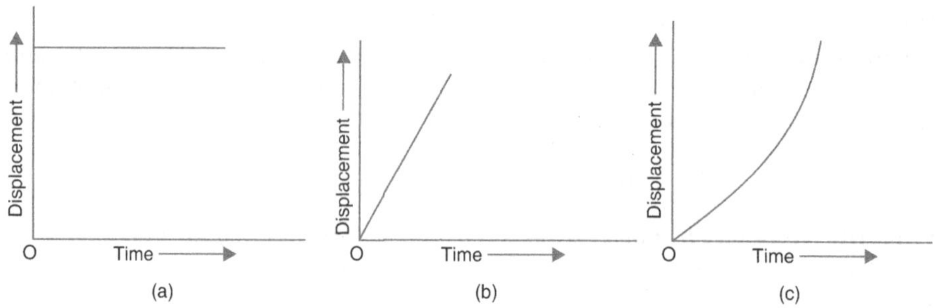

Fig. 7.1 Displacement-time graphs

Refer to Fig. 7.1 (*c*). The displacement time graph is a *curve*. This means that the displacement is not changing by equal amounts in equal intervals of time. The slope of the graph is different at different times. In other words, the velocity of the body is changing with time. The motion of the body is accelerated.

7.4. VELOCITY-TIME GRAPHS

Refer to Fig. 7.2 (*a*). The velocity of the body increases linearly with time. The slope of the graph is constant, *i.e.*, velocity changes by equal amounts in equal intervals of time. In other words, the *acceleration of the body is constant*. Also, at time $t = 0$, the velocity is finite. Thus, the body, *moving with a finite initial velocity, is having a constant acceleration*.

Refer to Fig.7.2 (*b*). The body has a finite initial velocity. As the time passes, the velocity decreases linearly with time until its final velocity becomes zero, *i.e.* it comes to rest. Thus, the body has a *constant deceleration* (or retardation) since the *slope of the graph is negative*.

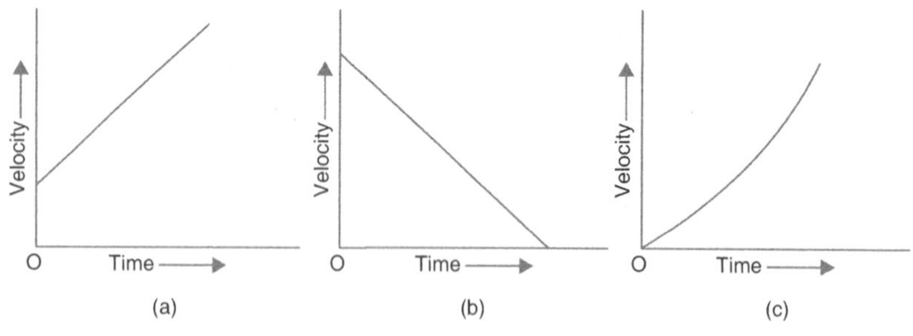

Fig. 7.2 Velocity-time graphs

Refer to Fig. 7.2 (*c*). The velocity-time graph is a *curve*. The slope is therefore, different at different times. In other words, the *velocity is not changing at a constant rate. The body does not have a uniform acceleration since the acceleration is changing with time.*

7.5. EQUATIONS OF MOTION UNDER UNIFORM ACCELERATION

7.5.1. First Equation of Motion. *Relation between u, v, a and t.*

Let us assume that a body starts with an initial velocity u and acceleration a. After time t, it attains a velocity v. Therefore, the change in velocity in t seconds $= v - u$. Hence, the change in velocity in one second $= \frac{v-u}{t}$. By definition, this is equal to the acceleration a.

Thus, $$a = \frac{v-u}{t}$$

or $$at = v - u$$

or $$v = u + at \qquad \text{...(7.4)}$$

7.5.2. Second Equation of Motion. *Relation between S, u, a and t.*

Let a body moving with an initial uniform velocity u is accelerated with a uniform acceleration a for time t. After time t its final velocity is v. The distance S which the body travels in time t is determined as follows :

Now, since the acceleration is uniform, *i.e.*, the velocity changes by an equal amount in equal intervals of time, it is obvious that the average velocity is just the average of initial and final velocities.

$$\text{Average velocity} = \left(\frac{u+v}{2}\right)$$

$\therefore$ Distance travelled = average velocity × time

$$S = \left(\frac{u+v}{2}\right) \times t$$

or $$S = \left(\frac{u+u+at}{2}\right) \times t \qquad (\because \ v = u + at)$$

$$= \left(u + \frac{at}{2}\right) \times t$$

or $$S = ut + \frac{1}{2}at^2 \qquad \text{...(7.5)}$$

7.5.3. Third Equation of Motion. *Relation u, v, a and S.*

We know, that

$$S = \text{average velocity} \times \text{time}$$

$$= \left(\frac{u+v}{2}\right) \times t$$

$$= \left(\frac{u+v}{2}\right) \times \left(\frac{v-u}{a}\right) \qquad \left(\because \ t = \frac{v-u}{a}\right)$$

$$= \frac{v^2 - u^2}{2a}$$

$\therefore$ $$v^2 - u^2 = 2aS \qquad \text{...(7.6)}$$

7.6. DISTANCE COVERED IN nth SECOND BY A BODY MOVING WITH UNIFORM ACCELERATION

Let u = initial velocity of the body

a = acceleration

S_{nth} = distance covered in nth second

then
$$S_{nth} = \begin{pmatrix}\text{distance covered} \\ \text{in } n \text{ second, } s_n\end{pmatrix} - \begin{pmatrix}\text{distance covered in } (n-1) \\ \text{second, } s_{n-1}\end{pmatrix}$$

Using the relation,

$$S_n = un + \frac{1}{2} an^2 \qquad (\because \ t = n)$$

and

$$S_{n-1} = u(n-1) + \frac{1}{2} a (n-1)^2$$

$$= u(n-1) + \frac{1}{2} a (n^2 - 2n + 1)$$

$\therefore$

$$S_{nth} = S_n - S_{n-1}$$

$$= \left(un + \frac{1}{2} an^2\right) - \left[u(n-1) + \frac{1}{2} a (n^2 - 2n + 1)\right]$$

$$= un + \frac{1}{2} an^2 - un + u - \frac{1}{2} an^2 + an - a/2$$

$$= u + an - a/2$$

$\therefore$

$$S_{nth} = u + a/2(2n - 1) \qquad \text{...(7.7)}$$

Example 7.1. *A car accelerates from a velocity of 36 km/hour to a velocity of 108 km/hour in a distance of 240 m. Calculate the average acceleration and time required.*

Sol. Initial velocity,

$$u = 36 \text{ km/hour}$$

$$= \frac{36 \times 1000}{60 \times 60} = 10 \text{ m/sec}$$

Final velocity, v = 108 km/hour

$$= \frac{108 \times 1000}{60 \times 60} = 30 \text{ m/sec}$$

Distance, S = 240 m.

Average acceleration, a = ?

Using the relation,

$$v^2 - u^2 = 2aS$$

$$(30)^2 - (10)^2 = 2 \times a \times 240$$

or

$$900 - 100 = 480\, a$$

or

$$a = \frac{800}{480} = \mathbf{1.67\ m/sec^2.\ (Ans.)}$$

Time required, t = ?

$$v = u + at$$

$$30 = 10 + 1.67 \times t$$

$\therefore$

$$t = \frac{(30 - 10)}{1.67} = \mathbf{11.97\ sec.\ (Ans.)}$$

Example 7.2. *A body has an initial velocity of 16 m/sec and an acceleration of 6 m/sec^2. Determine its speed after it has moved 120 metres distance. Also calculate the distance the body moves during 10th second.*

Sol. Initial velocity, $u = 16$ m/sec

Acceleration, $a = 6$ m/sec^2

Distance, $S = 120$ metres

Speed, $v = ?$

Using the relation,

$$v^2 - u^2 = 2aS$$

$$v^2 - (16)^2 = 2 \times 6 \times 120$$

or

$$v^2 = (16)^2 + 2 \times 6 \times 120$$

$$= 256 + 1440 = 1696$$

$$\mathbf{v = 41.18\ m/sec.\ (Ans.)}$$

Distance travelled in 10th sec ; $S_{10th} = ?$

Using the relation,

$$S_{nth} = u + \frac{a}{2}(2n - 1)$$

$$S_{10th} = 16 + \frac{6}{2}(2 \times 10 - 1) = 16 + 3(20 - 1)$$

$$= \mathbf{73\ m.\ (Ans.)}$$

Example 7.3. *On turning a corner, a motorist rushing at 15 m/sec, finds a child on the road 40 m ahead. He instantly stops the engine and applies brakes, so as to stop the car within 5 m of the child, calculate : (i) retardation, and (ii) time required to stop the car.*

Sol. Initial velocity, $u = 15$ m/sec

Final velocity, $v = 0$

Distance, $S = 40 - 5 = 35$ m.

(*i*) **Retardation,** $a = ?$

Using the relation,

$$v^2 - u^2 = 2aS$$

$$0^2 - 15^2 = 2 \times a \times 35$$

$$\therefore \quad \mathbf{a = -3.21\ m/sec^2.\ (Ans.)}$$

[– ve sign indicates that the acceleration is negative, *i.e.*, retardation]

(*ii*) **Time required to stop the car, $t = ?$**

Using the relation,

$$v = u + at$$

$$0 = 15 - 3.21 \times t \qquad (\because\ a = -3.21\ \text{m/sec}^2)$$

$$\therefore \quad t = \frac{15}{3.21} = \mathbf{4.67\ s.\ (Ans.)}$$

Example 7.4. *A burglar's car had a start with an acceleration 2 m/sec^2. A police vigilant party came after 5 seconds and continued to chase the burglar's car with a uniform velocity of 20 m/sec. Find the time taken, in which the police will overtake the car.*

Sol. Let the police party overtake the burglar's car in t seconds, after the instant of reaching the spot.

Distance travelled by the burglar's car in t seconds, S_1 :

Initial velocity, $u = 0$

Acceleration, $a = 2 \text{ m/sec}^2$

Time, $t = (5 + t)$ sec.

Using the relation,

$$S = ut + \frac{1}{2} at^2$$

$$S_1 = 0 + \frac{1}{2} \times 2 \times (5 + t)^2$$

$$= (5 + t)^2 \quad \text{...(i)}$$

Distance travelled by the police party, S_2 :

Uniform velocity, $v = 20$ m/sec.

Let t = time taken to overtake the burglar's car

∴ Distance travelled by the party,

$$S_2 = v \times t = 20t \quad \text{...(ii)}$$

For the police party to overtake the burglar's car, the two distances S_1 and S_2 should be equal.

i.e.,

$$S_1 = S_2$$

$$(5 + t)^2 = 20t$$

$$25 + t^2 + 10t = 20t$$

$$t^2 - 10t + 25 = 0$$

∴

$$t = \frac{+10 \pm \sqrt{100 - 100}}{2}$$

or $\quad$ **$t = 5$ sec. (Ans.)**

Example 7.5. *A car starts from rest and accelerates uniformly to a speed of 80 km/hour over a distance of 500 metres. Calculate the acceleration and time taken.*

If a further acceleration raises the speed to 96 km/hour in 10 seconds, find the acceleration and further distance moved.

The brakes are now applied and the car comes to rest under uniform retardation in 5 seconds. Find the distance travelled during braking.

Sol. Considering the **first period of motion :**

Initial velocity, $u = 0$

Velocity attained, $v = \frac{80 \times 1000}{60 \times 60} = 22.22$ m/sec.

Distance covered, $S = 500$ m

If a is the acceleration and t is the time taken,

Using the relation :

$$v^2 - u^2 = 2aS$$

$$(22.22)^2 - 0^2 = 2 \times a \times 500$$

$\therefore \quad a = \frac{(22.22)^2}{2 \times 500} = \mathbf{0.494\ m/sec^2}.$ **(Ans.)**

Also, $\quad v = u + at$

$22.22 = 0 + 0.494 \times t$

$\therefore \quad t = \frac{22.22}{0.494} = \mathbf{45\ sec.}$ **(Ans.)**

Now considering the **second period of motion,**

Using the relation,

$$v = u + at$$

where $\quad v = 96 \text{ km/hour} = \frac{96 \times 1000}{60 \times 60} = 26.66 \text{ m/sec}$

$u = 80 \text{ km/hour} = 22.22 \text{ m/sec}$

$t = 10 \text{ sec}$

$\therefore \quad 26.66 = 22.22 + a \times 10$

$\therefore \quad a = \frac{26.66 - 22.22}{10} = \mathbf{0.444\ m/sec^2}.$ **(Ans.)**

To calculate distance covered, using the relation

$$S = ut + \frac{1}{2} at^2$$

$$= 22.22 \times 10 + \frac{1}{2} \times 0.444 \times 10^2$$

$$= 222.2 + 22.2 = 244.4$$

$\therefore \quad \mathbf{S = 244.4\ m.}$ **(Ans.)**

During the period when brakes are applied :

Initial velocity, $\quad u = 96 \text{ km/hour} = 26.66 \text{ m/sec}$

Final velocity, $\quad v = 0$

Time taken, $\quad t = 5 \text{ sec.}$

Using the relation,

$$v = u + at$$

$$0 = 26.66 + a \times 5$$

$\therefore \quad a = \frac{-26.66}{5} = -5.33 \text{ m/sec}^2.$

(–ve sign indicates that acceleration is negative *i.e.*, retardation)

Now using the relation,

$$v^2 - u^2 = 2aS$$

$$0^2 - (26.66)^2 = 2 \times -5.33 \times S$$

$\therefore \quad S = \frac{26.66^2}{2 \times 5.33} = 66.67 \text{ m.}$

$\therefore$ **Distance travelled during braking = 66.67 m. (Ans.)**

Example 7.6. *Two trains A and B moving in opposite directions pass one another. Their lengths are 100 m and 75 m respectively. At the instant when they begin to pass, A is moving at 8.5 m/sec with a constant acceleration of 0.1 m/sec² and B has a uniform speed of 6.5 m/sec. Find the time the trains take to pass.*

Sol. Length of train $A = 100$ m

Length of train $B = 75$ m

$\therefore$ Total distance to be covered

$$= 100 + 75 = 175 \text{ m}$$

Imposing on the two trains A and B, a velocity equal and opposite to that of B.

Velocity of train $A = (8.5 + 6.5) = 15.0$ m/sec

and velocity of train $B = 6.5 - 6.5 = 0.$

Hence the train A has to cover the distance of 175 m with an acceleration of 0.1 m/sec² and an initial velocity of 15.0 m/sec.

Using the relation,

$$S = ut + \frac{1}{2} at^2$$

$$175 = 15t + \frac{1}{2} \times 0.1 \times t^2$$

$$3500 = 300t + t^2$$

or $$t^2 + 300t - 3500 = 0$$

$$t = \frac{-300 \pm \sqrt{90000 + 14000}}{2} = \frac{-300 \pm 322.49}{2}$$

$$= 11.24 \text{ sec.}$$

Hence the trains take 11.24 seconds to pass one another. (Ans.)

Example 7.7. *The distance between two stations is 2.6 km. A locomotive starting from one station, gives the train an acceleration (reaching a speed of 40 km/h in 0.5 minutes) until the speed reaches 48 km/hour. This speed is maintained until brakes are applied and train is brought to rest at the second station under a negative acceleration of 0.9 m/sec². Find the time taken to perform this journey.*

Sol. Considering the motion of the locomotive *starting from the first station*.

Initial velocity $u = 0$

Final velocity $v = 40$ km/hour

$$= \frac{40 \times 1000}{60 \times 60} = 11.11 \text{ m/sec.}$$

Time taken, $t = 0.5$ min or 30 sec.

Let 'a' be the resulting acceleration.

Using the relation,

$$v = u + at$$

$$11.11 = 0 + 30a$$

$\therefore$ $$a = \frac{11.11}{30} = 0.37 \text{ m/sec}^2.$$

Let t_1 = time taken to attain the speed of 48 km/hour

$$\left(\frac{48 \times 1000}{60 \times 60} = 13.33 \text{ m/sec.}\right)$$

Again, using the relation,

$$v = u + at$$

$$13.33 = 0 + 0.37t_1$$

$$\therefore \quad t_1 = \frac{13.33}{0.37} = 36 \text{ sec.} \qquad ...(i)$$

and the distance covered in this interval is given by the relation,

$$S_1 = ut_1 + \frac{1}{2}at_1^2$$

$$= 0 + \frac{1}{2} \times 0.37 \times 36^2 = \mathbf{240\ m.}$$

Now, considering the motion of the *retarding period* before the locomotive comes to rest at the second station (*i.e.*, portion *BC* in Fig. 7.3).

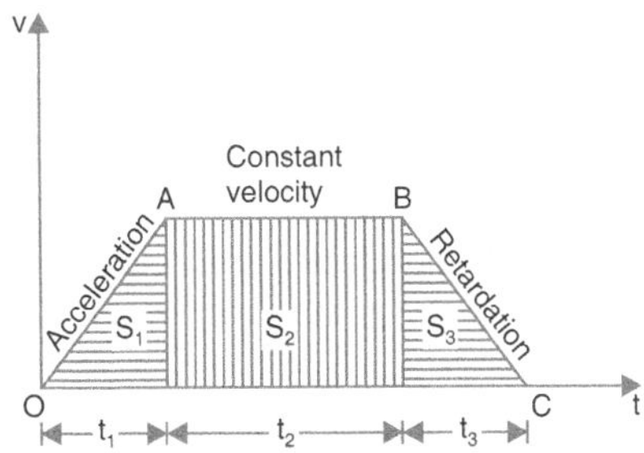

Fig. 7.3

Now, $u = 13.33$ m/sec

$v = 0$

$a = -0.9$ m/sec^2

Let $t = t_3$ be the time taken

Using the relation,

$$v = u + at$$

$$0 = 13.33 - 0.9t_3$$

$$\therefore \quad t_3 = \frac{13.33}{0.9} = 14.81 \text{ sec} \qquad ...(ii)$$

and distance covered,

$$S_3 = \text{average velocity} \times \text{time}$$

$$= \left(\frac{13.33 + 0}{2}\right) \times 14.81 = 98.7 \text{ m}$$

$\therefore$ Distance covered with constant velocity of 13.33 m/sec,

$$S_2 = \text{total distance between two stations} - (S_1 + S_2)$$

$$= (2.6 \times 1000) - (240 + 98.7) = 2261.3 \text{ m.}$$

$\therefore$ Time taken to cover this distance,

$$t_2 = \frac{2261.3}{13.33} = 169.6 \text{ sec} \qquad ...(iii)$$

Adding (*i*), (*ii*) and (*iii*)

Total time taken,

$$T = t_1 + t_2 + t_3$$

$$= 36 + 169.6 + 14.81$$

$$= \mathbf{220.41\ sec.} \quad \textbf{(Ans.)}$$

Example 7.8. *Two trains A and B leave the same station on parallel lines. A starts with a uniform acceleration of 0.15 m/sec^2 and attains a speed of 24 km / hour when the steam is required to keep speed constant. B leaves 40 seconds after with uniform acceleration of 0.30 m/sec^2 to attain a maximum speed of 48 km/hour. When will B overtake A ?*

Sol. Motion of train A :

Uniform acceleration, $a_1 = 0.15\ \text{m/sec}^2$

Initial velocity, $u_1 = 0$

Final velocity, $v_1 = 24\ \text{km/hour}$

$$= \frac{24 \times 1000}{60 \times 60} = \frac{20}{3}\ \text{m/sec.}$$

Let t_1 be the time taken to attain this velocity (in seconds).

Using the relation,

$$v = u + at$$

$$\frac{20}{3} = 0 + 0.15t_1$$

$$\therefore \quad t_1 = \frac{20}{3 \times 0.15} = 44.4\ \text{sec.}$$

Also, distance travelled during this interval,

$$S_1 = ut_1 + \frac{1}{2}at_1^2$$

$$= 0 + \frac{1}{2} \times 0.15 \times 44.4^2$$

$$= 148\ \text{m.}$$

Motion of train B :

Initial velocity, $u_2 = 0$

Acceleration, $a_2 = 0.3\ \text{m/sec}^2$

Final velocity, $v_2 = 48\ \text{km/hr}$

$$= \frac{48 \times 1000}{60 \times 60} = \frac{40}{3}\ \text{m/sec.}$$

Let t_2 be the time taken to travel this distance, say S_2.

Using the relation,

$$v = u + at$$

$$\frac{40}{3} = 0 + 0.3t_2$$

$$\therefore \quad t_2 = \frac{40}{3 \times 0.3} = 44.4\ \text{sec}$$

and

$$S_2 = u_2t_2 + \frac{1}{2}a_2 t_2^2$$

$$= 0 + \frac{1}{2} \times 0.3 \times (44.4)^2$$

$$= 296\ \text{m.}$$

Let the train B overtake the train A when they have covered a distance S from the start. And let the train B take t seconds to cover the distance.

Thus, time taken by the train $A = (t + 40)$ sec.

Total distance moved by train A,

$$S = 148 + \text{distance covered with constant speed}$$

$$S = 148 + [(t + 40) - t_1]\ \frac{20}{3}$$

$$= 148 + [t + 40 - 44.4] \times \frac{20}{3}$$

$$= 148 + (t - 4.4) \times \frac{20}{3} \quad ...(i)$$

[$\{(t + 40) - t_1\}$ is the time during which train A moves with constant speed]

Similarly, total distance travelled by the train B,

$$S = 296 + \text{distance covered with constant speed}$$

$$= 296 + (t - 44.4) \times \frac{40}{3} \quad ...(ii)$$

Equating (i) and (ii),

$$148 + (t - 4.4)\ \frac{20}{3} = 296 + (t - 44.4) \times \frac{40}{3}$$

$$148 + \frac{20}{3}\ t - \frac{88}{3} = 296 + \frac{40}{3} t - \frac{1776}{3}$$

$$\left(\frac{40}{3} - \frac{20}{2}\right) t = 148 - 296 + \frac{1776}{3} - \frac{88}{3}$$

or $$t = 62.26 \text{ sec.}$$

Hence, the train B overtakes the train A after 62.26 sec. of its start. (Ans.)

Example 7.9. *Two stations A and B are 10 km apart in a straight track, and a train starts from A and comes to rest at B. For three quarters of the distance, the train is uniformly accelerated and for the remainder uniformly retarded. If it takes 15 minutes over the whole journey, find its acceleration, its retardation and the maximum speed it attains.*

Sol. Refer to Fig. 7.4.

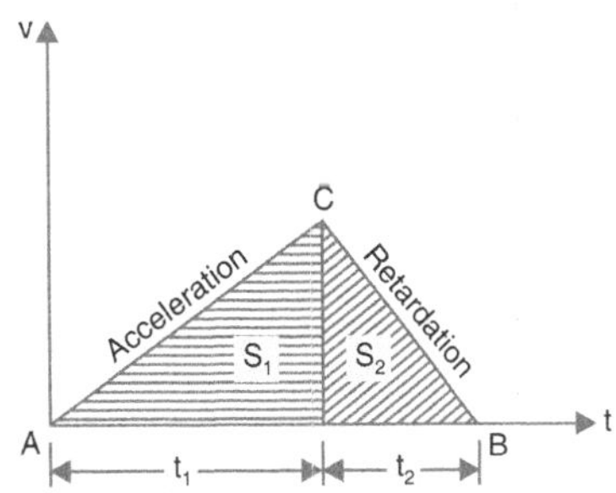

Fig. 7.4

Distance between A and B,

$$S = 10 \text{ km} = 10{,}000 \text{ m}$$

Considering the motion in the first part :

Let u_1 = initial velocity = 0

a_1 = acceleration

t_1 = time taken

S_1 = distance travelled.

Using the relation,

$$S = ut + \frac{1}{2}\ at^2$$

$$S_1 = 0 + \frac{1}{2}\ a_1 t_1^2 = \frac{1}{2}\ a_1 t_1^2 \quad ...(i)$$

$$7500 = \frac{1}{2}\ a_1 t_1^2 \quad ...(ii)$$

[$\because$ $S_1 = 3/4 \times 10{,}000 = 7500$ m]

Also, for the second retarding part

$$u_2 = \text{initial velocity}$$
$$= \text{final velocity at the end of first interval}$$
$$= 0 + a_1t_1 = a_1t_1$$

Hence v_2 = final velocity at the end of second part

$$= u_2 - a_2t_2$$
$$= a_1t_1 - a_2t_2$$
$$= 0, \text{ because the train comes to rest}$$

$\therefore$ $$a_1t_1 = a_2t_2$$

or $$\frac{a_1}{a_2} = \frac{t_2}{t_1} \quad ...(iii)$$

Also, S_2 = distance travelled in the second part

$$= \text{average velocity} \times \text{time}$$
$$= \left(\frac{a_1t_1 + 0}{2}\right) \times t_2$$
$$= \frac{a_1t_1}{2} \cdot t_2 \quad ...(iv)$$

Adding (*i*) and (*iv*),

$$S_1 + S_2 = \frac{a_1t_1^2}{2} + \frac{a_1t_1}{2} \cdot t_2$$
$$= \frac{a_1t_1}{2}(t_1 + t_2)$$

or $$S_1 + S_2 = \frac{a_1t_1}{2} \times 900 \quad (\because \; t_1 + t_2 = 15 \text{ min.} = 900 \text{ sec})$$

$$10{,}000 = \frac{a_1t_1}{2} \times 900 \quad (\because \; S_1 + S_2 = 10 \text{ km} = 10{,}000 \text{ m})$$

or $$a_1t_1 = \frac{20{,}000}{900} = \frac{200}{9}$$

But a_1t_1 = maximum velocity

Hence **max. velocity** $= \frac{200}{9}$ **= 22.22 m/sec (Ans.)**

Also, from eqn. (*ii*)

$$7500 = \frac{1}{2} \times 22.22 \times t_1$$

or $$t_1 = \frac{7500}{11.11} = 675 \text{ sec}$$

$\therefore$ $$t_2 = 900 - 675 = 225 \text{ sec}$$

Now from eqn. (*iii*),

$$\frac{a_1}{a_2} = \frac{t_2}{t_1} = \frac{225}{675} = \frac{1}{3}$$

$\therefore$ $$3a_1 = a_2.$$

Also, $v_{max} = 22.22 = a_1 t_1$

$\therefore$ $$\boldsymbol{a_1} = \frac{22.22}{675} = \mathbf{0.0329\ m/sec^2}. \quad \textbf{(Ans.)}$$

and $$\boldsymbol{a_2} = 3a_1 = 3 \times 0.0329 = \mathbf{0.0987\ m/sec^2}. \quad \textbf{(Ans.)}$$

7.7. MOTION UNDER GRAVITY

It has been seen that bodies falling to earth (through distances which are small as compared to the radius of the earth) and entirely unrestricted, increase in their velocity by about 9.81 m/sec for every second during their fall. This acceleration is called the acceleration due to gravity and is conventionally denoted by '*g*'. Though the value of this acceleration varies a little at different parts of the earth's surface but the generally adopted value is 9.81 m/sec^2.

For downward motion	**For upward motion**
$a = +g$	$a = -g$
$v = u + gt$	$v = u - gt$
$h = ut + \frac{1}{2} gt^2$	$h = ut - \frac{1}{2} gt^2$
$v^2 - u^2 = 2gh$	$v^2 - u^2 = -2gh.$

7.8. SOME HINTS ON THE USE OF EQUATIONS OF MOTION

(*i*) If a body starts from rest, its initial velocity, $u = 0$

(*ii*) If a body comes to rest ; its final velocity, $v = 0$

(*iii*) When a body is thrown upwards with a velocity u, time taken to reach the maximum height $= \frac{u}{g}$ and velocity on reaching the maximum height is zero (*i.e.*, $v = 0$) . This value of t is obtained by equating $v = u - gt$ equal to zero.

(*iv*) Greatest height attained by a body projected upwards with a velocity $u = \frac{u^2}{2g}$, which is obtained by substituting $v = 0$ in the equation $v^2 - u^2 = -2gh$.

(*v*) Total time taken to reach the ground $= \frac{2u}{g}$, the velocity on reaching the ground being $\sqrt{2gh}$.

$$(\because \quad v^2 - u^2 = 2gh \text{ or } v^2 - 0^2 = 2gh \text{ or } v = \sqrt{2gh})$$

(*vi*) *The velocity with which a body reaches the ground is same with which it is thrown upwards.*

Example 7.10. *A stone is dropped from the top of tower 100 m high. Another stone is projected upward at the same time from the foot of the tower, and meets the first stone at a height of 40 m. Find the velocity, with which the second stone is projected upwards.*

Sol. Motion of the first particle :

Height of tower $= 100$ m

Initial velocity, $u = 0$

Height, $h = 100 - 40 = 60$ m.

Let t be the time (in seconds) when the two particles meet after the first stone is dropped from the top of the tower.

Refer to Fig. 7.5.

Using the relation,

$$h = ut + \frac{1}{2} gt^2$$

$$60 = 0 + \frac{1}{2} \times 9.81\, t^2$$

$$\therefore \quad t = \sqrt{\frac{120}{9.81}} = 3.5 \text{ sec.}$$

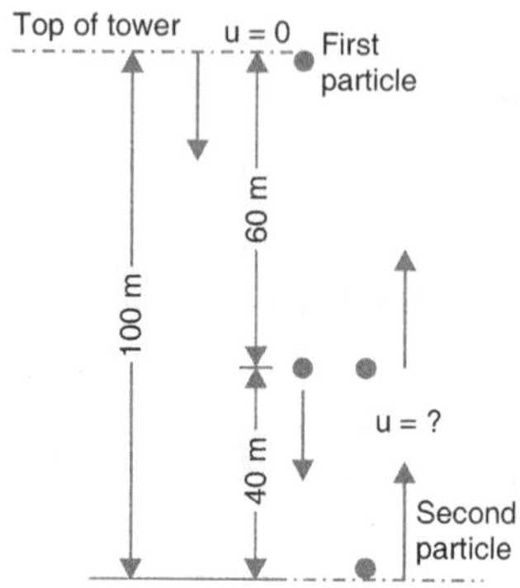

Fig. 7.5

Motion of the second particle :

Height, $h = 40$ m

Time, $t = 3.5$ sec.

Let u be the initial velocity with which the second particle has been projected upwards.

Using the relation,

$$h = ut - \frac{1}{2} gt^2 \qquad (\because \text{ Particle is projected upwards})$$

$$40 = u \times 3.5 - \frac{1}{2} \times 9.81 \times 3.5^2$$

$$3.5u = 40 + \frac{1}{2} \times 9.81 \times 3.5^2$$

$$\boldsymbol{u = 28.6 \text{ m/sec.}} \quad \textbf{(Ans.)}$$

Example 7.11. *A body projected vertically upwards attains a maximum height of 450 m. Calculate the velocity of projection and compute the time of flight in air. At what altitude will this body meet a second body projected 5 seconds later with a speed of 140 m/sec ?*

Sol. Maximum height attained by the body

$= 450$ m

Let u = initial velocity of the body

v = final velocity of the body = 0

Using the relation,

$$v^2 - u^2 = -\,2gh \qquad (\because \text{ body is thrown upwards})$$

$$0^2 - u^2 = -\,2 \times 9.81 \times 450$$

$$\boldsymbol{u = 94 \text{ m/sec.}} \quad \textbf{(Ans.)}$$

Let 't' be the time taken by the body in reaching the highest point from the point of projection.

Then, using the relation,

$$v = u - gt$$

$$0 = 94 - 9.81t$$

$$\therefore \quad t = \frac{94}{9.81} = 9.6 \text{ sec.}$$

$\therefore$ Total time of flight in air

$= 2 \times 9.6 =$ **19.2 sec. (Ans.)**

($\because$ The body will take the same time in returning also)

Let the second body meet the first body at a height 'h' from the ground. Let 't' be the time taken by the first body.

Then, time taken by the second body

$$= (t - 4) \text{ sec.}$$

Considering the motion of first body

$$h = ut - \frac{1}{2} gt^2$$

$$= 94t - \frac{1}{2} \times 9.81t^2 \qquad ...(i)$$

Considering the motion of the second body

$$h = 140 (t - 5) - \frac{1}{2} \times 9.81 (t - 5)^2 \qquad ...(ii)$$

Equating (*i*) and (*ii*), we get

$$94t - \frac{1}{2} \times 9.81t^2 = 140 (t - 5) - \frac{1}{2} \times 9.81 (t - 5)^2$$

$$188t - 9.81t^2 = 280 (t - 5) - 9.81 (t - 5)^2$$

$$188t - 9.81t^2 = 280t - 1400 - 9.81 (t - 5)^2$$

$$188t - 9.81t^2 = 280t - 1400 - 9.81t^2 + 98.1t - 245.25$$

From which $t = 8.65$ sec.

Putting this in eqn. (*i*), we get

$$h = 94 \times 8.65 - \frac{1}{2} \times 9.81 \times 8.65^2$$

$$= 813.3 - 367 = 446.3 \text{ m.}$$

Hence, the second body will meet the first one at a height of 446.3 m from the ground. (Ans.)

Example 7.12. *Two stones are thrown vertically upwards one from the ground with a velocity of 30 m/sec and another from a point 40 metres above with a velocity of 10 m/sec. When and where will they meet ?*

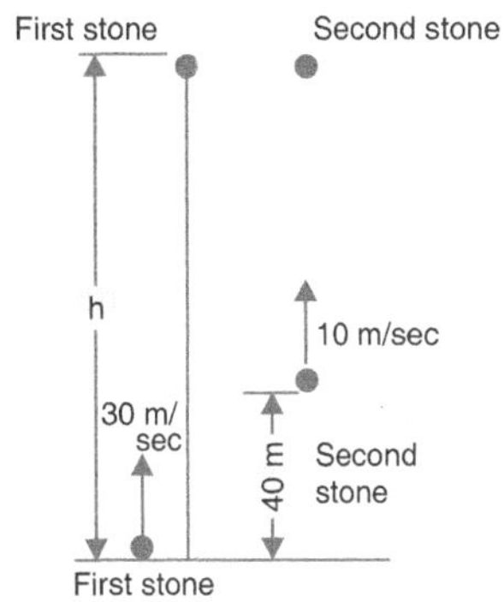

Fig. 7.6

Sol. Refer to Fig. 7.6.

Let the two stones meet after 't' seconds from their start at a height of 5 metres from the ground.

Motion of first stone :

u = initial velocity = 30 m/sec

h = vertical distance travelled

t = time taken

Using the relation, $h = ut - \frac{1}{2} gt^2$

($\because$ stone is thrown upwards)

$$h = 30t - \frac{1}{2} \times 9.81t^2 \qquad ...(i)$$

Motion of second stone :

Vertical distance travelled

$$h' = h - 40$$
$$u = 10 \text{ m/sec.}$$

Again using the relation,

$$h = ut + \frac{1}{2} gt^2$$

$$(h - 40) = 10t - \frac{1}{2} \times 9.8t^2 \quad ...(ii)$$

Subtracting (*ii*) from (*i*),

$$40 = 20t$$

$$\boldsymbol{t = 2 \textbf{ sec.}} \quad \textbf{(Ans.)}$$

Substituting this value in eqn. (*i*), we get

$$h = 30 \times 2 - \frac{1}{2} \times 9.81 \times 2^2 = \mathbf{40.38 \text{ m.} \text{ (Ans.)}}$$

Hence, the two stones meet after 2 seconds at 40.38 m from the ground.

Example 7.13. *A stone is thrown from the ground vertically upwards, with a velocity of 40 m/sec. After 3 seconds another stone is thrown in the same direction and from the same place. If both of the stones strike the ground at the same time, compute the velocity with which the second stone was thrown.*

Sol. Motion of first stone :

u = velocity of projection = 40 m/sec

v = velocity at the maximum height = 0

t = time taken to reach the maximum height = ?

Using the relation,

$$v = u - gt \quad (\because \text{ stone is moving upward})$$

$$0 = 40 - 9.81t$$

or

$$t = \frac{40}{9.81} = 4 \text{ sec.}$$

Therefore, total time taken by the first stone to return to the earth = 4 + 4 = 8 sec (because the time taken to reach the maximum height is same as that to come down to earth).

Therefore, the time taken by the second stone to return to the earth = 8 – 3 = 5 sec.

or time taken to reach the maximum height = 5/2 = 2.5 sec.

Motion of second stone :

u = velocity of projection = ?

v = final velocity at max. height = 0

t = time taken to reach the max. height

Using the relation,

$$v = u - gt$$

$$0 = u - 9.81 \times 2.5$$

$$\therefore \quad u = 9.81 \times 2.5 = 24.5 \text{ m/sec.}$$

Hence, the velocity of projection of second stone

= 24.5 m/sec. (Ans.)

Example 7.14. *A body, falling freely under the action of gravity passes two points 15 metres apart vertically in 0.3 seconds. From what height, above the higher point, did it start to fall.*

Sol. Refer to Fig. 7.7.

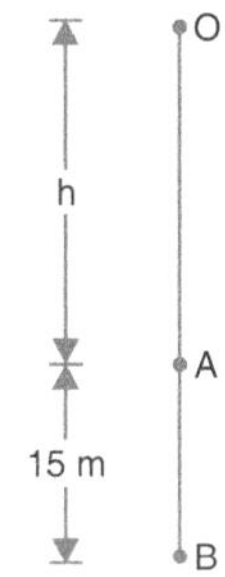

Fig. 7.7

Let the body start from O and pass two points A and B, 15 metres apart in 0.3 second after traversing the distance OA.

Let $OA = h$

Considering the motion from O to A,

Initial velocity, $u = 0$

Using the relation,

$$h = ut + \frac{1}{2}gt^2 \quad (\because \text{ the body is falling downward})$$

$$h = 0 + \frac{1}{2}gt^2 \qquad ...(i)$$

Considering the motion from O to B.

Initial velocity, $u = 0$

Time taken, $t = (t + 0.3)$ sec.

Again, using the relation, $h + 15 = 0 + \frac{1}{2}g\,(t + 0.3)^2 \qquad ...(ii)$

Subtracting, (i) from (ii),

$$15 = \frac{1}{2}g(t + 0.3)^2 - \frac{1}{2}gt^2$$

$$30 = g(t^2 + 0.6t + 0.09) - gt^2$$

$$30 = gt^2 + 0.6\,gt + 0.09\,g - gt^2$$

$$\therefore \quad 0.6gt = 30 - 0.09g$$

$$t = \frac{30}{0.6g} - \frac{0.09g}{0.6g} = 5.1 - 0.15 = 4.95 \text{ sec.} \qquad ...(iii)$$

Substituting the value of t in eqn. (i), we get

$$\boldsymbol{h} = \frac{1}{2} \times 9.81 \times (4.95)^2 = \mathbf{120.2\ m.} \quad \textbf{(Ans.)}$$

Example 7.15. *A stone dropped into a well is heard to strike the water after 4 seconds. Find the depth of the well, if the velocity of sound is 350 m/sec.*

Sol. Initial velocity of stone, $u = 0$

Let t = time taken by stone to reach the bottom of the well,

and h = depth of the well

Using the relation,

$$h = ut + \frac{1}{2}gt^2$$

$$h = 0 + \frac{1}{2} \times 9.8t^2 = 4.9t^2 \qquad ...(i)$$

Also, the time taken by the sound to reach the top

$$= \frac{\text{Depth of the well}}{\text{Velocity of sound}}$$

$$= \frac{h}{350} = \frac{4.9t^2}{350} \qquad ...(ii)$$

Total time taken = time taken by the stone to reach the bottom of the well
+ time taken by sound to reach the ground
= 4 seconds (given)

$$\therefore \qquad t + \frac{4.9t^2}{350} = 4$$

or $$4.9t^2 + 350t - 1400 = 0$$

or $$t = \frac{-350 \pm \sqrt{(350)^2 + 4 \times 4.9 \times 1400}}{2 \times 4.9}$$

$$= \frac{-350 + 387.2}{9.8} = 3.8 \text{ sec}$$

$$\therefore \qquad t = 3.8 \text{ sec.}$$

Substituting the value in eqn. (*i*), we get

$$h = 4.9 \times (3.8)^2 = 70.8 \text{ m}$$

Hence, the depth of well = 70.8 m. (Ans.)

VARIABLE ACCELERATION

Example 7.16. *The equation of motion of a particle is*

$$S = -6 - 5t^2 + t^3$$

where S is in metres and t in seconds.

Calculate : (i) The displacement and the acceleration when the velocity is zero.

(ii) The displacement and the velocity when the acceleration is zero.

Sol. The equation of motion is

$$S = -6 - 5t^2 + t^3 \qquad ...(\text{given}) \quad ...(i)$$

Differentiating both sides,

$$\frac{ds}{dt} \quad \text{or} \quad v = -10t + 3t^2$$

$$\therefore \qquad v = -10t + 3t^2 \qquad ...(ii)$$

Again, differentiating both sides,

$$\frac{dv}{dt} \quad \text{or} \quad a = -10 + 6t$$

$$\therefore \qquad a = -10 + 6t \qquad ...(iii)$$

Now, **(*i*) When the velocity is zero,**

$$v = -10t + 3t^2 = 0$$

$$\therefore \qquad t(3t - 10) = 0$$

$$t = \frac{10}{3} = 3.33 \text{ sec.} \qquad (\text{ignoring } t = 0 \text{ which means start})$$

Substituting this value in eqns. (*i*) and (*iii*),

$$S = \text{displacement}$$
$$= -6 - 5 \times 3.33^2 + 3.33^3$$
$$= -6 - 55.44 + 36.92$$
$$= \mathbf{-24.52\ m.}\quad \textbf{(Ans.)}$$

The negative sign indicates that distance is travelled in the other direction.

Also, a = acceleration

$$= -10 + 6 \times \frac{10}{3} = \mathbf{10\ m/sec^2.}\quad \textbf{(Ans.)}$$

(*ii*) When the acceleration is zero

$$a = -10 + 6t = 0$$

$\therefore$ $$6t = 10$$

or $$t = \frac{10}{6} = \frac{5}{3} = 1.67 \text{ sec.}$$

Substituting this value in eqns. (*i*) and (*ii*), we get

$$S = \text{displacement}$$
$$= -6 - 5t^2 + t^3 = -6 - 5 \times (1.67)^2 + (1.67)^3$$
$$= -6 - 13.94 + 4.66 = \mathbf{-15.28\ m.}\quad \textbf{(Ans.)}$$

The –ve sign again means that the distance is travelled in the other direction.

Also, $$v = -10t + 3t^2$$
$$= -10 \times 1.67 + 3 \times (1.67)^2 = -16.7 + 8.36$$
$$= \mathbf{-8.34\ m/sec.}\quad \textbf{(Ans.)}$$

Example 7.17. *If a body be moving in a straight line and its distance S in metres from a given point in the line after t seconds is given by the equation*

$$S = 20t + 3t^2 - 2t^3.$$

Calculate : (a) The velocity and acceleration at the start.

(b) The time when the particle reaches its maximum velocity.

(c) The maximum velocity of the body.

Sol. The equation of motion is

$$S = 20t + 3t^2 - 2t^3 \quad ...(i)$$

Differentiating both sides

$$\frac{dS}{dt} = v = 20 + 6t - 6t^2 \quad ...(ii)$$

Again, differentiating

$$\frac{d^2S}{dt^2} = \frac{dv}{dt} = a = 6 - 12t \quad ...(iii)$$

(*a*) At start, *t* = 0

Hence from eqns. (*ii*) and (*iii*),

$$v = 20 + 0 - 0 = \mathbf{20\ m/sec.}\quad \textbf{(Ans.)}$$
$$a = 6 - 12 \times 0 = \mathbf{6\ m/sec.}\quad \textbf{(Ans.)}$$

(b) When the particle reaches its maximum velocity

$$a = 0$$

i.e., $$6 - 12t = 0$$

$\therefore$ $$\mathbf{t = 0.5\ sec.}\quad \textbf{(Ans.)}$$

(b) The maximum velocity of the body

When $t = 0.5$ sec.

$$v_{max} = 20 + 6t - t^2$$
$$= 20 + 6 \times 0.5 - 6 \times 0.5^2$$
$$= 20 + 3 - 1.5$$
$$= \mathbf{21.5\ m/sec.}\quad \textbf{(Ans.)}$$

SELECTED QUESTIONS EXAMINATION PAPERS

Example 7.18. *Two trains A and B leave the same station on parallel lines. A starts with uniform acceleration of 0.15 m/s² and attains a speed of 24 km/hour when the steam is reduced to keep the speed constant. B leaves 40 seconds after with a uniform acceleration of 0.30 m/s² to attain a maximum speed of 48 km/hour. When will B overtake A ?*

Sol. Motion of train A:

Uniform acceleration, $a_1 = 0.15\ m/s^2$

Initial velocity, $u_1 = 0$

Final velocity, $v_1 = 24$ km/h

$$= \frac{24 \times 1000}{60 \times 60} = \frac{20}{3}\ \text{m/sec}$$

Let t_1 be the time taken to attain this velocity (in seconds)

Using the relation:

$$v = u + at$$

$$\frac{20}{3} = 0 + 0.15 \times t_1$$

$\therefore$ $$t_1 = \frac{20}{3 \times 0.15} = 44.4\ \text{sec}$$

Also, distance travelled during this interval,

$$s_1 = u_1 t_1 + \frac{1}{2} a_1 t_1^2$$

$$= 0 + \frac{1}{2} \times 0.15 \times 44.4^2 = 148\ \text{m}$$

Motion of train B:

Initial velocity, $u_2 = 0$

Acceleration, $a_2 = 0.3\ m/sec^2$

Final velocity, $v_2 = 48$ km/h

$$= \frac{48 \times 1000}{60 \times 60} = \frac{40}{3}\ \text{m/sec}$$

Let t_2 be taken to travel this distance, say s_2

Using the relation:

$$v = u + at$$

$$\frac{40}{3} = 0 + 0.3 \times t_2$$

$$\therefore \qquad t_2 = \frac{40}{3 \times 0.3} = 44.4 \text{ s}$$

and

$$s_2 = u_2 t_2 + \frac{1}{2} a_2 t_2^2$$

$$= 0 + \frac{1}{2} \times 0.3 \times (44.4)^2 = 296 \text{ m}$$

Let the train B overtake the train A when they have covered a distance s from the start. And let the train B take t seconds to cover the distance.

Thus, time taken by the train $A = (t + 40)$ sec.

Total distance moved by train A.

$$s = 148 + \text{distance covered with constant speed}$$
$$= 148 + [(t + 40) - t_1] \times 20/3$$
$$= 148 + [t + 40 - 44.4] \times 20/3$$
$$= 148 + (t - 4.4) \times 20/3 \qquad ...(i)$$

[$\{(t + 40) - t_2\}$ is the time during which train A moves with constant speed].

Similarly, total distance travelled by the train B,

$$s = 296 + \text{distance covered with constant speed}$$
$$= 296 + (t - 44.4) \times 40/3 \qquad ...(ii)$$

Equating (*i*) and (*ii*)

$$148 + (t - 4.4) \times 20/3 = 296 + (t - 44.4) \times 40/3$$

$$148 + \frac{20}{3}t - \frac{88}{3} = 296 + \frac{40}{3}t - \frac{1776}{3}$$

$$\left(\frac{40}{3} - \frac{20}{3}\right)t = 148 - 296 + \frac{1776}{3} - \frac{88}{3}$$

$$\mathbf{t = 62.26 \text{ s}}$$

Hence, *train B, overtakes train A after 62.26 s of its start.* **(Ans.)**

Example 7.19. *A cage descends a mine shaft with an acceleration of 1 m/s*^{2}. *After the cage has travelled 30 m, stone is dropped from the top of the shaft. Determine: (i) the time taken by the stone to hit the cage, and (ii) distance travelled by the cage before impact.*

Sol. Acceleration of cage,

$$a = 1 \text{ m/s}^2$$

Distance travelled by the shaft before dropping of the stone = 30 m

(*i*) Time taken by the stone to hit the cage = ?

Considering motion of the stone.

Initial velocity, $u = 0$

Let t = time taken by the stone to hit the cage, and

h_1 = vertical distance travelled by the stone before the impact.

Using the relation,

$$h = ut + \frac{1}{2} gt^2$$

$$h_1 = 0 + \frac{1}{2} \times 9.8\, t^2 = 4.9\, t^2 \qquad ...(i)$$

Now let us consider motion of the cage for 30 m

Initial velocity, $u = 0$

Acceleration, $a = 1.0 \text{ m/s}^2$.

Let t' = time taken by the shaft to travel 30 m

Using the relation,

$$s = ut + \frac{1}{2} at^2$$

$$30 = 0 + \frac{1}{2} \times 1 \times (t')^2$$

$$t' = 7.75 \text{ s.}$$

It means that cage has travelled for 7.75 s before the stone was dropped. Therefore total time taken by the cage before impact = $(7.75 + t)$.

Again using the relation:

$$s = ut + \frac{1}{2} at^2$$

$$s_1 = 0 + \frac{1}{2} \times 1 \times (7.75 + t)^2 \quad ...(ii)$$

In order that stone may hit the cage the two distances must be equal *i.e.*, equating (*i*) and (*ii*).

$$4.9\, t^2 = \frac{1}{2} \times (7.75 + t)^2$$

$$4.9 = 0.5\,(60 + t^2 + 15.5\, t)$$

or $$9.8 = t^2 + 15.5\, t + 60$$

or $$t^2 + 15.5\, t - 50.2 = 0$$

$$t = \frac{-15.5 \pm \sqrt{(15.5)^2 + 4 \times 50.2}}{2} = \frac{-15.5 \pm \sqrt{441.05}}{2}$$

$$= \frac{-15.5 \pm 21.0}{2} = 2.75 \text{ s} \quad \text{(neglecting –ve sign)}$$

∴ $\mathbf{t = 2.75 \text{ s.}}$ **(Ans.)**

(*ii*) Distance travelled by the cage before impact = ?

Let s_2 = distance travelled by the cage before impact.

We know total time taken by the cage before impact.

$$= 7.75 + 2.75 = 10.5 \text{ s.}$$

Now using the relation,

$$s_2 = ut + \frac{1}{2} at^2$$

$$= 0 + \frac{1}{2} \times 1 \times (10.5)^2 = 55.12 \text{ m}$$

Hence distance travelled by the cage before impact = **55.12 m. (Ans.)**

Example 7.20. *A body moving in a straight line has the equation of motion given by $s = 2t^3 - 4t + 10$, where s is measured in metres and t in seconds.*

Determine :

(i) The time required for the body to reach a velocity of 60 m/s starting from rest.

(ii) The acceleration of the body when velocity is equal to 32 m/s.

(iii) The net displacement of the body between the time interval of t = 0 to t = 4 s.

Solution. $$s = 2t^3 - 4t + 10 \quad \text{(Given) ...(1)}$$

By successive differentiation, we get the equations for velocity-time and acceleration-time.

∴ Differentiating the above equation, we get

$$\frac{ds}{dt} = v = 6t^2 - 4 \quad ...(2)$$

(*i*) Time required for the body to reach a velocity of 68 m/s, t = ?

Now, $v = 6t^2 - 4$

∴ $68 = 6t^2 - 4$

or $\boldsymbol{t = 3.46}$ **s. (Ans.)**

(*ii*) Acceleration of the body when velocity is equal to 32 m/s, a = ?

Differentiating eqn. (2), we get

$$\frac{dv}{dt} = a = 12t \quad ...(3)$$

To find out acceleration we should know t when $v = 32$ m/s

∴ From eqn. (2), $v = 6t^2 - 4$

or $32 = 6t^2 - 4$

or $t = 2.45$ s

Substituting this value of t in eqn. (3), we get

$$a = 12t = 12 \times 2.45 = \mathbf{29.4\ m/s^2}.\ \textbf{(Ans.)}$$

(*iii*) Net displacement of the body between the time $t = 0$ and $t = 4$ s, S_{net} = ?

Substituting these values in eqn. (1), we get

$$S_{(t=0)} = S_1 = 2 \times 0 - 4 \times 0 + 10 = 10 \text{ m, s} = 10 \text{ m.}$$

It means that the body is 10 m ahead of starting point. Hence, 10 m is to be deducted from the final displacement when $t = 4$ s.

$$S_{(t=4)} = S_2 = 2 \times (4)^3 - 4 \times 4 + 10 = 122 \text{ m}$$

∴ $$S_{net} = (S_2 - S_1) = 122 - 10 = \mathbf{112\ m.}\ \textbf{(Ans.)}$$

HIGHLIGHTS

1. A body is said to be in motion if it changes its position with respect to its surroundings. The motions may be of the following types :
   (*i*) Rectilinear motion (*ii*) Curvilinear motion
   (*iii*) Rotary or circular motion.
2. *Speed* of a body is defined as the rate of change of its position with respect to its surroundings irrespective of its directions. It is a *scalar* quantity.
3. *Velocity* of a body is the rate of change of its position with respect to its surroundings in a particular direction. It is a *vector* quantity.
4. If a body travels equal distances in equal intervals of time in the same direction, it is said to be moving with a *uniform or constant velocity.*

5. If a body travels unequal distances in equal intervals of time, in the same direction then its is said to be moving with a *variable velocity* or if it changes either its speed or its direction or both shall again be said to be moving with a *variable velocity.*
6. The *average or mean velocity* of a body is the velocity with which the distance travelled by the body in the same interval of time, is the same as that with the variable velocity.
7. The rate of change of velocity of a body is called its acceleration. When the velocity is increasing the acceleration is reckoned as *positive*, when decreasing as *negative*.
8. If the velocity of a body changes by equal amounts in equal intervals of time, the body is said to move with *uniform acceleration.*
9. If the velocity of a body changes by unequal amounts in equal intervals of time, the body is said to move with *variable acceleration.*
10. First equation of motion $v = u + at$.
11. Second equation of motion $S = ut + \frac{1}{2}at^2$.
12. Third equation of motion $v^2 - u^2 = 2aS$.
13. Distance covered in nth second $S_{nth} = u + a/2\,(2n - 1)$.
14. Motion under gravity

For downward motion

$a = +g$

$v = u + gt$

$h = ut + \frac{1}{2}gt^2$

$v^2 - u^2 = 2gh$

For upward motion

$a = -g$

$v = u - gt$

$h = ut - \frac{1}{2}gt^2$

$v^2 - u^2 = -2gh.$

OBJECTIVE TYPE QUESTIONS

Fill in the blanks :

(*i*) A body is said to be in motion if it ...... its position with respect to its surroundings.

(*ii*) A body is said to be at ...... at an instant if its position with respect to the surrounding objects remains unchanged during that instant.

(*iii*) The ...... of a body is defined as the rate of change of its position with respect to its surroundings irrespective of direction.

(*iv*) Speed is a ...... quantity.

(*v*) The ...... of a body is the rate of change of its position with respect to its surroundings in a particular direction.

(*vi*) Velocity is ...... quantity.

(*vii*) If a body travels equal distances in equal intervals of time in the same direction it is said to be moving with a ...... velocity.

(*viii*) The rate of change of velocity of a body is called its ....

(*ix*) The ...... velocity of a body is the velocity with which the distance travelled by the body in the same interval of time is the same as with variable velocity.

(*x*) If the velocity of a body changes by equal amounts in equal intervals of time, the body is said to move with ....

(*xi*) If the velocity of a body changes by unequal amounts in equal intervals of time, the body is said to move with ....

(*xii*) $v^2 - u^2 = 2$ ....

(*xiii*) $h = ut +$ ....

(*xiv*) $v^2 - u^2 = 2g$ ....

(*xv*) $S_{nth} = u + a/2$(......).

Answers

(*i*) changes (*ii*) rest (*iii*) speed (*iv*) scalar (*v*) velocity
(*vi*) vector (*vii*) uniform (*viii*) acceleration (*ix*) average velocity
(*x*) uniform acceleration (*xi*) variable acceleration
(*xii*) aS (*xiii*) $\frac{1}{2}gt^2$ (*xiv*) h (*xv*) $2n - 1$.

EXERCISES

Theoretical Questions

1. Define the term 'Motion' and state its types.
2. Define the following :
   Displacement, Velocity, Acceleration, Variable velocity and Variable acceleration.
3. Derive the following relations
$$v = u + at$$
$$S = ut + \frac{1}{2}at^2.$$
4. Find an expression for distance travelled in nth second by a body moving with a uniform acceleration.

Unsolved Examples

1. A body has an initial velocity of 10 cm/sec and is being accelerated at a constant rate of 5 cm/sec^2. What will be the velocity after 2 seconds ? What displacement does it travel during this time ?
[**Ans.** 20 cm/sec, 30 cm]
2. A train starting from rest and moving with a uniform acceleration attains a speed of 108 km/hr in 2.5 min. Find
   (*i*) Acceleration, and
   (*ii*) The distance it travels in this time. [**Ans.** (*i*) 0.2 m/sec^2, (*ii*) 2.25 km]
3. The brakes applied to a car produce a uniform retardation of 0.5 m/sec^2. The car covers a distance of 16 m before stopping. How long were the brakes applied ? [**Ans.** 8 sec.]
4. A particle has an initial velocity of 8 m/sec and an acceleration of 3 m/sec^2. Compute its speed after it has moved 60 metres distance. Compute the distance the particle moves during the sixth second.
[**Ans.** 20.59 m/sec; 24.5 m]
5. An automobile accelerates from a velocity of 24 km/hr to a velocity of 72 km/hr in a distance of 160 m. Compute the average acceleration and the time required. [**Ans.** 1.11 m/sec^2; 12 sec]
6. A cyclist riding at 36 km/hr passes a car which just starts to move with an acceleration of 0.2 m/sec^2. How far will he ride before the car overtakes him ? [**Ans.** 1000 m]
7. Two trains A and B leave the same station on parallel lines. A starts with uniform acceleration of $\frac{1}{6}$ m/sec^2 and attains a speed of 24 km/hr when steam is reduced to keep the speed constant. B leaves 40 seconds after with a uniform acceleration of $\frac{1}{3}$ m/sec^2 to attain a maximum speed of 48 km/hr. When will B overtake A ? [**Ans.** 100 sec]
8. A car starting from rest moves with a uniform acceleration for 4 km, attains a maximum speed 'v' and then retards uniformly till it comes to rest after moving further a distance of 2 km taking 15 minutes from start to stop. Find : (*i*) the maximum velocity attained by the car ; (*ii*) its acceleration ; (*iii*) its retardation; and (*iv*) time occupied in each operation.
[**Ans.** $\frac{4}{5}$ km/min, 80 m/min^2, – 160 m/min^2, 5 mts]
9. The distance between two stations is 2.5 km. A train starting from one station receives an acceleration (reaching a speed of 40 km/hr, in 0.5 minutes) until the speed reaches 48 km/hr. This speed is maintained

until the brakes are applied and train is brought to rest at the second station under a negative acceleration of 0.9 m/sec^2. Find the time to perform this journey. [**Ans.** 212.86 sec]

10. A stone is dropped from a cliff 500 m high. After what time will it strike the ground ? What will be its final velocity ? [**Ans.** 10.09 sec., 98.98 m/sec]

11. A stone is dropped from the top of a cliff 120 m high and 1 second afterwards another stone is thrown down and strikes the first stone when it has just reached the foot of the cliff. Find the speed with which the second stone was thrown, neglecting air resistance. [**Ans.** 10.8 m/s]

12. A cage goes down a mine shaft 675 m deep in 45 seconds. For the first quarter of the distance only the speed is being uniformly accelerated and during the last quarter uniformly retarded, the acceleration and retardation being equal. Find the uniform speed of cage while traversing the centre portion of the shaft. [**Ans.** 225 m/s]

13. A ball is projected vertically upwards with a velocity of 20 m/sec. Three seconds later a second ball is projected vertically upward with a velocity of 30 m/sec. At what height will they meet above the earth's surface ? [**Ans.** 11.35 m]

14. A train starting from rest, acceleration uniformly for 2 minutes, run at constant speed for 4 minutes. After this the train comes to stop with uniform retardation for 3 minutes and this covers a distance of 5 km. Find (*i*) its constant speed, (*ii*) its acceleration, and (*iii*) its retardation.

[**Ans.** (*i*) 46.15 km/hr, (*ii*) 1384 km/hr^2, (*iii*) 923 km/hr^2]

15. A stone is dropped into a well and the sound of splash is heard on the top after 4 seconds. If the sound travels at 340 m/sec, determine the depth of the well. [**Ans.** 70.53 m]

16. A particle moves along a straight line so that its displacement in metres from a fixed point is given by, $S = 2t^3 + 4t^2 - 6t + 8$. Find :

(*i*) Velocity at start

(*ii*) Velocity after 5 seconds

(*iii*) Acceleration at start

(*iv*) Acceleration after 5 seconds. [**Ans.** (*i*) 6 m/s, (*ii*) 184 m/s, (*iii*) 8 m/s^2, (*iv*) 68 m/s^2]

17. A car moves along a straight line whose equation of motion is given by $S = 12t + 3t^2 - 2t^3$, where S is in metres and t is in seconds. Calculate (*i*) velocity and acceleration at start, and (*ii*) acceleration, when velocity is zero. [**Ans.** (*i*) 12 m/s, (*ii*) 6 m/s^2, (*iii*) – 18 m/s^2]

18. A particle starts from rest with an acceleration a given by the relation $a = 2/(2 + 2v)$ m/s^2, where v is the velocity of particle in m/s. Determine the distance covered by the particle, when it acquires a velocity of 72 km/hr. [**Ans.** 1533.3 m]

8

Laws of Motion

8.1. INTRODUCTION

The relation between the motion and the force is defined by the laws of dynamics. Among other laws, Newton's laws of motion are most prominent. In 1686, Isacc Newton enunciated the following three laws of motion :

1. Newton's first law of motion
2. Newton's second law of motion
3. Newton's third law of motion.

8.2. MOMENTUM

The quantity of motion possessed by a body moving in a straight line is known as "momentum". It is product of mass of a body and its velocity. Momentum is a vector quantity because of velocity involved in it. If the mass is in kg and velocity in m/sec then its units are kg m/sec.

8.3. NEWTON'S FIRST LAW OF MOTION

It states : *"Every body continues in its state of rest or of uniform motion, in a straight line, unless it is acted upon by some external force to change that state".*

It is a matter of every day experience that a body at rest remains at rest unless a force is applied to it. An experimental determination of the law when a body is moving is almost impossible for when a body moves on a horizontal surface, there is always the force of friction and air resistance acting on the body which brings it to rest after some time ; had these forces been absent body would have gone on moving in a straight line with uniform velocity indefinitely. As a partial demonstration, it would be seen that the smoother the surface, the longer will be the distance travelled by the body.

8.4. NEWTON'S SECOND LAW OF MOTION

It states : *"The rate of change of momentum of a body is directly proportional to the impressed force and takes place in the direction of the straight line in which the force acts".* This law enables us to measure a force and establish the fundamental equation of dynamics.

Let m = mass of a body,
u = initial velocity,
v = final velocity,
a = constant acceleration,
t = time required to change the velocity from u to v ; and
P = force required to change velocity from u to v in t time.

Initial momentum of the body = mu

Final momentum of the body = mv

Change of momentum of the body = $mv - mu$

Rate of change of momentum of the body

$$= \frac{mv - mu}{t} = \frac{m(v-u)}{t}$$

$$= ma \qquad \left(\because \frac{v-u}{t} = a\right)$$

According to Newton's second law of motion, the rate of change of momentum is directly proportional to the impressed force.

$$\therefore \qquad P \propto ma$$

or $$P = kma$$

where k is a constant of proportionality.

If we define unit force as that which produces unit acceleration in unit mass, we get by putting

$$P = 1, m = 1, a = 1$$

$$k = 1$$

Hence $$P = ma \qquad ...(8.1)$$

If m = 1 kg, a = 1 m/sec^2, then

$$P = 1 \text{ Newton} = \frac{1}{9.81} \text{ kg wt or kgf}$$

If m = 1 gm, a = 1 cm/sec^2, then

$$P = 1 \text{ dyne} = \frac{1}{981} \text{ gm wt or gmf.}$$

A *dyne* (unit of force in C.G.S. system) *may be defined as the force, which while acting upon a mass of one gram produces an acceleration of 1 cm/sec^2 in the direction in which it acts.*

A *newton* (unit of force in S.I. units) *may be defined as the force, which while acting upon a mass of 1 kg produces as acceleration of 1 m/sec^2 in the direction in which it acts.*

Thus, $$1 \text{ N} = 10^5 \text{ dynes.}$$

8.5. NEWTON'S THIRD LAW OF MOTION

It states : "*To every action there is an equal and opposite reaction*".

Action means the force, which a body exerts on another, whereas reaction means the equal and opposite force, which a second body exerts on the first. For example, when a swimmer tries to swim, he pushes the water backwards and the reaction of the water, in the opposite direction, pushes the swimmer forward. Similarly, when a horse draws a stone tied by a rope the horse will be equally drawn back towards the stone.

8.6. GRAVITATIONAL AND ABSOLUTE UNITS OF FORCE

C.G.S. System of Units :

Gravitational unit of force : 1 gm wt (or gmf)

Absolute unit of force : 1 dyne

1 gm wt (or gmf) : 9.81 dynes.

A dyne is defined as the force, which while acting upon a mass of 1 gram produces an acceleration of 1 cm/sec^2 in the direction in which it acts.

M.K.S. System of Units :

Gravitational unit of force : 1 kg wt (or kgf)

Absolute unit of force : 1 newton

A newton (N) is that much force which produces an acceleration of 1 m/sec^2 in a mass 1 kg

$$\therefore \quad 1\text{ N} = 1\text{ kg} \times 1\text{ m/sec}^2$$
$$= 1000\text{ gm} \times 100\text{ cm/sec}^2$$
$$= 10^5\text{ gm cm/sec}^2$$
$$= 10^5\text{ dynes} \qquad \left[\begin{array}{r}\because\ 1\text{ dyne} = 1\text{ gm} \times 1\text{ cm/sec}\\ = 1\text{ gm cm/sec}\end{array}\right]$$

Now, 1 newton means that a body of mass 1 kg is moving with an acceleration of 1 m/sec^2. If the body is moving with an acceleration of 9.81 m/sec^2, then the force acting on the body will be 9.81 newtons. But 1 kg of mass attracted towards the earth with an acceleration of 9.81 m/sec^2 is denoted as 1 kg wt (or kgf).

$$\therefore \quad 1\text{ kg wt (or kgf)} = 9.81\text{ newtons}$$
$$= g\text{ newtons}$$

Similarly, 1 gm wt (or gmf) = 981 dynes.

Important Note. (*i*) *The units of force i.e., gm wt or kg wt (or gmf or kgf), generally written as gm or kg are referred as gravitational or engineer's units of force. Newton or dyne are absolute units of force. Therefore, gravitational units are 'g' times the absolute units.*

(*ii*) *Mass of a body is equal to the weight of the same body in gravitational units.*

Illustration. Let the mass of the body is equal to 300 kg, P *i.e.,* force with which it is attracted towards earth

$$= ma$$
$$= 300 \times 9.81 = 2943\text{ kg m/sec}^2 = 2943\text{ N.}$$

Now weight of the body is the force by which it is attracted towards the earth centre,

$$W = 2943\text{ N}$$
$$= \frac{2943}{9.81}\text{ kg wt (or kgf)} \qquad \left[\begin{array}{l}\because\ 1\text{ kg wt} = 9.81\text{ N}\\ \therefore\ 1\text{ N} = \dfrac{1}{9.81}\text{ kg wt}\end{array}\right]$$
$$= 300\text{ kg wt (or kgf)}$$

Points to Remember

1. If the weight of the body is given, then it means that the units are gravitational. The mass in absolute units will be represented by its numerical value. Suppose that the weight of the body is given as 50 kg, then

W = 50 kg wt (or kgf) .... in gravitational units.

m = 50 kg ...... in absolute units.

2. In the equation, $P = ma$, the value of m is taken in absolute units.

3. *Mass if given, is always in absolute units.*

In order to preserve the equation, $P = ma$ mass of the body is considered in *metric slugs* [weight of body in kg wt (or kgf) is divided by gravitational acceleration to give the mass in slugs], *i.e.*, 1 slug

$$= \frac{\text{kg}}{9.81} \text{ or } \frac{W}{g}.$$

If weight of the body is W, then

$$P = \frac{W}{g} \times a \qquad \text{... in gravitational units}$$

where $$m = \frac{W}{g}.$$

From above it may be concluded : (*i*) If mass is in slugs *i.e.*, $\frac{W}{g}$, then force will be in kg. (*i.e.*, kg wt of kgf) , (*ii*) If mass is in kg, then, the force will be in newtons. In both the cases the value of acceleration is in m/sec^2.

8.7. LAW OF CONSERVATION OF MOMENTUM

The law of conservation of momentum states : "*Total momentum of any group of objects always remains the same if no external force acts on them*".

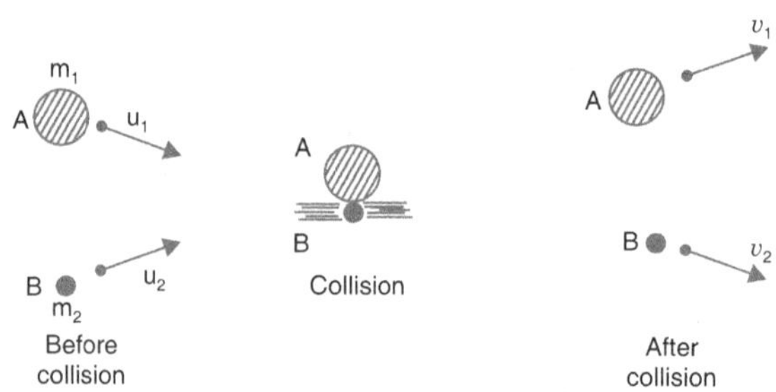

Fig. 8.1

Refer to Fig. 8.1. Let us take the case of collision of two balls A and B. Before collision the ball A of mass m_1 is moving with a velocity u_1 and the ball B of mass m_2 is moving with a velocity u_2. After collision let the velocity of ball A be v_1 and that of ball B be v_2.

Total momentum before collision :

Momentum of mass, $m_1 = m_1u_1$

Momentum of mass, $m_2 = m_2u_2$

Total momentum of masses m_1 and $m_2 = m_1u_1 + m_2u_2$

It may be remembered that momentum is a vector quantity and therefore momenta add vectorially.

Total momentum after collision

Momentum of mass, $m_1 = m_1v_1$

Momentum of mass, $m_2 = m_2v_2$

$\therefore$ Total momentum of masses m_1 and $m_2 = m_1v_1 + m_2v_2$

The law of conservation of momentum requires that the *total momentum before collision = total momentum after collision.*

Thus, $$m_1u_1 + m_2u_2 = m_1v_1 + m_2v_2 \quad ...(8.2)$$

Examples of conservation of momentum

1. Propulsion of a jet plane. In a jet engine, the gases at a high speed are ejected through a nozzle at the rear of the engine. The escaping gases have momentum. Since the total momentum must remain constant, the engine acquires an equal and opposite momentum. The momentum gained by the engine provides the force needed to push the plane forward.

2. Recoil of a gun. Consider, a gun loaded with a bullet. Before firing, the gun and the bullet are at rest. Therefore, the total momentum of the system is zero. After firing, the bullet moves in one direction but gun recoils *i.e.,* moves in the opposite direction. The momentum of the bullet is equal in magnitude but opposite in direction to that of the gun. The total momentum, being the vector sum of the momentum of the bullet and that of the gun, is zero. Since the mass of the gun is very large compared to that of the bullet the velocity of recoil of the gun will be much smaller than that of the bullet.

8.8. IMPULSE AND IMPULSIVE FORCE

Suppose a force P acts on a body for a time interval 't'. The product Pt is called the *impulse* of the force, *i.e.,*

$$\text{Impulse} = \text{force} \times \text{time interval}$$

According to Newton's second law,

$$\text{Force} = \text{mass} \times \text{acceleration}$$

$$P = ma$$

$$= m\left(\frac{v-u}{t}\right)$$

$$Pt = m(v-u) \quad ...(8.3)$$

i.e., Force × time interval = mass × change in velocity

or Impulse = mass × change in velocity

Impulse of a force is a useful concept when a force acts for a short duration of time. It tells how much the velocity of a body of a given mass will change if a given impulse is applied to it. When we hit a hockey stick we give it an impulse.

The units of impulse are newton second (Ns) in absolute units and kgf-sec in gravitational M.K.S. system.

Example 8.1. *A body has a mass of 250 kg. Find its weight in :*

(*i*) *Absolute units,* (*ii*) *Gravitational units.*

Sol. Mass of the body, m = 250 kg

(*i*) Weight of the body in absolute units

$= mg$

$= 250 \times 9.81 =$ **2452.5 N (Newton). (Ans.)**

(*ii*) Weight of the body in gravitational units

$= \frac{2452.5}{9.81} =$ **250 kg wt or kgf. (Ans.)**

Example 8.2. *Determine the force necessary to produce an acceleration of 5 m / sec² in a mass of 200 kg.*

Sol. Mass of the body, $m = 200$ kg

Acceleration, $a = 5$ m/sec^2

Using the relation,

$$P = ma$$
$$= 200 \times 5 = \textbf{1000 N (S.I. units). (Ans.)}$$

Example 8.3. *A body of mass 150 kg has its velocity changed from 10 m / sec to 20 m / sec in the same direction in 50 seconds.*

Find (i) Change in its momentum, and

(ii) Force responsible for this change in

(a) Absolute units, and *(b) Engineer's (or Gravitational) units.*

Sol. Mass of the body, $m = 150$ kg

Initial velocity of the body, $u = 10$ m/sec

Final velocity of the body, $v = 20$ m/sec

Time taken, $t = 50$ sec.

(*a*) **Absolute units**

(*i*) Change in momentum $= m\,(v - u) = 150(20 - 10)$

$$= \textbf{1500 kg m/sec. (Ans.)}$$

(*ii*) Force,
$$P = ma = m \times \left(\frac{v-u}{t}\right)$$
$$= 150 \times \left(\frac{20-10}{50}\right) = \textbf{30 N. (Ans.)}$$

(*b*) **Engineer's or Gravitational units**

(*i*) Change in momentum $= \dfrac{m}{g}\,(v - u)$

$$= \frac{150}{9.81}\,(20 - 10) = \textbf{152.9 kgf sec. (Ans.)}$$

(*ii*) Force,
$$P = \frac{m}{g}\,.\,a = \frac{m}{g}\left(\frac{v-u}{t}\right)$$
$$= \frac{150}{9.81}\left(\frac{20-10}{50}\right) = \textbf{3.058 kgf. (Ans.)}$$

Example 8.4. *The weight of the body on earth is 490 N. If the acceleration due to gravity on earth is 9.8 m / s², what will be the weight of the body on (i) the moon where the gravitational acceleration is 1.4 m / s², and (ii) the sun, where the gravitational acceleration is 270 m / s².*

Sol. Weight of the body on earth,

$$W = 490 \text{ N}$$

Acceleration due to gravity on earth,

$$g = 9.8 \text{ m/s}^2$$

Now, mass of body, $m = \frac{W}{g} = \frac{490}{9.8} = 50$ kg $(\because \ W = m.g)$

(*i*) **Weight of the body on moon where g = 1.4 m/s² = ?**

$$W = m.g$$
$$= 50 \times 1.4 = \mathbf{70\ N.\ (Ans.)}$$

(*ii*) **Weight of the body on sun where g = 270 m/s² = ?**

$$W = m.g$$
$$= 50 \times 270 = \mathbf{13500\ N.\ (Ans.)}$$

Example 8.5. *A body of mass 200 kg, is found to move with a velocity of 20 m/s when a force of 100 N acts on it for 90 s. Determine the velocity of the body :*

(*i*) *When the force acts in the direction of motion.*

(*ii*) *When the force acts in the opposite direction of motion.*

Sol. Mass of the body, $m = 200$ kg

Initial velocity of the body, $u = 20$ m/s

Magnitude of the force, $P = 100$ N

Time, $t = 90$ s

Now, using the relation,

$$P = ma$$
$$100 = 200 \times a \quad \text{(where } a \text{ is the acceleration of the body)}$$
$$\therefore \quad a = \frac{100}{200} = 0.5 \text{ m/s}^2.$$

(*i*) **Velocity of the body when the force acts in the direction of motion = ?**

$$v = u + at$$
$$= 25 + 0.5 \times 90 = \mathbf{70\ m/s.\ (Ans.)}$$

(*ii*) **Velocity of the body when the force acts in the opposite direction of motion = ?**

Here $a = -0.5$ m/s²

$$\therefore \quad v = u + at$$
$$= 25 + (-0.5 \times 90) = \mathbf{-20\ m/s.\ (Ans.)}$$

Minus sign means that the body is moving in the reverse direction.

Example 8.6. *A train of mass 20000 kg is moving at 10 km/h and after 20 seconds it is moving at 50 km/h. What is the average force acting upon it during this time, in the direction of motion ?*

Sol. Change in velocity $(v - u)$

$$= 50 - 10 = 40 \text{ km/h}$$
$$= \frac{40 \times 1000}{60 \times 60} = 11.11 \text{ m/s}$$

Change in momentum $= m\,(v - u) = 20000 \times 11.11$

$= 2.222 \times 10^5$ kg m/s

Average force = change in momentum per sec

$$\frac{2.222 \times 10^5}{20} = \mathbf{11.11 \times 10^3\ N\ or\ 11.11\ kN.\ (Ans.)}$$

Example 8.7. *An electric train travelling at 36 km/h is pulled up gradually, the retardation being 0.5 m/s².*

(*i*) *How far will it travel before coming to rest ?*

(*ii*) *If the retarding force is 600 kN, what is the mass of the train ?*

Sol. Initial velocity, $u = 36$ km/h

$$= \frac{36 \times 1000}{60 \times 60} = 10 \text{ m/s}$$

Retardation, $a = -0.5 \text{ m/s}^2$

Final velocity, $v = 0$

Retarding force $= 600$ kN

(*i*) Using the relation, $v^2 = u^2 + 2as$

$$0^2 = 10^2 + 2(-0.5) \times s$$

or $\boldsymbol{s = 100}$ **m. (Ans.)**

(*ii*) **Mass of the train,** $\boldsymbol{m = ?}$

Retarding force, $P = ma$

or $600 \times 1000 = m.a = m \times 0.5$

$$\therefore \quad \boldsymbol{m} = \frac{600 \times 1000}{0.5} = \boldsymbol{12 \times 10^5} \textbf{ kg. (Ans.)}$$

Example 8.8. *A train weighing 3500 kN has a frictional resistance of 5 N/kN. What average pull will be required if it is to attain a speed of 72 km/h from rest in 5 minutes on the level ?*

Sol. Weight of the train, $W = 3500$ kN

Frictional resistance, $R = 5$ N/kN

Initial velocity, $u = 0$

Final velocity, $v = 72$ km/h

$$= \frac{72 \times 1000}{60 \times 60} = 20 \text{ m/s.}$$

Time, $t = 5 \text{ min} = 5 \times 60 = 300$ s

Using the relation, $v = u + at$

$$20 = 0 + a \times 300 \quad \text{or} \quad a = \frac{20}{300} \text{ m/s}^2$$

Accelerating force, $P = m.a$

$$= \frac{3500 \times 1000}{9.81} \times \frac{20}{300} = 23785 \text{ N}$$

Force required to overcome frictional resistance,

$$F = 3500 \times 5 = 17500 \text{ N}$$

Total pull, $P_t = P + F$

$= 23785 + 17500$

$= 41285$ N $=$ **41.28 kN. (Ans.)**

Example 8.9. *A train has a weight of 3500 kN. If the frictional resistance amounts to 5 N per kN. What steady pull must the locomotive exert in order to increase the speed on a level track from 36 km/h to 72 km/h within a period of 2 minutes ?*

Sol. Weight of train $= 3500$ kN

Frictional resistance $= 5$ N/kN

Initial velocity, $u = 36$ km/h

$$= \frac{36 \times 1000}{60 \times 60} = 10 \text{ m/s}$$

Final velocity, $v = 72$ km/h

$$= \frac{72 \times 1000}{60 \times 60} = 20 \text{ m/s}$$

Time, $t = 2$ min $= 120$ seconds

Using the relation, $v = u + at$

$$20 = 10 + a \times 120$$

$$a = \frac{10}{120} = \frac{1}{12} \text{ m/s}^2$$

Let P = constant pull exerted by the locomotive on the train

Total frictional resistance $= 5 \times 3500 = 17500$ N

$\therefore$ Net force which causes motion of the train,

$$P' = P - 17500$$

Now using the relation,

$$P' = m.a$$

$$P - 17500 = \frac{3500 \times 1000}{9.81} \times \frac{1}{12} = 29731$$

$\therefore$ **P = 47231 N. (Ans.)**

Example 8.10. *A man of mass 80 kg dives into a swimming pool from a tower of height 18 m. He was found to go down in water by 2.2 m and then started rising. Find the average resistance of water. Neglect the resistance of air. Take the value of $g = 9.8$ m/s^2.*

Sol. Mass of man, $m = 80$ kg

Height of tower, $h = 18$ m

Distance travelled by man from water surface into the water, $s = 2.2$ m

Average resistance of water = ?

Motion of the man from top of the tower to the water surface :

Initial velocity, $u = 0$

Acceleration due to gravity, $g = 9.8$ m/s^2

Let the final velocity when the man reaches the water surface $= v$

Now using the relation,

$$v^2 - u^2 = 2gh$$

$$v^2 - 0^2 = 2 \times 9.8 \times 18$$

$$\therefore \quad v = \sqrt{2 \times 9.8 \times 18} = 18.78 \text{ m/s}$$

Motion of the man from the water surface upto the point from where he started rising :

Distance travelled, $s = 2.2$ m

Initial velocity of the man on the water surface,

$$u = 18.78 \text{ m/s}$$

Final velocity, $v = 0$

Let the retardation due to water resistance = a

Using the relation, $v^2 - u^2 = -2as$ (–ve sign is taken due to retardation)

$$0^2 - (18.78)^2 = -2 \times a \times 2.2$$

$$\therefore \quad a = \frac{(18.78)^2}{2 \times 2.2} = 80.15 \text{ m/s}^2$$

Hence, average resistance of water

$$\boldsymbol{P} = m.a. = 80 \times 80.15 = \mathbf{6412\ N.\ (Ans.)}$$

Example 8.11. *A gun of mass 3 × 10⁴ kg fires projectile of mass 456 kg with a velocity of 305 m/s. With what initial velocity will the gun recoil ? If the recoil is overcome by an average force of 60 kN, how far will the gun travel ? How long will it take ?*

Sol. Refer to Fig. 8.2.

Mass of the gun = 3×10^4 kg

Mass of the projectile , m = 456 kg

Velocity of the projectile, v = 305 m/s

Let V = velocity of recoil of gun in m/s.

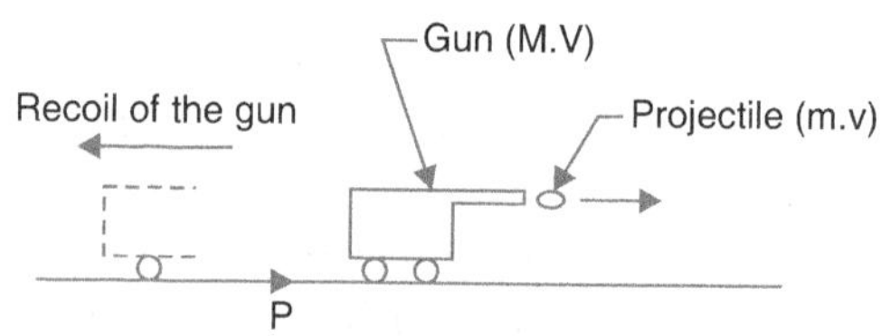

Fig. 8.2

Applying principle of conservation of momentum,

$$MV + mv = 0$$

$$V = -\frac{mv}{M} = -\frac{456 \times 305}{3 \times 10^4} = -4.64 \text{ m/s}$$

The negative sign here indicates that the direction in which the canon moves is *opposite* to the direction in which the shell is fired.

Hence the *velocity of recoil* = **4.64 m/s. (Ans.)**

Let t seconds be the time taken by the impulsive force, P, of 600 kN to overcome the *recoil.*

Using the relation

$$P = M\left(\frac{v - u}{t}\right)$$

$$600 \times 10^3 \times t = 3 \times 10^4 \times 4.64 \qquad \left[\because \begin{array}{l} \text{Here } u = 4.64 \text{ m/s} \\ P = 600 \text{ kN and } v = 0 \end{array}\right]$$

From which, $\quad$ **t = 0.232 s. (Ans.)**

Distance covered s = ?

$$a = \frac{v - u}{t} = \frac{0 - 4.64}{0.232} = -20 \text{ m/s}^2 \qquad \text{(–ve sign means retardation)}$$

Using the relation, $v^2 - u^2 = 2as$

$$0^2 - (4.64)^2 = 2 \times (-20)\, s$$

or

$$s = \frac{(4.64)^2}{2 \times 20} = \mathbf{0.538\ m.\ (Ans.)}$$

Example 8.12. *A canon of mass 18144 kg fires a shell of mass 91 kg with a muzzle velocity of 762 m/s. Calculate the velocity with which canon recoils and the uniform force required to stop it within 375 mm. In how much time will it stop ?*

Sol. Refer to Fig. 8.3

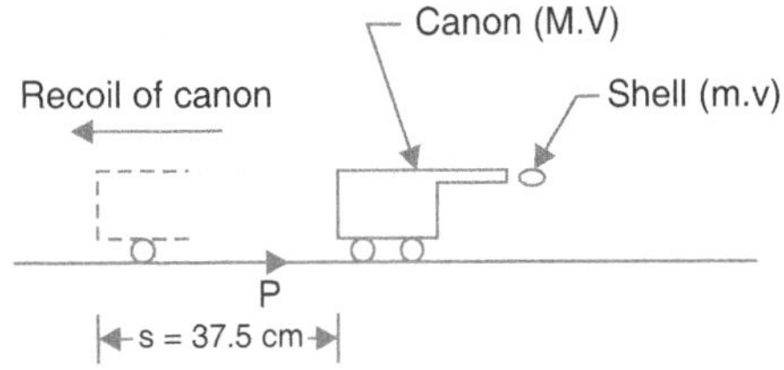

Fig. 8.3

Mass of canon, $M = 18144$ kg

Mass of the shell, $m = 91$ kg

Velocity of the shell, $v = 762$ m/s.

Velocity of the canon, V = ?

As per the principle of conservation of momentum,

Initial momentum of system = final momentum of system

i.e.,
$$MV + mv = 0$$

or
$$V = -\frac{mv}{M} = -\frac{91 \times 762}{18144} = -3.82 \text{ m/s}$$

The negative sign here indicates that the direction in which the canon moves is *opposite* to the direction in which the shell is fired.

Hence, velocity of recoil of the canon = **3.82 m/s. (Ans.)**

Now considering the motion of the canon,

u = initial velocity = 3.82 m/s

v = final velocity

= 0, as the recoil has to be stopped.

s = distance travelled by canon before stopping

Let a = acceleration of the canon

t = time taken to stop the canon.

Force required to stop the canon, P = ?

Using the relation :

$$v^2 - u^2 = 2as$$
$$0^2 - (3.82)^2 = 2 \times a \times 0.375$$
$$a = -\frac{(3.82)^2}{2 \times 0.375} = -19.46 \text{ m/s}^2$$

Hence, force (retarding) required to stop the canon,

$$P = M.a = 18144 \times 19.46 = \mathbf{3.53 \times 10^6\ N.\ (Ans.)}$$

Time required to stop the canon, t = ?

Using the relation, $v = u + at$

$$0 = 3.82 - 19.46\,t$$

or
$$t = \frac{3.82}{19.46} = \mathbf{0.196\ s.\ (Ans.)}$$

Example 8.13. *Two railway trucks of masses 10000 kg 8000 kg moving in the same direction at speeds of 10 m/s and 15 m/s respectively collide and subsequently move together. Calculate common speed due to impact.*

Sol. Mass of first truck, $m_1 = 10000$ kg

Mass of second truck, $m_2 = 8000$ kg

Speed of the first truck, $v_1 = 10$ m/s

Speed of the second truck, $v_2 = 15$ m/s

Let V = common speed in m/s after the collision.

As per law of conservation of momentum :

Momentum before collision = momentum after collision

$$m_1v_1 + m_2v_2 = (m_1 + m_2) V$$

$$10000 \times 10 + 8000 \times 15 = (10000 + 8000) V$$

$\therefore$ $\mathbf{V = 12.22}$ **m/s. (Ans.)**

Example 8.14. *A pile driver of mass 250 kg falls 3.0 m on a pile of mass 1200 kg. The pile is driven 0.2 m into the ground. Find the common velocity of the pile and the driver after the impact and average resistance of the ground to the penetration of the pile.*

Sol. Refer to Fig. 8.4.

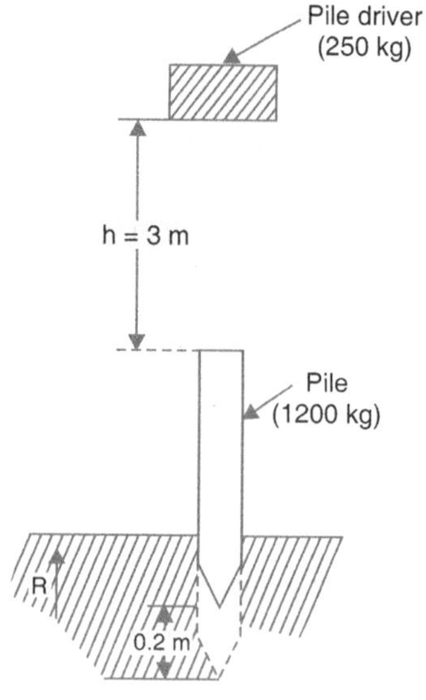

Fig. 8.4

Mass of the pile driver, $m = 250$ kg

Mass of the pile, $M = 1200$ kg

Velocity of the pile driver after failing 3 m,

$$v = \sqrt{2gh}$$

$$= \sqrt{2 \times 9.81 \times 3}$$

$$= 7.67 \text{ m/s}$$

After the impact total mass in motion is $(M + m)$. Let V be the common velocity.

As per *principle of conservation of momentum* :

Momentum before impact

$$= \text{momentum after impact}$$

$$mv = (M + m) V$$

$$250 \times 7.67 = (1200 + 250) V$$

$$\therefore \quad V = \frac{250 \times 7.67}{(1200 + 250)} = 1.32 \text{ m/s.}$$

Now, velocity is destroyed in 20 cm or 0.2 m and if 'a' is the *retardation*,

$$0^2 - (1.32)^2 = 2 \times a \times 0.2 \qquad (\because \ v^2 - u^2 = 2as)$$

$$\therefore \quad a = -\frac{(1.32)^2}{2 \times 0.2} = 4.36 \text{ m/s}^2 \text{ (Retardation)}$$

Resulting retarding force,

$$P = (M + m) \cdot a$$

$$= (1200 + 250) \times 4.36 = 6322 \text{ N}$$

The resistance of the ground (R)

= retarding force + weight of driver and the pile

$$\therefore \quad \boldsymbol{R} = 6322 + (250 + 1200) \times 9.81$$

$$= \mathbf{20546.5 \text{ N. (Ans.)}}$$

8.9. D' ALEMBERT'S PRINCIPLE

D' Alembert, a French mathematician, was the first to point out that on the lines of *equation of static equilibrium, equation of dynamic equilibrium can also be established by introducing inertia force in the direction opposite the acceleration in addition to the real forces on the plane.*

Static equilibrium equations are :

$$\Sigma H \text{ (or } P_x) = 0, \Sigma V \text{ (or } \Sigma P_y) = 0, \Sigma M = 0$$

Similarly when different external forces act on a system in motion, the algebraic sum of all the forces (including the *inertia force*) is zero. This is explained as under :

We know that, $P = ma$ (Newton's second law of motion)

or $P - ma = 0$ or $P + (-ma) = 0$

The expression in the block $(-ma)$ is the *inertia force* and negative sign signifies that it acts in a direction opposite to that of acceleration/retardation a.

It is also known as the "*principle of kinostatics*".

Example 8.15. *Two bodies of masses 80 kg and 20 kg are connected by a thread and move along a rough horizontal surface under the action of a force 400 N applied to the first body of mass 80 kg as shown in Fig. 8.6. The co-efficient of friction between the sliding surfaces of the bodies and the plane is 0.3.*

Determine the acceleration of the two bodies and the tension in the thread, using D' Alembert's principle.

Sol. Refer to Figs. 8.5 and 8.6

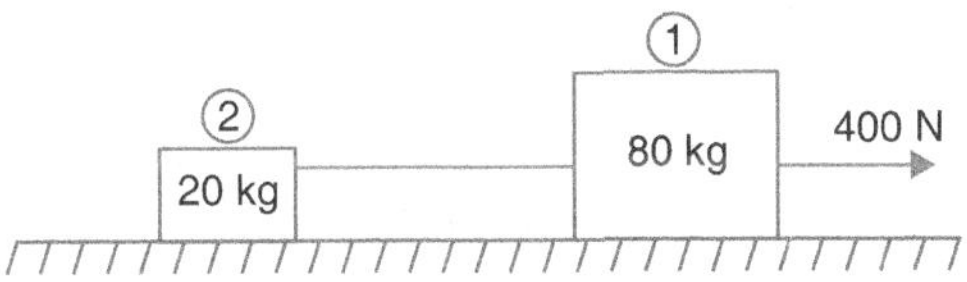

Fig. 8.5

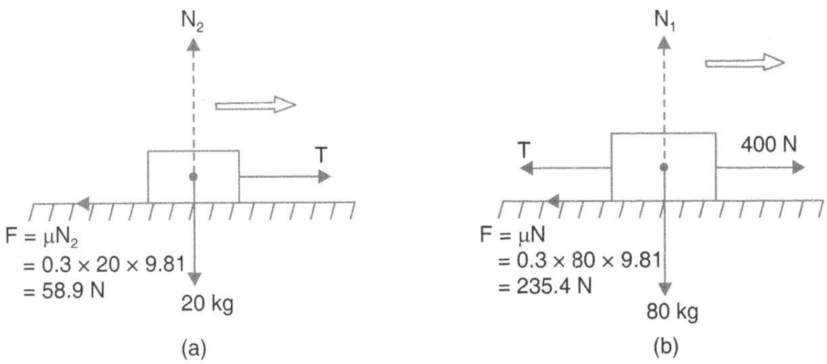

Fig. 8.6

Acceleration of the bodies, a :

As per *D' Alembert's principle for dynamic equilibrium condition the algebraic sum of all the active forces acting on a system should be zero.*

The various forces acting on the bodies are :

(*i*) Force applied $= 400$ N

(*ii*) Inertia force $= (80 + 20)\ a$

(*iii*) Frictional force $= 0.3 \times 80 \times 9.81 + 0.3 \times 20 \times 9.81$

$= 235.4 + 58.9 = 294.3$ N

$\therefore\quad 400 - (80 - 20)\ a = 294.3 = 0$

or
$$a = \frac{400 - 294.3}{(80 + 20)} = \mathbf{1.057\ m/s^2.\ (Ans.)}$$

Tension in the thread between the two masses, *T* :

Considering free body diagrams of the masses 80 kg and 20 kg separately as shown in Fig. 8.6 (*a*) and (*b*).

Applying D' Alembert's principle for Fig. 8.6 (*a*), we get

$$400 - T - 80 \times 1.057 - 0.3 \times 80 \times 9.81 = 0$$

$\therefore$ **T = 80 N. (Ans.)**

Now, applying D' Alembert's principle for Fig. 8.6 (*b*), we get

$$T - 0.3 \times 20 \times 9.81 - 20 \times 1.057 = 0$$

$\therefore$ **T = 80 N. (Ans.)**

It may be noted that the same answer is obtained by considering the two masses separately.

8.10. MOTION OF A LIFT

Consider a lift (elevator or cage etc.) carrying some mass and moving with a uniform acceleration.

Let m = mass carried by the lift in kg,

$W (= m.g)$ = weight carried by the lift in newtons,

a = uniform acceleration of the lift, and

T = tension in the cable supporting the lift.

There could be the following *two* cases :

(*i*) When the lift is moving *upwards*, and

(*ii*) When the lift is moving *downwards*.

1. Lift moving upwards :

Refer to Fig. 8.7.

The net upward force, which is responsible for the motion of the lift

$$= T - W = T - m.g \quad ...(i)$$

Also, this force = mass × acceleration

$$= m.a \quad ...(ii)$$

Equating (*i*) and (*ii*), we get

$$T - m.g = m.a$$

$$\therefore\quad T = m.a + m.g = m(a + g) \quad ...(8.4)$$

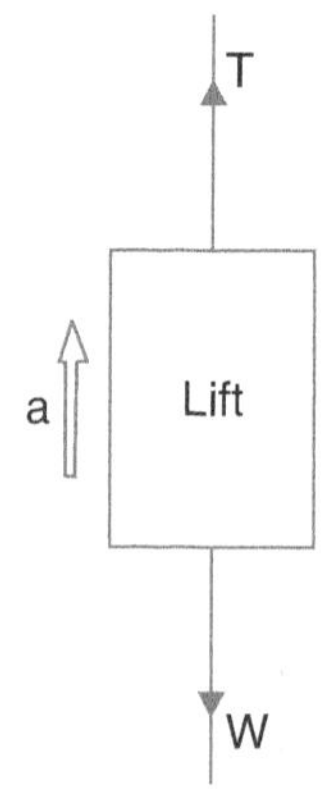

Fig. 8.7. Lift moving upwards.

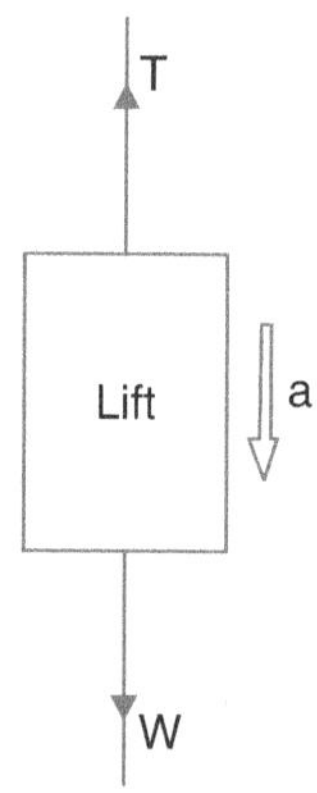

Fig. 8.8. Lift moving downwards.

2. Lift moving downwards :

Refer to Fig. 8.8.

Net downward force responsible for the motion of the lift

$$= W - T = m.g - T \quad ...(i)$$

Also, this force $= \text{mass} \times \text{acceleration}$

$$= m.a \quad ...(ii)$$

Equating (*i*) and (*ii*), we get

$$m.g - T = m.a$$

$$\therefore \quad T = m.g - m.a = m\,(g - a) \quad ...(8.5)$$

Example 8.16. *An elevator cage of mass 900 kg when empty is lifted or lowered vertically by means of a wire rope. A man of mass 72.5 kg is standing in it. Find :*

(*a*) *The tension in the rope,*

(*b*) *The reaction of the cage on the man, and*

(*c*) *The force exerted by the man on the cage, for the following two conditions :*

(*i*) *when the cage is moving up with an acceleration of 3 m/s² and*

(*ii*) *when the cage is moving down with a uniform velocity of 3 m/s.*

Sol. Mass of the cage, $M = 900$ kg

Mass of the man, $m = 72.5$ kg.

(*i*) **Upward acceleration, $a = 3$ m/s²**

(*a*) Let T be the tension in the rope in newtons

The various forces acting on the cage are :

1. Tension, T of the rope acting vertically upwards.
2. Total mass $= M + m$, of the cage and the man acting vertically downwards.

As the cage moves upwards, $T > (M + m)\,g$

$\therefore$ Net accelerating force $= T - (M + m)\,g = (m + m)\,a$

$$\therefore \quad T - (M + m)\,g = (M + m)\,a \quad ...(i)$$

Substituting the given values, we get

$$T - (900 + 72.5)\,9.81 = (900 + 72.5) \times 3$$

$$\therefore \quad \mathbf{T = 12458\ N.\ (Ans.)}$$

(*b*) Let 'R' be the reaction of the cage on the man in newtons.

Considering the various forces, the equation of motion is

$$R - mg = m.a \quad ...(ii)$$

or

$$R = mg + ma = m\,(g + a)$$

$$= 72.5\,(9.81 + 3) = \mathbf{928.7\ N.\ (Ans.)}$$

(*c*) The force exerted by the man on the cage must be equal to the force exerted by the cage on the man (Newton's third law of motion).

$\therefore$ Force exerted by the man on the cage = **928.7 N. (Ans.)**

(*ii*) **When the cage moves with a uniform velocity 3 m/s :**

When the cages moves with a uniform velocity, acceleration is equal to zero.

(*a*) Tension in the rope, T :

Putting $a = 0$ in eqn. (*i*), we get

$$T - (M + m)\,g = (M + m) \times 0 = 0$$

$$\therefore \quad T = (M + m)\,g$$

$$= (900 + 72.5) \times 9.81 = \mathbf{9540\ N.\ (Ans.)}$$

(*b*) Also from equation (*ii*)

When $a = 0$,

$$R = mg + m \times 0 = mg$$
$$= 72.5 \times 9.81 = \textbf{711.2 N. (Ans.)}$$

(*c*) Force exerted by the man on the cage

= force exerted by the cage on the man

= **711.2 N. (Ans.)**

Example 8.17. *An elevator of mass 500 kg is ascending with an acceleration of 3 m/s². During this ascent its operator whose mass is 70 kg is standing on the scales placed on the floor. What is the scale reading ? What will be total tension in the cables of the elevator during his motion ?*

Sol. Mass of the elevator, $M = 500$ kg

Acceleration, $a = 3 \text{ m/s}^2$

Mass of the operator, $m = 70$ kg

Pressure (R) exerted by the man, when the lift moves upward with an acceleration of 3 m/s²,

$$R = mg + ma = m\,(g + a)$$
$$= 70\,(9.81 + 3) = \textbf{896.7 N. (Ans.)}$$

Now, tension in the cable of elevator

$$T = M\,(g + a) + m\,(g + a)$$
$$= (M + m)\,(g + a)$$
$$= (500 + 70)\,(9.81 + 3) = \textbf{7301.7 N. (Ans.)}$$

8.11. MOTION OF TWO BODIES CONNECTED BY A STRING PASSING OVER A SMOOTH PULLEY

Fig. 8.9 shows two bodies of weights W_1 and W_2 respectively hanging vertically from a *weightless* and *inextensible* string, passing over a *smooth pulley.* Let T be the common tension in the string. If the *pulley were not smooth, the tension would have been different in the two sides of the string.*

Let W_1 be greater than W_2 and a be the acceleration of the bodies and their motion as shown.

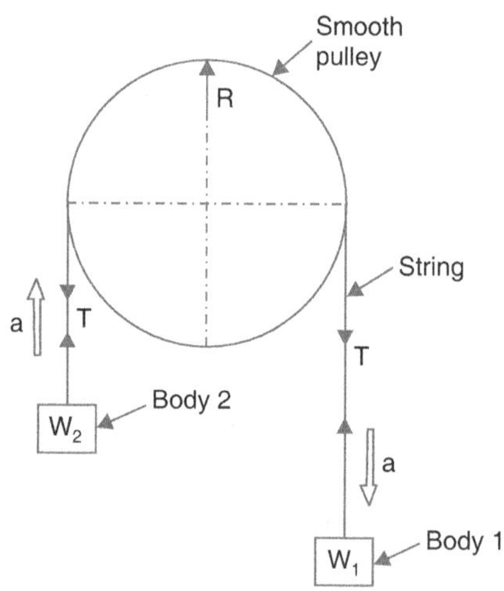

Fig. 8.9

Consider the motion of body 1:

Forces acting on it are : W_1 (downwards) and T (upwards).

$\therefore$ Resulting force $= W_1 - T$ (downwards) ...(*i*)

Since this weight is moving downward, therefore, force acting on this weight

$$= \frac{W_1}{g} \cdot a \qquad ...(ii)$$

Equating (*i*) and (*ii*)

$$W_1 - T = \frac{W_1}{g}\, a \qquad ...(1)$$

Now consider the motion of body 2:

Forces acting on it are : T (upwards) W_2 (downwards)

$\therefore$ Resultant force $= T - W_2$ ...(*iii*)

Since the body is moving upwards therefore force acting on the body

$$= \frac{W_2}{g} \,.\, a \qquad ...(iv)$$

Equating (*iii*) and (*iv*)

$$T - W_2 = \frac{W_2}{g} \,.\, a \qquad ...(2)$$

Now adding eqns. (1) and (2), we get

$$W_1 - W_2 = \left(\frac{W_1 + W_2}{g}\right) a$$

from which,

$$a = \left(\frac{W_1 + W_2}{W_1 + W_2}\right) . g \qquad ...(8.6)$$

From equation (2),

$$T - W_2 = \frac{W_2}{g}\, a$$

$$T = W_2 + \frac{W_2}{g}\, a = W_2 \left(1 + \frac{a}{g}\right)$$

Substituting the value of '*a*' from equation (8.6), we get

$$T = W_2 \left[1 + \left(\frac{W_1 - W_2}{W_1 + W_2}\right) . \frac{g}{g}\right]$$

from which,

$$T = \frac{2\, W_1\, W_2}{W_1 + W_2} \qquad ...(8.7)$$

Reaction of the pulley,

$$R = T + T = 2T$$

$$= \frac{4 W_1\, W_2}{W_1 + W_2} \qquad ...(8.8)$$

Example 8.18. *Two bodies weighing 45 N and 60 N are hung to the ends of a rope, passing over a frictionless pulley. With what acceleration the heavier weight comes down ? What is the tension in the string ?*

Sol. Weight of heavier body, $W_1 = 60$ N

Weight of lighter body, $W_2 = 45$ N

Acceleration of the system, *a* = ?

Using the relation,

$$a = \frac{g\,(W_1 - W_2)}{(W_1 + W_2)} = \frac{9.81\,(60 - 45)}{(60 + 45)} = \mathbf{1.4\ m/s^2.\ (Ans.)}$$

Tension in the string, T = ?

Using the relation,

$$T = \frac{2\, W_1\, W_2}{W_1 + W_2} = \frac{2 \times 60 \times 45}{(60 + 45)} = \mathbf{51.42\ N.\ (Ans.)}$$

Example 8.19. *A system of frictionless pulleys carries two weights hung by inextensible cords as shown in Fig. 8.10. Find :*

(*i*) *The acceleration of the weights and tension in the cords.*

(*ii*) *The velocity and displacement of weight '1' after 5 seconds from start if the system is released from rest.*

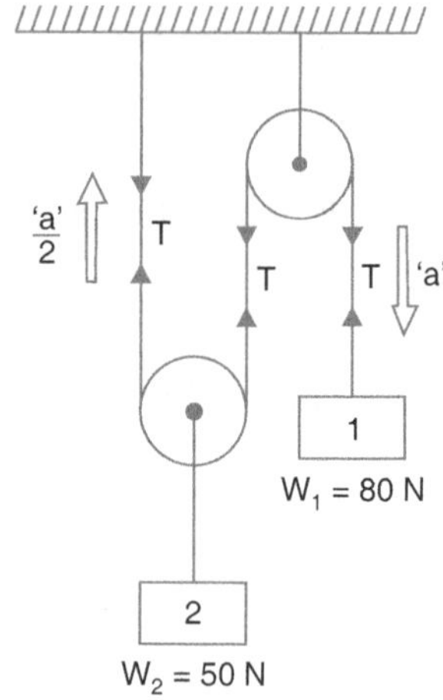

Fig. 8.10

Sol. Weight, $W_1 = 80$ N

Weight, $W_2 = 50$ N

Let T = tension (constant throughout the cord, because pulleys are *frictionless*, and *cord is continuous*).

When weight W_1 travels unit distance then weight W_2 travels half the distance. Acceleration is proportional to the distance.

$\therefore$ If a = acceleration of weight W_1

then, $a/2$ = acceleration of weight W_2.

It is clear from the figure that weight W_1 moves downward and weight W_2 moves upward.

(*i*) **Acceleration of weights, T = ?**

Consider the motion of weight W_1 :

$$W_1 - T = \frac{W_1}{g} a$$

$$80 - T = \frac{80}{g} \times a \quad ...(i)$$

Consider the motion of weight W_2 :

$$2T - W_2 = \frac{W_2}{g}$$

$$2T - 50 = \frac{50}{g} \times \frac{a}{2} \quad ...(ii)$$

Multiplying eqn. (*i*) by 2 and adding eqns. (*i*) and (*ii*), we get

$$110 = \frac{185}{g} a$$

$$\therefore \quad a = \frac{110 \times 9.81}{185} = 5.8 \text{ m/s}^2$$

Hence **acceleration of W_1 = 5.8 m/s². (Ans.)**

and **acceleration of W_2** = 5.8/2 = **2.9 m/s². (Ans.)**

Substituting the value of 'a' in eqn. (*i*), we get

$$80 - T = \frac{80}{9.81} \times 5.8$$

$$\therefore \quad \mathbf{T = 32.7 \text{ N. (Ans.)}}$$

(*ii*) **Velocity and displacement of weight W_1 after 5 sec. = ?**

$$u = 0,\ a = 5.8 \text{ m/s}^2,\ t = 5 \text{ s}$$

$$\therefore \quad v = u + at = 0 + 5.8 \times 5 = \mathbf{29 \text{ m/s. (Ans.)}}$$

and

$$s = ut + \frac{1}{2} at^2 = 0 + \frac{1}{2} \times 5.8 \times 5^2 = \mathbf{72.5 \text{ m. (Ans.)}}$$

8.12. MOTION OF TWO BODIES CONNECTED AT THE EDGE OF A HORIZONTAL SURFACE

Fig. 8.11 shows two bodies of weights W_1 and W_2 respectively connected by a light inextensible string. Let the body 1 hang free and body 2 be placed on a rough horizontal surface. Let the body 1 move downwards and the body 2 move along the surface of the plane. We know that the velocity and acceleration of the body will be the same as that of the body 2, therefore tension will be same throughout the string. Let μ be the co-efficient of friction between body 2 and the horizontal surface.

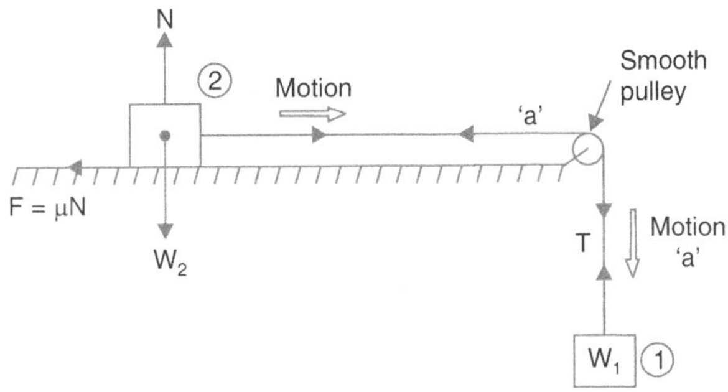

Fig. 8.11

Normal reaction at the surface, $N = W_2$

and force of friction, $F = \mu N = \mu W_2$

Let a = acceleration of the system

T = tension in the string.

Consider the motion of body 1 :

Forces acting on it are : W_1 (downwards) and T (upwards)

Resultant force $= W_1 - T$ ...(*i*)

Since the body is moving downwards, therefore force acting on this body

$$= \frac{W_1}{g} . a \quad ...(ii)$$

Equating (*i*) and (*ii*), $W_1 - T = \dfrac{W_1}{g} a$ ...(1)

Now consider the motion of body 2 :

Forces acting on it are : T (towards right), Force of friction F (towards left).

∴ Resultant force $= T - F = T - \mu W_2$ ...(*iii*)

Since, the body is moving horizontally with acceleration, therefore force acting on this body

$$= \frac{W_2}{g} . a \quad ...(iv)$$

Equating (*iii*) and (*iv*), we get

$$T - \mu W_2 = \frac{W_2}{g} a \quad ...(2)$$

Adding equations (1) and (2), we get

$$W_1 - \mu W_2 = \frac{W_1}{g} a + \frac{W_2}{g} a$$

or
$$W_1 - \mu W_2 = \frac{a}{g} (W_1 + W_2)$$

or
$$a = \left(\frac{W_1 - \mu W_2}{W_1 + W_2} \right) g \quad ...(8.9)$$

Substituting this value of 'a' in equation (1), we get

$$W_1 - T = \frac{W_1}{g} \left(\frac{W_1 - \mu W_2}{W_1 + W_2} \right) g$$

$$T = W_1 - W_1 \left(\frac{W_1 - \mu W_2}{W_1 + W_2} \right)$$

$$T = W_1 \left[1 - \frac{W_1 - \mu W_2}{W_1 + W_2} \right]$$

$$= W_1 \left[\frac{W_1 + W_2 - W_1 + \mu W_2}{W_1 + W_2} \right]$$

i.e.,
$$T = \frac{W_1 W_2 (1 + \mu)}{W_1 + W_2} \quad ...(8.10)$$

For *smooth horizontal surface* ; putting $\mu = 0$ in equations (8.9) and (8.10), we get

$$a = \frac{W_1 . g}{W_1 + W_2} \quad ...(8.11)$$

and
$$T = \frac{W_1 W_2}{W_1 + W_2} \quad ...(8.12)$$

Example 8.20. *Find the acceleration of a solid body A of weight 8 N, when it is being pulled by another body of weight 6 N along a smooth horizontal plane as shown in Fig. 8.12.*

Sol. Refer to Fig. 8.12.

Weight of body B, $W_1 = 6$ N

Weight of body A, $W_2 = 8$ N

Acceleration of body, a = ?

Tension in the string, T = ?

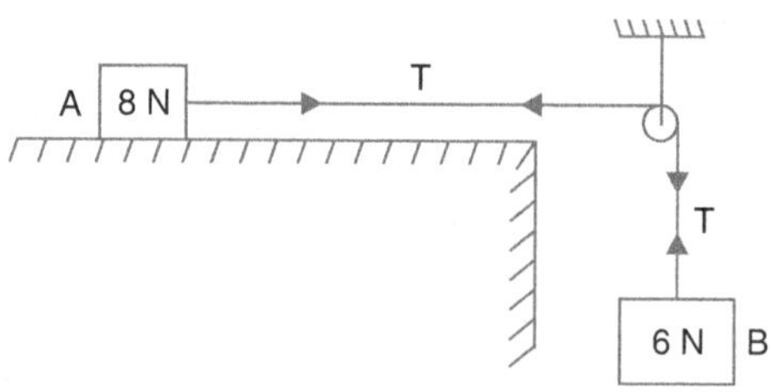

Fig. 8.12

Equation of motion for body B

$$6 - T = \frac{6}{g} . a \quad ...(i)$$

Equation of motion for body A

$$T = \frac{8}{g} . a \quad ...(ii)$$

Adding (*i*) and (*ii*), we get

$$6 = \frac{14}{g} \cdot a$$

$$\therefore \qquad \boldsymbol{a} = \frac{6 \times 9.81}{14} = \mathbf{4.2\ m/s^2. \ (Ans.)}$$

Substituting this value of a in (*i*), we get

$$6 - T = \frac{6}{9.81} \times 4.2$$

$$\therefore \qquad \boldsymbol{T} = \mathbf{3.43\ N. \ (Ans.)}$$

Example 8.21. *Two blocks shown in Fig. 8.13, have weights A = 8 N and B = 10 N and co-efficient of friction between the block A and horizontal plane, μ = 0.2.*

If the system is released, from rest and the block A falls through a vertical distance of 1.5 m, what is the velocity acquired by it ? Neglect the friction in the pulley and extension of the string.

Sol. Refer to Fig. 8.13.

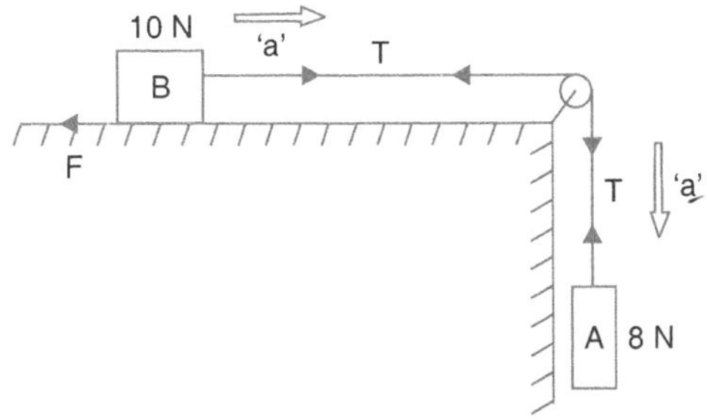

Fig. 8.13

Considering vertical string portion:

$$8 - T = \frac{8}{g} \cdot a \qquad ...(i)$$

Considering horizontal string portion :

$$T - F = \frac{10}{g} \cdot a$$

or $$T - \mu N_B = \frac{10}{g} \cdot a$$

or $$T - 0.2 \times 10 = \frac{10}{g} a \qquad (\because \ N_B = W_B = 10 \text{ newtons})$$

or $$T - 2 = \frac{10}{g} a \qquad ...(ii)$$

Adding (*i*) and (*ii*)

$$6 - \frac{18a}{g}$$

$$\therefore \qquad a = \frac{6 \times 9.81}{18} = 3.27 \text{ m/s}^2$$

Now using the relation :

$$v^2 - u^2 = 2as \quad \text{or} \quad v^2 - u^2 = 2 \times 3.27 \times 1.5$$

$$\therefore \qquad v = 3.13 \text{ m/s}$$

Hence the velocity acquired by weight **$\boldsymbol{A}$ = 3.13 m/s. (Ans.)**

Example 8.22. *A body '1' of weight 20 N is held on a rough horizontal table. An elastic string connected to the body '1' passes over a smooth pulley at the end of the table and then under a second smooth pulley carrying a body '2' of weight 10 N as shown in Fig. 8.14. The other end of the string is fixed to a point above the second pulley. When the 20 N body is released, it moves with an acceleration of g/5. Determine the value of co-efficient of friction between the block and the table.*

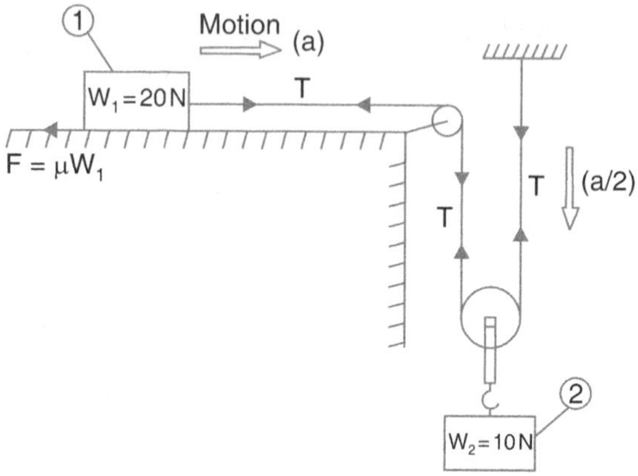

Fig. 8.14

Sol. Weight of body '1', $W_1 = 20$ N

Weight of body '2', $W_2 = 10$ N

Acceleration of body '1' $a = g/5$

Let T = tension in string in newtons, and

μ = co-efficient of friction between block and the table.

Considering the motion of body '1' :

$$T - \mu W_1 = \frac{W_1}{g} a$$

or $$T - \mu \times 20 = \frac{20}{g} \times \frac{g}{5} = 4 \quad ...(i)$$

Considering the motion of body '2' :

A little consideration will show that the acceleration of the body '2' will be half of that of the body '1' *i.e.*, $g/10$.

Now, $$W_2 - 2T = \frac{W_2}{g} \times \frac{a}{2}$$

or $$10 - 2T = \frac{10}{g} \times \frac{g}{10} = 1 \quad ...(ii)$$

Now multiplying eqn. (*i*) by 2 and adding eqns. (*i*) and (*ii*), we get

$$10 - 40\mu = 9$$

$\therefore$ $$40\mu = 1 \text{ or } \mu = \mathbf{0.025}. \textbf{ (Ans.)}$$

Example 8.23. *A string passing across a smooth table at right angle to two opposite edges has two masses M_1 and M_2 ($M_1 > M_2$) attached to its ends hanging vertically as shown in Fig. 8.15. If a mass M be attached to the portion of the string which is on the table, find the acceleration of the system when left to itself.*

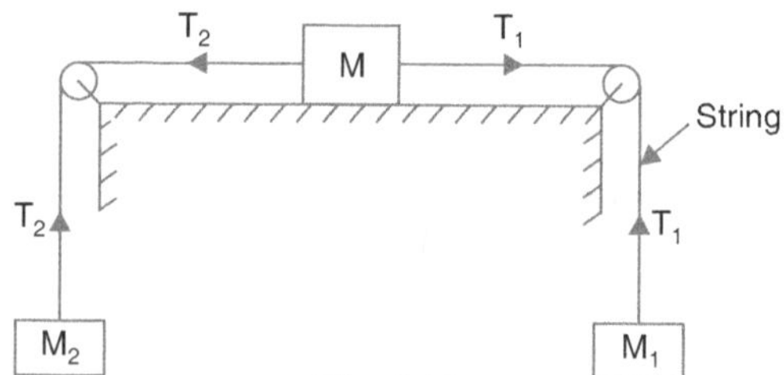

Fig. 8.15

Sol. Refer to Fig. 8.15.

Let T_1 and T_2 be the tensions in the two portions of the strings.

Acceleration of the system, *a* = ?

We know that

$$W_1 = M_1 g,\ W_2 = M_2 g$$

$\therefore$ Equations of motion are :

$$M_1 g - T_1 = M_1 a \quad ...(i)$$

$$T_1 - T_2 = M \cdot a \quad ...(ii)$$

$$T_2 - M_2 g = M_2 \cdot a \quad ...(iii)$$

Adding (*i*), (*ii*) and (*iii*), we get

$$M_1 g - M_2 g = a (M_1 + M + M_2)$$

$$\boldsymbol{a} = \left[\frac{\mathbf{M_1 - M_2}}{\mathbf{M_1 + M + M_2}}\right] \times \mathbf{g}. \quad \textbf{(Ans.)}$$

8.13. MOTION OF TWO BODIES CONNECTED BY A STRING ONE END OF WHICH IS HANGING FREE AND THE OTHER LYING ON A ROUGH INCLINED PLANE

Fig. 8.16 shows two bodies of weight W_1 and W_2 respectively connected by a light inextensible string. Let the body 1 of weight W_1 hang free and body 2 of weight W_2 be placed on an inclined rough surface. The velocity and acceleration of the body 1 will be the same as that of body 2. Since the string is inextensible, therefore, tension will be same throughout.

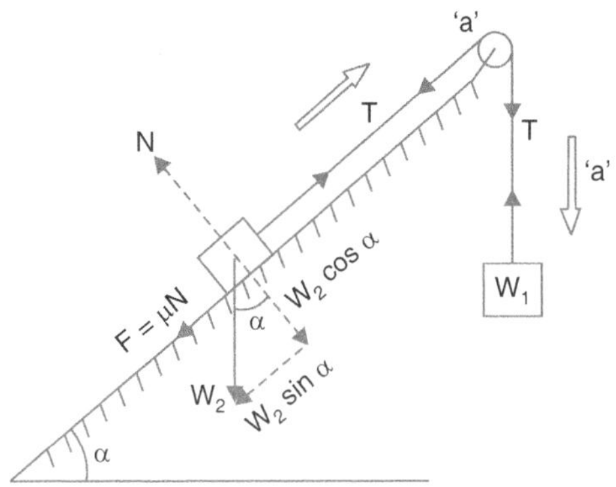

Fig. 8.16

Let a = acceleration of the system

α = inclination of the plane

μ = co-efficient of friction between body and the inclined surface

T = tension in the string.

Consider the motion of body 1 :

Forces acting on it are : W_1 (downwards), T (upwards)

Resultant force = $W_1 - T$ ...(*i*)

Since the body is moving downwards, therefore force acting on the body

$$= \frac{W_1}{g} . a \quad ...(ii)$$

Equating (*i*) and (*ii*)

$$W_1 - T = \frac{W_1}{g} . a \quad ...(1)$$

Now consider the motion of body 2 :

Normal reaction at the surface,

$$N = W_2 \cos \alpha$$

$\therefore$ Force of friction, $F = \mu N = \mu W_2 \cos \alpha$

The forces acting on the body 2 as shown are :

T (upwards), $W \sin \alpha$ (downwards)

and $F = \mu W_2 \cos \alpha$ (downwards)

$\therefore$ Resultant force = $T - W_2 \sin \alpha - \mu W_2 \cos \alpha$ ...(*iii*)

Since, this body is moving along the inclined surface with acceleration therefore force acting on this body

$$= \frac{W_2}{g} a \quad ...(iv)$$

Equating (*iii*) and (*iv*), we get

$$T - W_2 \sin \alpha - \mu W_2 \cos \alpha = \frac{W_2}{g} a \quad ...(2)$$

Adding equations (1) and (2), we get

$$W_1 - W_2 \sin \alpha - \mu W_2 \cos \alpha = \frac{a}{g} (W_1 + W_2)$$

$$\therefore \qquad a = \frac{g\,(W_1 - W_2 \sin \alpha - \mu W_2 \cos \alpha)}{W_1 + W_2} \qquad ...(8.13)$$

Substituting this value of 'a' in equation (1), we get

$$W_1 - T = \frac{W_1}{g}\, a$$

$$T = W_1 - \frac{W_1}{g}\, a = W_1 \left(1 - \frac{a}{g}\right)$$

$$= W_1 \left[1 - \frac{W_1 - W_2 \sin \alpha - \mu W_2 \cos \alpha}{W_1 + W_2}\right]$$

$$= W_1 \left[\frac{W_1 + W_2 - W_1 + W_2 \sin \alpha + \mu W_2 \cos \alpha}{W_1 + W_2}\right]$$

$$= W_1 W_2 \left[\frac{1 + \sin \alpha + \mu \cos \alpha}{W_1 + W_2}\right]$$

i.e.,

$$T = \frac{W_1 W_2 (1 + \sin \alpha + \mu \cos \alpha)}{W_1 + W_2} \qquad ...(8.14)$$

For *smooth inclined surface* ; putting $\mu = 0$ in equations (8.13) and (8.14).

$$a = \frac{g\,(W_1 - W_2 \sin \alpha)}{W_1 + W_2} \qquad ...(8.15)$$

and

$$T = \frac{W_1 W_2 (1 + \sin \alpha)}{W_1 + W_2} \qquad ...(8.16)$$

Example 8.24. *A body weighing 8 N rests on a rough plane inclined at 15° to the horizontal. It is pulled up the plane, from rest, by means of a light flexible rope running parallel to the plane. The portion of the rope, beyond the pulley hangs vertically down and carries a weight of 60 N at the end. If the co-efficient of friction for the plane and the body is 0.22, find:*

(*i*) *The tension in the rope,*

(*ii*) *The acceleration in m/s², with which the body moves up the plane, and*

(*iii*) *The distance in metres moved by the body in 2 seconds, starting from rest.*

Sol. Refer to Fig. 8.17.

Let T newton be the tension in the string and a m/s² the acceleration of the system.

Considering motion of 60 N weight (W_1) :

$$60 - T = \frac{60}{g} \cdot a \qquad ...(i)$$

Considering motion of 8 N weight (W_2) :

$$T - W_2 \sin \alpha - F = \frac{W_2}{g} \cdot a$$

$$T - 8 \sin \alpha - \mu N = \frac{8}{g} \cdot a$$

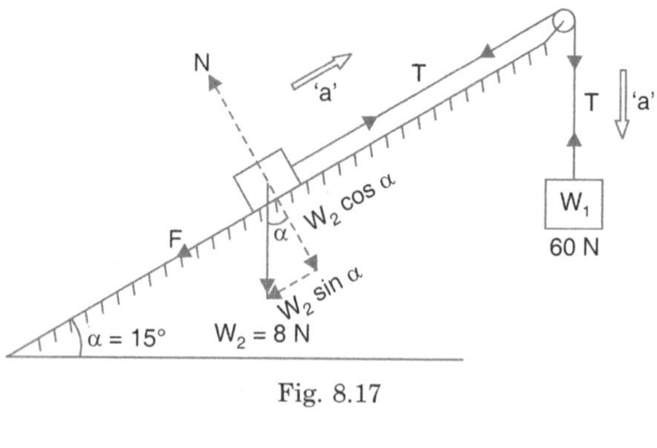

Fig. 8.17

$$T - 8 \sin \alpha - 0.22 \times 8 \cos \alpha = \frac{8}{g} . a \qquad (\because \ N = W_2 \cos \alpha = 8 \cos \alpha) \quad ...(ii)$$

Adding (*i*) and (*ii*)

$$60 - 8 \sin \alpha - 0.22 \times 8 \cos \alpha = \frac{68}{g} . a$$

$$60 - 8 \sin 15° - 1.76 \cos 15° = \frac{68}{9.81} \times a$$

$$60 - 2.07 - 1.7 = \frac{68}{9.81} \times a$$

$\therefore$ $\quad$ **a = 8.11 m/s². (Ans.)**

Substituting this value of 'a' in equation (*i*), we get

$$T = 60 - \frac{60}{9.81} \times 8.11 = \textbf{10.39 N. (Ans.)}$$

Distance moved in 5 seconds, s = ?

Initial velocity, $u = 0$

Time, $t = 2$ s.

Using the relation : $s = ut + \frac{1}{2} at^2$

$\therefore$ $$s = 0 + \frac{1}{2} \times 8.11 \times 2^2 = \textbf{16.22 m. (Ans.)}$$

Example 8.25. *Determine the resulting motion of the body '1' assuming the pulleys to be smooth and weightless as shown in Fig. 8.18. If the system starts from rest, determine the velocity of the body '1' after 5 seconds.*

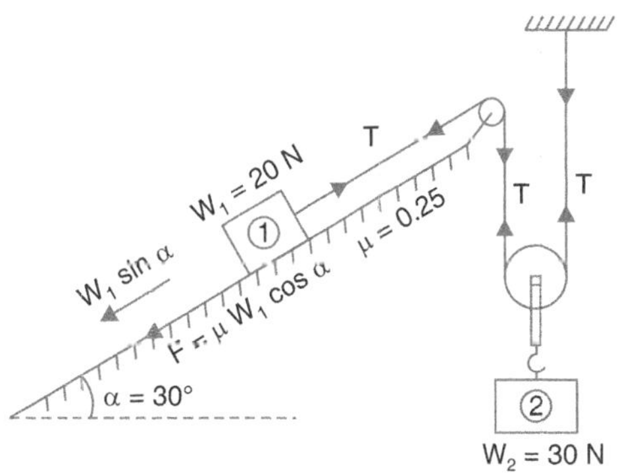

Fig. 8.18

Sol. Weight of body '1', W_1 = 20 N

Weight of body '2', W_2 = 30 N

Let T = tension in the string, and

a = acceleration of the body '1'.

Considering the motion of body '1' :

$$T - W_1 \sin \alpha - \mu W_1 \cos \alpha = \frac{W_1}{g} a$$

or $$T - 20 \sin 30° - 0.25 \times 20 \cos 30° = \frac{20}{g} \times a$$

or $$T - 10 - 4.33 = \frac{20}{g} a$$

or $$T - 14.33 = \frac{20}{g} a \quad ...(i)$$

Considering the motion of body '2' :

A little consideration will show that the acceleration of body '2' will be half the acceleration of body '1' (*i.e.*, $a/2$).

$$\therefore \quad 30 - 2T = \frac{30}{g} \times \frac{a}{2} = \frac{15}{g} a \quad ...(ii)$$

Multiplying eqn. (*i*) by 2 and adding eqns. (*i*) and (*ii*), we get

$$1.34 = \frac{55}{g} a$$

$$\therefore \quad a = \frac{1.34 \times g}{55} = \frac{1.34 \times 9.81}{55} = 0.239 \text{ m/s}^2$$

$\therefore$ Velocity of body '1' after 5 sec., if the system starts from rest,

$$\text{v} = u + at = 0 + 0.239 \times 5 = \mathbf{1.195 \text{ m/s. (Ans.)}}$$

8.14. MOTION OF TWO BODIES CONNECTED OVER ROUGH INCLINED PLANES

Fig. 8.19 shows two bodies of weight W_1 and W_2 respectively resting on the two inclined planes with inclinations α_1 and α_2 respectively.

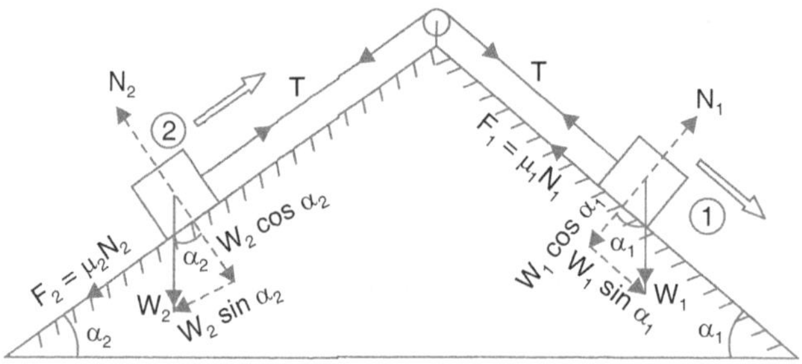

Fig. 8.19

Let a = acceleration of the system

μ_1 = co-efficient of friction between body 1 and the inclined plane 1 and

μ_2 = co-efficient of friction between body 2 and the inclined plane 2.

Consider the motion of body 1 :

Normal reaction at the surface,

$$N_1 = W_1 \cos \alpha_1$$

$\therefore$ Force of friction, $\quad F_1 = \mu_1 N_1 = \mu_1 W_1 \cos \alpha_1$

The forces acting on body 1 are :

T (upwards), force of friction F_1 (upwards) and $W_1 \sin \alpha_1$ (downwards) as shown in Fig. 8.19.

$\therefore$ Resultant force $= W_1 \sin \alpha_1 - T - \mu_1 W_1 \cos \alpha_1$ ...(*i*)

Since this body is moving downwards, the force acting on this body

$$= \frac{W_1}{g} . a \quad ...(ii)$$

Equating (*i*) and (*ii*)

$$W_1 \sin \alpha_1 - T - \mu_1 W_1 \cos \alpha_1 = \frac{W_1}{g} . a \quad ...(1)$$

Now consider motion of body 2 :

Normal reaction at the surface,

$$N_2 = W_2 \cos \alpha_2$$

$\therefore$ Force of friction, $F_2 = \mu_2 N_2 = \mu_2 W_2 \cos \alpha_2$

The forces acting on body 2 are :

T (upwards), force of friction of F_2 (downwards) and $W_2 \sin \alpha_2$ (downwards) as shown in Fig. 8.19.

Resultant force $= T - W_2 \sin \alpha_2 - \mu_2 W_2 \cos \alpha_2$ ...(*iii*)

Since the body is moving upwards, the force acting on the body

$$= \frac{W_2}{g} a \quad ...(iv)$$

Equating (*iii*) and (*iv*)

$$T - W_2 \sin \alpha_2 - \mu_2 W_2 \cos \alpha_2 = \frac{W_2}{g} a \quad ...(2)$$

Adding eqns. (1) and (2), we get

$$W_1 \sin \alpha_1 - W_2 \sin \alpha_2 - \mu_1 W_1 \cos \alpha_1 - \mu_2 W_2 \cos \alpha_2 = \frac{a}{g} (W_1 + W_2)$$

$$\therefore \quad a = \frac{g (W_1 \sin \alpha_1 - W_2 \sin \alpha_2 - \mu_1 W_1 \cos \alpha_1 - \mu_2 W_2 \cos \alpha_2)}{W_1 + W_2} \quad ...(8.17)$$

Substituting this value of 'a' in equation (1), we get

$$W_1 \sin \alpha_1 - T - \mu_1 W_1 \cos \alpha_1 = \frac{W_1 \times g}{g} \times \frac{(W_1 \sin \alpha_1 - W_2 \sin \alpha_2 - \mu_1 W_1 \cos \alpha_1 - \mu_2 W_2 \cos \alpha_2)}{W_1 + W_2}$$

$$\therefore \quad T = (W_1 \sin \alpha_1 - \mu_1 W_1 \cos \alpha_1) - \frac{W_1 (W_1 \sin \alpha_1 - W_2 \sin \alpha_2 - \mu_1 W_1 \cos \alpha_1 - \mu_2 W_2 \cos \alpha_2)}{W_1 + W_2}$$

or

$$T = \frac{1}{(W_1 + W_2)} [(W_1 + W_2)(W_1 \sin \alpha_1 - \mu W_1 \cos \alpha_1) - W_1 (W_1 \sin \alpha_1 - W_2 \sin \alpha_2 - \mu_1 W_1 \cos \alpha_1 - \mu_2 W_2 \cos \alpha_2)]$$

$$= \frac{1}{(W_1 + W_2)} \times [W_1^2 \sin \alpha_1 - \mu_1 W_1^2 \cos \alpha_1 + W_1 W_2 \sin \alpha_1 - \mu_1 W_1 W_2 \cos \alpha_1 - W_1^2 \sin \alpha_1 + W_1 W_2 \sin \alpha_2 + \mu_1 W_1^2 \cos \alpha_1 + \mu_2 W_1 W_2 \cos \alpha_2]$$

$$= \frac{1}{W_1 + W_2} (W_1 W_2 \sin \alpha_1 + W_1 W_2 \sin \alpha_2 - \mu_1 W_1 W_2 \cos \alpha_1 + \mu_2 W_1 W_2 \cos \alpha_2)$$

$$= \left[\frac{W_1 W_2 (\sin\alpha_1 + \sin\alpha_2) - W_1 W_2 (\mu_1 \cos\alpha_1 - \mu_2 \cos\alpha_2)}{W_1 + W_2}\right]$$

$$= \frac{1}{W_1 + W_2} [W_1 W_2 (\sin\alpha_1 + \sin\alpha_2) - W_1 W_2 (\mu_1 \cos\alpha_1 - \mu_2 \cos\alpha_2)]$$

i.e.,

$$T = \frac{W_1 W_2}{W_1 + W_2} (\sin\alpha_1 + \sin\alpha_2 - \mu_1 \cos\alpha_1 + \mu_2 \cos\alpha_2) \quad ...(8.18)$$

For *smooth inclined plane* : putting $\mu_1 = 0$ and $\mu_2 = 0$ in equations (8.17) and (8.18), we get

$$a = \frac{g (W_1 \sin\alpha_1 - W_2 \sin\alpha_2)}{W_1 + W_2} \quad ...(8.19)$$

and

$$T = \frac{W_1 W_2}{W_1 + W_2} (\sin\alpha_1 + \sin\alpha_2) \quad ...(8.20)$$

Example 8.26. *Blocks A and B weighing 10 N and 4 N respectively are connected by a weightless rope passing over a frictionless pulley and are placed on smooth inclined planes making 60° and 45° with the horizontal as shown in Fig. 8.20. Determine :*

(i) The tension in the string and

(ii) Velocity of the system 3 seconds after starting from rest.

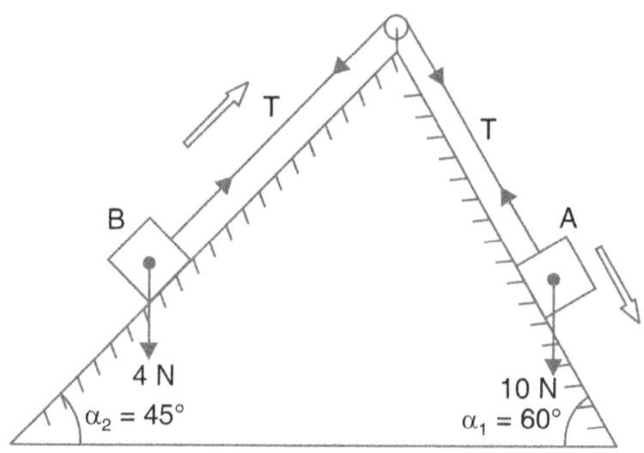

Fig. 8.20

Sol. Refer to Fig. 8.20.

Let 'T' be the tension in the rope and 'a' the acceleration of the system.

(*i*) Tension, T = ?

For block A :

Resolving forces *parallel* to the plane :

$$10 \sin 60° - T = \frac{10}{g} . a \quad ...(i)$$

For block B :

Resolving forces parallel to the plane,

$$T - 4 \sin 45° = \frac{4}{g} . a \quad ...(ii)$$

Adding (*i*) and (*ii*), we get

$$10 \sin 60° - 4 \sin 45° = \frac{14}{g} . a$$

$$8.66 - 2.83 = \frac{14}{9.81} \times a$$

$$a = 4.08 \text{ m/s}^2$$

Substituting this value of equations 'a' in (*i*), we get

$$10 \sin 60° - T = \frac{10}{9.81} \times 4.08$$

$$\therefore \quad T = 10 \sin 60° - \frac{10}{9.81} \times 4.08$$

$$= 8.66 - 4.16 = \mathbf{4.5\ N.\ (Ans.)}$$

(*ii*) Velocity after 3 seconds, v = ?

Using the relation : $v = u + at$

$$= 0 + 4.08 \times 3 \qquad (\because \; u = 0)$$

$$= \mathbf{12.24\ m/s.\ (Ans.)}$$

HIGHLIGHTS

1. *Newton's First Law of Motion* : "Every body continues in its state of rest or of uniform motion, in a straight line, unless it is acted upon by some external force to change that state".
2. Newton's Second Law of Motion : "The rate of change of momentum is directly proportional to the impressed force and takes place in the direction of the straight line in which the force acts".
3. Newton's Third Law of Motion : "To every action there is an equal and opposite reaction".
4. *Law of Conservation of Momentum* : "Total momentum of any group of objects always remains the same if no external force acts on them."
5. Impulse = mass × change in velocity.

OBJECTIVE TYPE QUESTIONS

Say 'Yes' or 'No' :

1. Kinematics deals with only the effect of force.
2. The relation between the motion and the force is defined by the laws of dynamics.
3. Momentum is the product of mass of a body and square of its velocity.
4. Momentum is a scalar quantity.
5. Inertia is an inherent property of a body which offers resistance to change in its state of rest or uniform motion.
6. Issac Newton enunciated laws of motion in 1699.
7. Mass of a body is equal to the weight of the same body in gravitational units.
8. Recoil of a gun is an example of conservation of momentum.
9. When we hit a hockey stick, we give it an impulse.

Choose the Correct Answer :

10. In case of motion of two bodies ($W_1 > W_2$) connected by a string passing over a smooth pulley, the tension (T) in the string is given by

(*a*) $\dfrac{W_1 W_2}{W_1 + W_2}$ (*b*) $\dfrac{2 W_1 W_2}{W_1 + W_2}$

(*c*) $\dfrac{3 W_1 W_2}{W_1 + W_2}$ (*d*) $\dfrac{4 W_1 W_2}{W_1 + W_2}$.

11. In case of motion of two bodies ($W_1 > W_2$) connected at the edge of a rough horizontal surface the tension (T) in the string is given by

(*a*) $\dfrac{W_1 W_2 (1 + \mu)}{W_1 + W_2}$ (*b*) $\dfrac{2 W_1 W_2 (1 + \mu)}{W_1 + W_2}$

(*c*) $\dfrac{W_1 W_2 (1 - \mu)}{W_1 + W_2}$ (*d*) $\dfrac{4 W_1 W_2 (1 - \mu)}{W_1 + W_2}$.

12. In case of motion of two bodies ($W_1 > W_2$) connected by a string one end of which is hanging free and the other lying on a smooth inclined plane, the tension, (T) in the string is given by

(*a*) $\dfrac{W_1 W_2 (1 + \sin \alpha)}{2 W_1 + W_2}$ (*b*) $\dfrac{W_1 W_2 (1 - \sin \alpha)}{2 W_1 + W_2}$

(*c*) $\dfrac{W_1 W_2 (1 + \sin \alpha)}{W_1 + W_2}$ (*d*) $\dfrac{2 W_1 W_2 (1 + \sin \alpha)}{2 W_1 + W_2}$.

13. In case of motion of two bodies ($W_1 > W_2$) connected over rough inclined planes, the tension (T) in the string is given by

(*a*) $\dfrac{W_1 W_2}{W_1 + W_2} (\sin \alpha_1 + \sin \alpha_2)$ (*b*) $\dfrac{2 W_1 W_2}{W_1 + W_2} (\sin \alpha_1 + \sin \alpha_2)$

(*c*) $\dfrac{2 W_1 W_2}{W_1 + W_2} (\sin \alpha_1 - \sin \alpha_2)$ (*d*) $\dfrac{4 W_1 W_2}{W_1 + W_2} (\sin \alpha_1 - \sin \alpha_2)$.

Answers

1. Yes	**2.** Yes	**3.** No	**4.** No	**5.** Yes
6. No	**7.** Yes	**8.** Yes	**9.** Yes	**10.** (*b*)
11. (*a*)	**12.** (*c*)	**13.** (*a*)		

Unsolved Examples

1. A body has a mass of 200 kg. Find its weight in (*i*) Absolute units ; (*ii*) Gravitational/Engineering units. [**Ans.** (*i*) 1962 N, (*ii*) 200 kgf]
2. A body of mass 100 kg has its velocity changed from 6 m/s to 10 m/s in the same direction in 40 seconds. Find (*i*) the change in momentum, (*ii*) the force responsible for this change in (*a*) Absolute units, (*b*) Engineer's/Gravitational units. [**Ans.** 400 kg m/s., 10 N ; 40.7 kgf sec., 102 kgf]
3. A body of mass 300 kg is acted upon by a force of 200 N for 90 seconds. If the initial velocity of the body is 20 m/s, determine the velocity of the body : (*i*) when the force acts in the direction of motion, and (*ii*) when the force acts in the opposite direction of motion.
[**Ans.** (*i*) 80 m/s, (*ii*) – 40 m/s (*i.e.,* body will be moving in the opposite direction)]
4. A body of mass 500 kg, initially at rest, is acted upon by a force of 200 N. Find the velocity of the body after the force has been acting for 5 sec. and the distance through which the body will travel in these 5 sec. [**Ans.** 2 m/s ; 5 m]
5. A man of mass 65 kg dives vertically downwards into a swimming pool from a tower of height 19.6 m. He was found to go down in water by 2 m and started rising. Find the average resistance of the water. Neglect the resistance of air. [**Ans.** 6243 N]
6. A bullet of mass 81 gm and moving with a velocity of 300 m/s is fired into a log of wood and it penetrates to a depth of 100 mm. If the bullet moving with the same velocity, were fired into a similar piece of wood 50 mm thick, with what velocity would it emerge ? Find also the force of resistance, assuming it to be uniform. [**Ans.** 212.13 m/s, 36450 N]
7. The tractive force, exerted by a railway car weighing 40 kN, is 1600 N. If the frictional resistance is 4 N per kN of the railway car's weight, determine the acceleration when the railway car is moving on a level track. [**Ans.** 0.353 m/s^2]
8. A body of mass 20 kg falls on the muddy ground from a height of 39.2 m. The body penetrates into the ground. Find the distance through which the body will penetrate into the ground, if the resistance offered by the ground to penetration is constant and equal to 980 N. Take g = 9.8 m/s^2. [**Ans.** 9.8 m]
9. A bullet of mass 200 gm, is fired horizontally with a velocity of 30 m/s, from a gun carried in a carriage, which together with the gun has mass of 100 kg. The resistance to sliding of the carriage over the ice on which it rests is 19.6 N. Find :
   (*i*) Velocity, with which the gun will recoil.
   (*ii*) Distance, in which it comes to rest.
   (*iii*) Time taken to do so. [**Ans.** (*i*) 0.06 m/s, (*ii*) 9 mm, (*iii*) 0.3 s]

10. A train of weight 2000 kN starts from rest and attains a speed of 100 km/h in 4 minutes. If the frictional resistance of the track is 8 N per kN of the train's weight, find the average pull required. Take $g = 9.8$ m/s^2. **[Ans.** 39.59 kN]

11. An elevator of weight 2000 N is moving vertically downwards with a uniform acceleration. Write the equation for the elevator cable tension. Starting from rest it travels a distance of 30 m during an interval of 12 seconds. Find the cable tension during this time. Neglect all other resistances to motion. What are the limits of cable tension ?

[Ans. 1915.2 N, at $a = 0$, $T = 2000$ N and at $a = 9.81$ m/s^2, $T = 0$]

12. Two bodies of weight 50 N and 30 N are connected to the two ends of a light inextensible string. The string is passing over a smooth pulley. Determine :

(*i*) The acceleration of the system, and

(*ii*) Tension in the string. Take $g = 9.8$ m/s^2. **[Ans.** (*i*) 2.45 m/s^2, (*ii*) 37.5 N]

13. A system of frictionless pulleys carries two masses hung by inextensible cords as shown in Fig. 8.21. Find the tension in the cords and acceleration of the weights.

[Ans. 630 N, 1.4 m/s^2 (mass *A*), 2.8 m/s^2 (mass *B*)]

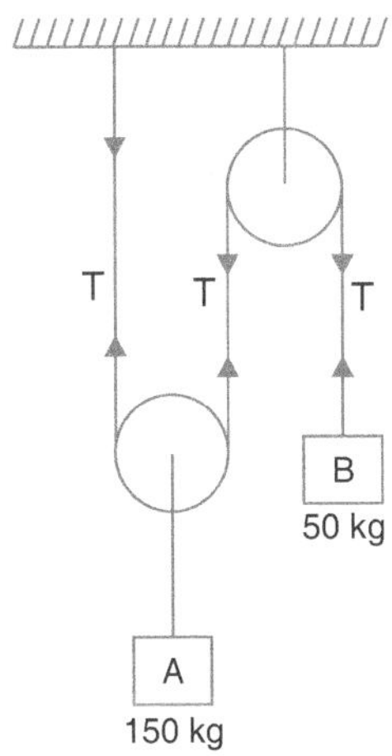

Fig. 8.21

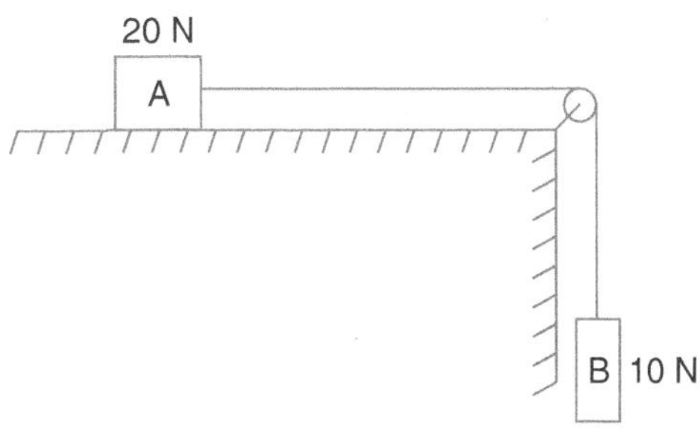

Fig. 8.22

14. Two blocks shown in Fig. 8.22, have weights $A = 20$ N and $B = 10$ N and the co-efficient of friction between the block *A* and the horizontal plane, $\mu = 0.25$.

If the system is released, from rest, and the block **B** falls through a vertical distance 2 m, what is the velocity attained by the block **B**. Neglect the friction in the pulley and the extension of the string.

[Ans. 2.56 m/s]

15. Two bodies of weight 40 N and 20 N are connected to the two ends of a light inextensible string, passing over a smooth pulley. The weight of 40 N is placed on a smooth horizontal surface while the weight of 20 N is hanging free in air. Find :

(*i*) The acceleration of the system, and

(*ii*) The tension in the string.

Take $g = 9.81$ m/s^2. **[Ans.** (*i*) 3.27 m/s^2, (*ii*) 13.33 N]

16. Two bodies of weights 60 N and 40 N are connected to the two ends of a light inextensible string, which passes over a smooth pulley. The weight 60 N is placed on a smooth inclined plane of angle of inclination of 10°, while the weight 40 N is hanging free in air. Determine :

(*i*) Acceleration of the system, and

(*ii*) Tension in the string.

Take $g = 9.81$ m/s^2. **[Ans.** (*i*) 2.89 m/s^2, (*ii*) 28.16 N]

17. A body of mass 150 kg, rests on a rough plane inclined at 10° to the horizontal. It is pulled up the plane, from rest, by means of a light flexible rope running parallel to the plane. The portion of the rope, beyond the pulleys hangs vertically down and carries a mass of 80 kg at the end. If the co-efficient of friction, for the plane and the body is 0.2, find :

(*i*) The tension in the rope.

(*ii*) The acceleration in m/s², with which the body moves up the plane.

(*iii*) The distance in metres moved by the body in 4 seconds starting from rest. [**Ans.** (*i*) 189.5 N, (*ii*) 1.04 m/s², (*iii*) 8.32 m]

18. Find the tension in the string during motion of the system shown in Fig. 8.23. μ = 0.2. Neglect the mass of the pulley. Take g = 9.81 m/s². [**Ans.** 63.38 N]

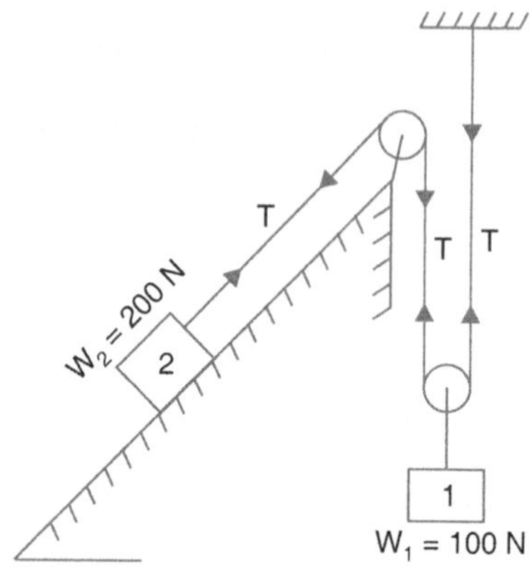

Fig. 8.23

19. Masses of 5 kg of each rest on two inclined planes each of elevation 30° and are connected by a string passing over a common apex. After two seconds the mass of 5 kg is removed. How far up the plane will the 3 kg mass continue ? [**Ans.** 0.647 m]

20. Two rough planes inclined at 30° and 15° to the horizontal and of the same height are placed back to back. Masses of 15 kg and 5 kg are placed on the faces and connected by a string over the top of the planes. If the co-efficient of friction be 0.3, find from fundamentals the resulting acceleration.

[**Ans.** 1.32 m/s²]

9

Work, Power and Energy

9.1. Concept of work. 9.2. Units of work. 9.3. Graphical representation of work. 9.4. Power—Units of power—Indicated power (I.P.)—Brake power (B.P.). 9.5. Law of conservation of energy—Highlights—Objective Type Questions—Unsolved Examples.

9.1. CONCEPT OF WORK

In 'Mechanics' work means accomplishment. A *force is said to have done work if it moves the body, on which it acts, through a certain distance.* A force is not able to produce any displacement-translational or rotational no work is said to have been done.

Work is measured by the *product of force (P) and displacement (S)* both being in the same direction. Work is positive or negative, according as the force acts in the same direction or in the direction opposite to the direction of displacement, Mathematically,

Work = Force × Displacement in the direction of force

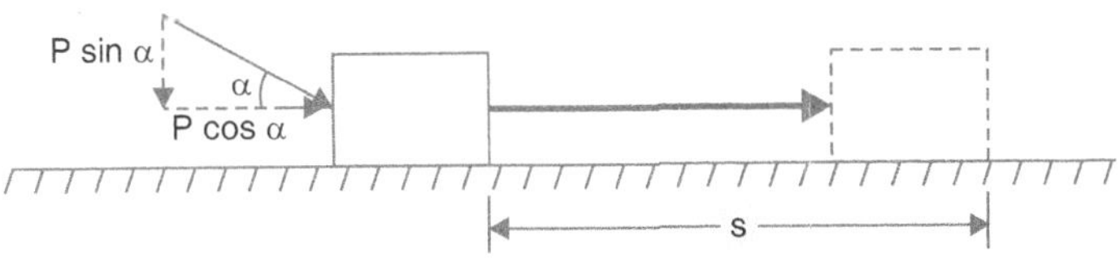

Fig. 9.1

Fig. 9.1 shows the force P acting at angle α to the direction of displacement. Then

Work done by the force

$$\text{`}P\text{'} = \text{Force} \times \text{displacement in the direction of force}$$

$$\therefore \quad W = P \cos \alpha \times s$$

or $$W = PS \cos \alpha \quad \text{...(9.1)}$$

If angle $\alpha = 0$ *i.e.,* the force P acts in the direction of displacement or motion, then

$$W = P \times s \quad \text{...(9.2)}$$

Note. It may be noted that *only the component of force in the direction of motion does the work.*

9.2. UNITS OF WORK

The units of work depend on the units of force and displacement.

In S.I. system, the force is expressed in newtons (N) whereas the distance is expressed in metres (m). Hence, the unit of work is Nm. When a force of 1 N displaces the body through 1 m, 1 Nm work is said to be done. It is also called *Joule* (briefly written as J). Mathematically.

$$1 \text{ Joule} = 1 \text{ Nm}$$

9.3. GRAPHICAL REPRESENTATIONS OF WORK

If we plot a graph between the force during any displacement along the Y-axis and the corresponding displacement along the X-axis, then the *area under the graph* (also called the *work diagram*) gives the work done. Figs. 9.2 and 9.3 show the graphs when the force is constant and variable respectively.

Example 9.1. *A trolley of mass 225 kg moves on a level track for a distance of 250 m. If the resistance of the track is 200 N, find the work done in moving the trolley.*

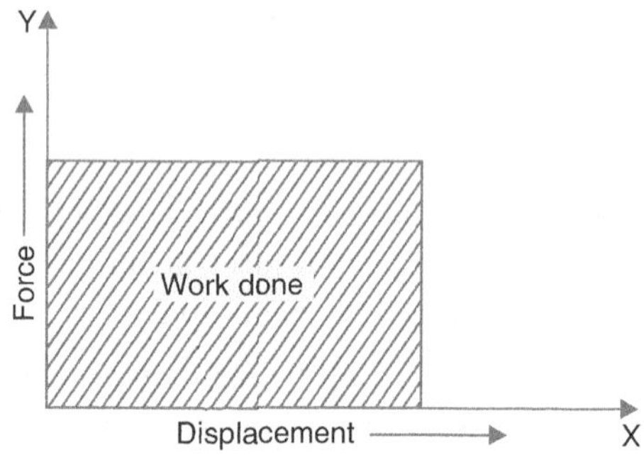

Fig. 9.2 When the force is constant.

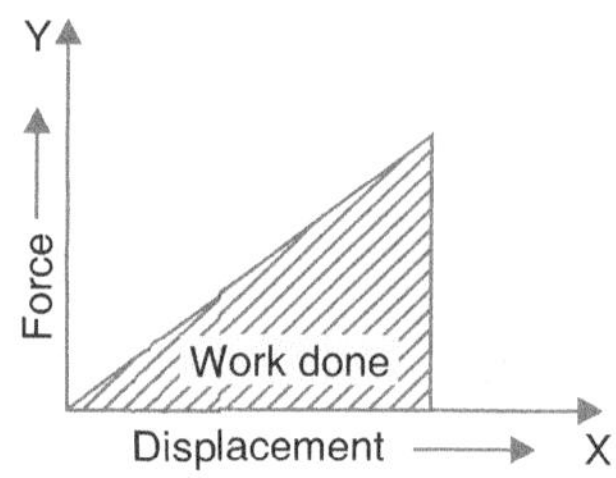

Fig. 9.3 When the force is variable.

Sol. Resistance of the track, $F\ (= P) = 200$ N

Distance moved, $s = 250$ m

$\therefore$ Work done in moving the trolley

$$= P \times s = 200 \times 250 \text{ Nm or J}$$

$$= \frac{200 \times 250}{1000} = \textbf{50 kJ.} \quad \textbf{(Ans.)}$$

Example 9.2. *When a force is applied to a certain spring it gets stretched by 25 mm. Calculate the work done, if the force required to stretch 1 mm of the spring is 20 N.*

Sol. Amount by which the spring is stretched, $s = 25$ mm.

Maximum force required to stretch the spring by 25 mm

$$= 20 \times 25 = 500 \text{ N.}$$

$\therefore$ Average force, $P = \dfrac{500 + 0}{2} = 250$ N.

Hence, work done = average force × distance

$$= 250 \times 25 = 6250 \text{ Nmm}$$

$$= 6.25 \text{ Nm} = \mathbf{6.25 \text{ J.}} \quad \textbf{(Ans.)}$$

9.4. POWER

Power is defined as *rate of doing work*. If an agent performs W units of work in t seconds, then

$$\text{Power} = \frac{\text{work done}}{\text{time taken}} = \frac{W}{t} \qquad ...(9.3)$$

$$= \frac{P \times s}{t} \qquad (\because \; W = P \times s)$$

$$= P \times v \qquad \left(\because \; v = \frac{s}{t}\right) \qquad ...(9.4)$$

Hence, power = force × velocity.

9.4.1. Units of Power

The unit of power is called *horse power* (H.P.). An agent is said to possess one H.P. If it can perform work at the rate of 75 m-kgf/sec or 4500 m-kgf/min. (**MKS System**). In S.I. unit the unit of power is watt (= 1 N m/s). Generally, a bigger unit of power *i.e.*, kW is used, which is equal to 1000 watts.

In MKS system the horse power developed by the force,

$$\text{H.P.} = \frac{P \times v}{75}, \text{ where } P \text{ is in kgf and } v \text{ is in m/sec.} \qquad ...[9.4(a)]$$

In S.I. system,

$$\text{Power} = \frac{P \times v}{1000} \text{ kW} \qquad ...[9.4\,(b)]$$

where, P is in newtons (N) and v is in m/sec.

9.4.2. Indicated Power (I.P.)

"Indicated power" is *the power developed in the cylinder of an engine.* It is usually measured from the indicator diagram drawn with the help of particular engine indicator. This represents the rate at which the work is being done on the head of the piston by the working agent.

9.4.3. Brake Power (B.P.)

"Brake power" is the power available at pulley or shaft of the engine. It is also called *shaft power.* It is called *brake power because it is measured by means of a brake drum.* B.P. is always *less* than I.P. because some of the power generated in the engine under is used up in overcoming friction of the living parts of the engine; this is known as *Frictional power* (F.P.)

$$\therefore \qquad \text{B.P.} = \text{I.P.} - \text{F.P.} \qquad ...(9.5)$$

Also, mechanical efficiency,

$$\eta_{mech.} = \frac{\text{B.P.}}{\text{I.P}} \times 100 \qquad ...(9.6)$$

Example 9.3. *A car weighing 8000 N accelerates from rest to a speed of 45 km / h in a distance of 50 m against a resistance of 100 N. Find the average driving force acting on the car. Using the average force, find the greatest power developed by the engine.*

Sol. *Given :* W = 800 N; $u = 0$,

$$v = 45 \text{ km/h} = \frac{45 \times 1000}{60 \times 60} = 12.5 \text{ m/s}; \; s = 50 \text{ m};$$

$$R = 100 \text{ N}.$$

Let P be the average driving force exerted by the engine.

Net force acting on the car = $(P - 100)$ N

Now, work done by force = gain in K.E.

or, $$(P - 100) \times 50 = \frac{1}{2} \times \frac{8000}{9.81} \times (12.5)^2$$

∴ $$P = \textbf{1374.21 N.} \quad \textbf{(Ans.)}$$

Power developed by the engine = $P \times v$

$$= 1374.21 \times 12.5 = 17177.6 \text{ Nm/s or J/s}$$

or, $$W = \textbf{17.177 kW.} \quad \textbf{(Ans.)}$$

Example 9.4. *A mechanical device can lift a weight of 24 kN suspended by a heavy rope weighting 60 N per metre. If the length of the rope is 360 metres and it takes 30 seconds in winding up the load to the top, find the power of the machine.*

Sol. *Given:* Weight to be lifted, W = 24 kN = 24000 N

Weight of the rope per metre length = 60 N/m

Length of rope = 360 m

Time taken to wind up the load = 30 s

Power of the machine:

Mean weight of the rope $= \dfrac{60 \times 360}{2} = 10800$ N

Total force against which work is performed

$$= P = 24000 + 10800 = 34800 \text{ N}$$

Distance, $s = 360$ m

∴ Work done per second

$$= \frac{P \times s}{t} = \frac{34800 \times 360}{30} \text{ W} = 417600 \text{ W}$$

or, **417.6 kW. (Ans.)**

Example 9.5. *Water from a tank of capacity 18000 litres it to be lifted in 20 minutes by a pump through a height of 12 m. If the efficiency of the pump is 65 percent, find the power of the pump.*

Sol. *Given:* Capacity of water tank = 18000 litres

Mass of water = 18000 kg (∵ 1 litre of water = 1 kg)

Time during which water is to be lifted = 20 minutes

Height through which water is to be lifted = 12 m

Efficiency of the pump = 65%

Power of the pump:

$$\text{Work done} = mgh$$

$$= 18000 \times 9.81 \times 12 = 2118960 \text{ J}$$

$$\text{Power} = \frac{\text{work done}}{\text{time taken}} = \frac{2118960}{20 \times 60} = 1765.8 \text{ W}$$

$$\eta = 0.65 = \frac{\text{power output}}{\text{power input}}$$

$\therefore$ Power input $= \dfrac{\text{power output}}{0.65} = \dfrac{1765.8}{0.65} = 2716.6$ W

or, **2.716 kW. (Ans.)**

9.5. LAW OF CONSERVATION OF ENERGY

In its most general form, the law of conservation of energy states:

"The total amount of energy in the universe is constant; energy cannot be created or destroyed although it may be converted into various forms".

An important particular case of the principle of conservation of energy is the following theorem.

"A body of mass 'm' fall from rest at a height 'h' above the ground. The sum of its potential and kinetic energies remains constant throughout the motion".

Proof. Refer to Fig. 9.4. Let a body of mass 'm' be lifted from the ground to a position A, so that it is at a height 'h' above the ground.

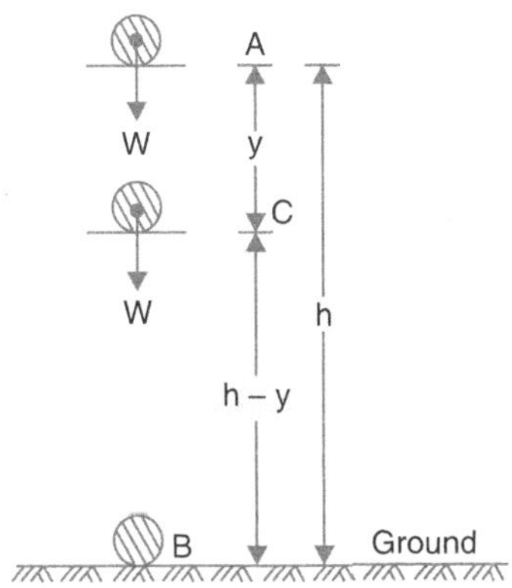

Fig. 9.4

Energy at A:

Since the body is at rest, at A

$$\text{P.E. at } A = mgh$$

$$\text{K.E. at } A = 0$$

$\therefore$ Total energy at $A = mgh + 0 = mgh$ ...(1)

Energy at C:

When the body is allowed to fall it begins to move, thus acquiring kinetic energy. Let us calculate its K.E. and P.E. at the instant it is at a position C so that it has fallen through a distance 'y' and is at a height $(h - y)$ above the ground. Therefore,

$$\text{P.E. at } C = mg\,(h - y)$$

Let 'v' be the velocity when the body *reaches* at C.

We know that $v^2 - u^2 = 2as$

Since $u = 0$, $s = y$ and $a = g$,

We have $v^2 = 2gy$

$\therefore \quad \text{K.E. at } C = \frac{1}{2} mv^2 = \frac{1}{2} m\,(2gy)$

$= mgy$

Hence total energy at $C \quad = mg\,(h - y) + mgy$

$= mgh$ (same as the energy at A) ...(2)

Thus, *a part of the potential energy is converted into K.E., but the sum of the two energies is the same.*

Energy at B:

When the body reaches at B

$$y = h$$

$$\text{P.E. at } B = mg\,(h - y) = mg\,(h - h) = 0$$

$$\text{K.E. at } B = mgy = mgh$$

$\therefore$ Total energy at $B = 0 + mgh$

$= mgh$ or Wh ...(3)

Thus, although the K.E. and P.E. individually change throughout the motion, *their sum always remains constant.*

Example 9.6. *A body of mass 5 kg is thrown up vertically with a velocity of 9.8 m/s. What is its kinetic energy?*

(*i*) *at the moment of propulsion ;* (*ii*) *after half second ;*

(*iii*) *after one second? Assume g = 9.81 m/s².*

Sol. *Given*: $m = 5$ kg ; $u = 9.8$ m/s

(*i*) **K.E. at the moment of propulsion:**

$$\text{K.E.} = \frac{1}{2} mu^2 = \frac{1}{2} \times 5 \times (9.8)^2$$

$$= \mathbf{240.1\ Nm.} \quad \textbf{(Ans.)}$$

(*ii*) **K.E. after half second:**

Velocity after half second,

$$v = u - gt = 9.8 - 9.81 \times 0.5 = 4.895 \text{ m/s}$$

$$\therefore \quad \text{K.E.} = \frac{1}{2} \times 5 \times (4.895)^2 = \mathbf{59.9\ Nm.} \quad \textbf{(Ans.)}$$

(*iii*) **K.E. after one second:**

Velocity after one second,

$$v = u - gt = 9.8 - 9.81 \times 1 = 0$$

$$\therefore \quad \text{K.E.} = \frac{1}{2} \times 5 \times 0^2 = \mathbf{0.} \quad \textbf{(Ans.)}$$

Example 9.7. *A bullet of mass 80 gm moving with a velocity of 250 m/s is fired into a block of wood and gets embedded in it to a depth of 20 cm. What will be the resistance offended by the wood ?*

Sol. *Given:* $m = 80$ gm $= 0.08$ kg ;

$v = 250$ m/s ; $s = 20$ cm $= 0.2$ m.

Resistance offered by the wood, R:

Loss of K.E. of the bullet = work done against resistance

or, $$\frac{1}{2} mv^2 = R \times s$$

or, $$\frac{1}{2} \times 0.08 \times 250^2 = R \times 0.2$$

$\therefore$ **R** = 12500 N or **12.5 kN.** **(Ans.)**

Example 9.8. *A hammer head weighing 70 ms arranged to swing downwards in a circular path. It is released at a point 1.2 m higher than the lowest point of the circle. Find the kinetic energy and the speed of the hammer head at the lowest point. If at that point it breaks a piece of metal and so parts with 42 Nm of energy, to what height will it rise on the other side ?*

Sol. Weight of the hammer, $W = 70$ N

Height above the lowest point of the circle = 1.2 m

$\therefore$ P.E. of the hammer $= Wh = 70 \times 1.2$

$= 84$ Nm (or Joule)

Because rise/gain in K.E. = loss of P.E.

$\therefore$ K.E. at the lowest point = **84 Nm. (Ans.)**

Speed of the hammer head at the lowest point, v :

Using the relation $v^2 - u^2 = 2gh$, we get

$$v^2 - 0^2 = 2 \times 9.81 \times 1.2$$

or, $$\mathbf{v} = \sqrt{2 \times 9.81 \times 1.2} = \mathbf{4.85\ m/s.}$$ **(Ans.)**

Energy lost in breaking the piece of metal = 42 Nm

$\therefore$ Balance energy of the hammer

$= 84 - 42 = 42$ Nm

As the hammer rises, this energy is converted back into potential energy. If h' is the height at which the hammer comes to rest, then

$$Wh' = 42$$

or, $$70h' = 42$$

or, $$\mathbf{h'} = \frac{42}{70} = \mathbf{0.6\ m.}$$ **(Ans.)**

Example 9.9. *A bullet of mass 0.03 kg is fired horizontally into a body of mass 10 kg which is suspended by the string of 1 m length. Due to this impact, the body swings through an angle of 30° with the vertical. Find the velocity of the bullet at impact.*

Sol. Refer to Fig. 9.5

Mass of the bullet, $m = 0.03$ kg

Mass of the body, $M = 10$ kg

Length of the string = 1 m

Angle of swing = 30°.

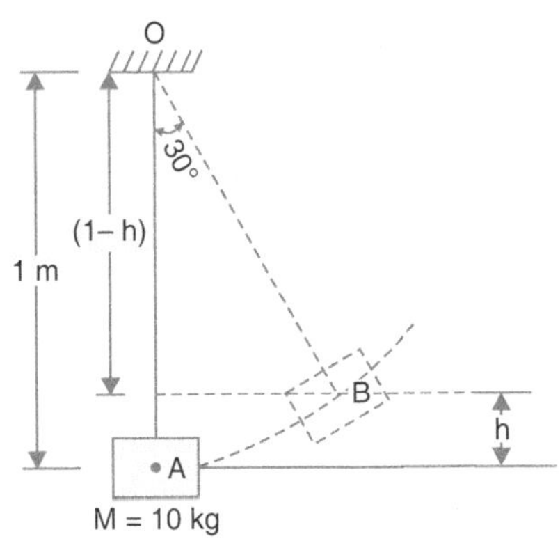

Fig. 9.5

Velocity of the bullet at impact, v :

Let h = height between A and B during the swing.

Now, momentum before impact = momentum after impact.

$$\therefore \quad m \times v = (m + M)\, V$$

[where, v = velocity of the bullet, V = common velocity of bullet and the body]

or, $$0.03\, v = (0.03 + 10)\, V$$

$$0.03\, v = 10.03\, V \quad ...(1)$$

After the bullet is fired the block takes the place at B at height h

$$\therefore \quad \text{K.E. at } A = \text{P.E. at B}$$

or, $$\frac{1}{2}(m + M)\, V^2 = (m + M)\, g \times h$$

or, $$h = \frac{V^2}{2g} = \frac{V^2}{2 \times 9.81} = 0.051\, V^2 \quad ...(2)$$

Now from Fig. 9.5,

$$\cos 30° = \frac{1 - h}{1} = 1 - 0.051\, V^2$$

or, $$0.051\, V^2 = 1 - \cos 30° = 0.134$$

or, $$V = \left(\frac{0.134}{0.051}\right)^{1/2} = 1.62 \text{ m/s}$$

Substituting this value of V in eqn. (1), we get

$$0.03\, v = 10.03 \times 1.62$$

$$\therefore \quad v = \mathbf{541.6 \text{ m/s.}} \quad \textbf{(Ans.)}$$

Example 9.10. *A pile of mass 500 kg is driven into ground by dropping freely a hammer of mass 318 kg through a height of 2.7 m. If the pile is driven into the ground by 0.15 m, calculate the average resistance of the soil.*

Sol. Refer to Fig. 9.6

Mass of hammer, m = 31.8 kg

Height of drop, h = 2.7 m

Let v = velocity attained by the falling weight.

$$\therefore \quad v = \sqrt{2gh} = \sqrt{2 \times 9.81 \times 2.7} = 7.28 \text{ m/s}$$

This is the velocity with which the hammer hits the pile. Let V be the common velocity with which the hammer and pile move together.

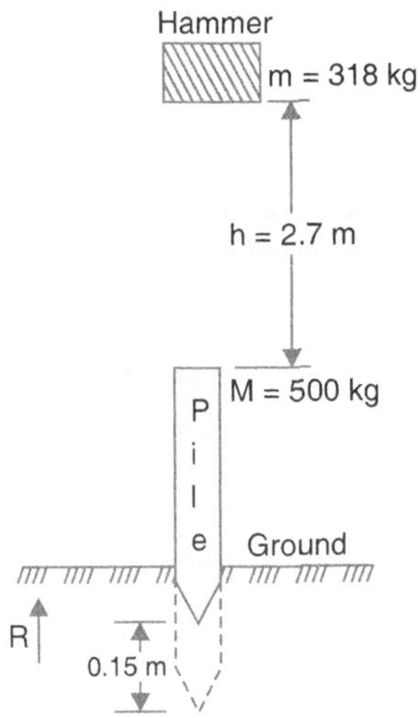

Fig. 9.6

By the principle of conservation of momentum, we have

$$mv = (M + m)\, V$$

or, $$318 \times 7.28 = (500 + 318) \times V$$

$\therefore$ $$V = \frac{318 \times 7.28}{818} = 2.83 \text{ m/s}$$

Let R be the resistance of the ground in newtons.

Now the kinetic energy of the two masses and the work done while falling through a distance $s = 0.15$ m will be absorbed by R during this distance of 0.15 m.

Now, K.E. at impact

$$= \frac{1}{2}(M + m)\, V^2$$

$$= \frac{1}{2}(500 + 318) \times 2.83^2 = 3275.6 \text{ Nm}$$

and *the work done by gravity*

$$= (500 + 318) \times 9.81 \times 0.15 = 1203.7 \text{ Nm}$$

$\therefore$ Total energy $= 3275.6 + 1203.7 = 4479.3$ Nm

Hence, by the principle of conservation of energy,

$$4479.3 = R \times 0.15$$

or, $$R = \frac{4479.3}{0.15} = 29862 \text{ N}$$

Hence, average resistance of the soil

= **29862 N. (Ans.)**

Example 9.11. *A pile of mass 1000 kg is driven vertically into the ground by the blows of pile hammer of mass 2000 kg and falling on the head of the pile from a height of 1 m, in each blow. If the average resistance of the ground is 200 kN, calculate:*

(*i*) *The pile penetration per blow,*

(*ii*) *Loss of energy at impact, and*

(*iii*) *Duration of time during which pile is in motion.*

Solution. Refer to Fig. 9.7

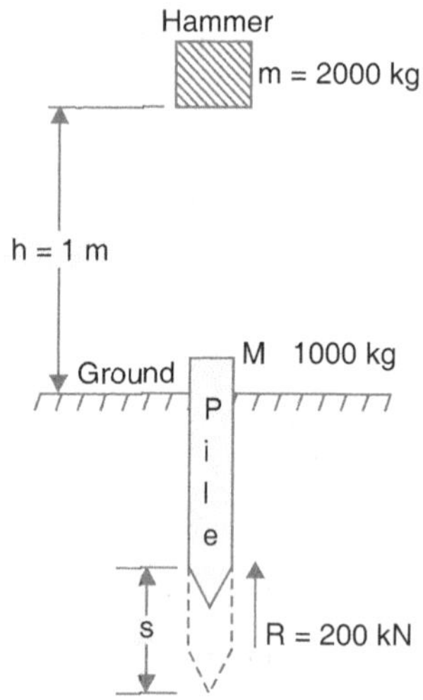

Fig. 9.7

Mass of the hammer, $m = 2000$ kg

Mass of the pile, $M = 1000$ kg

Height of fall, $h = 1$ m

Average resistance of the ground, $R = 200$ kN

(*i*) **Pile penetration per blow, *s*:**

The action of driving the pile is having potential energy of $(m + M)\, g \times s$ and kinetic energy of $\frac{1}{2}(m + M)\, V^2$; this energy is utilised in overcoming the resistance of the ground $R \times s$.

$$\therefore \qquad (m + M)\, g \times s + \frac{1}{2}(m + M)\, V^2 = R \times s \qquad ...(1)$$

where V = velocity after the impact, when hammer and pile move together.

Let us consider momentum just before and after impact.

$$mv = (m + M)\, V \qquad ...(2)$$

where, v = velocity before impact and is given by the equation $v^2 - u^2 = 2gh$

$$\therefore \qquad v = \sqrt{2gh} = \sqrt{2 \times 9.81 \times 1} = 4.43 \text{ m/s} \qquad (\because \; u = 0)$$

Now, let us find the value of V from equation (2),

$$2000 \times 4.43 = (2000 + 1000)\, V$$

$$\therefore \qquad V = 2.95 \text{ m/s}$$

Let us find '*s*' from equation (1),

$$(2000 + 1000) \times 9.81 \times s$$

$$+ \frac{1}{2}(2000 + 1000) \times (2.95)^2 = 200 \times 1000 \times s$$

or, $$29430\, s + 13053.7 = 200000\, s$$

or, $$s = 0.0765 \text{ m or } \mathbf{76.5 \text{ mm.}} \quad \textbf{(Ans.)}$$

(*ii*) **Loss of energy at impact :**

The loss of energy at impact

$$= \frac{1}{2} mv^2 - \frac{1}{2}(m + M)V^2$$

$$= \frac{1}{2} \times 2000 \times (4.43)^2 - \frac{1}{2}(2000 + 1000) \times (2.95)^2$$

$$= \mathbf{6571 \text{ Nm.}} \quad \textbf{(Ans.)}$$

(*iii*) **Duration of time during which pile is in motion, *t* :**

Using the relation : $v^2 - u^2 = 2as.$

$$0^2 - (2.95)^2 = 2 \times a \times 0.0765 = a = -\,56.88 \text{ m/s}^2$$

Also, $$v = u + at$$

$\therefore$ $$0 = 2.95 - 56.88 \times t$$

or, $$t = \mathbf{0.052 \text{ s.}} \quad \textbf{(Ans.)}$$

Example 9.12. *A block of wood weighing 6.8 N is placed on a rough horizontal floor, co-efficient of friction between the block and the floor being 0.4. A bullet of weight 0.284 N is fired with a velocity of 180 m/s into block. Find :*

(*i*) *The velocity with which the block and the bullet begin to move together after the impact,*

(*ii*) *The distance which the block moves along the floor, and*

(*iii*) *The ratio of energy lost during the impact to that lost through the friction with the floor.*

Sol. Weight of wooden block, $W = 6.8$ N

Coefficient of friction between block and the floor, $\mu = 0.4$.

Weight of the bullet, $w = 0.284$ N

velocity of the bullet $= 180$ m/s.

(*i*) Let '*u*' be the common velocity of the bullet and the block after the impact.

As per law of conservation of momentum,

Momentum of bullet before impact = momentum of bullet and block after impact

or, $$0.284 \times 180 = (6.8 + 0.284) \times u$$

$\therefore$ $$\mathbf{u} = \frac{0.284 \times 180}{(6.8 + 0.284)}$$

$$= \mathbf{7.22 \text{ m/s.}} \quad \textbf{(Ans.)}$$

(*ii*) Weight of the bullet and block = 6.8 + 0.284 = 7.084 N

Force of friction on the two $= \mu N = 0.4 \times 7.084 = 2.834$ N

Also, $$2.834 = \frac{7.084}{9.81} \times a, \textit{ a being retardation}$$

$\therefore$ $$a = \frac{2.834 \times 9.81}{7.084} = 3.92 \text{ m/s}^2$$

The distance travelled by the block is given by

$$0^2 - (7.22)^2 = 2 \times (-\,3.92) \times s \qquad [\because\ a = -\,3.92 \text{ m/s}^2]$$

$\therefore$ $$\mathbf{s} = \frac{(7.22)^2}{2 \times 3.92} = \mathbf{6.65 \text{ m.}} \quad \textbf{(Ans.)}$$

(*iii*) K.E. of the bullet before impact

$$= \frac{1}{2} \times \frac{0.284}{9.81} \times (180)^2$$
$$= 469 \text{ Nm}$$

K.E. of the bullet and block immediately after impact

$$= \frac{1}{2} \times \frac{7.084}{9.81} \times (7.22)^2$$
$$= 18.82 \text{ Nm}$$

K.E. lost during impact

$$= 469 - 18.82 = 450.18 \text{ Nm (app.)} = U_1 \text{ (say)}$$

The K.E. which the block and bullet possessed immediately after the impact is lost through friction with the floor.

K.E. lost through friction

$$= \frac{1}{2} \times \frac{7.084}{9.81} \times (7.22)^2 = 18.82 = U_2 \text{ (say)}$$

$$\therefore \quad \frac{U_1}{U_2} = \frac{450.2}{18.82} = \mathbf{23.9.} \quad \textbf{(Ans.)}$$

SELECTED QUESTIONS FROM EXAMINATION PAPERS

Example 9.13. *A locomotive draws a train of mass 400 tonnes, including its own mass, on a level ground with a uniform acceleration, until it acquires a velocity of 54 km/hour in 5 minutes. If the frictional resistance is 40 newtons/tonne of mass and the air resistance varies with the square of the velocity, find the power of the engine. Take air resistance as 500 newtons at 18 km/hour.*

Sol. Mass of the train, $m = 400$ tonnes

Velocity attained, $v = 54$ km/hour

$$= \frac{54 \times 1000}{3600} = 15 \text{ m/s}$$

Time taken, $t = 5$ min. $= 300$ s.

Frictional resistance $= 40$ N/tonne

$\therefore$ Total frictional resistance $= 40 \times 400$

$$= 16000 \text{ N or } 16 \text{ kN}$$

Let a = acceleration of the train.

To find 'a' using the relation:

$$v = u + at$$

or,
$$15 = 0 + a \times 300$$

$$\therefore \quad a = \frac{15}{300} = 0.05 \text{ m/s}^2$$

Force required for this acceleration,

$$F = m.a = 400 \times 0.05 = 20 \text{ kN}$$

As the air resistance varies with the square of the velocity, therefore air resistance at 54 km/hour

$$= 500 \times \left(\frac{54}{18}\right)^2 = 4500 \text{ N or } 4.5 \text{ kN}$$

$\therefore$ Total resistance $= 16 + 20 + 4.5 = 40.5$ kN

Hence, power of the engine $= P \times v = 40.5 \times 15$

$$= \mathbf{607.5 \text{ kW.}} \quad \textbf{(Ans.)}$$

Example 9.14. *A train of weight 1500 kN is drawn up an incline of 1 in 150 with a uniform speed of 24 km/h. If the resistance due to friction, air etc., is 4 N/kN, calculate the power of the engine.*

Sol. Weight of the train = 1500 kN

Speed of the train = 24 km/h

$$= \frac{24 \times 1000}{60 \times 60} = \frac{20}{3} \text{ m/s}$$

Frictional resistance,

$$F = 4 \times 1500 = 6000 \text{ N}$$

Resolved part of the weight of the train parallel to the track

$$= W \sin \alpha = 1500 \times 10^3 \times \frac{1}{150} \quad \left[\because \tan \alpha = \sin \alpha = \frac{1}{50}, \text{ when } \alpha \text{ is very small}\right]$$

$$= 10000 \text{ N}$$

Total resistance to be overcome

$$= F + W \sin \alpha = 6000 + 10000$$

$$= 16000 \text{ N}$$

Engine must apply a force of 16000 N for moving the train with a uniform speed.

Power = work done/sec. = force × speed

$$= 16000 \times \frac{20}{3} \text{ watts} = 106667 \text{ watts}$$

$$= \frac{106667}{1000} = \mathbf{106.67 \text{ kW.}} \quad \textbf{(Ans.)}$$

Example 9.15. *A locomotive starts from rest and pulls a train weighing 10000 kN down an incline of 1 in 100. The tractive resistance is 3 N/kN. If the constant engine pull is 50 kN, find the speed of the train after it has moved 2000 m. Find also the final power developed by the engine.*

Sol. Refer to Fig. 9.8

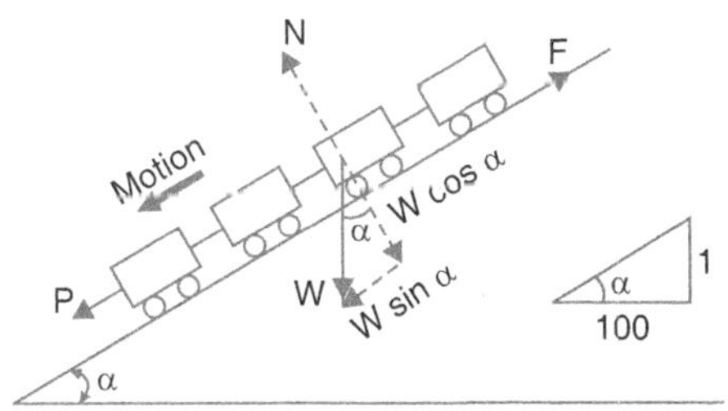

Fig. 9.8

Weight of train, $W = 10000$ kN

Tractive resistance, $F = 3 \times 10000 = 30000$ N

Slope = 1 in 100

Constant engine pull = 50 kN or 50000 N

Distance moved, $s = 2000$ m

Speed of the train, v :

Let a = acceleration of the locomotive.

The equation of motion is:

$$P + W \sin \alpha - F = \frac{W}{g} . a$$

$$50000 + 10000000 \times \frac{1}{100} - 30000 = \frac{10000000}{9.81} \times a$$

$$\left[\because \tan \alpha = \sin \alpha = \frac{1}{100} \text{ when } \alpha \text{ is very small}\right]$$

or,
$$120000 = \frac{10000000}{9.81} \times a$$

$$\therefore \quad a = \frac{1200000 \times 9.81}{10000000} = 0.1177 \text{ m/s}^2$$

Also, $u = 0, s = 2000$ m

Using the relation : $v^2 - u^2 = 2as$, we get

$$v^2 - 0^2 = 2 \times 0.1177 \times 2000$$

or,
$$\mathbf{v} = \sqrt{2 \times 0.1177 \times 2000}$$

$$= \mathbf{21.7 \text{ m/s.}} \quad \textbf{(Ans.)}$$

Power developed by the engine:

Work done/sec. against resistance

$$= 30000 \times 21.7 \text{ watts}$$

$$\therefore \text{ Power developed} = \frac{30000 \times 21.7}{1000}$$

$$= \mathbf{651 \text{ kW.}} \quad \textbf{(Ans.)}$$

Example 9.16. *A 5 kg mass drops from 1.5 m upon a string whose modulus is 10 N/mm. What will be the speed of the block when the spring is compressed 60 mm?*

Sol. *Given:* $m = 5$ kg; $h = 1.5$ m;

$\delta = 60$ mm $= 0.06$ m; $k = 10$ N/mm $= 10000$ N/m.

Potential energy of mass $= mg\,(h + \delta)$

$$= 5 \times 9.81 \times (1.5 + 0.06) = 76.52 \text{ Nm}$$

The work done by the spring on the mass is negative.

$$\text{Work done by spring} = -\frac{1}{2} k\delta^2$$

$$= -\frac{1}{2} \times 10000 \times (0.06)^2$$

$$= -18 \text{ Nm}$$

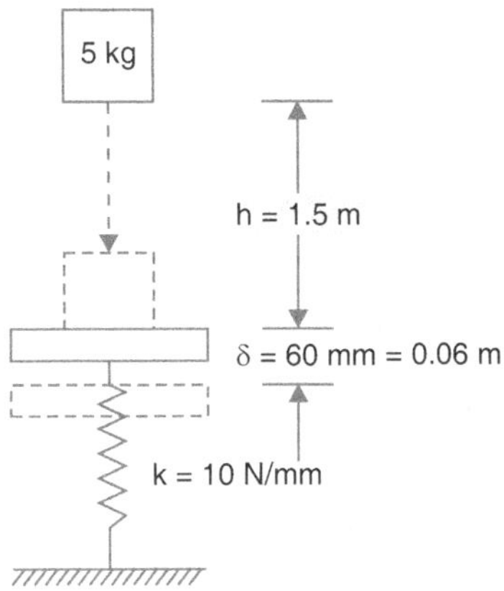

Fig. 9.9

Let v be the velocity/speed of the block. Then, K.E. of the mass

$$= \frac{1}{2} mv^2 = \frac{1}{2} \times 5 \times v^2 = 2.5\, v^2$$

Now, K.E. = net P.E.

i.e., $2.5\, v^2 = 76.52 - 18 = 58.52$

$\therefore$ v = **4.838 m/s. (Ans.)**

Example 9.17. *A 45 N weight is dropped on a pan 450 mm below it. Pan is mounted on spring (spring constant 1800 N/m). Assume impact to be perfect plastic, weight of pan, W_p = 113 N as shown in Fig. 9.10. Determine:*

(*i*) *The velocity of the weight and pan immediately after impact.*

(*ii*) *The maximum compression of the spring where system comes to rest.*

Sol. *Given:* $W = 45$ N; $h = 450$ mm $= 0.45$ m; $k = 1800$ N/m; $W_p = 113$ N

(*i*) **The velocity of the weight and pan immediately after impact v_1 :**

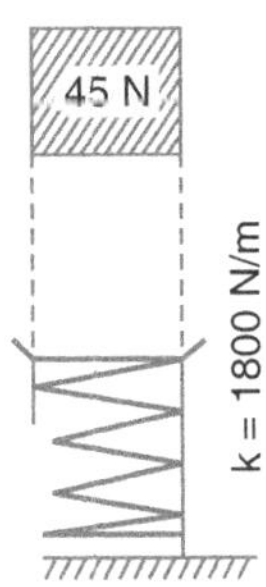

Fig. 9.10

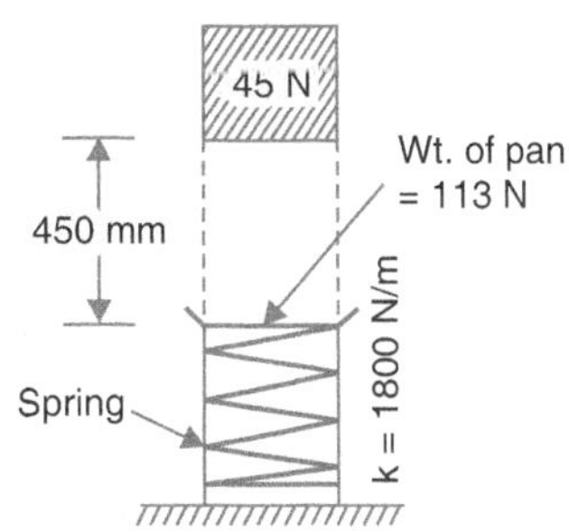

Fig. 9.11

The velocity of the weight W after moving done a height of 0.45 m,

$$= \sqrt{2gh} = \sqrt{2 \times 9.81 \times 0.45} = 2.971 \text{ m/s}$$

$$\left[\because\ v^2 - u^2 = 2gh \text{ or, } v^2 - 0 = 2gh \quad \therefore\ v = \sqrt{2gh}\right]$$

Combined velocity of weight and pan immediately after impact v_1 is found as follows:

Using law of conservation of momentum, we get

$$\frac{45}{g} \times v = \left(\frac{45 + 113}{g}\right) v_1$$

$$\therefore \qquad v_1 = \frac{45 \times 2.971}{(45 + 113)} = \mathbf{0.846\ m/s.} \quad \textbf{(Ans.)}$$

(*ii*) **The maximum compression of the system, δ :**

Loss of P.E.

$$= 45 \times 0.45 - \frac{1}{2}\left(\frac{45 + 113}{9.81}\right) \times 0.846^2$$

$$= 20.25 - 5.77 = 14.48$$

Let spring be compressed by δ metres.

Work done = Loss in P.E.

or, $$\frac{1}{2} k\delta^2 = 14.48$$

or, $$\frac{1}{2} \times 1800 \times \delta^2 = 14.48$$

$$\therefore \qquad \delta = \mathbf{0.1268\ m.} \quad \textbf{(Ans.)}$$

Example 9.18. *A block of weight 654 N is compressed against a spring as shown in Fig. 9.12. How far from 'B' will it strike on place at A ? Take k = 4000 N/m and μ = 0.2.*

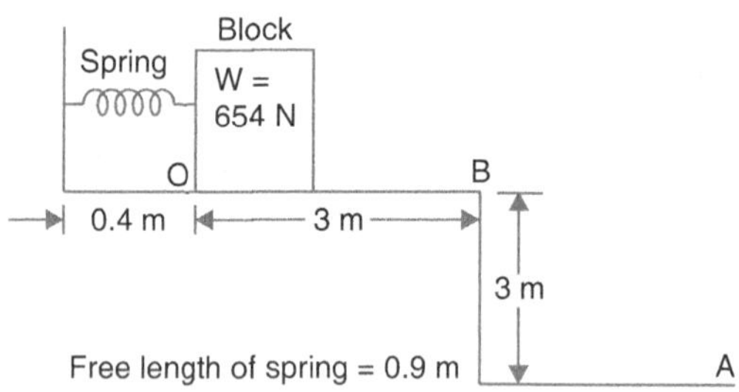

Fig. 9.12

Sol. *Given:* $W = 654$ N; $k = 4000$ N/m; $\mu = 0.2$

Free length of the spring = 0.9 m

∴ Compression of the spring,

$$\delta = 0.9 - 0.4 = 0.5 \text{ m}$$

Work done by the spring

$$= \frac{1}{2} k\delta^2 = \frac{1}{2} \times 4000 \times 0.5^2 = 500 \text{ Nm}$$

Work done by the friction upto the surface *B*

$$= 0.2 \times 654 \times 3 = 392.4 \text{ Nm}$$

Applying *work-energy principle,* we get

Total work done = change in K.E.

or,
$$(500 - 392.4) = \frac{1}{2} \cdot \frac{W}{g} \cdot v^2 = \frac{1}{2} \times \frac{654}{9.81} \times v^2$$

$$\therefore \quad v = 1.797 \text{ m/s}$$

From the point '*B*', the block will undergo projectile motion.

Let '*t*' be the time taken to reach at *A* in seconds.

Using the equation of motion for projectile, we have

$$h = ut + \frac{1}{2} gt^2$$

or,
$$3 = 0 + \frac{1}{2} + 9.81\, t^2 \qquad \therefore \quad t = 0.782 \text{ s}$$

$\therefore$ Horizontal distance moved by the block from the point *B*,

$$s = v \times t = 1.797 \times 0.782 = \mathbf{1.405\ m.} \quad \textbf{(Ans.)}$$

Example 9.19. *A box of mass 2 kg is projected down a slope of 15° with an initial velocity of 4 m/s. Assuming the coefficient of friction during its motion along the plane to be 0.4, find the velocity of the box when it covers a distance of 3 m along the plane. Also, find the time and the distance in which it will come to rest.*

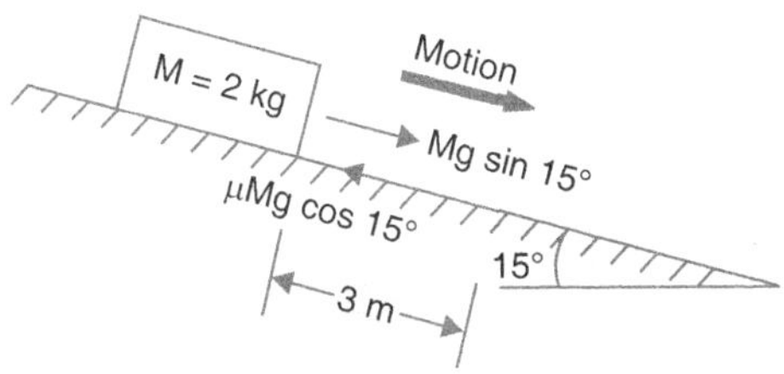

Fig. 9.13

Sol. Mass of the box, $M = 2$ kg

Inclination of the plane = 15°

Initial velocity, $U = 4$ m/s

Coefficient of friction, $\mu = 0.4$

Velocity after covering a distance of 3 m, *V* :

The box is having initial velocity 4 m/s, this *contributes K.E.* When the box travels 3 m distance there is fall in height. Hence there is *potential energy. The K.E. and P.E. are utilised doing work against friction and retaining part of K.E.*

$\therefore$ Initial K.E. + work done by gravity (P.E.)

= Work done against friction + K.E. retained.

i.e.,
$$\frac{1}{2} MU^2 + Mg \times 3 \sin 15^\circ = \mu\, Mg \cos 15^\circ \times s + \frac{1}{2} MV^2$$

or,
$$\frac{1}{2} \times 2 \times (4)^2 + 2 \times 9.81 \times 3 \sin 15^\circ = 0.4 \times 2 \times 9.81 \cos 15^\circ \times 3 + \frac{1}{2} \times 2 \times V^2$$

or, $16 + 15.23 = 22.74 + V^2$

or, $V = \mathbf{2.9\ m/s.}$ **(Ans.)**

Time and distance required to come to rest:

Let s_1 = the distance travelled by box along the inclined plane, and

t = time taken to come to rest.

Now, K.E. + Work done by gravity (P.E.) = Work done against friction.

$$\therefore \quad \frac{1}{2} MU^2 + Mg\, s_1 \sin 15° = \mu\, Mg \cos 15° \times s_1$$

or, $$\frac{1}{2} \times 2 \times (4)^2 + 2 \times 9.81\, s_1 \sin 15°$$

$$= 0.4 \times 2 \times 9.81 \times \cos 15° \times s_1$$

or, $16 + 5.08\, s_1 = 7.58\, s_1$

$\therefore$ $s_1 = \mathbf{6.4\ m.}$ **(Ans.)**

To find time 't', let us find out the acceleration 'a'.

Using the relation,

$$V^2 - U^2 = 2as$$

Here, $U = 4$ m/s

$V = 0$ (Because the box comes to rest)

$s_1 = 6.4$ m

$a = ?$

Substituting the various values, we get

$$0^2 - (4)^2 = 2 \times a \times 6.4$$

$\therefore$ $a = -\ 1.25\ \text{m/s}^2$

(–ve sign means *retardation.*)

Now, using the equation,

$$V = U + at$$

or $0 = 4 - 1.25 \times t$

$\therefore$ $t = \mathbf{3.2\ s.}$ **(Ans.)**

Example 9.20. *A small block starts from A and slides down the inclined plane AB as shown in Fig. 9.14. What distance along the horizontal plane BC will it travel before coming to rest ? Given that μ = 0.3.*

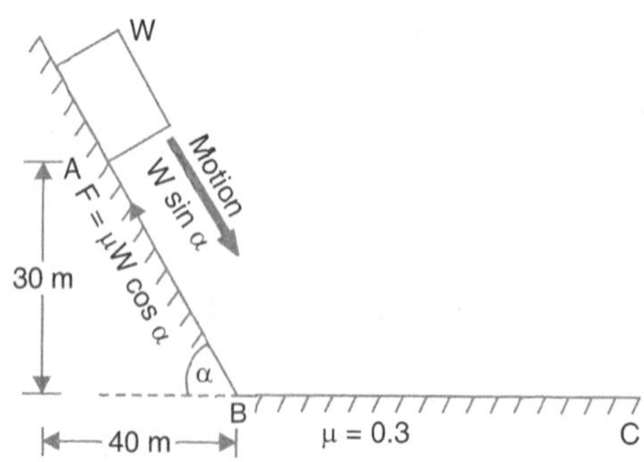

Fig. 9.14

Sol. Let, W = weight of the block

Now, $\tan \alpha = 30/40$, *i.e.*, $\alpha = 36.86°$

Consider the motion of the block:

$$W \sin \alpha - \mu W \cos \alpha = \frac{W}{g} a$$

or,
$$W \times \sin 36.86° - 0.3\, W \cos 36.86° = \frac{W}{9.81} . a$$

or,
$$0.6\, W - 0.24\, W = \frac{W}{9.81} . a$$

or,
$$a = 3.53 \text{ m/s}^2$$

We are to find out distance (s) moved by the block along BC. By virtue of inclination of AB the block will have velocity developed at point B and is given by

$$v^2 = u^2 + 2as$$

$$v^2 = 0 + 2 \times 3.53 \times 50 \qquad \left[\because AB = \sqrt{(30)^2 + (40)^2} = 50 \text{ m}\right]$$

or,
$$v = 18.79 \text{ m/s} \qquad \text{(Block starts from rest } i.e., u = 0)$$

The block at B possesses K.E. and it is utilised in overcoming the frictional resistance along BC.

$$\therefore \qquad \text{K.E.} = F \times s$$

or,
$$\frac{1}{2} . \frac{W}{g} v^2 = \mu W \times S$$

or,
$$\frac{1}{2} \times \frac{(18.79)^2}{9.81} = 0.3 \times s$$

or, s = **60 m. (Ans.)**

HIGHLIGHTS

1. A force is said to have done work if it moves the body, on which it acts, through a certain distance.
2. Work is measured by the product of force (P) and displacement (s) both being in the same direction.
3. Power is defined as the *rate of doing work*.

   Power = force × velocity.

   In M.K.S. system, the horse power developed by the force,

   $$\text{H.P.} = \frac{P \times v}{75},$$

   where P is in *kgf* and v is m/sec.

   In *S.I. system,*

   $$\text{Power} = \frac{P \times v}{1000} \text{ kW}$$

   where P is in newtons (N) and v is in m/s.

4. *Indicated power* is the power developed in the cylinder of an engine. It is usually measured from the indicator diagram drawn with the help of particular engine indicator.
5. *Break power* is the power available at the pulley or shaft of the engine. It is also called shaft power.
6. Mechanical efficiency,

$$\eta_{mech.} = \frac{\text{B.P.}}{\text{I.P.}} \times 100.$$

7. *Energy* of a body is its capacity of doing work. The energy may exist in the following forms: Mechanical energy, electrical energy, light energy, heat energy, chemical energy, sound energy, magnetic energy and nuclear energy.
8. *Potential energy (P.E.)*: The energy which a body possesses by virtue of its position or configuration is called 'potential energy' P.E. = mgh.

   Kinetic energy (K.E.): The energy which a body possesses by virtue of its motion is called kinetic energy.

$$\text{K.E.} = \frac{1}{2} mv^2$$

9. *Law of conservation of energy* states: "The total amount of energy in the universe is constant; energy cannot be created or destroyed although it may be converted into various forms".

OBJECTIVE TYPE QUESTIONS

Say 'Yes' or 'No'/Choose Correct Answer:

1. Work is measured by the product of force (P) and displacement (s) both being in the opposite direction.
2. Power is defined as the rate of doing work.
3. In SI units the unit of power is watt.
4. Brake power (B.P.) is the power developed in the cylinder of an engine.
5. Indicated power (I.P.) is the power available at pulley or shaft of the engine.
6. Mechanical efficiency = $\frac{\text{B.P.}}{\text{I.P.}}$.
7. The energy which a body possesses by virtue of its position or configuration is called,
   (*a*) potential energy (*b*) kinetic energy
   (*c*) sound energy (*d*) chemical energy.
8. The energy which a body possesses by virtue of its motion is called,
   (*a*) electromagnetic energy (*b*) chemical energy
   (*c*) potential energy (*d*) kinetic energy.
9. The potential energy of a body of mass '*m*' and lying at a height '*h*' is given by
   (*a*) mgh (*b*) m^2gh
   (*c*) mgh^2 (*d*) mg^2h.
10. The kinetic energy of a body of mass '*m*' and moving with velocity '*v*' is given by
    (*a*) $\frac{1}{2} mv^3$ (*b*) $\frac{1}{3} mv^2$
    (*c*) $\frac{1}{2} mv^2$ (*d*) mv^2.

Answers

1. No	**2.** Yes	**3.** Yes	**4.** No	**5.** No
6. No	**7.** (*a*)	**8.** (*d*)	**9.** (*a*)	**10.** (*c*).

Unsolved Examples

1. A train of mass 500 × 10^3 kg starts from rest and accelerates uniformly to a speed of 90 km/hour in 50 seconds. The total frictional resistance to motion is 15 kN. Determine:
   (*i*) The maximum power required.
   (*ii*) The power required to maintain the speed of 90 km/h. [**Ans.** (*i*) 6.625 MW, (*ii*) 375 kW]
   [**Hint.** (*i*) Total force required, P' = force required to overcome friction + force required to accelerate the train
   Power = $P \times b$
   (*ii*) Total force required, P' = force required to overcome the frictional resistance only power = $P' \times v$]
2. A block of wood of weight 1600 N is placed on a smooth inclined plane which makes an angle of 30° with the horizontal. Find the work done in pulling the block up for a length of 5 m. [**Ans.** 4000 Nm]
3. A train of weight 1500 kN, is ascending a slope of 1 in 100 with a uniform speed of 36 km/hour. Find the power exerted by the engine, if the load resistance is 5 N per kN weight of the train. [**Ans.** 225 kW]
4. A bullet of mass 0.025 kg is fired with a velocity of 400 m/s. What is the K.E. of the bullet ? If the bullet can penetrate 200 mm in a block of wood, what is the average resistance of wood ? What will be the exit velocity of the bullet if fired into a similar block of wood 100 mm thick? [**Ans.** 2000 Nm; 1000 N; 282.8 m/s]
5. A hammer of mass 1500 kg drops from a height of 0.6 m on a pile of mass 750 kg. Find the depth of penetration of the pile into the ground, if the average resistance of the ground is 140 kN. Assume the impact between the hammer and the pile to be plastic. [**Ans.** 50 mm]
6. A hammer of mass 400 kg falls through a height of 3 m on a pile of negligible mass. If it drives the pile 1 m into the ground, find the average resistance of the ground for penetration. [**Ans.** 15696 N]
7. A bullet of mass 0.01 kg is fired into a body of mass 1 kg, which is suspended by a string 1 m long. Due to this impact, the body swings through an angle of 18.2°. Find the velocity of the bullet. [**Ans.** 100 m/s]
8. A bullet of mass 0.025 kg, moving horizontally with a velocity of 600 m/s strikes a wooden block of mass 5 kg resting on a rough horizontal surface. The bullet after striking the block remains burried in the block and both travel for a distance of 0.9 m, before coming to rest.
   Determine:
   (*i*) Average resistance between the block and horizontal surface.
   (*ii*) Co-efficient of friction between between block and horizontal surface. [**Ans.** (*i*) 28.87 N, (*ii*) 0.5]
9. An engine of mass 50 tonnes pulls a train of mass 250 tonnes up a gradient of 1 in 125 with a uniform speed of 36 km/h. Find the power transmitted by the engine, if the tractive resistance is 60 N/tonne. Also find the power transmitted by the engine, if acceleration of the engine is 0.2 m/s^2. Take g = 9.8 m/s^2. [**Ans.** 415 kW, 1015 kW]
10. A wagon of mass 50 tonnes, starting from rest, travels 30 m down a 1% grade and strikes a post with bumper spring as shown in Fig. 9.15.

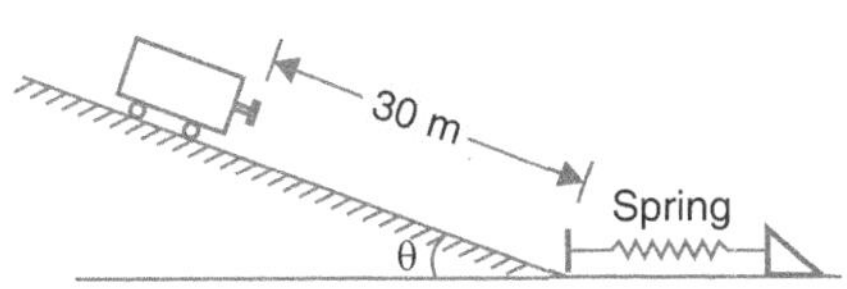

Fig. 9.15

If the rolling resistance of the track is 50 N/t, find the velocity with which the wagon strikes the post. Also, find the amount by which the spring will be compressed, if the bumper spring compresses 1 mm per 15 kN. [**Ans.** 1.7 m/s, 98.1 mm]

11. A railway wagon of mass 20 tonnes runs into a buffer stop having two buffer springs each of 10 kN/mm stiffness. Find the maximum compression of the springs, if the wagon is travelling at 18 km/h. [**Ans.** 158 mm]

12. A pendulum consists of a bob of 4 kg mass, supported by a thin rod pivoted 1 m from the centre of the bob. The bob is moved, so that the rod makes an angle of 20° with the vertical and then released. What is the linear and angular velocity of the bob, when it swings back to the vertical position ?
Neglect the weight of the rod and frictional resistance. [**Ans.** 1.09 m/s; 1.09 rad/s]

13. Determine the constant force P that will give the system of bodies shown in Fig. 9.16 velocity of 3 m/s after moving 4.5 m from rest. Take μ_k = 0.2 for all surfaces in contact. [**Ans.** 97.9 N]

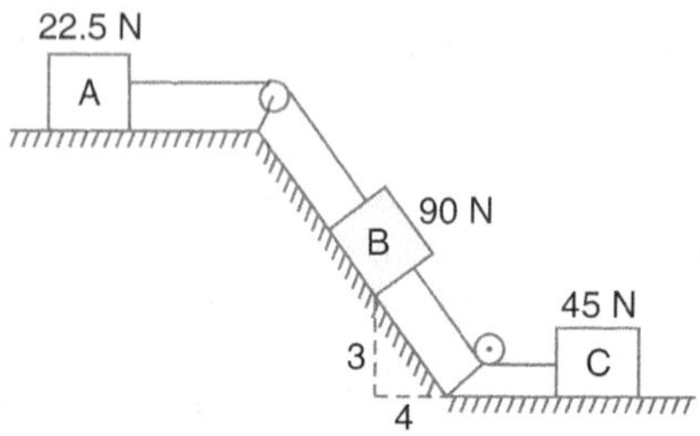

Fig. 9.16

10

Simple Machines

10.1. GENERAL CONCEPT OF A MACHINE

A 'machine' may be defined as a contrivance or a device which receives energy in some available form and uses it for doing a particular useful work.

Example. The steam engine converts the energy provided by the steam into motion of translation.

A *'lifting machine'* may be defined as contrivance to overcome a force or load (W) applied at one point by means of another force called effort (P). In most of the cases, P is smaller than W, but sometimes it may be greater than W.

A machine has the following three features :

(*i*) The point of application of effort

(*ii*) Mechanism—the working part that makes up a machine

(*iii*) The point where useful work is done or load is lifted.

The machines may be of the following two types :

1. Simple machines
2. Compound machines.

A *simple machine has only one point* for the *application of effort* and *one point for load. Its mechanism* is *simple. Examples : Lever, inclined plane, screw jack, sewing machine etc.*

A *compound machine has more than one point* for *application of effort and for load. Its mechanism is compound or complicated. Examples : Printing machine, lathe machine, milling machine, planer, shaper etc.*

10.2. IMPORTANT DEFINITIONS

(*i*) **Mechanical advantage (M.A.).** It is the defined as the ratio of weight lifted (W) to the effort applied (P). It is always expressed in pure number.

$$\text{M.A.} = \frac{W}{P} \qquad \text{...(10.1)}$$

(*ii*) **Velocity ratio (V.R.).** It is the ratio of distance (y) moved by the effort to the distance (x) moved by the load

$$\text{V.R.} = \frac{y}{x} \qquad ...(10.2)$$

(*iii*) **Input of a machine.** It is the work done *on the machine.* In a lifting machine, it is measured by the product of effort and the distance through which it has moved (*i.e.*, $P \times y$).

(*iv*) **Output of a machine.** It is the actual work done *by the machine*. In a lifting machine it is measured by the product of the weight lifted and the distance through which it has been lifted (*i.e.*, $W \times x$).

(*v*) **Efficiency of a machine (η).** It is ratio of output to the input of a machine.

i.e.,
$$\eta = \frac{\text{Output}}{\text{Input}} \qquad ...(10.3)$$

It is generally expressed as a *percentage.*

(*vi*) **Ideal machine.** A machine is said to be ideal if its efficiency is 100%. In this case, output is equal to input.

10.3. RELATION BETWEEN M.A., V.R. AND η

Let W = load lifted by the machine

P = effort required to lift the load

y = distance moved by the effort, in lifting the load

x = distance moved by the load, and

η = efficiency of the machine.

Then, mechanical advantage of the machine,

$$\text{M.A.} = \frac{W}{P}$$

and velocity ratio of the machine,

$$\text{V.R.} = \frac{y}{x}$$

Input of a machine = effort applied × distance through which the effort has been moved

$= P \times y$

Output of a machine = load lifted × distance through which the load has been lifted

$= W \times x$

∴ Efficiency,
$$\eta = \frac{\text{Output}}{\text{Input}} = \frac{W \times x}{P \times y} = \frac{W/P}{y/x}$$

$$= \frac{\text{M.A.}}{\text{V.R.}} \qquad ...(10.4)$$

10.4. CONCEPT OF FRICTION IN A MACHINE

Some losses always occur in every machine and mostly they are due to *friction.* One way of determining the same is by calculating the efficiency of the machine. The alternate method is to express the frictional losses in the following ways :

(*i*) in terms of loss of effort

(*ii*) in terms of loss of weight/load lifted.

Let P = actual effort (considering the machine friction) required to lift a load.

W = actual load (considering the machine friction) lifted or resistance overcome.

W_i = ideal load (neglecting machine friction) lifted or resistance overcome.

F_p = loss of effort due to friction

F_w = loss of load due to friction

Loss of effort due to friction.

$$F_P = P - P_i \qquad \text{...(10.5)}$$

Loss of load due to friction

$$F_W = W_i - W \qquad \text{...(10.6)}$$

To calculate, P_i consider an *ideal machine* in which $\eta = 1$.

We know, $\text{M.A.} = \eta \times \text{V.R.}$

$$\frac{W}{P_i} = 1 \times \text{V.R.}$$

or

$$P_i = \frac{W}{\text{V.R.}} \qquad \text{...(10.7)}$$

Substituting the value of P_i in eqn. (10.5)

$$F_P = P - \frac{W}{\text{V.R.}} \qquad \text{...(10.8)}$$

To calculate, W_i again consider an *ideal machine*

$$\text{M.A.} = \eta \times \text{V.R.}$$

$$\frac{W_i}{P} = 1 \times \text{V.R.}$$

$$\therefore \quad W_i = P \times \text{V.R.} \qquad \text{...(10.9)}$$

Substituting the value of W_i in eqn. (10.6)

$$F_W = P \times \text{V.R.} - W \qquad \text{...(10.10)}$$

From eqns. (10.8) and (10.10),

$$F_W = F_P \times \text{V.R.} \qquad \text{...(10.11)}$$

If for a machine P, P_i, W and W_i are known, the efficiency of a machine

$$\eta = \frac{P_i}{P} = \frac{W}{W_i} \qquad \text{...[10.12 (a)]}$$

Hence,

$$\eta = \frac{\text{Output}}{\text{Input}} = \frac{\text{M.A.}}{\text{V.R.}} = \frac{P_i}{P} = \frac{W}{W_i} \qquad \text{...[10.12 (b)]}$$

10.5. CONDITION FOR REVERSIBILITY OF A MACHINE

When the machine is capable of doing work in the reverse direction due to its load, as the effort is totally removed, the machine is said to be *reversible*. As soon as the effort is removed the work done *on the load* will be input rather than output and here the input will be equal to frictional losses.

The *condition for reversibility* of a machine is determined as follows :

Let W = load lifted by the machine,

P = effort required to lift the load,

y = distance moved by the effort, and

x = distance moved by the load.

Then, input of the machine

$$= P \times y$$

and output of the machine

$$= W \times x$$

Frictional losses in the machine

$$= \text{input} - \text{output}$$
$$= P \times y - W \times x$$

In a *reversible machine, the output of the machine should be more than the machine friction when the effort (P) is zero.*

i.e.,
$$W \times x > (P \times y - W \times x) \quad \text{...[10.13 }(a)]$$
$$2W \times x > P \times y$$
$$\frac{W \times x}{P \times y} > \frac{1}{2}$$
$$\frac{W/P}{y/x} > \frac{1}{2}$$
$$\frac{\text{M.A.}}{\text{V.R.}} > \frac{1}{2} \qquad \left(\because \frac{W}{P} = \text{M.A. and } \frac{y}{x} = \text{V.R.}\right)$$
$$\eta > \frac{1}{2} > 0.5$$
$$> 50\% \quad \text{...[10.13 }(b)]$$

Hence, for a *machine to be reversible the efficiency of the machine is to be greater than 50%.*

10.6. NON-REVERSIBLE/IRREVERSIBLE OR SELF-LOCKING MACHINE

When the machine is not capable of doing any work in the reversed direction, after the effort is removed, then the machine is called a non-reversible/irreversible or self-locking machine. The *condition* for a machine to be self *locking* is that *its efficiency should not be more than 50%.*

10.7. LAW OF A MACHINE

The *law of a machine gives the relationship between the effort applied and the load lifted.*

For any machine if a graph is plotted between effort P and load lifted W it will be found that it follows a straight line relationship

$$P = mW + C \quad \text{...(10.14)}$$

This graph does not pass through O but makes an intercept C on the P-axis which shows the effort is lost in overcoming frictional resistance only (refer to Fig. 10.1). The constant 'm' *shows the slope of the graph.*

The equation $P = mW + C$ is known as the *law of machine.* W_1 and W_2 are noted the values of constants 'm' and 'C' can be found as follows :

$$P_1 = mW_1 + C \qquad \text{...[10.15 (a)]}$$

$$P_2 = mW_2 + C \qquad \text{...[10.15 (b)]}$$

Subtracting eqn. [10.15 (*b*)] from eqn. [10.15 (*a*)]

$$P_1 - P_2 = m(W_1 - W_2)$$

$$m = \frac{P_1 - P_2}{W_1 - W_2} \qquad \text{...(10.16)}$$

Substituting the value of m in eqn. [10.15 (*a*)], we get

$$P_1 = \left(\frac{P_1 - P_2}{W_1 - W_2}\right) W_1 + C$$

or

$$C = P_1 - \left(\frac{P_1 - P_2}{W_1 - W_2}\right) W_1 \qquad \text{...(10.17)}$$

10.8. MAXIMUM MECHANICAL ADVANTAGE AND EFFICIENCY

We know that,

Mechanical advantage, $\quad \text{M.A.} = W/P$

$$= \frac{W}{mW + C} \qquad (\because \; P = mW + C)$$

$$= \frac{1}{m + \frac{C}{W}} \qquad \text{...(10.18)}$$

If W increases, the denominator $\left(m + \frac{C}{W}\right)$ will decrease and M.A. will increase. When the value of W is extremely high, $\frac{C}{W}$ will tend to be zero and corresponding M.A. will be maximum.

$$\therefore \qquad (\text{M.A.})_{max} = \frac{1}{m} \qquad \text{...(10.19)}$$

We know that, $\quad \text{M.A.} = \eta \times \text{V.R.}$

$$\eta = \frac{\text{M.A.}}{\text{V.R.}} = \frac{1}{m + \frac{C}{W}} \times \frac{1}{\text{V.R.}} \qquad \text{...(10.20)}$$

The velocity ratio in any machine is fixed. Thus, η is directly proportional to M.A. For *maximum M.A., the efficiency will also be maximum.*

Maximum efficiency, $\quad \eta_{max} = \frac{(\text{M.A.})_{max}}{\text{V.R.}}$

$$= \frac{1}{m \times \text{V.R.}} \qquad \text{...(10.21)}$$

If η-W (*i.e.*, efficiency-load) graph is plotted the curve will be as shown in Fig. 10.1.

From eqns. (10.18) to (10.21), it is concluded that

(*i*) *M.A. increases with load,*

(*ii*) η *also increases with load,*

(*iii*) *M.A. will be maximum at a very high value of load, and*

(*iv*) η *will also be maximum at a very high value of load.*

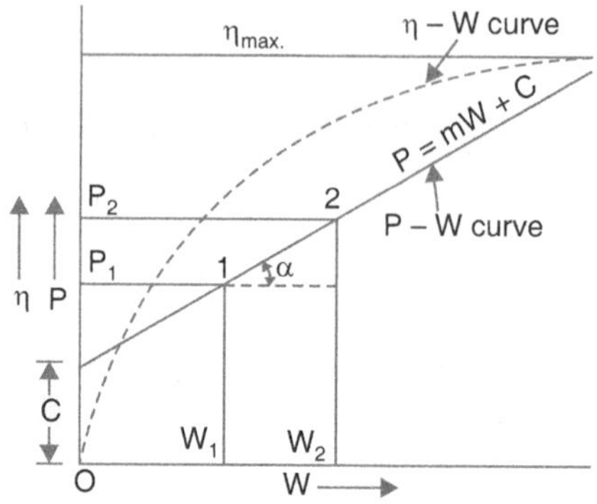

Fig. 10.1

Example 10.1. *In a lifting machine a load of 100 N is moved through a distance of 5 cm, by an effort P moving through a distance of 80 cm. Find the value of P if the efficiency of the machine is 60%.*

Sol. Load lifted $W = 100$ N

Distance moved by the effort $= 80$ cm

Distance moved by the load $= 5$ cm

Efficiency, $\eta = 60\%$

Effort, value of P = ?

We know that :

Mechanical advantage,

$$\text{M.A.} = \frac{W}{P} = \frac{100}{P}.$$

Velocity ratio,
$$\text{V.R.} = \frac{y}{x} = \frac{\text{distance moved by the effort}}{\text{distance moved by the load}}$$

$$= \frac{80}{5} = 16.$$

Efficiency,
$$\eta = \frac{\text{M.A}}{\text{V.R.}} \times 100$$

$$60 = \frac{100/P}{16} \times 100 = \frac{100 \times 100}{P \times 16}$$

or
$$P = \frac{100 \times 100}{60 \times 16}$$

$$= \mathbf{10.42\ N.}\quad \textbf{(Ans.)}$$

Example 10.2. *A machine raised a load of 360 N through a distance of 200 mm. The effort, a force of 60 N moved 1.8 m during the process. Calculate :*

(*i*) *velocity ratio;*

(*ii*) *mechanical advantage;*

(*iii*) *effect of friction; and*

(*iv*) *efficiency at this load.*

Sol. Load raised, $W = 360$ N

Effort applied, $P = 60$ N

Distance moved by the effort, $y = 1.8$ m $= 1800$ mm

Distance moved by the load, $x = 200$ mm

(*i*) **Mechanical advantage, M.A. = ?**

We know that mechanical advantage,

$$\text{M.A.} = \frac{W}{P} = \frac{360}{60} = \mathbf{6.} \quad \textbf{(Ans.)}$$

(*ii*) **Velocity ratio, V.R. = ?**

Velocity ratio, $$\text{V.R.} = \frac{y}{x} = \frac{1800}{200} = \mathbf{9.} \quad \textbf{(Ans.)}$$

(*iii*) **Efficiency = ?**

Efficiency at this load, $$\eta = \frac{\text{M.A}}{\text{V.R.}} \times 100 = \frac{6}{9} \times 100$$

$$= \mathbf{66.66\%.} \quad \textbf{(Ans.)}$$

(*iv*) **Effect of friction = ?**

We know that,

Effort lost in friction, $$F_p = P - \frac{W}{\text{V.R.}} = 60 - \frac{360}{9}$$

$$= \mathbf{20\ N.} \quad \textbf{(Ans.)}$$

Friction in terms of load,

$$F_w = P \times \text{V.R.} - W$$

$$= 60 \times 9 - 360 = \mathbf{180\ N.} \quad \textbf{(Ans.)}$$

Example 10.3. *The law of a certain lifting machine is* $P = \frac{W}{50} + 1.5$. *The velocity ratio of the machine is 100. Find the maximum possible mechanical advantage and maximum possible efficiency of the machine. Determine the effort required to overcome friction while lifting a load of 150 N. Also calculate the efficiency at this load.*

Sol. Mechanical advantage,

$$\text{M.A.} = \frac{W}{P} = \frac{W}{\frac{W}{50} + 1.5} = \frac{1}{\frac{1}{50} + \frac{1.5}{W}}$$

$$(\text{M.A.})_{max} = \underset{W \to \infty}{\text{Lt}} \frac{1}{\frac{1}{50} + \frac{1.5}{W}} = \mathbf{50.} \quad \textbf{(Ans.)}$$

$$\eta_{max} = \frac{(\text{M.A.})_{max}}{\text{V.R.}} = \frac{50}{100} = \mathbf{50\%.} \quad \textbf{(Ans.)}$$

The effort required to lift a load of 150 N,

$$P = \frac{150}{50} + 1.5 = 4.5 \text{ N},$$

The ideal effort P_i required to lift a load of 150 N

$$P_i = \frac{W}{\text{V.R.}} = \frac{150}{100} = 1.5 \text{ N}.$$

$\therefore$ Effort required to overcome friction at a load of 150 N is

$$= P - P_i$$
$$= 4.5 - 1.5 = \mathbf{3\ N.}\quad \textbf{(Ans.)}$$

The efficiency at this load is given by,

$$\eta = \frac{W}{P \times \text{V.R.}} = \frac{150}{4.5 \times 100} = 0.3333$$
$$= \mathbf{33.33\%.}\quad \textbf{(Ans.)}$$

☞**Example 10.4.** *In a lifting machine an effort of 15 N raised a load of 770 N. What is the mechanical advantage ? Find the velocity ratio, if the efficiency at this load is 60%.*

If on the machine effort of 25 N raised a load of 1320 N, what is the efficiency ? What will be the effort required to raise a load of 500 N ? Calculate the maximum mechanical advantage and maximum efficiency ?

Sol. Case I

Load raised, $W = 770$ N

Effort applied, $P = 15$ N

Efficiency, $\eta = 60\%$

$$\therefore \quad \textbf{M.A.} = \frac{W}{P} = \frac{770}{15} = \mathbf{51.33.}\quad \textbf{(Ans.)}$$

Also $$\eta = \frac{\text{M.A.}}{\text{V.R.}} = \frac{W}{P \times \text{V.R.}}$$

or $$0.6 = \frac{770}{15 \times \text{V.R}}.$$

$$\therefore \quad \textbf{V.R.} = \frac{770}{0.6 \times 15} = \mathbf{85.55.}\quad \textbf{(Ans.)}$$

Case II

Velocity ratio, V.R. = 85.55

Load raised, $W = 1320$ N

Effort applied, $P = 25$ N

$$\therefore \quad \text{M.A.} = \frac{1320}{25} = 52.8$$

Also, $$\eta = \frac{\text{M.A.}}{\text{V.R.}} = \frac{52.8}{85.55} = 0.6172 = \mathbf{61.72\%.}\quad \textbf{(Ans.)}$$

Case III

When $P = 15$ N $\quad W = 770$ N

and when $P = 25$ N $\quad W = 1320$ N

Using these values in the law of machine,

i.e., $$P = mW + C$$

$$15 = m \times 770 + C \quad ...(i)$$

$$25 = m \times 1320 + C \quad ...(ii)$$

Subtracting (*i*) from (*ii*),

$$10 = 550 \text{ m}$$

$$\therefore \quad m = \frac{10}{550} = \frac{1}{55} = 0.0182$$

Substituting this value of m in eqn. (*i*), we get

$$15 = 770 \times 0.0182 + C$$

$$\therefore \quad C = 1$$

Thus, the law of machine is

$$P = 0.0182\ W + 1$$

∴ Effort required to raise a load of 500 N

$$P = 0.0182 \times 500 + 1$$

$$= \mathbf{10.1\ N.} \quad \textbf{(Ans.)}$$

Maximum mechanical advantage :

Using the relation,

$$(\text{M.A.})_{max} = \frac{1}{m} = \frac{1}{0.0182} = \mathbf{55.} \quad \textbf{(Ans.)}$$

Maximum efficiency :

Using the relation,

$$\eta_{max} = \frac{1}{m \times \text{V.R.}}$$

$$= \frac{1}{0.0182 \times 85.55} = 0.642$$

or, $$= \mathbf{64.2\%.} \quad \textbf{(Ans.)}$$

☞ **Example 10.5.** *(a) Explain the difference between a reversible and a self-locking machine.*

(b) What load will be lifted by an effort of 12 N if the velocity ratio is 18 and the efficiency of the machine at this load is 60%. If the machine has a constant frictional resistance, determine the law of machine and find the effort required to run the machine at (i) no load and (ii) load of 90 N.

Sol. (*a*) Refer to text.

(*b*) Effort applied, $P = 12$ N

Velocity ratio, V.R. = 18

Efficiency, $\eta = 60\%$

Load lifted, $W = ?$

Using the relation,

$$\eta = \frac{\text{M.A.}}{\text{V.R.}} = \frac{W}{P \times \text{V.R.}}$$

$$0.6 = \frac{W}{12 \times 18}$$

$$\therefore \quad \mathbf{W = 129.6\ N.} \quad \textbf{(Ans.)}$$

Effort lost in friction,

$$F_P = P - \frac{W}{\text{V.R.}} = 12 - \frac{129.6}{18} = \mathbf{4.8\ N.}$$

As the frictional resistance is constant the effort required to run the machine with no load must be equal to 4.8 N.

Let the law of machine is

$$P = mW + C \qquad ...(i)$$

When $P = 4.8$ N, $W = 0$

When $P = 12$ N, $W = 129.6$ N.

Substituting the values in (*i*), we get

$$12 = 129.6\ m + C \qquad ...(ii)$$

$$4.8 = 0 + C \qquad ...(iii)$$

Solving (*ii*) and (*iii*), we get

$$m = \frac{1}{18} \text{ and } C = 4.8$$

Hence, the **law of machine is**

$$\mathbf{P = \frac{1}{18}\ W + 4.8.} \quad \textbf{(Ans.)}$$

At no load, *i.e.*, $W = 0$

$$\mathbf{P = \frac{1}{18} \times 0 + 4.8 = 4.8\ N.} \quad \textbf{(Ans.)}$$

At $W = 90$ N

$$\mathbf{P} = \frac{1}{18} \times 90 + 4.8 = \mathbf{9.8\ N.} \quad \textbf{(Ans.)}$$

Example 10.6. *In a lifting machine, the efforts required to lift loads of 20 N and 30 N were 5 N and 6 N respectively. If the velocity ratio of the machine is 20, determine.*

(i) Law of the machine,

(ii) Efficiencies corresponding to loads of 20 N and 30 N,

(iii) Effort lost in friction in both cases, and

(iv) The maximum efficiency which can be expected from this machine.

Sol. (*i*) **Law of the machine = ?**

Let the law of machine be

$$P = mW + C$$

where P = effort applied

W = load lifted

m and C being two constants.

Now, when $W = 20$ N $\quad P = 5$ N

and when $W = 30$ N $\quad P = 6$ N.

Putting these values in the law of machine,

$$5 = 20\ m + C \qquad ...(i)$$

$$6 = 30\ m + C \qquad ..(ii)$$

Subtracting (*i*) from (*ii*), we get

$$1 = 10\,m$$

$$\therefore \quad m = \frac{1}{10}$$

Putting this value in eqn. (*i*),

$$5 = 20 \times \frac{1}{10} + C$$

$$\therefore \quad C = 3$$

Hence, the machine follows the law

$$P = \frac{1}{10}\,\mathbf{W + 3.} \quad \textbf{(Ans.)}$$

(*ii*) **Efficiencies = ?**

When $W = 20$ N, $P = 5$ N

$$\therefore \quad \text{M.A.} = \frac{W}{P} = \frac{20}{5} = 4$$

Also V.R. = 20

$\therefore$ Efficiency at this load

$$\eta = \frac{4}{20} \times 100 = \mathbf{20\%.} \quad \textbf{(Ans.)}$$

When $W = 30$ N, $P = 6$ N

$$\therefore \quad \text{M.A.} = \frac{W}{P} = \frac{30}{6} = 5.$$

$\therefore$ Efficiency at this load

$$= \frac{5}{20} \times 100 = \mathbf{25\%.} \quad \textbf{(Ans.)}$$

(*iii*) **Effort lost in friction = ?**

When $W = 20$ N, $P = 5$ N

Effort lost in friction,

$$F_P = P - \frac{W}{\text{V.R}} = 5 - \frac{20}{20} = \mathbf{4\ N.} \quad \textbf{(Ans.)}$$

When $W = 30$ N, $P = 6$ N

Effort lost in friction $F_P = P - \dfrac{W}{\text{V.R.}}$

$$= 6 - \frac{30}{20} = \mathbf{4.5\ N.} \quad \textbf{(Ans.)}$$

(*iv*) **Maximum efficiency = ?**

Maximum possible efficiency of any machine

$$= \frac{1}{m \times \text{V.R.}} = \frac{1}{\frac{1}{10} \times 20}$$

$$= 0.5 = \mathbf{50\%.} \quad \textbf{(Ans.)}$$

Example 10.7. *The law of a machine taken from the load (W) versus effort (P) graph, where W and P are in N is*

$$W = 3P - 16$$

Find: (i) the effort which can raise a load of 200 N,

(ii) the load which an effort of 20 N will lift,

(iii) the effort required to operate the machine at no load, and

(iv) the efficiency under a load or 110 N if the velocity ratio is 4.

Sol. The law of machine is

$$W = 3P - 16$$

where W = load lifted in N

P = effort applied in N

(*i*) When $W = 200$ N

Let P be the corresponding value of the effort

$$\therefore \quad 200 = 3P - 16$$

$$\therefore \quad \mathbf{P = 72\ N.} \quad \textbf{(Ans.)}$$

(*ii*) When $P = 20$ N

Let W = load lifted

$$\therefore \quad W = 3 \times 20 - 16 = \mathbf{44\ N.} \quad \textbf{(Ans.)}$$

(*iii*) When there is no load on the machine

i.e., $W = 0$

$$\therefore \quad 0 = 3P - 16$$

or
$$P = \frac{16}{3} = \mathbf{5.33\ N.} \quad \textbf{(Ans.)}$$

(*iv*) When $W = 110$ N

$$110 = 3P - 16$$

$$P = 42 \text{ N}$$

$$\therefore \quad \text{M.A.} = \frac{W}{P} = \frac{110}{42} = 2.62$$

and
$$\eta = \frac{\text{M.A.}}{\text{V.R.}} = \frac{2.62}{4} = 0.655$$

or
$$= \mathbf{65.5\%} \quad \textbf{(Ans.)}$$

10.9. SOME LIFTING MACHINES

Some important lifting machines are given below :

1. Simple wheel and axle
2. Wheel and differential axle
3. Pulleys :
   (*i*) First system of pulleys
   (*ii*) Second system of pulleys
   (*iii*) Third system of pulleys
4. Weston's differential pulley block

5. Worm and worm wheel
6. Single purchase crab winch
7. Double purchase crab winch
8. Simple screw jack
9. Differential screw jack.

10.10. SIMPLE WHEEL AND AXLE

Fig. 10.2 shows a simple wheel and axle lifting machine. The wheel and axle are keyed to the same spindle which is mounted on ball bearings to reduce the frictional resistance to a minimum. Two ropes are taken. Fixing its one end, one rope is wound round the axle a few turns and the load is attached to its free end. Similarly, one end of the other rope is fixed on the wheel and it is wound round the wheel in the *opposite direction* in such a way that as the rope on the axle is wound up, the rope on the wheel is unwound and *vice versa*. Effort is applied on the free end of the rope wound round the wheel.

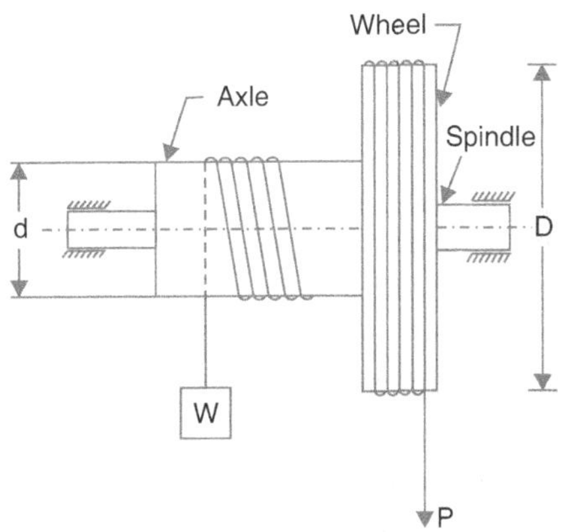

Fig. 10.2. Simple wheel and axle.

Let D = diameter of the wheel

d = diameter of the axle.

In one revolution of the wheel;

the displacement of the effort $= \pi D$

and the displacement of the load $= \pi d$

$$\text{V.R.} = \frac{\pi D}{\pi d} = \frac{D}{d} \qquad \text{...(10.22)}$$

If the diameter of the rope is also to be considered, then

$$\text{V.R.} = \frac{D + dr_1}{d + dr_2}$$

where, dr_1 = diameter of the rope on the wheel

dr_2 = diameter of the rope on the axle

$$\text{M.A.} = \frac{W}{P} \qquad \text{...as usual}$$

$$\eta = \frac{\text{M.A}}{\text{V.R.}} \qquad \text{...as usual}$$

Example 10.8. *In a wheel and axle, the diameter of the wheel is 400 mm and that of axle 100 mm. The thickness of the rope is 10 mm. What is the velocity ratio ? If the efficiency when lifting a load of 200 N at 10 m/min is 85%, find the power that must be supplied to the machine.*

Sol. Diameter of the wheel, $D = 400$ mm

Thickness of the rope, $d_r = 10$ mm

Diameter of the axle, $d = 100$ mm

$$\therefore \qquad \text{V.R.} = \frac{D + d_r}{d + d_r} = \frac{400 + 10}{100 + 10} = 3.73$$

Work done per minute by the machine in lifting 200 N at 10 m/min.

$$= 200 \times 10 = 2000 \text{ Nm}$$

Power obtained from the machine

$$= \frac{2000}{60} = 33.3 \text{ W}$$

This is 85% of the power supplied

$$\therefore \qquad \text{Power } \textbf{supplied} = \frac{33.3}{0.85} = \textbf{39.2 W.} \quad \textbf{(Ans.)}$$

Example 10.9. *In a wheel and axle, the diameter of the wheel is 55 cm and the diameter of the axle is 20 cm. The thickness of the cord on the wheel is 4 mm and that on the drum is 8 mm. Find the velocity ratio of the machine. If the efficiency when lifting a load of 150 N with a velocity of 20 m/min is 80 percent how many Nm of work must be supplied to the machine per minute ?*

Sol. Diameter of the wheel, $D = 55$ cm

Diameter of the axle, $d = 20$ cm

Thickness (or dia.) of the cord on the wheel,

$$dr_1 = 4 \text{ mm} = 0.4 \text{ cm}$$

Thickness (or dia.) of the cord on the axle,

$$dr_2 = 8 \text{ mm} = 0.8 \text{ cm}$$

Hence velocity ratio,

$$\textbf{V.R.} = \frac{D + dr_1}{d + dr_2} = \frac{55 + 0.4}{20 + 0.8} = \frac{55.4}{20.8} = \textbf{2.66.} \quad \textbf{(Ans.)}$$

Now, W = load lifted = 150 N

η = efficiency at this load = 80%

Also

$$\eta = \frac{W}{P \times \text{V.R.}}.$$

$$0.8 = \frac{150}{P \times 2.66}$$

$$P = \frac{150}{0.8 \times 2.66} = 70.5 \text{ N}$$

But velocity ratio $= \dfrac{\text{distance covered by the effort } (P) \text{ in certain interval of time}}{\text{distance moved by the load } (W) \text{ in the same interval of time}}$

$$= \frac{y}{x}$$

Hence, x = distance moved by the load
= 20 m/min

But $\quad$ V.R. $= 2.66 = \dfrac{y}{20}$

$\therefore$ $\quad$ y = distance moved by the effort per min
= 2.66 × 20 = 53.2 metres

Hence maximum work to be supplied
= effort × distance moved by the effort
= $P \times y$
= 70.5 × 53.2 = 3750.6 Nm

$\therefore$ **Work to be supplied per min**
= 3750.6 Nm. (Ans.)

10.11. WHEEL AND DIFFERENTIAL AXLE

Refer to Fig. 10.3. In this lifting machine the axle used is a compound axle consisting of a larger diameter B and a smaller diameter axle C. The rope passing round the axle C further passes round the movable pulley carrying the load, and then is wound round the axle B, but in *opposite direction*. One end of the rope is fixed on the axle C and the other end on the axle B.

Let $\quad D$ = diameter of the wheel
d_1 = diameter of the larger axle
d_2 = diameter of the smaller axle.

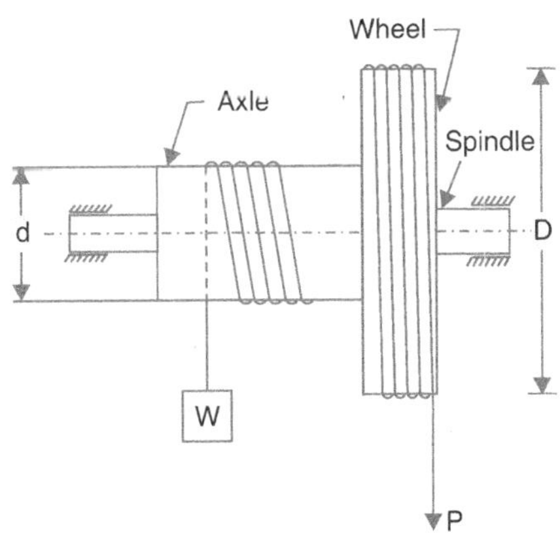

Fig. 10.3. Wheel and differential axle.

With one revolution of the wheel A

Distance moved by the effort $= \pi D$

Rope unwound by the axle, $C = \pi d_2$

Rope wound by axle, $B = \pi d_1$

∴ Length of rope which will wound in one revolution

$$= \pi d_1 - \pi d_2 \qquad (\because \; d_1 > d_2)$$

and displacement of weight

$$= \frac{\pi d_1 - \pi d_2}{2}$$

$$\therefore \qquad \text{V.R.} = \frac{\text{Distance moved by the effort}}{\text{Distance moved by the weight}}$$

$$= \frac{\pi D}{\left(\dfrac{\pi d_1 - \pi d_2}{2}\right)}$$

$$= \frac{2D}{d_1 - d_2} \qquad \text{...[10.23(}a\text{)]}$$

In case the diameter of the rope is also taken into account, then

$$\text{V.R.} = \frac{2(D + dr_1)}{(d_1 + dr_2) - (d_2 + dr_2)} \qquad \text{...[10.23 (}b\text{)]}$$

where dr_1 and dr_2 are the diameters of ropes on wheel and the axles respectively.

Example 10.10. *In a differential wheel and axle, the diameter of the effort wheel is 40 cm. The radii of the axles are 15 cm and 10 cm respectively. The diameter of the rope is 1 cm. Find the load which can be lifted by an effort of 10 N, assuming an efficiency of the machine to be 75%.*

Sol. Diameter of the effort wheel, $D = 40$ cm

Diameter of the larger axle, $d_1 = 2 \times 15 = 30$ cm

Diameter of the smaller axle, $d_2 = 2 \times 10 = 20$ cm

Diameter of the rope, $d_r = 1$ cm

$$\therefore \qquad \text{V.R.} = \frac{2(D + d_r)}{(d_1 + d_r) - (d_2 + d_r)}$$

$$= \frac{2(40 + 1)}{(30 + 1) - (20 + 1)} = \frac{82}{10} = 8.2$$

Effort $P = 10$ N

$\eta = 75\%$

Let W = load which can be lifted by the machine using the relation,

$$\eta = \frac{\text{M.A.}}{\text{V.R.}}$$

$$0.75 = \frac{W}{P \times 8.2} = \frac{W}{10 \times 8.2}$$

or $\qquad$ **W = 61.5 N. (Ans.)**

10.12. PULLEYS

10.12.1. Simple Pulley. A simple pulley is a wheel of metal or wood having a groove on its circumference, to receive a rope or chain, capable of rotation about an axle passing through its centre and perpendicular to its plane. The axle is supported in a frame work of metal or wood known as the block. The pulley is known as *fixed pulley* if the frame work supporting the axle of pulley is fixed. If the frame work or the block is movable the pulley is termed as *movable pulley*.

Following *assumptions* are made in the study of pulley systems which are quite reasonable from practical point of view.

1. The weight of the pulley block is small, as compared the weight lifted, and thus may be neglected in calculations.

2. The friction between pulley surface and string is negligible and thus the tension in the two sides of the rope, passing round the pulley, may be taken to be equal.

10.12.2. A Single Fixed Pulley. In a single fixed pulley the weight W is attached to one end of string and the effort P is applied at the other end. Under the assumed condition mentioned above, in the case of a single pulley :

$$\text{M.A.} = \text{V.R.} = 1$$

The only advantage of a single fixed pulley is that the effort can be applied in any convenient direction as shown in Fig. 10.4 (*a*), (*b*) and (*c*).

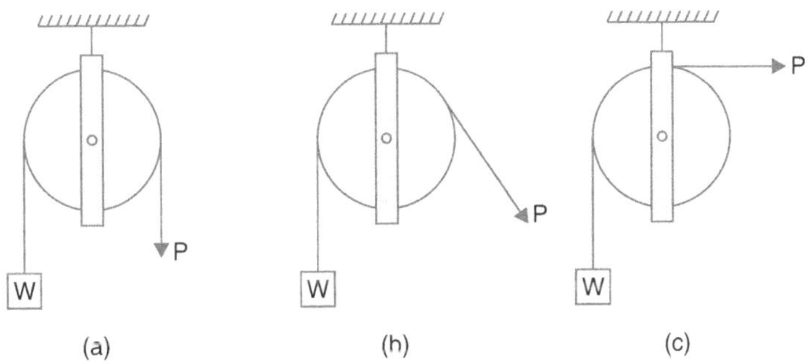

Fig. 10.4. Single fixed pulley.

10.12.3. A Movable and a Fixed Pulley. In a single movable pulley (Refer to Fig. 10.5) one end of the string is attached to the fixed support and effort P is applied at the other end. Assuming pulley to be frictionless and string and pulley to be of negligible weight, the weight W will be equally shared by the two portions of the string on fixed and moving portions as shown.

Thus,
$$P = \frac{W}{2} \text{ or } \frac{W}{P} = 2$$

If weight of the pulley is w, then

$$P = \frac{1}{2}(W + w)$$

$$\frac{W}{P} + \frac{w}{P} = 2$$

$$\frac{W}{P} = 2 - \frac{w}{P} \qquad ...(10.24)$$

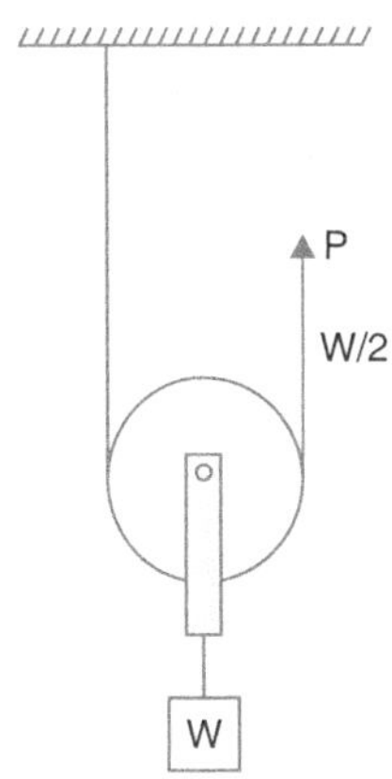

Fig. 10.5 Movable pulley.

The M.A. in the case of movable pulley is greater than unity but application of force is not easy. So to make force of application to be applied easily a fixed pulley is introduced which will not add to the mechanical advantage but will help the operator to use the force conveniently (Refer to Fig. 10.6). A little consideration will show that to raise the load W through a certain distance, the effort P will have to traverse double the distance. So velocity ratio in this case will be equal to 2.

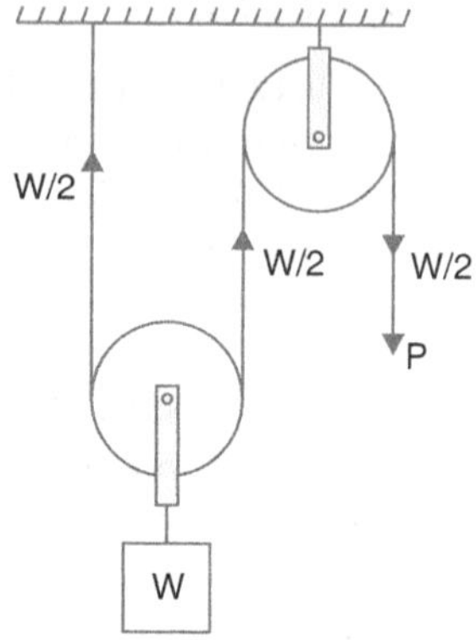

Fig. 10.6. A movable and a fixed pulley.

10.12.4. Combination of Pulleys. In order to get an increased mechanical advantage combination of pulleys is very often used. There are usually three systems of pulleys in use:

1. First system of pulleys (or separate string system)
2. Second system of pulleys (or the pulley block system)
3. Third system of pulleys (or separate string attached to weight system).

10.12.4.1. First system of pulleys (or separate string system). In this system a number of *movable pulleys* are used to give a greater mechanical advantage and *convenience in application of effort is applied by using last of all a fixed pulley.* In the system, one end of each string is tied to the fixed support while the other end passing round the periphery of the bottom pulley is fastened to the block of the next higher pulley (refer to Fig. 10.7).

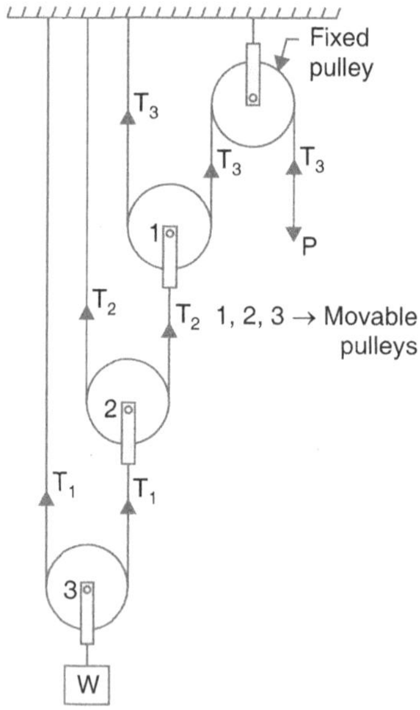

Fig. 10.7. First system of pulleys.

Let 1, 2, 3 etc., be the movable pulleys and w be the weight of each pulley ; W is the weight to be lifted. Also T_1, T_2, T_3 be the tensions in the different strings as shown.

Considering the equilibrium of pulleys 1, 2, 3 etc.

For movable pulley 1 : $2T_1 = W + w$

$$T_1 = \frac{W}{2} + \frac{w}{2}$$

For movable pulley 2 :

$$2T_2 = T_1 + w = \frac{W}{2} + \frac{w}{2} + w$$

$$T_2 = \frac{W}{2^2} + \frac{w}{2^2} + \frac{w}{2}$$

For movable pulley 3 :

$$2T_3 = T_2 + w$$

$$T_3 = \frac{W}{2^3} + \frac{w}{2^3} + \frac{w}{2^2} + \frac{w}{2}$$

If there are only three pulleys, then

$$T_3 = P = \frac{W}{2^3} + \frac{w}{2^3} + \frac{w}{2^2} + \frac{w}{2}$$

If there are n pulleys, then

$$T_n = P = \frac{W}{2^n} + \frac{w}{2^n} + \frac{w}{2^{n-1}} + \frac{w}{2^{n-2}} + \ldots\ldots$$

Multiplying both sides by 2^n

$$2^nP = W + w + 2w + 2^2 w + \ldots + 2^{n-1} w$$

$$= W + w(1 + 2 + 2^2 + \ldots + 2^{n-1})$$

$$= W + w(2^n - 1)$$

But mechanical advantage,

$$\text{M.A.} = \frac{W}{P}$$

or $$\text{M.A.} = \frac{2^n W}{2^n P}$$ (multiplying both numerator and denomenator by 2^n)

or $$\text{M.A.} = \frac{2^n W}{W + w(2^{2^n -1})} \qquad \ldots(10.25)$$

$[\because \; 2^nP = W + w(2^n - 1)]$

If the pulleys are weightless and frictionless, then

$$w = 0$$

and $$\text{M.A.} = \frac{W}{P}$$

or $$\text{M.A.} = \frac{2^n W}{W + 0} = 2^n \qquad \ldots(10.26)$$

For finding the velocity ratio of the system let us assume that P moves a distance y.

The distance by which the movable pulley 1 will move

$$= \frac{y}{2}$$

The distance by which the movable pulley 2 will move

$$= \frac{y}{2} \times \frac{1}{2} = \frac{y}{2^2}$$

Similarly, the third pulley will move

$$= \frac{y}{2^3}$$

Hence, with n movable pulleys, the lower one, and therefore the weight will move a distance

$$= \frac{y}{2^n}$$

$$\text{Velocity ratio, V.R.} = \frac{\text{Distance moved by } P}{\text{Distance moved by } W}$$

or

$$\text{V.R.} = \frac{y}{\frac{y}{2^n}} = 2^n \quad ...(10.27)$$

10.12.4.2. Second system of pulleys (or pulley block system). Second system of pulleys consists of two blocks each containing a number of pulleys. The *upper block is fixed to a fixed support and the lower block is movable.* The same string is passed round all the pulleys starting from the top pulley of the fixed block where effort is applied, to the bottom pulley of the movable block going round the next top pulley of the fixed block and next bottom pulley of the movable block and so on. The *end is finally attached to fixed block if the number of pulleys in each block is same* [Fig. 10.8 (*a*)]. *If the number of pulleys in the movable block is one less than that in the fixed block then the end of the string is attached to the movable block* [Fig. 10.8 (*b*)].

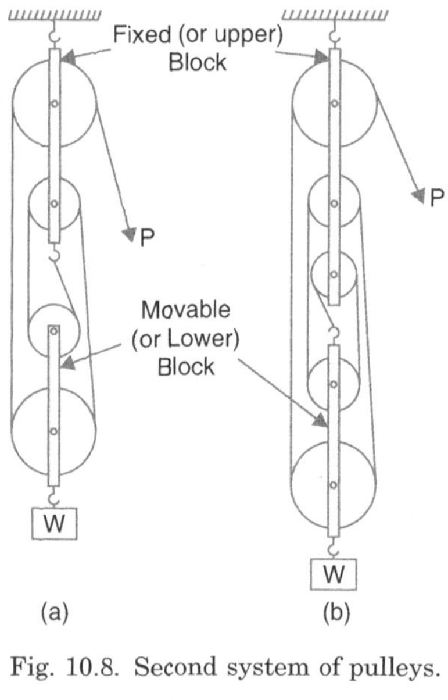

Fig. 10.8. Second system of pulleys.

The pulley being smooth the tension in the string throughout will remain equal to the applied effort P. If the number of segments of the string supporting the lower block is 'n', then the force supporting the lower block will be nP. When lower block is in equilibrium

$$W = nP$$

$$\frac{W}{P} = \text{M.A.} = n \quad ...(10.28)$$

= number of segments supporting lower block

If w is the weight of the lower block

$$W + w = nP$$

Mechanical advantage,

$$\text{M.A.} = \frac{W}{P} = n - \frac{w}{P} \qquad \text{...(10.29)}$$

It may be remembered that *number of segments supporting the movable block is always equal to total number of pulleys in the two blocks.*

10.12.4.3. Third system of pulleys (or separate string attached to the weight system). In this system, several movable pulleys are arranged keeping the top most pulley as fixed as shown in Fig. 10.9. The string passing over any pulley is fastened to load on one end and attached to the block of the lower moving on the other end. Effort is applied at the other end than that attached to the load of the string passing round the lower most pulley.

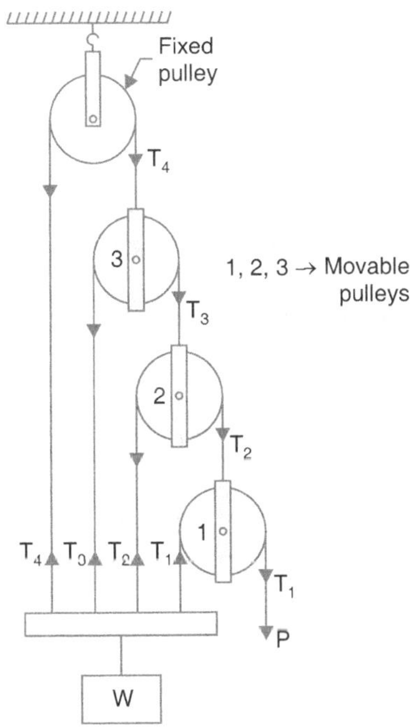

Fig. 10.9. Third system of pulleys.

Let the movable pulleys be 1, 2, 3 etc., beginning from the lowest end and let T_1, T_2, T_3 etc., be the tensions in the strings passing round them respectively.

Let w be the weight of each pulley. Considering equilibrium of pulleys 1, 2, 3 etc.

$$T_1 = P$$

$$T_2 = 2T_1 + w = 2P + w$$

$$T_3 = 2T_2 + w = 2^2P + 2w + w$$

$$T_4 = 2^3P + 2^2w + 2w + w$$

If there are n pulleys *including the fixed pulley.*

$$T_n = 2^{n-1} P + 2^{n-2} w + 2^{n-3} w + ... + 2w + w$$

But

$$\begin{aligned} W &= T_1 + T_2 + T_3 + T_4 + ... + T_n \\ &= P + (2P + w) + (2^2P + 2w + w) + (2^3P + 2^2w + 2w + w) \\ &\qquad + ... + (2^{n-1} P + 2^{n-2} w + 2^{n-3} w + ... + 2w + w) \\ &= P (1 + 2^1 + 2^2 + 2^3 + ... + 2^{n-1}) + w(1 + 2^1 + 2^2 + ... + 2^{n-2}) \\ &\qquad + w(1 + 2 + 2^2 + ... + 2^{n-3}) + ... + 2w + w \\ &= P(2^n - 1) + w(2 + 2^2 + 2^3 + ... + 2^{n-2} + 2^{n-1}) - w(n - 1) \\ &= P(2^n - 1) + w(2^n - n - 1) \end{aligned}$$

Dividing both sides by P

$$\frac{W}{P} = \text{M.A.} = (2^n - 1) + \frac{w}{P}(2^n - n - 1) \qquad ...(10.30)$$

This result *shows that the weight of pulleys assist the effort P, increasing thereby the mechanical advantage of the system.*

Assuming weights of pulleys as negligible *i.e.*, $w = 0$,

then $\quad \text{M.A.} = 2^n - 1$

For finding the V.R. of the system let the load moves x distance.

This will lower down pulley 3 = $x = (2^1 - 1)\, x$.

With the lowering down of pulley '3' by distance x and moving up of load by x, the slagness $2x$ produced thus will lower pulley '2' = $2x + x = 3x$ which is equal to $(2^2 - 1)\, x$. With the lowering down of pulley '2' by $3x$ and load moving up by x the slagness $4x$ will lower pulley '1' by $4x + 3x = 7x = (2^3 - 1)x$.

If there are n pulleys in the system then the distance moved by the last pulley will be $(2^n - 1)x$.

$$\therefore \qquad \text{V.R.} = \frac{\text{Distance moved by } P}{\text{Distance moved by } W} = \frac{(2^n - 1)\,x}{x}$$

or

$$\text{V.R.} = 2^n - 1 \qquad ...(10.31)$$

Example 10.11. *Four movable pulleys are arranged in the first system. If the weight of each pulley is 0.5 N, calculate the effort which can lift a load of 1000 N.*

Sol. We know that

$$\text{M.A.} = \frac{2^n\, W}{W + w\,(2^n - 1)}$$

where W = load to be lifted,

w = weight of each pulley

n = number of movable pulleys.

$$\therefore \qquad \text{M.A.} = \frac{2^4 \times 1000}{1000 + 0.5\,(2^4 - 1)} = \frac{1000}{P}$$

$$\therefore \qquad P = \frac{1000 + 0.5\,(2^4 - 1)}{2^4}$$

$$= \frac{1000 + 0.5 \times 15}{16} = \textbf{62.96 N.} \quad \textbf{(Ans.)}$$

Example 10.12. *A person weighing 600 N sits on a platform attached to the lower block of a system of 5 pulleys arranged in the second system. The platform and the lower block weigh 100 N. The man supports himself by exerting a downward pull at the free end of the rope. Neglecting friction, find the minimum pull the man exerts.*

Sol. Given n = total number of pulleys in the system

$= 5$

$W = 600$ N

Weight of the lower block and platform

$= 100$ N

Total weight $= 600 + 100 = 700$ N

Let the pull exerted by the man be P newtons.

Due to this pull the effective load on the lower platform will reduce to $(700 - P)$.

$\therefore$ nP = effective load

$= 700 - P$

or $5P = 700 - P$ or $6P = 700$

$\therefore$ $\mathbf{P = 116.7}$ **N. (Ans.)**

Example 10.13. *Five pulleys are arranged in the second system of pulleys. When not loaded the effort required to raise the movable block is 3.5 N. Further wastage in friction increases the pull at the rate of 3% of load lifted. Find :*

(i) the effort required to raise a load of 200 N

(ii) efficiency of the system at this load.

Sol. n = number of pulleys in the system = 5

Frictional effort at zero loading

$= 3.5$ N

Frictional effort at 200 N loading

$$= 3.5 + \frac{200 \times 3}{100} = 9.5 \text{ N}$$

When the system is considered frictionless

$nP = W$

or $5P = 200$

or $P = 40$ N

Due to friction an extra effort of 9.5 N will be required to lift the load.

Hence the total effort $= 40 + 9.5 = 49.5$ N

Efficiency at this load $= \dfrac{\text{effort without friction}}{\text{effort with friction}}$

$$= \frac{40}{49.5} \times 100 = \mathbf{80.81\%}. \quad \textbf{(Ans.)}$$

Example 10.14. *In the third system of pulleys there are in all seven pulleys including the fixed pulley. If the weight of each pulley is 0.5 N, how much weight can be supported by applying an effort equal to 8 N. What weight will be supported by the system without the application of any effort ?*

Sol. (*i*) We know that in third system of pulleys,

$$\frac{W}{P} = (2^n - 1) + \frac{w}{P}(2^n - n - 1)$$

where W = load lifted

P = effort applied

n = number of pulleys in the system

$$\therefore \quad W = P(2^n - 1) + w(2^n - n - 1)$$
$$= 80(2^7 - 1) + 0.5(2^7 - 7 - 1)$$
$$= 1016 + 60 = \mathbf{1076\ N.} \quad \textbf{(Ans.)}$$
$$= \mathbf{10.76\ kN.} \quad \textbf{(Ans.)}$$

(*ii*) $W = P(2^n - 1) + w(2^n - n - 1)$

As $P = 0$

$$\therefore \quad W = 0.5(2^7 - 7 - 1) = \mathbf{60\ N.} \quad \textbf{(Ans.)}$$

10.13. WESTON'S DIFFERENTIAL PULLEY BLOCK

Refer to Fig. 10.10. Weston's differential pulley block consists of two blocks A and B. The upper block A has two pulleys (1 and 2), one having its diameter a little larger than that of the other. These pulleys run together on one pulley the both of them *have one pulley with two grooves.* The lower block B also carries a pulley, to which the load W is attached. An endless chain passes round the pulley '1' then round the lower block pulley and then finally round the pulley '2'. The remaining chain hangs slack and is joined to the first portion of the chain. The effort P is applied to the chain passing over the pulley '1'. To *prevent the chain from slipping, projections are provided in the grooves of the pulley.*

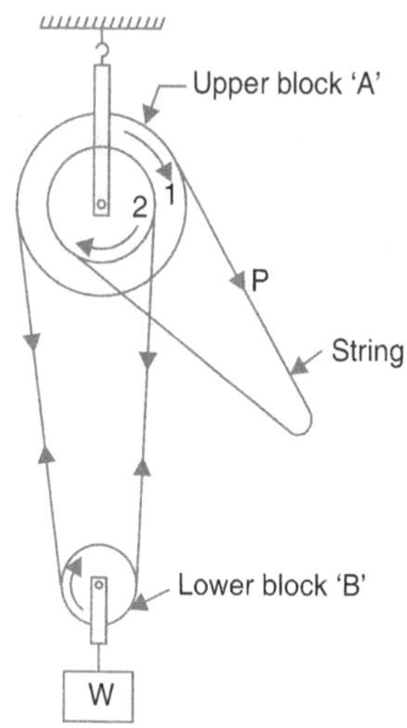

Fig. 10.10. Weston's differential pulley block.

Let D = diameter of the pulley '1',

d = diameter of the pulley '2',

W = weight lifted, and

P = the effort applied, to lift the weight.

For one complete revolution of the upper pulley block by the application of the effort P, the length of the chain pulled over the larger pulley = πD. This is *equal to the distance through which effort has moved.* Since, the smaller pulley also turns with the larger one, therefore, length of the chain released by the smaller pulley = πd.

$\therefore$ Net shortening of the chain

$$= \pi D - \pi d = \pi(D - d)$$

This shortening of chain will be equally divided between the two portions of the chain, supporting the load. Therefore, the distance through which the load will move up

$$= \frac{\pi}{2}(D - d)$$

$$\therefore \quad \text{V.R.} = \frac{\text{distance moved by } P}{\text{distance moved by } W}$$

or
$$V.R. = \frac{\pi D}{\frac{\pi}{2}(D - d)}$$

or
$$V.R. = \frac{2D}{D - d} \quad ...(10.32)$$

When the pulleys are provided with recesses it is convenient to express the velocity ratio in terms of the number of recesses.

Let n_1 = number of recesses on pulley '1' (large pulley)

n_2 = number of recesses on pulley '2' (smaller pulley)

Then
$$V.R. = \frac{2n_1}{n_1 - n_2} \quad ...(10.33)$$

$$M.A. = \frac{W}{P} \quad ...\text{as usual}$$

$$\eta = \frac{M.A.}{V.R.} \quad ...\text{as usual}$$

Example 10.15. *A weston differential pulley block consists of a lower block and upper block. The upper block has two cogged grooves, one of which has a radius of 15 cm and the other a radius of 12.50 cm. If the efficiency of a machine is 50 per cent, calculate the effort required to raise a load of 150 N.*

Sol. We know that in case of a weston differential pulley block,

$$V.R. = \frac{2D}{D - d} = \frac{2 \times 30}{30 - 25} = \frac{60}{5} = 12$$

Using the relation,

$$\text{Efficiency} = \frac{M.A.}{V.A.} \times 100$$

$$50 = \frac{M.A.}{12} \times 100$$

$\therefore$
$$M.A. = 6$$

But
$$M.A. = \frac{W \text{ (load lifted)}}{P \text{ (effort applied)}}$$

$\therefore$
$$6 = \frac{150}{P}$$

$\therefore$ **P = 25 N. (Ans.)**

Example 10.16. *In a weston differential pulley block, the number of recesses in the smaller wheel is 9/10 of that of the larger wheel. If the efficiency of the machine is 60%, find the load lifted by an effort of 30 N.*

Sol. Let the recesses in the larger wheel, $n_1 = 10$

$\therefore$ Recesses in the smaller wheel,

$$n_2 = \frac{9}{10} \times 10 = 9$$

Using the relation,

$$\text{V.R.} = \frac{2n_1}{n_1 - n_2} = \frac{2 \times 10}{10 - 9} = 20$$

and, mechanical advantage

$$\text{M.A.} = \frac{W}{P} = \frac{W}{30}$$

Also, efficiency, $$\eta = \frac{\text{M.A.}}{\text{V.R.}}$$

$\therefore$ $$0.5 = \frac{W}{30 \times 20}$$

or $$W = \mathbf{300\ N.\ (Ans.)}$$

Example 10.17. *The initial frictional resistance of an unloaded pulley block is 12 N. The friction increases at the rate of 15 N per 1000 N lifted by the block. The velocity ratio is 20. Find the effort required to lift a load of 4000 N and calculate the efficiency of the block at this load.*

Sol. Initial frictional resistance of an unloaded pulley block

$$F_1 = 12 \text{ N}$$

Rate of increase of friction = 15 N for every 1000 N lifted

$\therefore$ When W (load lifted) = 4000 N

Total amount of frictional resistance

$$= 12 + \frac{4000 \times 15}{1000} = 72 \text{ N}$$

In an ideal machine, $$\eta = 1 = \frac{W}{P_i \times \text{V.R.}} \qquad (\because\ P_i = \text{ideal effort})$$

If P denotes the actual effort required, then effort lost in friction,

$$F_P = P - P_i$$

$$= P - \frac{W}{\text{V.R.}} = P - \frac{4000}{20}$$

$$= P = -\ 200$$

But effort lost in friction = 72 N (from above)

$\therefore$ $$72 \text{ N} = P - 200 \text{ N}$$

$$\mathbf{P = 272\ N.\quad (Ans.)}$$

Also efficiency, $$\eta = \frac{\text{M.A.}}{\text{V.R.}} = \frac{W}{P \times \text{V.R.}}$$

$$= \frac{4000}{272 \times 20} = 0.735 = \mathbf{73.5\%.\quad (Ans.)}$$

Example 10.18. *A load of 400 N is lifted up by a pulley system as shown in Fig. 10.11. The co-efficient of friction on journal bearings supporting the larger pulley is 0.2. The friction in the moving pulley may be neglected. Determine :*

(*i*) *Magnitude of the effort*

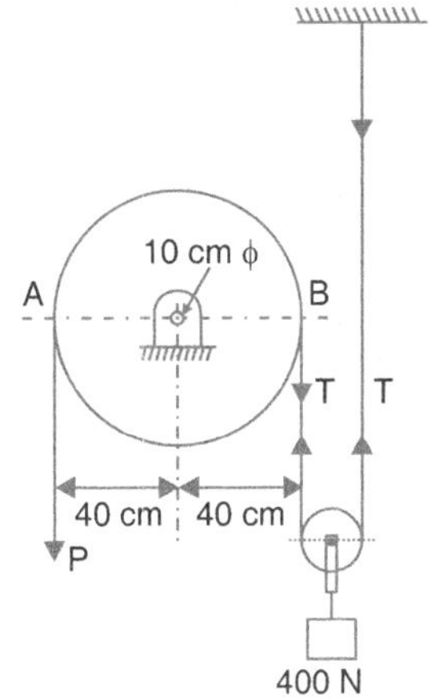

Fig. 10.11

Sol. Let P = effort applied

T = tension in the cord

Load W is-shared equally by $(T + T) = 2T$

$\therefore \quad W = 2T$

or $\quad 400 = 2T$

$\therefore \quad T = 200$ N

(*i*) **Magnitude of effort P = ?**

Pressure on the journal bearing

$$= P + T = P + 200$$

and frictional force at the journal bearing

$$= \mu \times \text{pressure} = 0.2\,(P + 200)$$

Taking moments about the centre of the bearing, we get

$$P \times 40 = T \times 40 + 0.2\,(P + 200) \times 5$$

$$40P = 200 \times 40 + P + 200$$

or $\quad$ $\mathbf{P = 210.2\ N.}$ **(Ans.)**

10.14. WORM AND WORM WHEEL

Refer to Fig. 10.12. It consists of a square threaded screw (known as *worm*) and a toothed wheel (known as *worm wheel*) geared with each other. A wheel is attached to the worm, over which passes a rope. Sometimes a handle is also fixed to the worm instead of the wheel. A load drum is securely mounted on the worm wheel.

Let l = radius of the wheel (or length of the handle),

r = radius of the load drum,

W = load lifted,

P = effort applied to lift the load, and

T = number of teeth on the worm wheel.

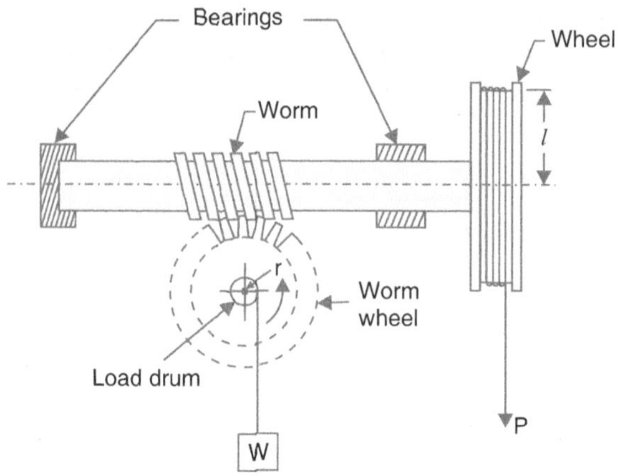

Fig. 10.12. Worm and worm wheel.

If the worm is single threaded (*i.e.,* for one revolution of the *worm,* the worm pushes the *worm wheel* through one teeth) then for one revolution of the wheel or handle the distance moved by the effort = $2\pi l$. The load drum will move through

$$= \frac{1}{T} \text{ revolution.}$$

$\therefore$ Distance, through which the load will move

$$= \frac{2\pi r}{T}$$

and $$\text{V.R.} = \frac{\text{distance moved by } P}{\text{distance moved by } W}$$

or $$\text{V.R.} = \frac{2\pi l}{\frac{2\pi r}{T}} = \frac{lT}{r} \qquad \text{...(10.34)}$$

If the worm is double threaded (*i.e.,* for one revolution of worm the worm pushes the worm wheel through half tooth, then

$$\text{V.R.} = \frac{lT}{2r} \qquad \text{...(10.35)}$$

or in general, if the worm is n threaded, then

$$\text{V.R.} = \frac{lT}{nr} \qquad \text{...(10.36)}$$

$$\text{M.A.} = \frac{W}{P} \qquad \text{...as usual}$$

$$\eta = \frac{\text{M.A.}}{\text{V.R.}} \qquad \text{...as usual}$$

Example 10.19. *In a single thread worm and worm wheel, the number of teeth on the worm wheel is 72. The diameters of the effort wheel is 18 cm and that of load drum 12 cm. Calculate the velocity ratio. If the efficiency of the machine is 60%, determine the effort requested to lift of load of 1000 N.*

Sol. Number of teeth on the worm wheel, $T = 72$

Radius of the effort wheel, $l = \frac{18}{2} = 9 \text{ cm}$

Radius of the load drum, $r = \frac{12}{2} = 6 \text{ cm}$

$$\textbf{V.R.} = \frac{lT}{r} = \frac{9 \times 72}{6} = \textbf{108.} \quad \textbf{(Ans.)}$$

$$\text{M.A.} = \frac{W}{P} = \frac{1000}{P}$$

Now, Efficiency $\eta = \frac{\text{M.A.}}{\text{V.R.}}$

$$\frac{60}{100} = \frac{100/P}{108}$$

$$\therefore \quad \boldsymbol{P} = \frac{1000}{108} \times \frac{100}{60} = \textbf{15.4 N.} \quad \textbf{(Ans.)}$$

10.15. SINGLE PURCHASE CRAB WINCH

Refer to Fig. 10.13. In a single purchase crab winch, a rope is fixed to the drum and is wound a few turns round it. The free end of the rope carries the load W. A large toothed wheel known as spur wheel is rigidly mounted on the load drum. Another small toothed wheel, called pinion, is geared with the spur-wheel.

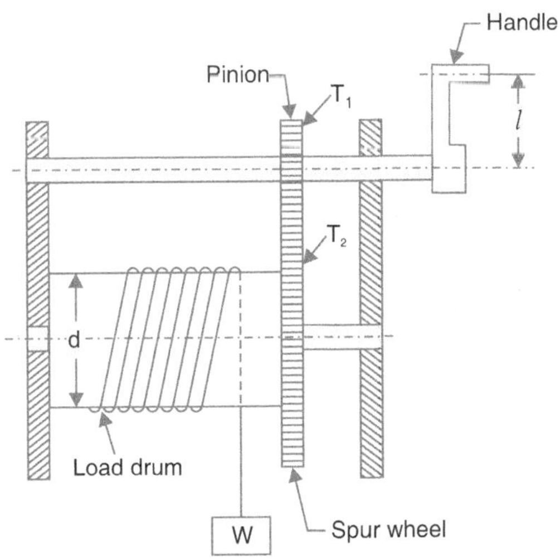

Fig. 10.13. Single purchase crab winch.

Let T_1 = number of teeth on the pinion,

T_2 = number of teeth on the spur wheel,

l = length of the handle,

W = load lifted, and

P = effort applied to lift the load, at the end of the handle.

Consider one revolution of the handle.

Then, distance moved by the effort = $2\pi l$.

Number of revolutions made by the pinion = 1

and number of revolutions made by the spur wheel

$$= \frac{T_1}{T_2}$$

Number of revolutions made by the load drum

$$= \frac{T_1}{T_2}$$

Distance moved by the load

$$= \pi d \cdot \frac{T_1}{T_2}$$

$$\therefore \qquad \text{V.R.} = \frac{\text{distance moved by } P}{\text{distance moved by } W}$$

$$= \frac{2\pi l}{\pi d \cdot \frac{T_1}{T_2}} = \frac{2l}{d} \times \frac{T_2}{T_1}$$

i.e.,

$$\text{V.R.} = \frac{2l}{d} \times \frac{T_2}{T_1} \qquad \text{...(10.37)}$$

If the diameter of the lifting rope is dr

Then

$$\text{V.R.} = \frac{2l}{d + dr} \times \frac{T_2}{T_1}$$

$$\text{M.A.} = \frac{W}{P} \qquad \text{...as usual}$$

$$\eta = \frac{\text{M.A.}}{\text{V.R.}} \qquad \text{...as usual}$$

☞ **Example 10.20.** *Following are the specifications of a single purchase crab :*

Diameter of the load drum, d = *0.2 m*

Length of lever, l = *1.2 m*

No. of teeth on the spur pinion, T_1 = *10*

No. of teeth on the spur wheel, T_2 = *100*

Find the velocity ratio of the machine. On this machine, efforts of 10 N and 16 N are required to lift the loads of 300 N and 900 N respectively. Find the law of machine and the efficiencies at the above loads.

Sol. Velocity ratio, V.R. = ?

Using the relation,

$$\text{V.R.} = \frac{2l}{d} \cdot \frac{T_2}{T_1} = \frac{2 \times 1.2}{0.2} \times \frac{100}{10} = \mathbf{120.\ Ans.}$$

Law of machine = ?

Let the law of machine be

$$P = mW + C \quad \text{...}(i)$$

When $P = 10$ N $\quad W = 300$ N

and when $P = 16$ N $\quad W = 900$ N

Substituting the above values in eqn. (*i*),

$$10 = m \times 300 + C \quad \text{...}(ii)$$

$$16 = m \times 900 + C \quad \text{...}(iii)$$

Subtracting (*ii*) from (*iii*), we get

$$6 = 600\, m$$

or $m = 6/600 = 0.01$

Substituting the value of m in eqn. (*ii*), we get

$$10 = 0.01 \times 300 + C$$

$\therefore$ $C = 7$

Hence law of machine is

$$P = 0.01\, W + 7$$

Efficiencies of the machine ?

Mechanical advantage in the first case,

$$\text{M.A.} = \frac{W}{P} = \frac{300}{10} = 30$$

Using the relation, $\eta = \frac{\text{M.A.}}{\text{V.R.}} = \frac{30}{120} = 0.25 = \mathbf{25\%}$. **(Ans.)**

Similarly, M.A. in the second case,

$$\text{M.A.} = \frac{W}{P} = \frac{900}{16} = 56.25.$$

Again, using the relation,

$$\eta = \frac{\text{M.A.}}{\text{V.R.}} = \frac{56.25}{120} = 0.469$$

or $= \mathbf{46.9\%}$. **(Ans.)**

10.16. DOUBLE PURCHASE CRAB WINCH

Refer to Fig. 10.14. In a double purchase crab winch the velocity ratio is obtained in two stages. In this lifting machine there are two spur wheels of teeth T_2 and T_4 and two pinions of teeth T_1 and T_3. The arrangement of spur wheels and pinions is such that the pinion with teeth T_1 gears meshes with the spur wheel of teeth T_2. Similarly, the pinion with teeth T_3 gears with meshes spur wheel of teeth T_4. The effort is applied at the handle.

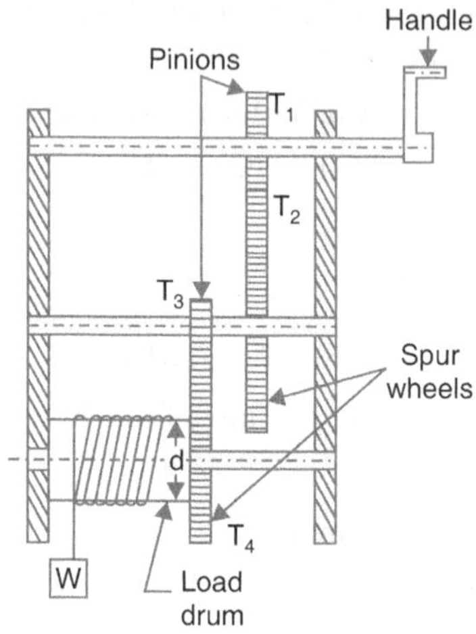

Fig. 10.14. Double purchase crab winch

Consider one revolution of the handle.

Then, distance moved by the effort = $2\pi l$

where l is the length of the handle.

Number of revolutions made by the pinion = 1

Number of revolutions made by the spur wheel 2 = $\dfrac{T_1}{T_2}$

Number of revolutions made by the pinion 3 = $\dfrac{T_1}{T_2}$

Number of revolutions made by the spur wheel 4 = $\dfrac{T_1}{T_2} \times \dfrac{T_3}{T_4}$

Distance moved by the load

$$= \pi d \times \frac{T_1}{T_2} \times \frac{T_3}{T_4}$$

$$\therefore \quad \text{V.R.} = \frac{\text{distance moved by the effort}}{\text{distance moved by the load}}$$

or
$$\text{V.R.} = \frac{2\pi l}{\pi d \times \dfrac{T_1}{T_2} \times \dfrac{T_3}{T_4}} \qquad ...(10.38)$$

or
$$\text{V.R.} = \frac{2l}{d}\left(\frac{T_2}{T_1} \times \frac{T_4}{T_3}\right) \qquad ...(10.39)$$

Here the velocity ratio is greatly increased by selecting suitable gears and the distances l and d.

If the diameter of rope d_r is also given, the formula for velocity ratio can be modified by taking the diameter of the drum from the centre of the rope on one side to the centre of the rope on the other side.

$$\therefore \quad \text{V.R.} = \frac{2l}{d + d_r} \times \left(\frac{T_2}{T_1} \times \frac{T_4}{T_3}\right) \qquad ...(10.40)$$

Example 10.21. *In a double purchase crab, the pinions have 15 and 20 teeth, while the spur wheels have 45 and 40 teeth. The effort handle is 400 mm, while the effective diameter of the drum is 150 mm. If the efficiency of this machine is 40%, what load will be lifted by an effort of 250 N applied at the end of the handle ?*

Sol. No. of teeth on the pinions,

$$T_1 = 15 \text{ and } T_3 = 20$$

No. of teeth on the spur wheels

$$T_2 = 45 \text{ and } T_4 = 40$$

Length of the effort handle,

$$l = 400 \text{ mm}$$

Effective diameter of the drum,

$$d = 150 \text{ mm}$$

Efficiency, $\eta = 40\%$

Effort, $P = 250$ N

Let W be the load lifted by the effort

We know that,
$$\text{V.R.} = \frac{2l}{d} \cdot \frac{T_2}{T_1} \times \frac{T_4}{T_3} = \frac{2 \times 400}{150} \times \frac{45 \times 40}{15 \times 20} = 32$$

Also
$$\eta = \frac{\text{M.A.}}{\text{V.R.}} = \frac{W}{P \times \text{V.R.}}$$

$$0.4 = \frac{W}{250 \times 32}$$

$W =$ **3200 N or 3.2 kN. (Ans.)**

Example 10.22. *A double purchase crab has the following dimensions :*

Effective diameter of load drum = *16 cm*

Length of handle = *36 cm*

Number of teeth on pinions = *20 and 30*

Number of teeth on spur wheels = *75 and 90*

When tested it was found that an effort of 9 N was required to lift a load of 180 N and effort of 13.5 N was required to lift a load of 315 N determine :

(*a*) *The law of machine.*

(*b*) *Probable effort to lift a load of 500 N.*

(*c*) *Efficiency in the above case.*

(*d*) *Maximum efficiency.*

Sol. Number of teeth on pinions,

$$T_1 = 20 \text{ and } T_3 = 30$$

Number of teeth on spur wheels

$$T_2 = 75 \text{ and } T_4 = 90$$

We know that
$$\text{V.R.} = \frac{2l}{d} \times \frac{T_2}{T_1} \times \frac{T_4}{T_3} = \frac{2 \times 36}{16} \times \frac{75 \times 90}{20 \times 30} = 50.6.$$

(*a*) **Law of machine ?**

Let the law of machine be

$$P = mW + C \quad ...(i)$$

When $P = 9$ N $\quad W = 180$ N

and when $P = 13.5$ N $\quad W = 315$ N

Substituting these values in eqn. (*i*), we get

$$9 = 180m + C \quad ...(ii)$$

$$13.5 = 315m + C \quad ...(iii)$$

Subtracting (*ii*) from (*iii*), we get

$$4.5 = 135m$$

or $$m = \frac{4.5}{135} = \frac{1}{30}$$

Substituting the value of m in eqn. (*ii*), we get

$$9 = 180 \times \frac{1}{30} + C$$

∴ $$C = 3$$

Hence the law of machine is

$$\mathbf{P = \frac{1}{30} W + 3.} \quad \textbf{(Ans.)}$$

(*b*) **Probable effort to lift a load of 500 N ?**

We know that, $$P = \frac{1}{30} W + 3$$

$$= \frac{1}{30} \times 500 + 3 = \mathbf{19.67\ N.} \quad \textbf{(Ans.)}$$

(*c*) **Efficiency in above case ?**

Using the relation, $$\eta = \frac{\text{M.A.}}{\text{V.R.}} = \frac{W}{P \times \text{V.R.}}$$

$$= \frac{500}{19.67 \times 50.6} = 0.50$$

or $$= \mathbf{50.2\%.} \quad \textbf{(Ans.)}$$

(*d*) **Maximum efficiency ?**

Using the relation, $$\eta_{max} = \frac{(\text{M.A.})_{max}}{\text{V.R.}} = \frac{1}{m \times \text{V.R.}} = \frac{1}{\frac{1}{30} \times 50.6}$$

$$= \frac{30}{50.6} = 0.593$$

or $$= \mathbf{59.3\%.} \quad \textbf{(Ans.)}$$

10.17. SIMPLE SCREW JACK

Refer to Fig. 10.15. An outstanding application of an inclined plane involving friction is found in simple screw jack. It is a device employed for lifting heavy loads which are usually centrally loaded upon it. Horizontal power is applied with the help of lever (or handle).

Let l = length of lever (or power arm),

P = the effort applied,

W = the load lifted

p = pitch of the screw.

Suppose, screw has taken one full revolution then, distance moved by the load = p

and distance moved by the effort = $2\pi l$

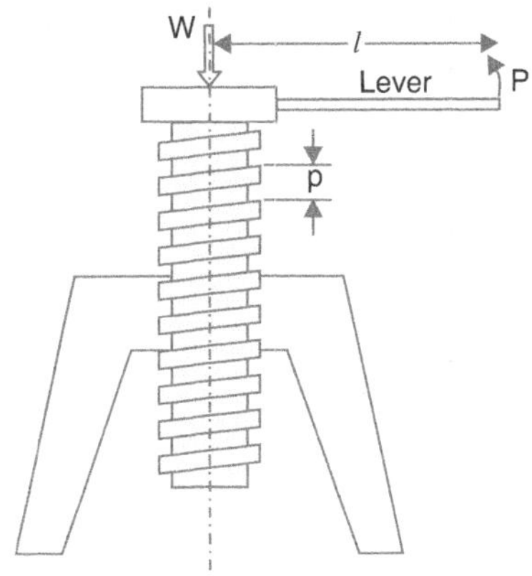

Fig. 10.15. Simple screw jack.

$$\therefore \qquad \text{V.R.} = \frac{\text{distance moved by } P}{\text{distance moved by } W}$$

i.e.,

$$\text{V.R.} = \frac{2\pi l}{p} \qquad \text{...(10.41)}$$

If the screw is *double threaded* then for one revolution of power arm the load will be lifted up through *twice the pitch*.

Hence, V.R. for double threaded screw

$$= \frac{2\pi l}{2p} = \frac{\pi l}{p} \qquad \text{...(10.42)}$$

$$\text{M.A.} = \frac{W}{P} \qquad \text{...as usual}$$

$$\eta = \frac{\text{M.A.}}{\text{V.R.}} \qquad \text{...as usual}$$

10.18. DIFFERENTIAL SCREW JACK

A differential screw jack is shown in Fig. 10.15. It has the same system of applying effort on the lever but has two threaded spindles B and C. The spindle B has both external and internal threads. The threads on spindle C mesh with internal threads of spindle B. Spindle B is screwed to a fixed base A. Nut O is provided at the top of the spindle B. The nut is rotated by the application of effort P at a distance l as shown.

When the nut is rotated spindle B is also rotated. While spindle B rotates, it acts as *a* nut for spindle C. If the nut is rotated such that spindle B rises, the spindle C will come down.

Let p_B = pitch of threads on spindle B

p_C = pitch of threads on spindle C.

For one revolution of the lever (or power arm)

Displacement of P – $2\pi l$

Spindle (screw) B rises by p_B while the spindle (or screw) C moves down by p_C.

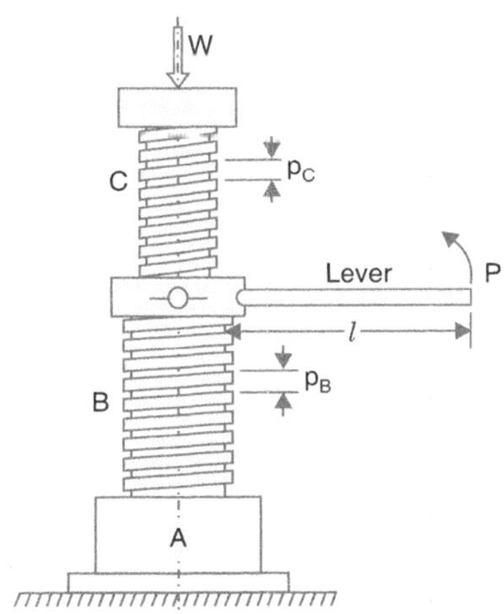

Fig. 10.16. Differential screw jack

Net raising up of load $W = p_B - p_C$

$$\therefore \quad \text{V.R.} = \frac{\text{displacement of effort}}{\text{displacement of load}}$$

i.e.,

$$\text{V.R.} = \frac{2\pi l}{p_B - p_C} \quad ...(10.43)$$

$$\text{M.A.} = \frac{W}{P} \quad \text{...as usual}$$

$$\eta = \frac{\text{M.A.}}{\text{V.R.}} \quad \text{...as usual}$$

Example 10.23. (*a*) *What is understood by 'pitch' and 'lead' in case of double thread ?*

(*b*) *A screw jack has a thread of 10 mm pitch. What effort applied at the end of the handle 400 mm long will be required to lift a load of 20 kN, if efficiency at this load is 45 per cent ?*

Sol. (*a*) **Pitch.** *Pitch of a screw is the distance from a point on a screw thread to a corresponding point on the next thread measured parallel to the axis of the screw.*

Lead. *Lead is the distance a screw thread advances axially in one turn. In a double thread, lead is equal to twice the pitch.*

(*b*) Load to be lifted, $W = 20$ kN

Pitch of the screw jack, $p = 10$ mm

Length of the handle, $l = 400$ mm

Velocity ratio of the screw jack,

$$\text{V.R.} = \frac{2\pi l}{p} = \frac{2\pi \times 400}{10} = 80\,\pi$$

Effort applied, P = ?

Using the relation,

$$\eta = \frac{\text{M.A.}}{\text{V.R.}}$$

$$0.45 = \frac{W}{P \times \text{V.R.}} = \frac{20}{P \times 80\pi}$$

$$P = \frac{20}{80\pi \times 0.45} = \mathbf{0.177\ kN.} \quad \textbf{(Ans.)}$$

HIGHLIGHTS

1. A '*machine*' may be defined as a contrivance or a device which receives energy in some available form and uses it for doing a particular useful work.
2. A '*lifting machine*' may be defined as contrivance to overcome a force or load (W) applied at one point by means of another force called effort (P).
3. Mechanical advantage,

$$\text{M.A.} = \frac{\text{load lifted } (W)}{\text{effort applied } (P)}$$

4. Velocity ratio,

$$\text{V.R.} = \frac{\text{distance moved by the effort } (y)}{\text{distance moved by the load } (x)}$$

5. Efficiency, $\eta = \frac{\text{output}}{\text{input}} = \frac{\text{M.A.}}{\text{V.R.}} = \frac{P_i}{P} = \frac{W}{W_i}$

6. Loss of effort due to friction,

$$F_P = P - P_i = P - \frac{W}{\text{V.R.}}$$

7. Loss of load due to friction,

$$F_W = W_i - W = P \times \text{V.R.} - W$$

8. A machine is said to be *reversible* when its efficiency is *greater* than 50%.

9. A machine is said to be *'non-reversible'* or *'self-locking'* when it is not capable of doing any work in the reversed direction (η < 50%).

10. The law of a machine gives the relationship between the effort applied and the load lifted. It is expressed as

$$P = mW + C$$

11. Maximum mechanical advantage, (M.A.) max = $\frac{1}{m}$ and maximum efficiency, η_{max}

$$= \frac{1}{m} \times \frac{1}{\text{V.R.}}.$$

12. Velocity ratio in case of :

(*i*) Simple wheel and axle : $\frac{D + dr_1}{d + dr_2}$

(*ii*) Wheel and differential axial :

$$\frac{2(D + dr_1)}{(d_1 + dr_2) - (d_2 + dr_2)}$$

(*iii*) *First system of pulleys* : 2^n, where n = number of movable pulleys.

(*iv*) *Second system of pulleys* : Total number of pulleys in the blocks.

(*v*) *Third system of pulleys* : $2^n - 1$, where n = number of pulleys including the fixed pulley.

(*vi*) Weston's differential pulley block : $\frac{2D}{D - d}$

(*vii*) Worm and worm wheel : $\frac{lT}{nr}$

(*viii*) Single purchase crab winch $\frac{2l}{d + dr} \times \frac{T_2}{T_1}$

(*ix*) Double purchase crab winch : $\frac{2l}{d + dr} \times \left(\frac{T_2}{T_1} \times \frac{T_4}{T_3}\right)$

(*x*) Simple screw jack : $\frac{2\pi l}{p}$ (when the screw is single threaded) $\frac{\pi l}{p}$ (when the screw is double threaded)

(*xi*) Differential screw jack : $\frac{2\pi l}{p_B p_C}$.

OBJECTIVE TYPE QUESTIONS

Fill in the blanks :

(*i*) A ...... is a contrivance or a device which receives energy in some available form and uses it for doing a particular useful work.
(*ii*) A simple machine has ...... point of application of effort and ...... point for load.
(*iii*) A ...... machine has more than one point for the application of effort and for load.
(*iv*) Mechanical advantage is the ratio of ...... lifted to ...... applied.
(*v*) ...... is the ratio of the distance moved by the effort to the distance moved by the load.
(*vi*) ...... of a machine is the work done on the machine.
(*vii*) ...... of a machine is the actual work done by the machine.
(*viii*) A machine is said to be ...... if its efficiency is 100 percent.
(*ix*) Loss of effort due to friction, $F_P = P -$ ......
(*x*) Loss of load due to friction, $F_W = P \times$ V.R. $-$ ......
(*xi*) A machine is said to be reversible if its efficiency is ...... than 50%.
(*xii*) The efficiency of a self–locking machine is ...... than 50%.
(*xiii*) The ...... of a machine gives the relationship between the effort applied and the load lifted.
(*xiv*) Maximum mechanical advantage of a machine is given as : $(\text{M.A.})_{max}$ = ...
(*xv*) V.R. of third system of pulleys = ...

Answers

(*i*) machine	(*ii*) one, one	(*iii*) compound	(*iv*) weight, effort
(*v*) Velocity ratio	(*vi*) Input	(*vii*) Output	(*viii*) ideal
(*ix*) $\dfrac{W}{\text{V.R.}}$	(*x*) W	(*xi*) greater	(*xii*) less
(*xiii*) law	(*xiv*) $\dfrac{1}{m}$	(*xv*) $2^n - 1$.	

EXERCISES

Theoretical Questions

1. What is a machine ? State its features.
2. Explain the following with examples :
   (*i*) Simple machines
   (*ii*) Compound machines.
3. Define the following :
   Mechanical advantage, velocity ratio, input of a machine and efficiency of a machine.
4. Establish relation between efficiency, mechanical advantage and velocity ratio of a machine.
5. Discuss briefly the effect of friction in a machine.
6. Prove that a machine will be reversible if its efficiency exceeds 50%.
7. What is a self-locking machine ?
8. What is the law of a machine ?
9. Discuss briefly with neat diagrams the following lifting machines :
   (*i*) Wheel and differential axle,
   (*ii*) Simple screw jack,
   (*iii*) Weston's differential pulley block,
   (*iv*) Single purchase crab winch,
   (*v*) Worm and worm wheel.

Unsolved Examples

1. In a lifting machine a load of 500 N is lifted by an effort of 12.5 N. While the load moves up by 5 mm the point of application of effort moves by 400 cm. Find :
   (*i*) Mechanical advantage
   (*ii*) Velocity ratio
   (*iii*) Efficiency of this machine. [**Ans.** (*i*) 40, (*ii*) 80, (*iii*) 50%]
2. The velocity ratio of a mechanic is 20. If the efficiency of this machine is 60% what effort will be required to lift a load of 240 N. [**Ans.** 20 N]
3. A machine can lift a load of 450 N with an effort of 30 N. The velocity ratio of a machine is 20. Determine the machine friction in terms of effort and load. Also find its efficiency. [**Ans.** F_p = 7.5 N ; F_W = 150 N ; 75%]
4. The velocity ratio of a certain machine is 75. The law of machine is $P = \frac{W}{45} + 3$. Determine the maximum mechanical advantage and maximum efficiency. Also state whether the machine is reversible or not. [**Ans.** 45 ; 60%, reversible]
5. In a certain machine it is observed that an effort of 10 N lifts a load of 250 N and an effort of 13 N lifts a load of 400 N through the same distance. What will be the effort required to lift a load of 550 N ? Find out the maximum mechanical advantage and maximum efficiency. Assume the V.R. of the machine as 75. [**Ans.** 16 N ; 50 ; 66.66%]
6. The velocity ratio of a machine is 150. It lifts loads of 240 N and 420 N with efforts of 5 N and 6.5 N respectively. Determine the efficiency of the machine when the load to be lifted is 300 N. [**Ans.** 36.36%]
7. In a machine it was found that efforts of 20 N and 15 N are required to raise load of 150 N and 250 N respectively. If the velocity ratio of the machine is 25, find :
   (*i*) Law of the machine
   (*ii*) Probable effort required to raise a load of 400 N
   (*iii*) Mechanical advantage
   (*iv*) Efficiency
   (*v*) Effort lost in friction. $\left[\textbf{Ans. } (i)\ P = \frac{W}{20} + 5, (ii)\ 25\text{ N}, (iii)\ 16, (iv)\ 64\%, (v)\ 9\text{ N}\right]$
8. The initial frictional resistance of an unloaded pulley block is 2.6 N. The friction increases at the rate of 1.4 N per 100 N lifted by the block. The velocity ratio is 18. Find the effort required to lift a load of 1200 N and calculate the efficiency of the block at this load. [**Ans.** 86.07 N ; 77.5%]
9. In a simple wheel and axle, the diameter of the wheel is 30 cm and that of axle is 15 cm. If the efficiency of the machine is 60%, determine the effort required to lift a load of 30 N. [**Ans.** 25 N]
10. In a simple wheel and axle, the radius of effort wheel is 240 mm and that of the axle is 400 mm. Determine the efficiency, if a load of 300 N can be lifted by an effort of 60 N. [**Ans.** 83.33%]
11. In a wheel and differential axle having a velocity ratio of 24, a load of 200 N can be lifted by an effort of 15 N and a load of 250 N can be lifted by an effort of 18 N.
    Find : (*i*) effort required to lift a load of 375 N, (*ii*) effort wasted in friction, and (*iii*) maximum efficiency of the machine. [**Ans.** (*i*) 25.5 N ; (*ii*) 9.875 N ; (*iii*) 69.4%]
12. Four movable pulleys are arranged in the first system of pulleys. If the efficiency of the system is 85%, find what load can be raised by an effort of 18 N. If the weight of each pulley is 1 N, calculate what effort can lift a load 1000 N. [**Ans.** 244.8 N; 68.43 N]
13. In a second system of pulleys, the weight of the lower movable block of pulleys is 100 N. Each block contains three pulleys. Find the greatest load that can be lifted with an effort of 250 N when one end of the string is tied to the movable block. [**Ans.** 1.4 kN]
14. In a system where same rope constantly passes round all the pulleys, the load to be lifted is 135 N and effort is of magnitude of 40 N. The weight of the lower block is 5 N. If the number of pulleys is 4, find the efficiency of the system. [**Ans.** 87.5%]
15. In a third system of pulleys there are 4 pulleys in all. Find the velocity ratio. Also find the total load lifted by an effort of 320 N, if the efficiency of the machine is 75%. [**Ans.** 3.6 kN]

16. In a Weston's differential pulley block there are 10 recesses on the larger pulley and 9 on the smaller one. Determine the efficiency of the machine, if an effort of 50 N, is required to lift a load of 600 N. [**Ans.** 60%]

17. In a differential pulley block, the radii of the concentric pulleys are 15 cm and 10 cm respectively. It has been found that an effort of 20 N just lifts a load of 100 N. Calculate

(*i*) efficiency of the machine (*ii*) effort lost in friction and

(*iii*) frictional load [**Ans.** (*i*) 83.33% ; (*ii*) 3.33 N ; (*iii*) 20 N]

18. In a single purchase crab winch, length of handle is 20 cm and gear ratio is 6. Find the velocity ratio and efficiency of the machine if a load of 540 N is lifted by an effort of 30 N. Diameter of the drum is 10 cm. [**Ans.** 32 ; 75%]

19. A single purchase crab has number of teeth on the pinion and spur wheel as 20 and 200 respectively. The handle is 300 mm long and the radius of the load drum is 150 mm. Find the velocity ratio. If an effort of 20 N can lift a load of 360 N, determine the efficiency of the machine. [**Ans.** 20 ; 90%]

20. In a double purchase crab winch, the following are the specifications :

Length of lever = 1 m; Number of teeth on the pinion on the effort axle = 10 ; Number of teeth on the spur wheel on the counter axle = 100 ; Number of teeth on the pinion of counter axle = 15 ; Number of teeth on the load axle = 150 ; Diameter of the load axle (drum) = 250 mm.

Find the velocity ratio and the load which can be lifted by an effort of 50 N if efficiency at this load is 50%. [**Ans.** 80 ; 2 kN]

21. Find the mechanical advantage and efficiency of a screw jack if an effort of 10 N is applied at the end of an arm 210 mm long to lift a load of 2500 N. The pitch of the screw being 5 mm. [**Ans.** 250 ; 94.7%]

22. In a differential screw jack the pitches of the outer and inner screws are 1 cm and 0.75 cm respectively. What load is raised by an effort of 10 N, if the efficiency of the machine is 75% ? Length of handle is given to be 50 cm. [**Ans.** 9424.5 N]

23. In a differential screw jack, pitch of the screws are 1 cm and 0.5 cm. Efficiency is 50%. What will be the load lifted by an effort of 100 N applied on a handle 40 cm long. [**Ans.** 25.13 kN]

11

Circular and Curvilinear Motion

11.1. INTRODUCTION AND DEFINITIONS

Introduction

In a rectilinear motion, a moving particle describes a *straight path* while in a curvilinear motion it describes a *curved path.* In *circular or rotating motion a moving particle describes a circular path,* its position at any instant can be defined by the angle θ covered by it w.r.t. X or Y-axis (generally it is done w.r.t. X-axis).

Definitions

1. Angular displacement. When a particle moves on a circular path its angle of rotation (or its angular displacement) θ, varies with time. Thus, we can say, $\theta = f(t)$.

Angular displacement (θ) is usually expressed in radians.

2. Angular velocity. The angular velocity is defined as *the rate of change of angular displacement denoted by ω (Greek letter omega).*

The particle is said to have *uniform rotation or uniform angular velocity if at any instant, having covered angular displacement θ in t time, receives equal increment of finite angular displacement.*

If θ = angular displacement,

t = time taken for the displacement to take place, and

ω = angular velocity

Then mathematically,

$$\omega = \frac{\theta}{t} \qquad \text{...(11.1)}$$

The particle is said to have *non-uniform rotation if it receives unequal increments of angular displacement in a circular path, during equal intervals of time.*

The unit of angular velocity is *rad/sec.*

If r is the radius of the circular path on which the particle is moving, the instantaneous velocity at that instant,

$$v = \omega r \qquad \text{...(11.2)}$$

If N is the uniform speed of a rotating part in revolutions per minute (r.p.m.) then its angular velocity is given by the relation,

$$\omega = \frac{2\pi N}{60} \text{ rad/sec.} \qquad ...(11.3)$$

3. Angular acceleration. It is defined as the *rate of change of angular velocity* denoted by α (Greek letter alpha). It is usually expressed in *rad / sec*^{2}.

Angular acceleration of a particle is uniform, if the changes in angular velocity are equal in equal intervals of time, however, small these intervals may be.

11.2. EQUATIONS OF ANGULAR MOTION

Let ω_0 = initial angular velocity in rad/sec^2

ω = final angular velocity in rad/sec^2

t = time taken in seconds

θ = angle covered/traversed (in t seconds) in radians

α = angular acceleration in rad/sec^2.

By definition, acceleration:

$$\alpha = \frac{\omega - \omega_0}{t}$$

$$\alpha t = \omega - \omega_0$$

or

$$\omega = \omega_0 + \alpha t \qquad ...(11.4)$$

Also, average angular velocity $= \left(\frac{\omega_0 + \omega}{2}\right)$

$\therefore$ Angular displacement,

$$\theta = \text{average angular velocity} \times \text{time}$$

$$= \left(\frac{\omega_0 + \omega}{2}\right) \times t$$

$$= \left(\frac{\omega_0 + \omega_0 + \alpha t}{2}\right) \times t \qquad (\because \ \omega = \omega_0 + \alpha t)$$

$$= \left(\frac{2\omega_0 + \alpha t}{2}\right) \times t$$

$$= \omega_0 t + \frac{1}{2} \alpha t^2$$

$\therefore$

$$\theta = \omega_0 t + \frac{1}{2} \alpha t^2 \qquad ...(11.5)$$

From equation (11.4), we have

$$\omega = \omega_0 + \alpha t$$

But

$$\theta = \left(\frac{\omega_0 + \omega}{2}\right) \times t$$

or $$t = \frac{2\theta}{\omega_0 + \omega}$$

Substituting this value of t in equation (11.4), we get

$$\omega = \omega_0 + \alpha\left(\frac{2\theta}{\omega_0 + \omega}\right)$$

$$\omega(\omega_0 + \omega) = \omega_0(\omega_0 + \omega) + 2\alpha\theta$$

$$\omega\omega_0 + \omega^2 = \omega_0^2 + \omega_0\omega + 2\alpha\theta$$

$$\omega^2 - \omega_0^2 = 2\alpha\theta \quad \text{...(11.6)}$$

11.3. EQUATIONS OF LINEAR MOTION AND ANGULAR MOTION

Following are the equations of the linear motion and angular motion :

Linear motion	*Angular motion*
$v = u + at$	$\omega = \omega_0 + \alpha t$
$S = ut + \frac{1}{2}at^2$	$\theta = \omega_0 t + \frac{1}{2}\alpha t^2$
$v^2 - u^2 = 2aS$	$\omega^2 - \omega_0^2 = 2\alpha\theta$

It may be calculated that u, v, a and S in linear motion correspond to ω_0, ω, α and θ respectively in angular motion.

11.4. RELATION BETWEEN LINEAR AND ANGULAR MOTION

Refer to Fig. 11.1. Consider a body rotating in a circular path. Let it moves from L to M after say t time.

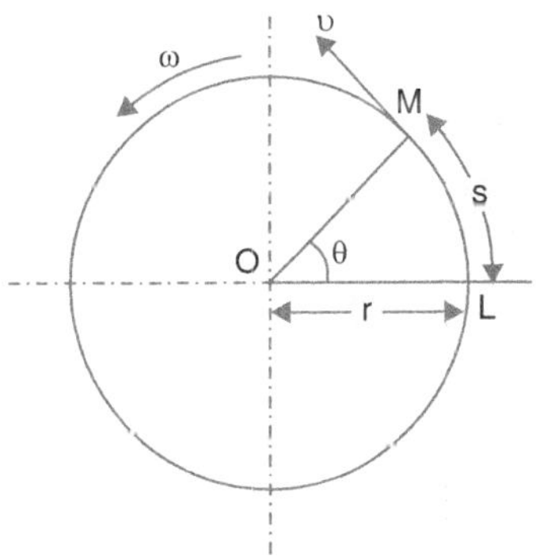

Fig. 11.1

Let ω = angular velocity of the body in rad/sec

r = radius of the circular path in metres

v = linear velocity of the body on the periphery in m/sec

θ = angular displacement in radians (from L to M in time t)

S = linear displacement in metres from L to M in time t

α = angular acceleration of the body and

a = linear acceleration.

(*i*) We know that

Length of the arc, LM = radius of the arc × angle subtended

or $$S = r \times \theta$$

$$\therefore \quad \theta = \frac{S}{r} \quad \text{...(11.7)}$$

(*ii*) Also, linear velocity,

$$v = \frac{\text{linear displacement from } L \text{ to } M}{\text{time taken}}$$

$$= \frac{\text{Arc } LM}{\text{time}} = \frac{S}{t}$$

or $$v = \frac{r \times \theta}{t}$$

or $$v = r\omega \qquad \left(\because \ \omega = \frac{\theta}{t}\right)$$

$\therefore$ $$\omega = \frac{v}{r} \qquad ...(11.8)$$

(*iii*) We know that, final angular velocity,

$$\omega = \omega_0 + \alpha t$$

or $$\frac{v}{r} = \frac{u}{r} + \alpha t$$

or $$\alpha t = \frac{v}{r} - \frac{u}{r}$$

or $$\alpha t = \frac{1}{r}(v - u)$$

or $$\alpha = \frac{1}{r}\left(\frac{v-u}{t}\right)$$

or $$\alpha = \frac{a}{r} \qquad \left(\because \ a = \frac{v-u}{t}\right) \qquad ...(11.9)$$

11.5. CENTRIFUGAL AND CENTRIPETAL FORCE

The word *centrifugal* means "*directed away from the centre*" and the word *centripetal* means "*directed towards the centre of curvature*".

When a body moves on a circular path, at every instant it has a tendency to move off tangentially as shown in Fig. 11.2 and to keep it moving in the circular path a *radial inward force* has to be applied, which is known as "*centrifugal force*".

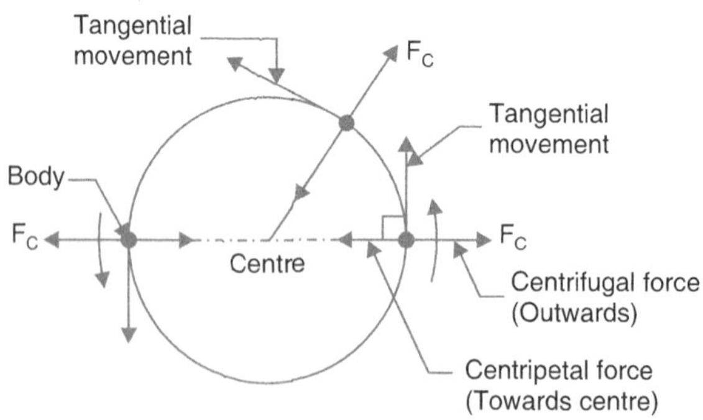

Fig. 11.2

When a particle moves on a curved or circular path of radius it is subjected to: (*i*) angular acceleration $\frac{d\omega}{dt}$ or tangential acceleration $r\frac{dw}{dt}$ and (*ii*) normal acceleration $\frac{v^2}{r}$ or $\omega^2 r$ directed towards the centre of the circle. If the particle is replaced by a body of weight W, a force known as "centripetal force", acting towards the centre of the circle, will have to be applied to keep the body moving along the circular path. This force denoted by

$$F_c = \text{mass} \times \text{acceleration} = \frac{W}{g} \times \frac{v^2}{r} = \frac{W}{g}\omega^2 r.$$

This inertia of body resisting change of motion exerts an equal outward force known as *"centrifugal force"*.

As *centripetal and centrifugal forces are equal* and *opposite*.

$$\therefore \quad \text{Centrifugal force} = \frac{W}{g}\frac{v^2}{r} = \frac{W}{g}\omega^2 r \quad ...(11.10)$$

The centrifugal force tends to throw the body radially outwards and to counter balance the effect of this force an inward force or centipetal force has got to be applied.

Useful applications of centripetal force are found in *clutches* and *governors*.

Centrifugal force entails many disadvantages, most commonly experienced are ; unbalanced dangerous force due to rotating masses, vehicles or trains rounding on curved path.

Example 11.1. *A wheel is rotating about its axis with a constant angular acceleration of 1 radian/sec^2. If the initial and final angular velocities are 2.5 radians/sec and 10.5 radians/sec, determine the total angle turned through during the time interval this change of angular velocity took place.*

Sol. Constant angular acceleration $\alpha = 1$ rad/sec^2

Initial angular velocity, $\omega_0 = 5.25$ rad/sec

Final angular velocity, $\omega = 10.5$ rad/sec

Total angle turned through by the wheel, θ = ?

Using the relation,

$$\omega^2 - \omega_0^2 = 2\alpha\theta$$

$$(10.5)^2 - (5.25)^2 = 2 \times 1 \times \theta$$

$$110.25 - 27.56 = 2\theta$$

$$\theta = \mathbf{41.34\ radians.} \quad \textbf{(Ans.)}$$

Example 11.2. *A wheel accelerates uniformly from rest to 1500 r.p.m. in 15 seconds. What is its angular acceleration ? How many revolutions does the wheel make in attaining the speed of 1500 r.p.m.*

Sol. Initial angular velocity, $\omega_0 = 0$

Final angular velocity, $\omega = \dfrac{2\pi \times 1500}{60} = 50\pi$ rad/sec

Time taken, $t = 15$ sec

Let α be the angular acceleration in rad/sec^2

Using the relation, $\omega = \omega_0 + \alpha t$

$$50\pi = 0 + \alpha \times 15$$

$$\therefore \quad \alpha = \frac{50\pi}{15} - 10.47 \text{ rad/sec}^2.$$

Let θ be the angular displacement moved in radians during this interval.

Now using the relation, $\theta = \omega_0 \,.\, t + \frac{1}{2}\alpha t^2$

$$= 0 \times 15 + \frac{1}{2} \times \frac{50\pi}{15} \times 15^2$$

$$\therefore \quad = 375\pi.$$

But 1 revolution = 2π

$\therefore$ Number of revolutions made by the wheel in attaining the speed of 1500 r.p.m.

$$= \frac{375\pi}{2\pi} = \mathbf{187.5.} \quad \textbf{(Ans.)}$$

Example 11.3. *A wheel rotating about a fixed axis at 20 r.p.m. is uniformly accelerated for 70 seconds during which time it makes 50 revolutions. Find :*

(i) Angular velocity at the end of this interval, and

(ii) Time required for the speed to reach 100 revolutions per minute.

Sol. Initial angular velocity,

$$\omega_0 = \frac{2\pi \times 20}{60} = 0.67\pi \text{ rad/sec}$$

Time, $t = 70$ sec

Angular displacement,

$$\theta = 50 \text{ revolutions}$$
$$= 50 \times 2\pi = 100\pi \text{ radians.}$$

(*i*) **Angular velocity at the end of 70 sec., ω = ?**

Let α = angular acceleration

Using the relation $\theta = \omega_0 \,.\, t + \frac{1}{2}\alpha t^2$

$$100\pi = 0.67\pi \times 70 + \frac{1}{2}\alpha \times 70^2 = 46.9\pi + 2450\alpha$$

$$\therefore \quad \alpha = 0.068 \text{ rad/sec}^2.$$

Now using the relation,

$$\omega = \omega_0 + \alpha t$$
$$= 0.67\pi + 0.068 \times 70$$
$$= 2.1 + 4.76 = \mathbf{6.86\ rad/sec.} \quad \textbf{(Ans.)}$$

(*ii*) **Time required for the speed to reach 100 r.p.m., t = ?**

Final velocity, $\omega = 100 \text{ r.p.m.} = \frac{2\pi \times 100}{60} = \frac{10}{3}\pi$ rad/sec.

Again, using the relation,

$$\omega = \omega_0 + \alpha t$$

$$\frac{10\pi}{3} = 0.67\pi + 0.068\, t$$

$$\therefore \quad t = 123 \text{ sec. approx.}$$
$$= \mathbf{2\ min\ 3\ sec.} \quad \textbf{(Ans.)}$$

Example 11.4. *A stationary flywheel of 0.5 m diameter is subjected to a constant torque of 200 Nm so that the wheel attains a speed of 100 r.p.m., in 3 minutes time. Determine :*

(*i*) *Angular acceleration*

(*ii*) *Number of revolutions made during this time*

(*iii*) *Word done to attain the speed, and*

(*iv*) *Circumferential (peripheral) speed of wheel at 100 r.p.m.*

Sol. Constant torque,

$$T = 200 \text{ Nm}$$

Initial angular speed,

$$\omega_0 = 0$$

Final angular speed,

$$\omega = \frac{2\pi \times 100}{60} = \frac{10\pi}{3} \text{ rad/s}$$

Time taken, $t = 3$ min or 180 sec.

(*i*) **Angular acceleration, α = ?**

Using the relation, $\omega = \omega_0 + \alpha t$

$$\frac{10\pi}{3} = 0 + \alpha \times 180$$

$$\alpha = \frac{10\pi}{3 \times 180} = \frac{\pi}{54} \text{ rad/s}^2. \quad \textbf{(Ans.)}$$

(*ii*) **Number of revolutions made N = ?**

Using the relation,

$$\omega^2 - \omega_0^2 = 2\alpha\theta$$

$$\left(\frac{10\pi}{3}\right)^2 - 0^2 = 2 \times \frac{\pi}{54} \times \theta$$

$$\theta = \frac{100\pi^2}{9} \times \frac{54}{2\pi} = 300\pi$$

or $$N = \frac{300\pi}{2\pi} = \textbf{150 revolutions.} \quad \textbf{(Ans.)}$$

(*iii*) **Work done = ?**

Using the relation,

$$\text{Work done} = \text{torque} \times \text{angle turned}$$
$$= 200 \times 300\,\pi$$
$$= \textbf{188495 Nm.} \quad \textbf{(Ans.)}$$

(*iv*) **Circumferential speed of the flywheel, v = ?**

Using the relation,

$$v = \omega r$$
$$= \frac{10\pi}{3} \times 0.25$$
$$= \textbf{2.62 m/s.} \quad \textbf{(Ans.)}$$

Example 11.5. *The relation between the angle of rotation and the time in the case of a rotating body is given by the equation,*

$\theta = 2t^3 + 3t^2 + 15$ *radians, where t is in seconds. Determine the angular velocity, displacement and acceleration when t = 10 seconds and state whether the acceleration is uniform.*

Sol. $$\theta = 2t^3 + 3t^2 + 15 \quad ...(i)$$

Differentiating both sides, we get

$$\frac{d\theta}{dt} = \omega = 6t^2 + 6t$$

or $$\omega = 6t^2 + 6t \quad ...(ii)$$

where ω is the angular velocity of the rotating body.

Again, differentiating the equation (*ii*), we get

$$\frac{d\omega}{dt} = \alpha = 12\,t + 6$$

$\therefore$ $$\alpha = 12\,t + 6 \quad ...(iii)$$

where α is the angular acceleration of the body.

(*i*) **Angular velocity, ω = ?**

Substituting t = 10 sec in equation (*ii*), we get

$$\omega = 6 \times 10^2 + 6 \times 10 = \mathbf{660\ rad/sec.}\quad \textbf{(Ans.)}$$

(*ii*) **Angular displacement, θ = ?**

Substituting t = 10 seconds in equation (*i*), we get

$$\theta = 2 \times 10^3 + 3 \times 10^2 + 15 = \mathbf{2315\ radians.}\quad \textbf{(Ans.)}$$

(*iii*) **Angular acceleration, α = ?**

Substituting t = 10 sec in equation (*iii*), we get

$$\alpha = 12 \times 10 + 6 = \mathbf{126\ rad/sec^2.}\quad \textbf{(Ans.)}$$

Example 11.6. *A pulley weighing 1000 N is supported on a horizontal shaft of negligible weight between two bearings, A and B, 5 metres apart. The pulley is fixed at a distance of 2 m from the bearing A. The mass centre of the polley is 20 mm from the axis of the shaft. Determine the maximum pressure on the two bearings when the shaft rotates at 200 r.p.m. Neglect the effect of deflection of the axis of the shaft.*

Sol. Refer to Fig. 11.3. Let R_A and R_B be the maximum pressures on the bearings.

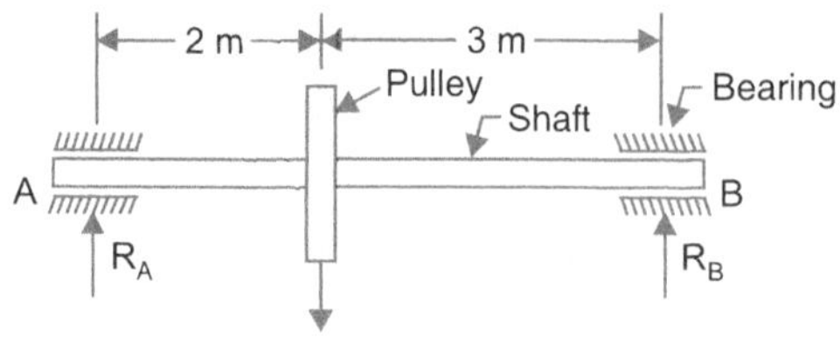

Fig. 11.3

Angular speed of the shaft,

$$\omega = \frac{2\pi \times 200}{60} = 20.94 \text{ rad/s.}$$

Centrifugal force exerted by the pulley on the shaft,

$$= \frac{W}{g}\,\omega^2 r = \frac{1000}{9.81} \times 20.94^2 \times \left(\frac{20}{1000}\right) = 894 \text{ N.}$$

The pressure on the shaft is the greatest at that instant when the mass centre of the pulley is *vertically below the axis of the shaft.*

The pressure will be

$$= \text{weight of pulley} + \text{centrifugal force exerted by the pulley}$$

$$= 1000 + 894 = 1894 \text{ N}$$

Taking moments about A, we get

$$R_B \times 5 = 1894 \times 2$$

$$\boldsymbol{R_B = 757.6 \text{ N}}$$

and $\boldsymbol{R_A = 1894 - 757.6 = 1136.4 \text{ N}}$. **(Ans.)**

Example 11.7. *How many revolutions per second must a stone weighing 0.2 N make when whirled round horizontally at the extremity of a string 0.5 m long in order to cause a tension of 10 N in the string ?*

Sol. Tension in the string, $T = 10$ N

Weight of the stone, $W = 0.2$ N

Length of the string, $r = 0.5$ m

Revolutions/sec, $n = ?$

We know that tension in the string

$$T = Fc = \frac{W}{g}\omega^2 r$$

or $$10 = \frac{0.2}{9.81} \times \omega^2 \times 0.5$$

or $$\omega^2 = \frac{9.81}{0.2 \times 0.5} = 981$$

or $$\omega = \sqrt{981} = 31.32 \text{ rad/sec.}$$

But $$\omega = 2\pi n$$

$\therefore$ $$31.32 = 2\pi n, \text{ where } n \text{ is the revolutions/sec}$$

or $$n = \frac{31.32}{2\pi} = \mathbf{4.98}. \quad \textbf{(Ans.)}$$

11.6. MOTION OF A CYCLIST ROUND A CURVE

Consider the motion of a cyclist of weight W which goes round a circular path of radius 'r' with an angular velocity 'ω' or a linear velocity 'v' such that $\omega = \frac{v}{r}$. When the cyclist moves round this circular path, he is acted upon by the centripetal force to keep him in the circular path. This centripetal force is provided by the frictional force exerted between the cycle tyre and the road surface as shown in Fig. 11.4. Thus, the cyclist has to lean inwards *i.e.*, towards to centre of the path to maintain equilibrium. The *angle of inclination with the vertical is more if the cyclist is riding at faster speed than that when he is riding at a slower speed.*

When rounding a circular path a cyclist is acted upon by the following forces :

(*i*) The centrifugal force $P = \frac{W}{g} \times \frac{v^2}{r}$ (acting horizontally or outwardly).

(*ii*) Weight of the cyclist 'W' and that of the bicycle, acting vertically downwards through the centre of gravity (G).

(*iii*) Frictional force $F_A = \mu N_A$ exerted between the cycle tyre and the road surface, and

(*iv*) Normal reaction N_A of the ground surface.

Now for equilibrium of system of forces shown in Fig. 11.4, the inclination (θ) of the cyclist with the vertical can be found out as under.

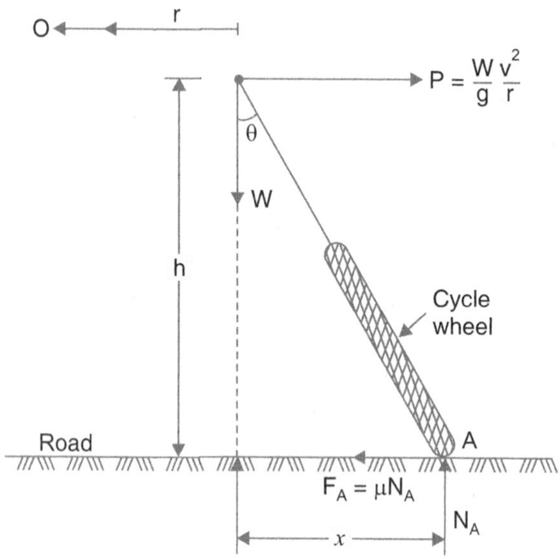

Fig. 11.4

Taking moments about A, we have

$$W \times x + F_A \times 0 - P \times h = 0$$

or $$Wx = Ph$$

or $$Wx = \frac{W}{g} \cdot \frac{v^2}{r} \cdot h$$

∴ $$x = \frac{v^2}{gr} \cdot h$$

Therefore, the cyclist should lean inwards such that the normal reaction N_A of the road acts at a distance $\frac{v^2}{gr} \times h$ from his centre of gravity (G) in a direction opposite to that of the centre of the circular path.

Also, $$\tan\theta = \frac{x}{h} = \frac{v^2}{gr} \cdot \frac{h}{h}$$

or $$\tan\theta = \frac{v^2}{gr} \qquad ...(11.11)$$

∴ $$\theta = \tan^{-1}\left(\frac{v^2}{gr}\right) \qquad ...[11.11\ (a)]$$

Therefore, the maximum speed 'v' at which the cyclist can drive, is such that

$$\tan\theta = \frac{v^2}{gr}$$

or $$v = \sqrt{\tan\theta \,.\, gr} \qquad ...[11.11\ (b)]$$

11.7. MOTION OF A VEHICLE ON A LEVEL CURVED PATH

Consider a vehicle of weight W, running with a velocity 'v', on a curved road of radius of curvature 'r' as shown in Fig. 11.5. Due to the centrifugal force (inertia force) $\frac{W}{g} \times \frac{v^2}{r}$, acting radially outward as shown, the wheel of the vehicle may tend to skid outwardly resulting in the frictional forces acting inwardly.

From Fig. 11.5 it is obvious that the vehicle is in equilibrium under the following six forces :

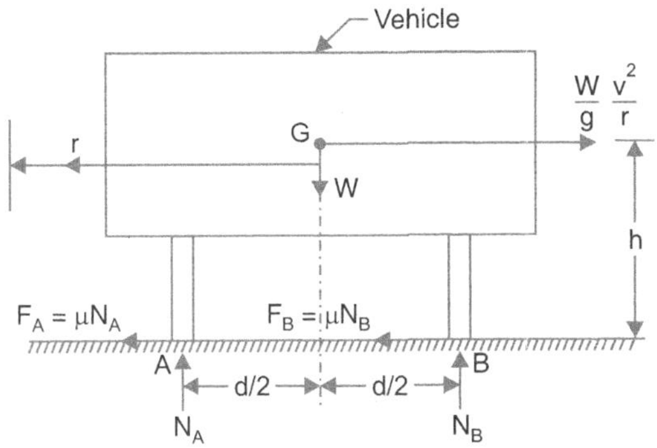

Fig. 11.5

(i) Centrifugal force : $\frac{W}{g} \cdot \frac{v^2}{r} \rightarrow$

(ii) Weight of the vehicle : $W \downarrow$

(iii) Normal reaction : $N_A \uparrow$

(iv) Normal reaction : $N_B \uparrow$

(v) Frictional force : $\mu N_A \leftarrow$

(vi) Frictional force : $\mu N_B \leftarrow$

(a) Maximum velocity to avoid skidding :

Considering vertical equilibrium ; $\Sigma V = 0$

$$W = N_A + N_B$$

Considering horizontal equilibrium ; $\Sigma H = 0$

$$\frac{W}{g} \cdot \frac{v^2}{r} = \mu N_A + \mu N_B$$

$$= \mu(N_A + N_B)$$

$$= \mu W \qquad (\because \; W = N_A + N_B)$$

$$\therefore \qquad \frac{v^2}{gr} = \mu$$

or

$$v = \sqrt{gr\mu} \qquad \text{...(11.12)}$$

This gives the maximum velocity against skidding of the vehicle. In fact the actual velocity, for *safety against skidding,* should be *less* than $\sqrt{gr\mu}$.

(*b*) Maximum speed to control overturing :

Taking moments of all the forces say about A ; $\Sigma M = 0$

$$N_B \times d = W \times \frac{d}{2} + \frac{W}{g} \cdot \frac{v^2}{r} \times h$$

$$N_B = \frac{W}{2} + \frac{W}{g} \frac{v^2}{r} \times \frac{h}{d}$$

i.e.,

$$N_B = \frac{W}{2}\left(1 + \frac{2v^2h}{grd}\right) \quad ...(11.13)$$

Also $W = N_A + N_B$

$\therefore$

$$N_A = W - N_B$$

$$= W - \frac{W}{2}\left(1 + \frac{2v^2h}{grd}\right)$$

$$= \frac{W}{2}\left[2 - \left(1 + \frac{2v^2h}{grd}\right)\right]$$

$$= \frac{W}{2}\left(2 - 1 - \frac{2v^2h}{grd}\right)$$

$$= \frac{W}{2}\left(1 - \frac{2v^2h}{grd}\right) \quad ...(11.14)$$

It may be seen from the equation (11.14) that if the velocity of the vehicle goes on increasing the expression in the bracket will go on decreasing. At one stage $\frac{2v^2h}{grd}$ may become more than one and the reaction N_A will become negative and the vehicle will *overturn.* The maximum speed at which the vehicle may be run without getting *over turned* can be calculated by equating $\frac{2v^2h}{grd} = 1$ in which case N_A will be zero

or

$$v = \sqrt{\frac{grd}{2h}} \quad ...(11.15)$$

(*c*) Condition for skidding and overturning :

The speeds for side slipping/skidding and overturning are given by the equations :

$$v^2 = gr\mu$$

and

$$v^2 = \frac{grd}{2h}.$$

The vehicle will be on the point of both skidding and overturning if these speeds are equal *i.e.*,

$$gr\mu = \frac{grd}{2h}$$

or

$$h = \frac{d}{2\mu} \quad ...(11.16)$$

which is the required condition.

11.8. MOTION OF A VEHICLE ON A BANKED CIRCULAR TRACK

(a) Super elevation or banking of roads. When a vehicle is moving in a circular path, a force is required towards the centre of the path. This force is provided by the frictional force between the wheels of the vehicle and the track, or in case of railed track, by the thrust of rails on the wheel flanges. This may *lead to wear-out of the wheels or rails rapidly* and, therefore, undesirable. To avoid this wear-out, the track is made *sloping* towards the centre so that the reaction of the track is inclined to the vertical and its horizontal components provide the necessary centrifugal force. A track which is made *sloping* in this manner is called a *banked* track. In case of track with rails, the outer rail is slightly raised above the inner rail ; this *excess elevation of the outer rail over the inner rail is called* **super elevation.**

Thus the process of slightly raising the outer edge of the road above the level of inner edge is called **'banking of roads' or 'super elevation'.**

The functions of banking the road or providing super-elevation to the railway track may be summarised as follows :

(*i*) To provide the necessary inward force (centripetal force) to counter balance the effect of centrifugal force to pull the vehicle outward to a horizontal curve.

(*ii*) To provide equal distribution of the load of the vehicle on outer and inner wheels of the vehicle.

(*iii*) To provide safety to the fast moving vehicles to negotiate the curved path without *overturning* or *skidding.*

(b) Angle of super-elevation and maximum speed of the vehicle. Let OM be the raised road on which the vehicle moves. Let θ be the angle of super-elevation as shown in Fig. 11.6. This angle θ should be such that the horizontal components of normal reaction N_A and N_B should provide the centripetal force

$$= \frac{W}{g} \frac{v^2}{r}$$

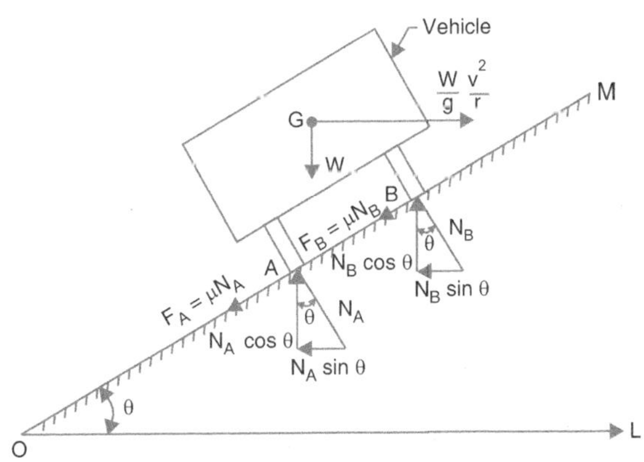

Fig. 11.6

and the vertical component of normal reactions should balance the weight W. Thus,

$$(N_A + N_B) \sin \theta = \frac{W}{g} \frac{v^2}{r} \qquad ...(i)$$

and $$(N_A + N_B) \cos \theta = W \qquad ...(ii)$$

Dividing eqns. (*i*) by (*ii*), we get

$$\frac{(N_A + N_B)\sin\theta}{(N_A + N_B)\cos\theta} = \frac{W}{g} \cdot \frac{v^2}{g} \times \frac{1}{W} \qquad ...(iii)$$

or
$$\tan\theta = \frac{v^2}{gr}$$

$\therefore$ Angle of super-elevation,

$$\theta = \tan^{-1}\left(\frac{v^2}{gr}\right)$$

Also, maximum speed of the vehicle,

$$v = \sqrt{gr\tan\theta}.$$

(*c*) Maximum speed to avoid skidding on a banked road. Refer to Fig. 11.7. Let N be the total normal reaction on the vehicle = $N_A + N_B$.

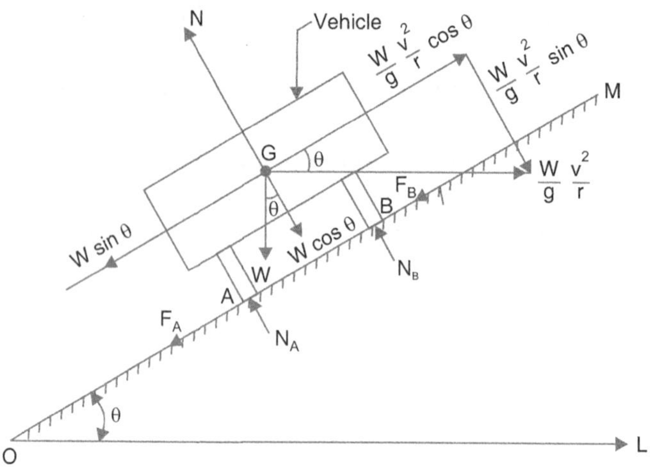

Fig. 11.7

Resolving all the forces along and perpendicular to the banked track *OM*, for equilibrium :

$$W\sin\theta + F_A + F_B = \frac{W}{g}\frac{v^2}{r}\cos\theta \qquad ...(i)$$

(along the plane *OM*)

and
$$N = \frac{W}{g}\frac{v^2}{r}\sin\theta + W\cos\theta \qquad \text{(perpendicular to plane } OM\text{)}$$

Also
$$F_A + F_B = \mu(N_A + N_B) = \mu N$$

$$= \mu\left(\frac{W}{g}\frac{v^2}{r}\sin\theta + W\cos\theta\right) \qquad ...(ii)$$

Substituting this in equation (*i*), we get

$$W \sin \theta + \mu W \left(\frac{v^2}{gr} \sin \theta + \cos \theta \right) = \frac{W}{g} \frac{v^2}{r} \cos \theta$$

Dividing both sides by $W \cos \theta$, we get

$$\tan \theta + \mu \left(\frac{v^2}{gr} \tan \theta + 1 \right) = \frac{v^2}{gr}$$

$$\therefore \qquad \tan \theta + \frac{\mu v^2}{gr} \tan \theta + \mu = \frac{v^2}{gr}$$

If μ is the co-efficient of friction between the wheels A and B and ϕ the angle of friction, then $\mu = \tan \phi$. Therefore, the above equation reduces to

$$\tan \theta + \frac{v^2}{gr} \tan \theta \,.\, \tan \phi + \tan \phi = \frac{v^2}{gr}$$

$$\frac{v^2}{gr} (1 - \tan \theta \tan \phi) = \tan \theta + \tan \phi$$

$$\therefore \qquad \frac{v^2}{gr} = \frac{\tan \theta + \tan \phi}{1 - \tan \theta \,.\, \tan \phi} = \tan (\theta + \phi)$$

Thus, to avoid skidding, $\qquad v = \sqrt{gr \tan (\theta + \phi)}$ ...(11.17)

(*d*) Maximum speed to avoid overturning on a banked road. In order to obtain maximum speed to avoid overturning it should be made sure that the *normal reaction* N_A *does not become negative*. This means that at the most N_A may be zero. Under these conditions the vehicle will be subjected to the following four forces:

(*i*) Centrifugal force $\frac{W}{g} \frac{v^2}{r}$

(*ii*) Weight W

(*iii*) Normal reaction at B, N_B

(*iv*) Frictional force F_B ($= \mu N_B$).

The resultant of N_B and F_B will be the total reaction at B, R_B inclined at an angle ϕ with N_B and meeting with $\frac{W}{g} \frac{v^2}{r}$ and W at the common point G for equilibrium as shown in Fig. 11.8.

$$\tan (\theta + \phi) = \frac{W/g \times v^2/r}{W} = \frac{v^2}{gr}$$

$$\therefore \qquad v = \sqrt{gr \tan (\theta + \phi)} \qquad \text{...}(i)$$

But from Fig. 11.8 (*a*),

$$\tan \phi = \frac{d/2}{h} = \frac{d}{2h} \qquad \text{...}(ii)$$

where, h = height of G above the track OM.

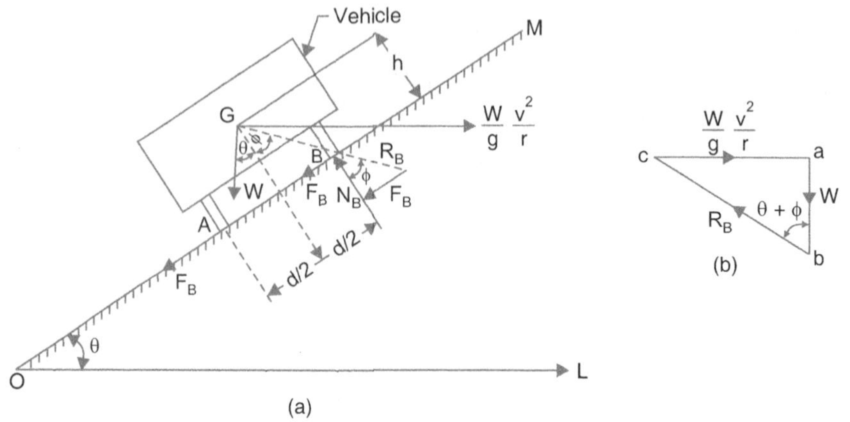

Fig. 11.8

Combining equations (*i*) and (*ii*), we get

$$v = \sqrt{gr\left(\frac{\tan\theta + \tan\phi}{1 - \tan\theta \,.\, \tan\phi}\right)}$$

$$v = \sqrt{gr\left(\frac{\tan\theta + d/2h}{1 - d/2h \,.\, \tan\theta}\right)} \quad ...(11.18)$$

which gives the **maximum speed for overturning.**

11.9. SUPER ELEVATION FOR RAILWAYS

Fig. 11.9 shows a super-elevated railways track and a wagon negotiating a curve of radius r. The wagon is in dynamic equilibrium under the following forces :

(*i*) Its weight, W

(*ii*) Centrifugal force, $F_c = \frac{W}{g}\frac{v^2}{r}$

(*iii*) Normal reaction at A, N_A

(*iv*) Normal reaction at B, N_B.

Let d = distance between the rails

and e = super-elevation

In railways, *super-elevation is so given that the inner wheel does not exert any end thrust on the rail.*

Resolving F_c and W into two components, one along the inclined plane and the other normal to it.

Considering components along with the plane, for zero end thrust.

$$F_C \cos\theta = W \sin\theta$$

$$\frac{W}{g}\frac{v^2}{r}\cos\theta = W\sin\theta$$

$$\tan\theta = \frac{v^2}{gr} \quad ...(i)$$

which is the same as expression as obtained previously for the roads.

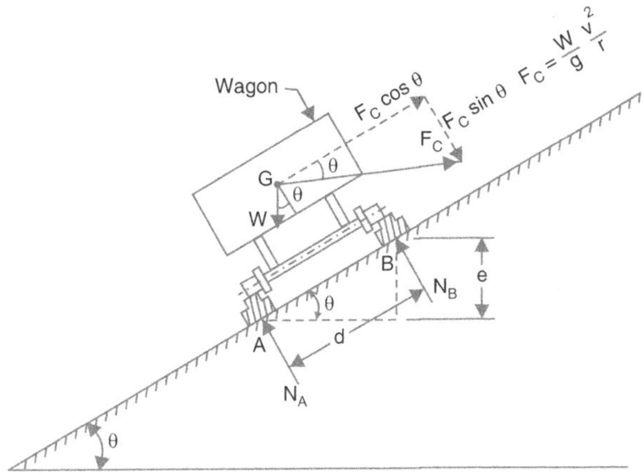

Fig. 11.9

∴ Angle of super-elevation

$$\theta = \tan^{-1}\left(\frac{v^2}{gr}\right) \qquad ...(ii)$$

And maximum speed, $v = \sqrt{gr \tan\theta}$ ...(*iii*)

If '*d*' is the distance between the center lines of the wheels, then as is clear from Fig. 11.9, the super elevation, '*e*', of the rails may be obtained from the relationship :

$$e = d \sin\theta$$

∴ Super-elevation, $e = d \sin\theta$

As angle θ is usually very small, from equation (*i*)

$$\sin\theta = \tan\theta = \frac{v^2}{gr}$$

∴ **Super-elevation,**

$$e = d\sin\theta = d\tan\theta$$

or $$e = d \cdot \frac{v^2}{gr} \qquad ...(iv)$$

Example 11.8. *A cyclist riding at a speed of 20 km/hr takes a turn around a circular road of radius 10 m. What is his inclination to the vertical ?*

Sol. Refer to Fig. 11.10.

$$v = 20 \text{ km/h}$$

$$= \frac{20 \times 1000}{60 \times 60} = 5.55 \text{ m/sec}$$

and $$r = 10 \text{ m}$$

Inclination to the vertical, θ = ?

Using the relation,

$$\tan\theta = \frac{v^2}{gr} = \frac{(5.55)^2}{9.81 \times 10} = 0.314$$

or $$\boldsymbol{\theta = 17.43° \text{ or } 17° \, 26' \text{ app.}} \quad \textbf{(Ans.)}$$

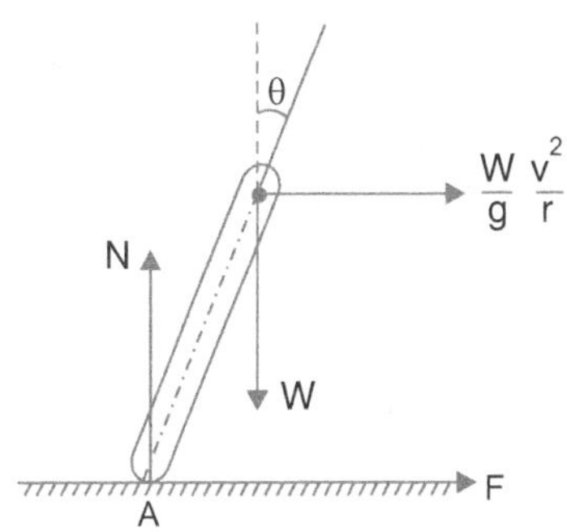

Fig. 11.10

Example 11.9. *A locomotive weighing 500 kN is rounding a curve of 250 m radius at 60 km/hr. Calculate the radial pressure between the wheels and the rails. To what angle would the track have to be banked in order to reduce the radial pressure to zero at the above speed ? Find also the vertical height at which the outer rail should be elevated above the inner wheel base, distance between two rails being 1.2 m.*

Sol. Refer to Fig. 11.11.

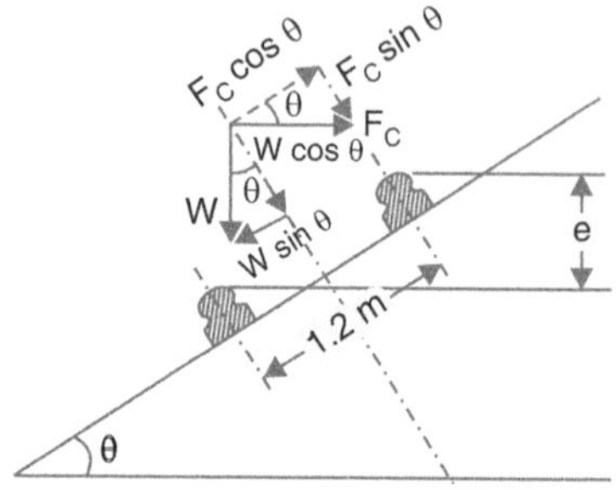

Fig. 11.11

Velocity of the locomotive

$$v = 60 \text{ km/hr} = \frac{60 \times 1000}{60 \times 60}$$

$$= 16.67 \text{ m/sec.}$$

Radial pressure is the centripetal force

$$= \frac{W}{g}\frac{v^2}{r} = \frac{500}{9.81} \times \frac{(16.67)^2}{250}$$

$$= \mathbf{56.65\ kN.} \quad \textbf{(Ans.)}$$

In order to eliminate the side thrust on the wheels, the track should be banked such that the components of W and F_C along the inclined plane may be equal and opposite.

Let θ = angle of banking

Then $F_C \cos\theta = W \sin\theta$

or $$\tan\theta = \frac{F_C}{W} = \frac{56.65}{500} = 0.1133$$

$\therefore$ $\theta = 6.46°$ or $6° \, 27'$

Height of the outer rail above the inner or super-elevation.

$$e = 1.2 \sin\theta = 1.2 \times 0.1133 \quad (\because \ \sin\theta = \tan\theta \text{ when } \theta \text{ is very small})$$

$$= \mathbf{0.136\ m\ or\ 13.6\ cm.} \quad \textbf{(Ans.)}$$

Example 11.10. *An what speed a vehicle should move round a curve of 80 m radius without any slip (i) on a level road, (ii) on a road banked to an angle of one vertical to 10 horizontal, and (iii) on the banked road without any lateral frictional support. The co-efficient of friction may be taken as 0.22.*

Sol. Radius of the curve, $r = 80$ m

Co-efficient of friction, $\mu = 0.22$ (= $\tan\phi$)

where ϕ = angle of friction

$$\tan\theta = \frac{1}{10} = 0.1, \text{ where } \theta \text{ is the angle of banking.}$$

(*i*) Speed of the vehicle on a level road without any slip, v = ?

Using the relation, $v = \sqrt{\mu g r}$

$$= \sqrt{0.22 \times 9.81 \times 80}$$

$$= \mathbf{13.14\ m/sec.} \quad \textbf{(Ans.)}$$

(*ii*) Speed of the vehicle on the banked road without any slip, v = ?

Using the relation, $v = \sqrt{gr \tan(\theta + \phi)}$

$$= \sqrt{gr\left(\frac{\tan\theta + \tan\phi}{1 - \tan\theta\tan\phi}\right)}$$

$$= \sqrt{9.81 \times 80\left(\frac{0.1 + 0.22}{1 - 0.1 \times 0.22}\right)}$$

$$= \mathbf{16.02\ m/sec.}\quad \textbf{(Ans.)}$$

(*iii*) **Speed of the vehicle without any lateral frictional support, v = ?**

Using the relation, $v = \sqrt{gr\tan\theta}$

$$= \sqrt{9.81 \times 80 \times 0.1}$$

$$= \mathbf{8.859\ m/sec.}\quad \textbf{(Ans.)}$$

Example 11.11. *In a circus show, a motor cyclist is moving inside a spherical cage of radius 3 m. The motorcycle and the man together weigh 7250 N. Find the least velocity with which he must pass the highest point of the cage without long contact with the inside of the cage.*

Sol. The motor cyclist can maintain the contact with the highest point only when the centrifugal force, F_C, balance the gravitational force W.

We are given that :

Radius of sphere, r = 3 m

Weight of man and motorcycle, W = 7250 N

Let v be the required velocity (in m/sec)

For the contact to be maintained,

$$\frac{W}{g}\frac{v^2}{r} = W$$

$$\frac{7250}{g} \times \frac{v^2}{r} = 7250$$

or $$v^2 = gr$$

$\therefore$ $$v = \sqrt{gr} = \sqrt{9.81 \times 3}$$

$$= \mathbf{5.425\ m/s.}\quad \textbf{(Ans.)}$$

HIGHLIGHTS

1. The *angular velocity* (ω) is defined as the rate of change of angular displacement (θ).

   Mathematically, $\omega = \dfrac{\theta}{t}$

   If N is the uniform speed of a rotating part in revolutions per minute (r.p.m.) then its angular velocity is given by the relation,

   $$\omega = \frac{2\pi N}{60}\ \text{rad/sec.}$$

2. Equations of angular motion are

   $$\omega = \omega_0 + \alpha t \quad ...(i)$$

   $$\theta = \omega_0 t + \frac{1}{2}\alpha t^2 \quad ...(ii)$$

   $$\omega^2 - \omega_0^2 = 2\alpha\theta \quad ...(iii)$$

where ω_0 = initial angular velocity, rad/sec
ω = final angular velocity, rad/sec
t = time taken in seconds
θ = angle covered traversed (in t seconds) in radians
α = angular acceleration, rad/sec^2.

3. Relation between linear and angular motion :

$$\omega = \frac{v}{r} \quad ...(i)$$

$$\alpha = \frac{a}{r} \quad ...(ii)$$

4. When a body moves on a circular path, at every instant it has a tendency to move off tangentially and to keep it moving in the circular path a radial inward force has to be applied which is known as "*centripetal force*". Inertia force, equal and opposite to the *centripetal force*, is known as "*centrifugal force*".

Centrifugal force $= \frac{W}{g}\frac{v^2}{r} = \frac{W}{g}\omega^2 r$

5. Motion of a cyclist round a curve,

$$v = \sqrt{\tan\theta \,.\, gr} \quad ...(i)$$

and $$\theta = \tan^{-1}\left(\frac{v^2}{gr}\right) \quad ...(ii)$$

6. Motion of a vehicle on a level curved path :

$$v = \sqrt{gr\mu} \quad ...(i)$$

The maximum speed at which vehicle may be run without getting overturned :

$$v = \sqrt{\frac{grd}{2h}} \quad ...(ii)$$

The vehicle will be on the point of both skidding and overturning, if

$$h = \frac{d}{2\mu} \quad ...(iii)$$

7. Motion of a vehicle on a banked circular track :
Angle of super-elevation,

$$\theta = \tan^{-1}\frac{v^2}{gr} \quad ...(i)$$

and maximum speed of the vehicle

$$v = \sqrt{gr\tan\theta} \quad ...(ii)$$

To avoid skidding.

$$v = \sqrt{gr\tan(\theta + \phi)} \quad ...(iii)$$

where ϕ is the angle of friction.

OBJECTIVE TYPE QUESTIONS

Fill in the blanks :

(*i*) The rate of change of angular displacement is termed as ...... velocity.

(*ii*) Angular velocity = $\frac{......}{\text{time}}$.

(*iii*) Linear velocity = Angular velocity × ......

(*iv*) ...... is defined as the rate of change of angular velocity.

(*v*) $\theta = ... + \frac{1}{2}\alpha t^2$.

(*vi*) Linear acceleration = ... × radius.

(*vii*) Centrifugal force = $\frac{W}{g} \times ......$

(*viii*) $\tan\theta = \frac{v^2}{......}$

(*ix*) Maximum velocity to avoid skidding.

$$v = \sqrt{gr ......}$$

(*x*) Excess elevation of the outer rail over the inner rail is called ......

Answers

(*i*) angular
(*ii*) angular displacement
(*iii*) radius
(*iv*) Angular acceleration
(*v*) $\omega_0 t$
(*vi*) angular acceleration
(*vii*) $\frac{v^2}{r}$ or $w^2 r$
(*viii*) gr
(*ix*) μ
(*x*) super-elevation.

EXERCISES

Theoretical Questions

1. Define the following :
Angular displacement, angular velocity and angular acceleration.

2. Derive the following equations of angular motion :
(*i*) $\omega^2 - \omega_0^2 = 2\alpha\theta$
(*ii*) $\theta = \omega_0 t + \frac{1}{2}\alpha t^2$.

3. Prove that, Linear velocity = angular velocity × radius of circular path.

4. Define "centrifugal force" and "centripetal force".

5. Explain why a cyclist has to lean inwards while negotiating a curve ?

6. What do you understand by the term super elevation ? Discuss the necessity of providing super-elevation on the roads.

7. Derive the expression for maximum speed to avoid skidding on a banked road.

Unsolved Examples

1. A flywheel is making 180 r.p.m. and after 20 seconds it is running at 150 r.p.m. How many revolutions will it make and what time will elapse before it stops, its retardation is uniform ? [**Ans.** 180, 2 min.]

2. Find the acceleration of a point on the rim of a flywheel 2 m diameter rotating at 100 r.p.m. [**Ans.** 109.7 m/sec^2]

3. A wheel accelerates uniformly from rest to a speed of 300 r.p.m. in 0.5 sec. It then rotates at that speed for 2 sec before retarding to rest in $\frac{1}{3}$ sec. How many revolutions does it make during the entire time intervals ? [**Ans.** 12.08 rev.]

4. A rotating wheel increases its speed from 200 to 500 r.p.m. in 35 seconds. What is the angular acceleration in rad/sec^2? [**Ans.** 0.89 rad/sec^2]

5. A wheel accelerates uniformly from rest to 2000 r.p.m. in 20 seconds. What is the angular acceleration ? How many revolutions does the wheel make in attaining the speed of 2000 r.p.m. ?

[**Ans.** 10.5 rad/sec^2, 333 revolutions]

6. Determine the tangential and centripetal acceleration of a crank pin at the instant the crank is rotating at 60 r.p.m. with an angular acceleration of 50 rad/sec^2. The radius of crank is 12 cm.

[**Ans.** a_n = 4.738 cm/sec^2, a_t = 600 cm/sec^2]

7. A wheel rotates for 5 seconds with a constant angular acceleration and describe during this time 100 radians. It then rotates with constant angular velocity, and during the next 5 seconds describes 80 radians. Find the initial angular velocity and angular acceleration.

[**Ans.** ω_0 = 24 rad/sec, α = – 1.6 rad/sec^2]

8. A particle having a constant speed of 4.5 m/sec moves around a circle 3 m in diameter. What is the normal acceleration ? [**Ans.** 13.5 m/sec^2]

9. A particle starting from rest moves along a circular path 20 cm radius with an angular acceleration of 12 rad/sec^2. Find tangential, centripetal and resultant acceleration after 0.5 second of taking the start.

[**Ans.** 2.4 m/sec^2, 7.2 m/sec^2, 7.59 m/sec^2]

10. A constant torque is applied to a stationary wheel of diameter 0.5 m which is mounted on a shaft between two frictionless bearings. Thus, the wheel acquires a speed of 150 r.p.m. in 5 minutes. Determine the following :

(*i*) Number of revolutions made by the wheel during the time;

(*ii*) Angular acceleration; and

(*iii*) Peripheral speed of the wheel at 150 r.p.m.

[**Ans.** (*i*) 375 revolutions, (*ii*) π/60 rad/sec^2, (*iii*) 3.93 m/sec]

11. A wheel is rotating about its axis with a constant angular acceleration of 1.0467 rad/s^2. If the initial and final angular velocities are 5.23 rad/sec and 10.47 rad/sec determine the total angle turned through during the time interval this change of angular velocity took place. [**Ans.** 39.298 rad]

12. The relation between the angle of rotation and the time in the case of a rotating body is given by the equation

$$\theta = 2t^3 + 3t^2 + 10 \text{ radians}$$

where t is in seconds.

Determine the angular velocity, displacement and acceleration when t = 5 seconds.

[**Ans.** 180 rad/sec, 235 rad., 66 rad/sec^2]

13. A cyclist riding at a speed of 24 km/h takes a turn round a circular road of radius 12 m. What is his inclination to the vertical ? [**Ans.** 20° 41′]

14. In a circus show a motor cyclist is moving inside a spherical cage of radius 3 m. The motorcycle and the man together weight 7.25 kN. Find the least velocity with which the motorcyclist must pass through the highest point on the cage without losing the contact inside the cage. [**Ans.** 19.53 km/hr]

15. Find the angle of the track made with the horizontal such that the car speeding at 120 km/hr around the track of 300 m radius is safe against skidding. [**Ans.** 20° 42′]

16. A four wheeled two axle vehicle travels on a flat road at a uniform speed of 72 km/hr around a curve of 240 m radius. The distance between wheel tracks is 1.7 m and c.g. of the vehicle is on the vertical through the centre of the wheel base and 1.5 m above the ground. The total weight of the vehicle is 160 kN. Determine :

(*i*) the vertical pressure on each of the inner and outer wheels, and

(*ii*) the maximum speed at which the vehicle can negotiate the curve without overturning.

[**Ans.** 28.01 kN ; 51.99 kN ; 131.5 km/hr]

17. A locomotive weighing 600 kN is rounding a curve of 200 metres radius at 54 km/h. Calculate the radial pressure between the wheels and the rails. To what angle would the truck have to be banked in order to reduce the radial pressure to zero at the above speed ? Find also the vertical height at which the outer rail should be elevated above the inner when the wheel base is 1 m.

[**Ans.** 68.8 kN ; 6° 33′ ; 11.4 cm]

12

Simple Stresses and Strains

12.1. CLASSIFICATION OF LOADS

A load may be defined as the combined effect of external forces acting on a body. The loads may be classified as : (*i*) dead loads, (*ii*) live or fluctuating loads, (*iii*) inertia loads or forces and (*iv*) centrifugal loads or forces.

The other way of classification is (*i*) tensile loads, (*ii*) compressive loads, (*iii*) torsional or twisting loads, (*iv*) bending loads and (*v*) shearing loads.

The load may be a 'point' (or concentrated) or 'distributed'.

Point load. A point load or concentrated load is one which is considered to act at a point. In actual practical, the load has to be distributed over a small area, because, such small knife-edge contacts are generally neither possible, nor desirable.

Distributed load. A distributed load is one which is distributed or spread in some manner over the length of the beam. If the spread is uniform, (*i.e.*, at the uniform rate, say w kN or N/metre run) it is said to be uniformly distributed load and is abbreviated as u.d.l. If the spread is not at uniform rate, it is said to be non-uniformly distributed load. *Triangulary* and *trapezoidally* distributed loads fall under this category.

12.2. STRESS

When a body is acted upon by some load or external force, it undergoes deformation (*i.e.*, change in shape or dimensions) which increases gradually. During deformation, the material of the body resists the tendency of the load to deform the body and when the load influence is taken over by the internal resistance of the material of the body, it becomes stable. This *internal resistance which the body offers to meet with the load is called* **stress.**

Stress can be considered either as total stress or unit stress. Total stress represents the total resistance to an external effect and is expressed in N, kN or MN. Unit stress represents the resistance developed by a unit area of cross-section and is expressed in kN/m^2 or MN/m^2 or N/mm^2. For the remainder of this text, the word stress will be used to signify unit stress.

The various types of stresses may be classified as :

1. Simple or direct stress

(*i*) Tension (*ii*) Compression (*iii*) Shear.

2. Indirect stress

(*i*) Bending (*ii*) Torsion.

3. Combined stress. Any possible combination of types 1 and 2.

This chapter deals with simple stresses only.

12.3. SIMPLE STRESS

Simple stress is often called *direct stress* because it develops under direct loading conditions. That is, simple tension and simple compression occurs when the applied force, called load is in line with the axis of the member (axial loading) (Figs. 12.1 and 12.2) and simple shear occurs, when equal, parallel and opposite forces tend to cause a surface to slide relative to the adjacent surface (Fig. 12.3).

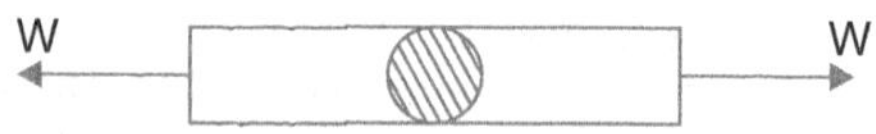

Fig. 12.1. Tensile stress

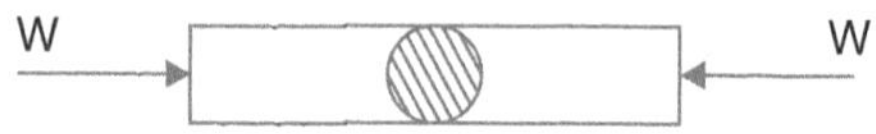

Fig. 12.2. Compressive stress

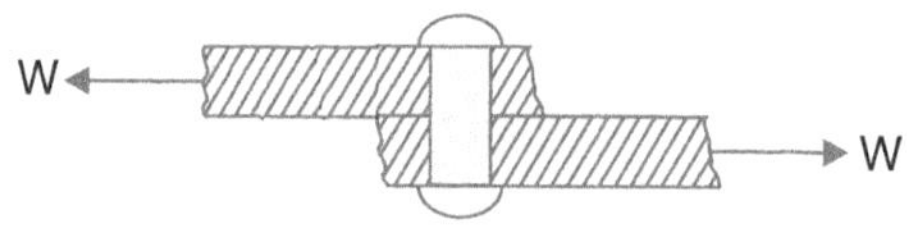

(*a*) Rivet resisting shear

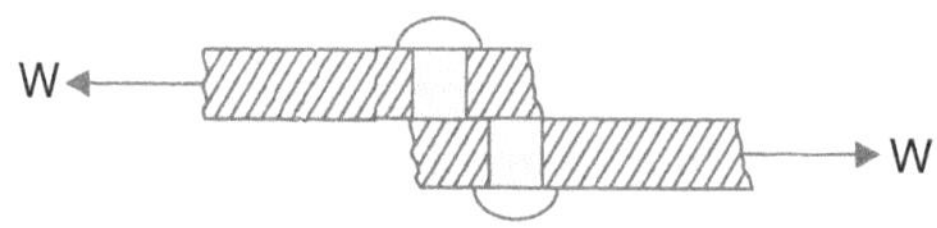

(*b*) Rivet failure due to shear

Fig. 12.3

In certain loading situations, the stresses that develop are not simple stresses. For example, referring to Fig. 12.4, the member is subjected to a load which is perpendicular to the axis of the member (*transverse loadings*) (Fig. 12.5). This will cause the member to bend, resulting in deformation of the material and stresses being developed internally to resist the deformation. All three types of stresses—*tension*, *compression* and *shear* will develop, but they will not be simple stresses, since they were not caused by direct loading.

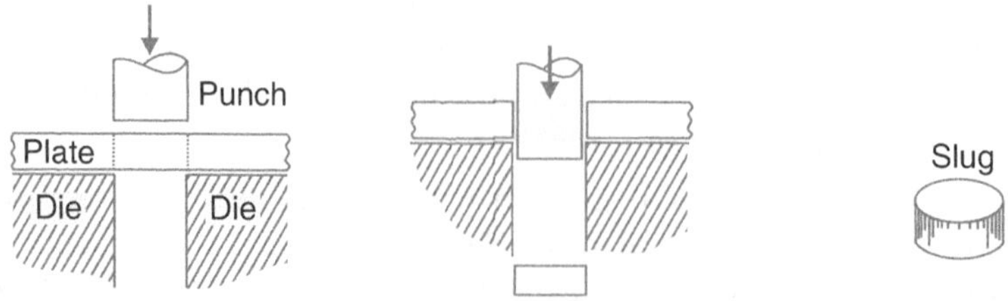

(*a*) Punch approaching plate (*b*) Punch shearing plate (*c*) Slug showing sheared area

Fig. 12.4

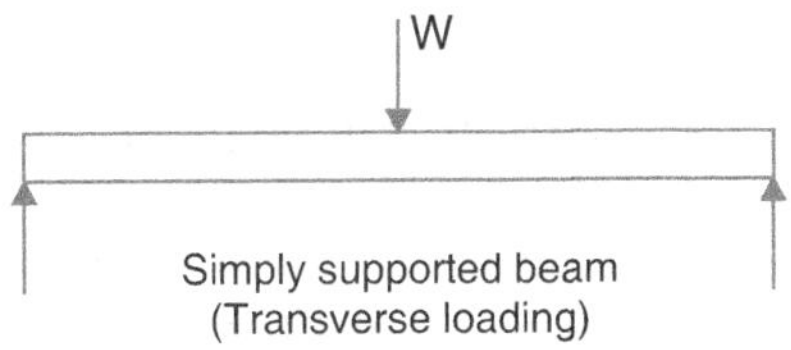

Fig. 12.5

When any type of *simple* stress develops, we can calculate the magnitude of the stress by,

$$f = \frac{W}{A} \quad ...(12.1)$$

where f = stress, N/cm^2 or kgf/cm^2

W = load $\left[\begin{array}{l}\text{external force causing}\\ \text{stress to develop}\end{array}\right]$, N or kgf

A = area over which stress develops, cm^2.

It may be noted that in cases of either simple tension or simple compression, the areas which resist the load are *perpendicular* to the direction of forces. When a member is subjected to simple shear, the resisting area is parallel to the direction of the force. Common situations causing shear stresses are shown in Figs. 12.3 and 12.4.

12.4. STRAIN

Any element in a material subjected to stress is said to be strained. *The strain (e) is the deformation produced by stress.* The various types of strains are explained below :

12.4.1. Tensile Strain

A piece of material, with uniform cross-section, subject to a uniform axial tensile stress, will increase its length from l to $(l + \delta l)$ (Fig. 12.6) and the increment of length δl is the actual deformation of the material. The fractional deformation or the tensile strain is given by

$$e_t = \frac{\delta l}{l} \quad ...(12.2)$$

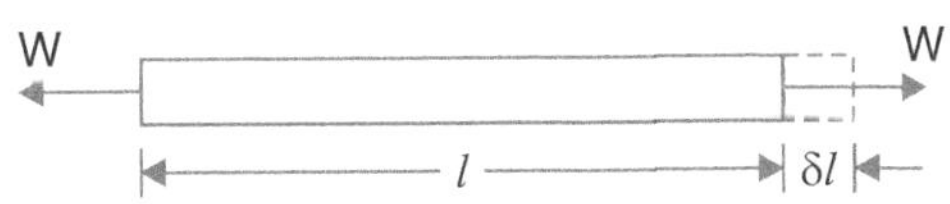

Fig. 12.6

12.4.2. Compressive Strain

Under compressive forces, a similar piece of material would be reduced in length (Fig. 12.7) from l to $(l - \delta l)$.

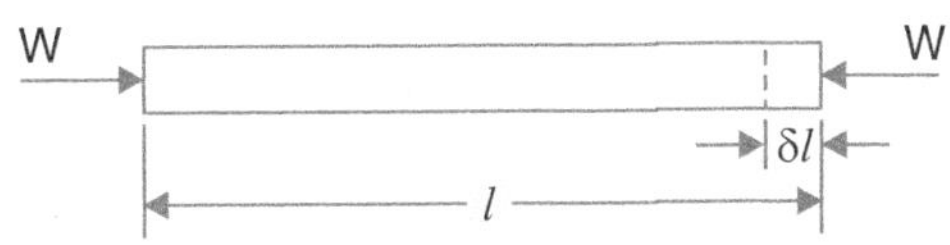

Fig. 12.7

The fractional deformation again gives the strain e,

where $$e = \frac{\delta l}{l} \quad ...(12.2\ a)$$

12.4.3. Shear Strain

In case of a shearing load, a shear strain will be produced which is measured by the angle through which the body distorts.

In Fig. 12.8 is shown a rectangular block *LMNP* fixed at one face and subjected to force *F*. After application of force, it distorts through an angle ϕ and occupies new position *LM′N′P*. The shear strain (e_s) is given by

$$e_s = \frac{NN'}{NP} = \tan \phi = \phi \text{ (radians) ...... (since } \phi \text{ is very small)}$$

The above result has been obtained by assuming *NN′* equal to arc (as *NN′* is very small) drawn with centre *P* and radius *PN*.

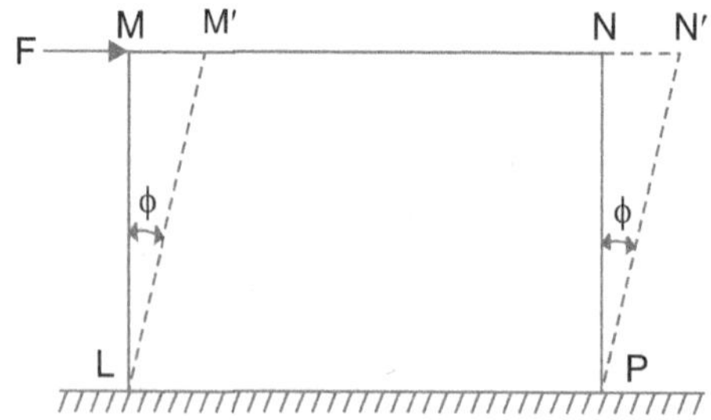

Fig. 12.8

12.4.4. Volumetric Strain

It is defined as the ratio between change in volume and original volume of the body, and is denoted by e_v.

$$\therefore \qquad e_v = \frac{\text{Change in volume}}{\text{Original volume}} = \frac{\delta V}{V} \qquad \text{...(12.3)}$$

The strains which disappear with the removal of load are termed as *elastic strains* and the body which *regains its original position on the removal of force is called an elastic body*. The body is said to be *plastic* if the *strains exist even after the removal of external force.* There is always a limiting value of load upto which the strain totally disappears on the removal of load, the stress corresponding to this load is called *elastic limit.*

12.5. HOOKE'S LAW

Robert Hooke discovered experimentally that within elastic limit, stress varies directly as strain

$$\text{Stress} \propto \text{Strain}$$

i.e., or
$$\frac{\text{Stress}}{\text{Strain}} = \text{a constant}$$

This constant is termed as Modulus of Elasticity.

(*i*) **Young's modulus.** It is the *ratio between tensile stress and tensile strain or compressive stress and compressive strain*. It is denoted by *E*. It is the same as modulus of elasticity

or
$$E = \frac{f}{e} \qquad \text{...(12.4)}$$

(*ii*) **Modulus of rigidity.** It is defined as the ratio of *shear stress to shear strain* and is denoted by *C*, *N* or *G*. It is also called *shear modulus of elasticity*.

or
$$\frac{f_s}{e_s} = C, N \text{ or } G \qquad \text{...(12.5)}$$

(*iii*) **Bulk or volume modulus of elasticity.** It may be defined as the ratio of normal stress (on each face of a solid cube) to volumetric strain and is denoted by the letter *k*.

or $$\frac{f_n}{e_v} = K \quad \text{...(12.6)}$$

12.6. MECHANICAL PROPERTIES OF METALS

Some important mechanical properties of metals are discussed below :

1. Strength. The *strength of a metal is its ability to withstand various forces to which it is subjected during a test or in service.* It is usually defined as tensile strength, compressive strength, proof stress, shear strength, etc. *Strength of materials* is general expression for the *measure of capacity* of resistance possessed by solid masses or pieces of various kinds to any causes tending to produce in them a permanent and disabling change of form or positive fracture. Materials of all kinds owe their strength to the action of the forces residing in and about the molecules of the bodies (the molecular forces) but mainly to that ones of these known as *cohesion*; certain modified results of cohesion as toughness or tenacity, hardness, stiffness and elasticity are also important elements and strength is in relation of the toughness and stiffness combined.

2. Elasticity. A material is said to be perfectly elastic if the whole strains produced by a load *disappears* completely on the removal of the load. The modulus of elasticity or Young's modulus (*E*) is the proportionally constant between stress and strain for eastic materials. Young's modulus is indicative of the property elastic materials. Young's modulus is indicative of the property called *stiffness ; small values of E indicate flexible materials* and *large value of E reflects stiffness and rigidity*. The property of *spring back* is a function of modulus of elasticity and refers to the extent to which metal springs back when an elastic deforming load is removed. In metal cutting, modulus of elasticity of the work-piece affects its rigidity, and modulus of elasticity of the cutting tools and tool-holder affect their rigidity.

3. Plasticity. *Plasticity is the property that enables the formation of a permanent deformation in a material*. It is the reverse of elasticity: a plastic material will retain exactly the shape it takes under load, even after the load is removed. Gold and lead are highly plastic materials. Plasticity is made use of in *stamping images on coins and ornamental works.*

During plastic deformation there is the displacement of atoms within the metallic grains and consequently the shapes of metallic components change.

It is because of this property that certain synthetic materials are given the name 'plastics'. These materials can be changed into required shape easily.

4. Ductility. *It is the ability of a metal to withstand elongation or bending.* Due to this property, wires are made by drawing out through a hole. The material shows a considerable amount of plasticity during the ductile extension. This is a valuable property in chains, ropes etc., because they do not snap off, while in service, without giving sufficient warning by elongation.

Note. The elongation of a test piece of metal which occurs when it is subjected to a sufficiently high tensile stress is a measure of the *ductility* of the material. The elongation is measured accurately by an extensometer etc. The value of this test as a measure of ductility has often been questioned however. The actual property measured is the ability of the metal to distort or 'flow' without breaking, this is the same as ductility, but there is no universal agreement regarding the value of the test to the engineer. For *sheet strip* and *wire material, a bend test* used as a substitute for *elongation test.* A simple bend test consists of bending the material through 180°; the radius of the bend depending on the gauge and composition of the material under test. The sample must withstand this test without cracking. If the *reverse-bend test* as its name suggests the test piece is bent through 90° and then back again; if necessary, this reversal is repeated a specified number of times, or until the sample breaks. This test is considerably more searching than the elongation test and reveals defects which the elongation test may not show up. For *tube material,* a bend test is carried out on a test strip cut from the wall of the tube while a length of the tube itself is usually flattened between two plates until its internal walls are a specified distance apart.

5. Malleability. This is the *property by virtue of which a material may be hammered or rolled into thin sheets without rupture*. This property generally increases with the increase of temperature.

The common metals in order of their ductility and malleability (at room temperature) are given below:

Ductility	*Malleability*
Gold	Gold
Silver	Silver
Platinum	Copper
Iron	Aluminium
Nickel	Tin
Copper	Platinum
Aluminium	Lead
Zinc	Zinc
Tin	Iron
Lead	Nickel

6. Tenacity or toughness. *Tenacity* or *toughness* is the *strength with which the material opposes rupture*. It is due to the attraction which the molecules have for each other ; giving them power to resist tearing apart.

The *area under the stress-strain curve indicates the toughness* (*i.e.*, energy which can be absorbed by the material upto the point of rupture). Although the engineering stress-strain curve is often used for this computation, a more realistic result is obtained from the *true stress-true strain* curve. Toughness is expressed as energy absorbed Nm per unit of volume of material participating in absorption (m^3), or Nm/m^3. This result is obtained by multiplying the ordinate by the abscissa (in appropriate units) of stress-strain plot.

7. Brittleness. *Lack of ductility is brittleness*. When a body breaks easily when subjected to shocks it is said to be brittle.

8. Hardness. *Hardness is usually defined as resistance of a material to penetration*. Hard materials resist scratches or being worn out by friction with another body.

Hardness is primarily a function of the elastic limit (*i.e.*, yield strength) of *the material* and to a *lesser extent a function of the work hardening coefficient*. The modulus of elasticity also exerts a slight effect on hardness.

In the most generally accepted tests, an indenter is pressed into the surface of the material by a slowly applied known load, and the extent of the resulting impression is measured mechanically or optically. A large impression for a given load and indenter indicates a soft material, and the opposite is true for a small impression.

The converse of hardness is known as *softness*.

12.7. TENSILE TEST

The tensile test is one of the most widely used of the mechanical tests. There are many variations of this test to accommodate the widely differing character of materials such as metals, elastomers, plastic and glass. The tensile test on a mild steel test piece is described below:

Fig. 12.9 (*a*, *b*, *c*) shows different types of tensile test pieces used for the test.

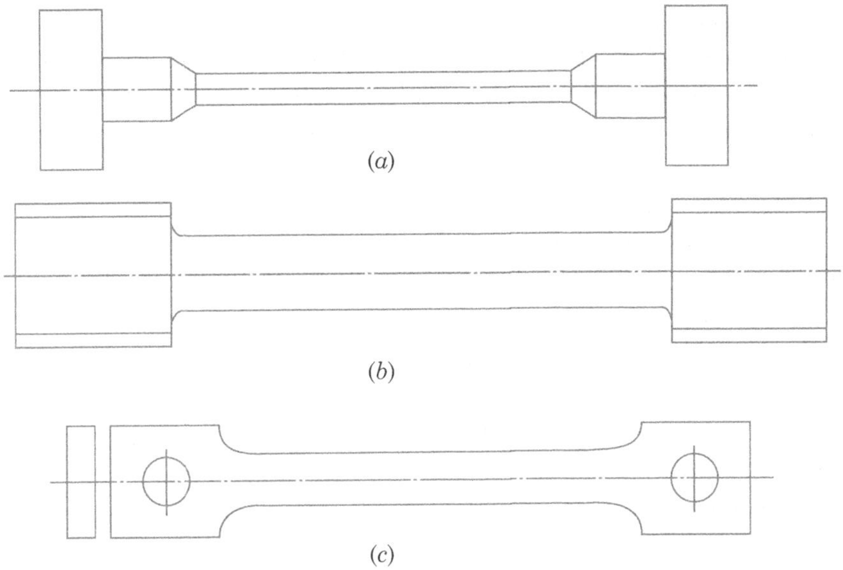

Fig. 12.9. Tensile test specimens.

The tensile test is carried out on a bar of uniform cross-section throughout the gauge length. The specimen is mounted in the jaw of a testing machine with which a gradually increasing load can be applied. The extension or elongation of the gauge length is recorded continuously and finally a graph is drawn between the load and extension or between stress and strain ; which is of the type shown in Fig. 12.10 (*a*).

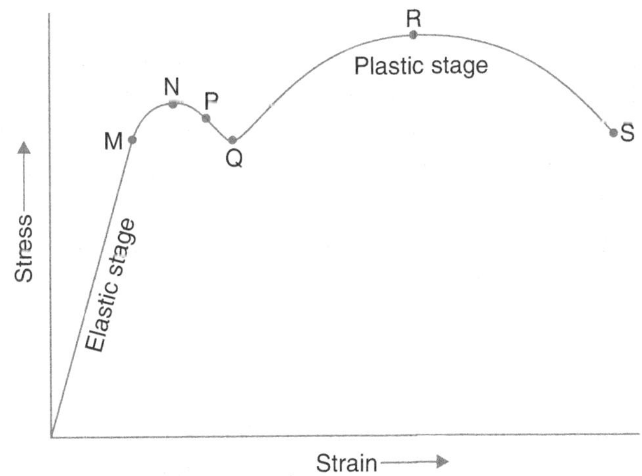

Fig. 12.10. (*a*) Stress-strain curve.

Upto point *M* Hookes' law holds good and this point is known as *limit of proportionality*. Beyond the point *M* Hooke's law is not obeyed although the material remains elastic *i.e.*, strain completely disappears after the removal of load. At the point *N* elastic limit is reached. If the material is loaded or stressed upto this point the material will regain its original shape on the removal of the load. Upto the point *P* strain increases more quickly than stress ; at this point the metal yields. In the mild steel yielding commences immediately and two points *P* and *Q*, the *upper* and *lower* yield *points* respectively are obtained. On further increasing the load slightly, the strain increases rapidly till R when neck or waist is formed. When this point (R) is reached the deformation or extension continues even with lesser load and ultimately fracture occurs.

The various properties connected with this test are given more elaborately in the following paragraphs:

(*i*) **Proportional limit.** *It is the maximum stress at which stress remains directly proportional to strain*. The proportional limit is determined from the stress-strain curve by drawing a straight line tangent at the origin and noting the first deviation of the plot from the line.

The proportional limit has *limited engineering significance because of its great dependence upon the precision available for its determination.*

(*ii*) **Elastic limit.** *The elastic limit is the maximum stress which the material can withstand without causing permanent deformation which remains after retnoval of stress.*

For engineering usage the elastic limit has *little significance*.

(*iii*) **Yield strength.** The *yield strength is the stress at which a material exhibits a specified limiting permanent set.*

The yield strength of a metal is a *property of considerable significance*. The tensile yield strength indicates resistance to permanent deformation produced by tensile loads. It is related to resistance to permanent deformation by shearing, bending, compressive and complex combination of forces. Because of this and the ease of its measurement the tensile yield strength is used widely as a *factor of design* ; it is preferable in most instances to the use of tensile strength. The *yield strength* also is *indicative of the ease of forming or shaping metals by mechanical stress*.

(*iv*) **Yield point.** The *yield paint is the stress at which there first occurs a marked increase in strain without an increase in stress*. The yield point can be determined by noting the first load at which there is visible increase in the distance between two gauge marks on a tensile specimen. This is conveniently accomplished by checking the length with a pair of dividers. If an extensometer is used, the length can be observed to increase rapidly without an increase in load. Still a third method is to coat the specimen with a brittle lacquer which cracks when the yield point is reached. The yield point most commonly is observed in mild steels, although it has been detected in a few other alloys.

(*v*) **Tensile strength (ultimate or maximum strength).** *It is calculated by dividing the maximum load carried by the specimen during a tension test by the original cross-sectional area of the specimen.*

Tensile strength is widely used design factor, although there is more justification for yield strength.

(*vi*) **Rupture strength.** *It is determined by dividing the load at the time of fracture by the original cross-sectional area*. If the rupture load is divided by the *actual cross-section* at the time of fracture, the *time rupture strength* is obtained.

The rupture strength is of indirect and limited interest to engineers. It provide the terminal point of the stress-strain curve and makes possible a computation of static toughness.

(*vii*) **Elongation.** Elongation of a specimen after fracture may be determined by placing the parts of the broken specimen closely together and holding them to place by a vice. The distance between gauge marks may be measured by means of dividers.

$$\text{Percentage elongation} = \frac{\text{Final length} - \text{Original length}}{\text{Original length}} \times 100$$

Elongation has considerable engineering significance because it indicates ductility, or the ability to deform appreciably without rupture. Ductility is essential in forming operations for metals, where it is desirable to achieve as much deformation as possible in one operation without danger of calling rupture. Ductility is also essential to avoid local failures leading to overall failures in metal members which are locally highly stressed as a result of design or fabricating techniques.

(*viii*) **Reduction of area.** After the metal has fractured the percentage reduction in area is calculated by measuring the test piece diameter at the point of fracture, calculating the cross-sectional area at this point, and expressing it as a percentage of original area.

$$\text{Percentage reduction of area} = \frac{\text{Original area} - \text{Area at fracture}}{\text{Original area}} \times 100$$

(*ix*) *Proof stress* is the stress at which the stress strain curve departs from a straight line by not more than 0.1 per cent of length of the test piece. The material is said to have passed the proof stress test if application of certain load for 15 seconds does not produce more than 0.1 per cent elongation.

(*x*) **Ultimate stress.** The *maximum load to which a bar is subjected to in a test divided by the original cross-sectional area, gives a nominal stress which is known as ultimate or maximum stress.* A tensile ultimate stress is called "*Tenacity*".

(*xi*) **Working stress.** The *maximum (or greatest) calculated stress to which a part of machine or structure is ever subjected to is called the working stress.* It should be well below the elastic limit.

(*xii*) **Factor of safety (F.O.S).** The ratio of ultimate stress to working stress is called factor of safety. It is also known as *factor ignorance.* In America it is sometimes called as he factor of stropidity. It depends upon the following factors:

(*i*) nature of loading,

(*ii*) frequency of variation of load,

(*iii*) degree of reliability required, and

(*iv*) decrease of section by corrosion etc.

Factor of safety is decided by experience.

Stress-strain curve for 'brittle materials'. *Structural steel is the only material that exhibits a marked yield point.* Most of the other materials show a gradual change from linear to the non-linear range. Brittle materials have a very low proportional point and do not show the yield point. Fig. 12.10 (*b*) shows a typical stress-strain curve for *cast iron.*

Note. The stress-strains curves for *compression* can similarly be plotted to determine the characteristic stresses such as proportional stress, yield stress and the ultimate stress. In case of *steel these stresses are the same both in tension and in compression.*

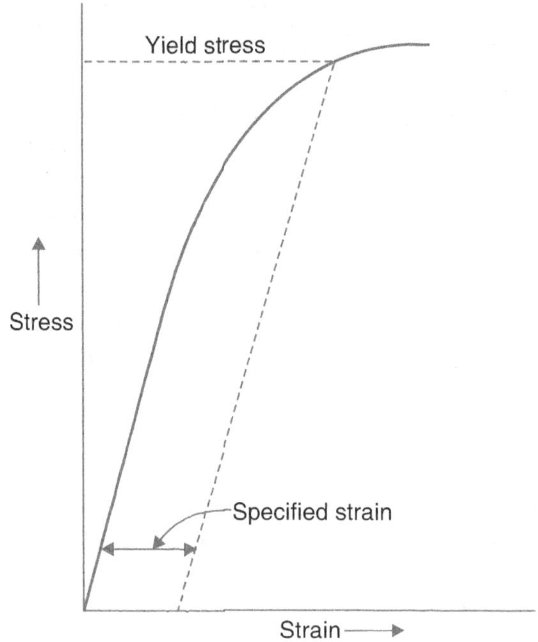

Fig. 12.10 (*b*)

12.8. STRAIN HARDENING (OR WORK HARDENING)

When a material is subjected to plastic deformation, a certain amount of work done on it is stored internally as strain energy. This *additional energy in a crystal results in strengthening or work hardening of solids*. Thus work hardening or strain hardening may be defined as *increased hardness accompanying plastic deformation*. This increase in hardness is accompanied by an *increase in both tensile and yield strength*. Work hardening *reduces ductility*.

Strain hardening is used in many manufacturing processes such as rolling of bars and drawing the tubes. It is also used to improve the elastic strength in the manufacture of many parts such as : (*i*) Pre-stretching of hoisting chains and cables, (*ii*) Initial pressurisation of pressure vessel, cylinders of hydraulic press and guns.

12.9. POISSON'S RATIO

If a body is subjected to a load its length changes ; ratio of this change in length to the original length is known as *linear* or *primary strain*. Due to this load, the dimensions of the body in all directions at right angles to its line of applications change ; the strains thus produced are called *lateral or* secondary or transverse strains and are of nature opposite to that of primary strains. For example, if the load is tensile, there will be an increase in length and a corresponding decrease in cross-sectional area of the body (Fig. 12.11). In this case, linear or primary strain will be tensile and secondary or lateral or transverse strain compressive.

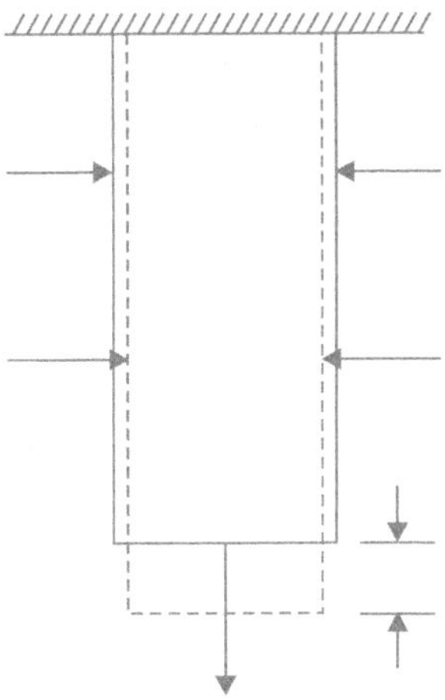

Fig. 12.11

The *ratio of lateral strain to linear strain is known as Poisson's ratio.*

i.e., Poisson's ratio, $\mu = \dfrac{\text{lateral strain or transverse strain}}{\text{linear or primary strain}} = \dfrac{1}{m}$

where m is a constant and its value varies between 3 and 4 for different materials.

Table 12.1 gives the average values of Poisson's ratio for common materials.

Table 12.1. Poisson's Ratio for Some of the Common Materials

S. No.	Material	Poisson's ratio
1	Aluminium	0.330
2	Brass	0.340
3	Bronze	0.350
4	Cast iron	0.270
5	Concrete	0.200
6	Copper	0.355
7	Mone	0.315
8	Steel	0.288
9	Stainless steel	0.305
10	Wrough iron	0.278

12.10. RELATIONS BETWEEN THE ELASTIC MODULII

Relations exist between the elastic constants for any specific materials and these relations hold good for all materials within the elastic range. The relations result from the fact that the application of any particular type of stress necessarily produces other types of stress on other places in the material. Further, each of the stresses produces its corresponding strain and all the strains produced must be consistent.

12.10.1. Relation between E and C

Refer to Fig. 12.12. *LMST* is a solid cube subjected to a shearing force *F*. Let f_s be the shear stress produced in the faces *MS* and *LT* due to this shearing force. The complementary shear stress consequently produced in the faces *ML* and *ST* is also f_s. Due to the shearing load the cube is distorted to *LM′S′ T*, and as such, the edge *M* moves to *M′*, *S* to *S′* and the diagonal *LS* to *L′S′*.

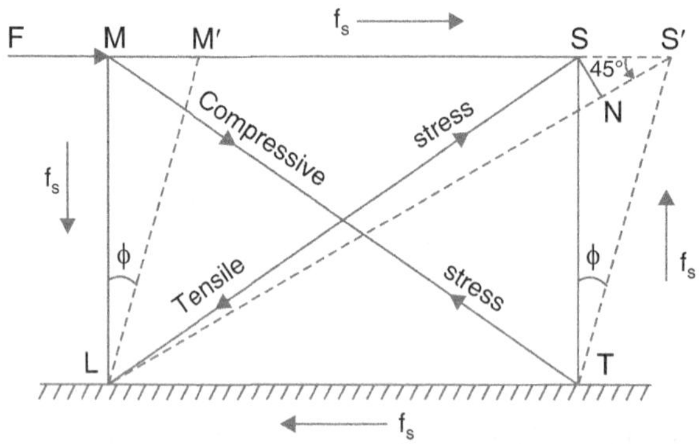

Fig. 12.12

$$\text{Shear strain} = \phi = \frac{SS'}{ST}$$

Also
$$\text{shear strain} = \frac{f_s}{C}$$

$$\therefore \quad \frac{SS'}{ST} = \frac{f_s}{C} \quad ...(i)$$

On the diagonal *LS′*, draw a perpendicular *SN* to *S*.

$$\text{Now diagonal strain} = \frac{NS'}{LN} = \frac{NS'}{LS} \quad ...(ii)$$

$$NS' = SS' \cos 45° = \frac{SS'}{\sqrt{2}}$$

[∠*LS′T′* is assumed to be equal to ∠*LST* since *SS′* is very small]

and
$$LS = ST \times \sqrt{2}$$

Putting the value of *LS* in (*ii*), we get

$$\text{Diagonal strain} = \frac{SS'}{\sqrt{2}ST \times \sqrt{2}} = \frac{SS'}{2ST}$$

But
$$\frac{SS'}{ST} = \frac{f_s}{C}$$

$$\therefore \quad \text{Diagonal strain} = \frac{f_s}{2C} = \frac{f_n}{2C} \quad ...(iii)$$

where f_n is the normal stress due to shear stress f_s. The net strain in the direction of diagonal *LS*

$$= \frac{f_n}{E} + \frac{f_n}{mE} \qquad \left[\begin{array}{l}\text{Since the diagonals } LS \text{ and } MT \text{ have normal} \\ \text{tensile and compressive stress } f_n \text{ respectively.}\end{array}\right]$$

$$= \frac{f_n}{E}\left(1+\frac{1}{m}\right) \qquad ...(iv)$$

Comparing (*iii*) and (*iv*), we get

$$\frac{f_n}{2C} = \frac{f_n}{m}\left(1+\frac{1}{m}\right) \quad i.e., \quad E = 2C\left(1+\frac{1}{m}\right) \qquad ...(12.7)$$

12.10.2. Relation between E and K

If the solid cube in question is subjected to f_n (normal compressive stress) on all the faces, the direct strain in each axis = $\frac{f_n}{mE}$ (compressive) and lateral strain in other axis = $\frac{f_n}{mE}$ (tensile).

$\therefore$ Net compressive strain in each axis

$$= \frac{f_n}{E} - \frac{f_n}{mE} - \frac{f_n}{mE} = \frac{f_n}{E}\left(1-\frac{2}{m}\right)$$

Volumetric strain (e_v) in this case will be,

$$e_v = 3 \times \text{linear strain} = 3 \times \frac{f_n}{E}\left(1-\frac{2}{m}\right)$$

But

$$e_v = \frac{f_n}{K}$$

$\therefore$

$$\frac{f_n}{K} = \frac{3f_n}{E}\left(1-\frac{2}{m}\right) \quad \text{or} \quad E = 3K\left(1-\frac{2}{m}\right) \qquad ...(12.8)$$

The relation between E, C and K can be established by eliminating m from the equations (12.7) and (12.8) as follows :

From equation (12.7), $m = \frac{2C}{E-2C}$.

Putting this value in equation (12.8), we get

$$E = 3K\left[1-\frac{2}{\frac{2C}{E-2C}}\right] \quad \text{or} \quad E = 3K\left(1-\frac{E-2C}{C}\right)$$

or

$$\frac{E}{3K} = \frac{C-E+2C}{C} = \frac{3C-E}{C} \quad \text{or} \quad \frac{E}{3K} + \frac{E}{C} = 3$$

or

$$EC + 3KE = 9KC$$

$$E(3K + C) = 9KC$$

$$E = \frac{9KC}{3K+C} \qquad ...(12.9)$$

Note. *When a square or rectangular block subjected to a shear load is in equilibrium, the shear stress in one plane is always associated with a complementary shear stress (of equal value) in the other plane at right angles to it.*

12.11. STRESSES INDUCED IN COMPOUND TIES OR STRUTS

Frequently ties consist of two materials, rigidly fastened together to prevent uneven straining of the two materials. In these cases, it is interesting to calculate the distribution of the load between the materials. It will be assumed that the two materials are symmetrically distributed about the axis of the bar, as with a cylindrical rod encased in a tube (Fig. 12.13). If then an axial load W is applied to the bar

$$W = f_1 A_1 + f_2 A_2 \quad \text{...(12.10)}$$

where f_1 and f_2 are the stresses induced and A_1 and A_2 are the cross sectional areas of the materials.

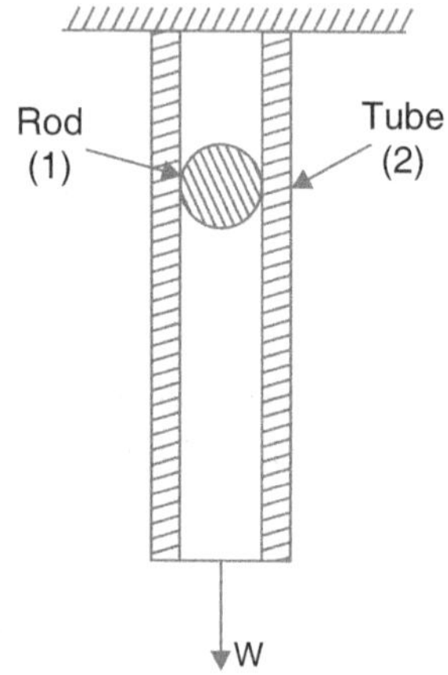

Fig. 12.13

The strains produced, e_1 and e_2 are equal.

$$\therefore \quad e_1 = e_2$$

$$\therefore \quad \frac{f_1}{E_1} = \frac{f_2}{E_2} \quad \therefore \quad \frac{f_1}{f_2} = \frac{E_1}{E_2} \quad \text{...[12.10 (a)]}$$

Hence the stresses are in the ratio of the elastic modulii.

12.12. THERMAL STRESSES AND STRAINS

If the temperature of a body is lowered or raised ; its dimensions will decrease or increase correspondingly. If these changes, however, are checked; the stresses thus developed in the body are called *temperature stresses* and corresponding strains are called *temperature strains*.

Let l = length of a bar of uniform cross-section

t_1 = initial temperature of the bar

t_2 = final temperature of the bar

α = co-efficient of linear expansion.

The extension in the bar due to rise in temperature will be = $\alpha (t_2 - t_1) l$.

If this elongation in bar is prevented by some external force or by fixing bar ends, temperature strain thus produced will be given by

$$\text{Temperature strain} = \frac{\alpha(t_2 - t_1)l}{l} = \alpha (t_2 - t_1) \text{ (compressive)} \quad \text{...(12.11)}$$

$$\therefore \text{ Temperature stress developed} = \alpha (t_2 - t_1)E \text{ (compressive)} \quad \text{...(12.12)}$$

If, however, the *temperature of the bar is lowered, the temperature strain and stress will be tensile in nature.*

STRESS-STRAIN AND YOUNG'S MODULUS

Example 12.1. *A square steel rod 20 mm × 20 mm in section is to carry an axial load (compressive) of 100 kN. Calculate the shortening in a length of 50 mm. $E = 2.14 \times 10^8$ kN/m².*

Sol. Area, $A = 0.02 \times 0.02 = 0.0004$ m² ; Length, l = 50 mm or 0.05 m

Load, $P = 100$ kN ; $E = 2.14 \times 10^8$ kN/m²

Shortening of the rod, δl :

Stress, $$\sigma = \frac{P}{A}$$

$\therefore$ $$\sigma = \frac{100}{0.0004} = 250000 \text{ kN/m}^2$$

Also $$E = \frac{\text{stress}}{\text{strain}} \quad \text{or} \quad \text{Strain} = \frac{\text{stress}}{E} = \frac{250000}{2.14 \times 10^8}$$

or $$\frac{\delta l}{l} = \frac{250000}{2.14 \times 10^8}$$

$\therefore$ $$\delta l = \frac{250000}{2.14 \times 10^8} \times l = \frac{250000}{2.14 \times 10^8} \times 0.05 = 0.0000584 \text{ m or } 0.0584 \text{ mm}$$

Hence the shortening of the rod = 0.0584 mm. (Ans.)

Example 12.2. *A hollow cast-iron cylinder 4 m long, 300 mm outer diameter and thickness of metal 50 mm is subjected to a central load on the top when standing straight. The stress produced is 75000 kN/m². Assume Young's modulus for cast iron as 1.5 × 10⁸ kN/m² and find (i) magnitude of the load, (ii) longitudinal strain produced and (iii) total decrease in length.*

Sol. Outer diameter, D = 300 mm = 0.3 m

Thickness, t = 50 mm = 0.05 m

Length, l = 4 m

Stress produced, $\sigma = 75000 \text{ kN/m}^2$

$E = 1.5 \times 10^8 \text{ kN/m}^2$

Here diameter of the cylinder, $d = D - 2t = 0.3 - 2 \times 0.05 = 0.2$ m

(*i*) **Magnitude of the load P :**

Using the relation, $$\sigma = \frac{P}{A}$$

or $$P = \sigma \times A = 75000 \times \frac{\pi}{4}(D^2 - d^2) = 75000 \times \frac{\pi}{4}(0.3^2 - 0.2^2)$$

or $$\mathbf{P = 2945.2 \text{ kN.}} \quad \textbf{(Ans.)}$$

(*ii*) **Longitudinal strain produced, e :**

Using the relation,

$$\text{Strain, } e = \frac{\text{stress}}{E} = \frac{75000}{1.5 \times 10^8} = \mathbf{0.0005.} \quad \textbf{(Ans.)}$$

(*iii*) **Total decrease in length, δl :**

Using the relation,

$$\text{Strain} = \frac{\text{change in length}}{\text{original length}} = \frac{\delta l}{l}$$

$$0.0005 = \frac{\delta l}{4}$$

$$\delta l = 0.0005 \times 4 \text{ m} = 0.002 \text{ m} = 2 \text{ mm}$$

Hence decrease in length = 2 mm. (Ans.)

Example 12.3. *The following observations were made during a tensile test on a mild steel specimen 40 mm in diameter and 200 mm long.*

Elongation with 40 kN load (within limit of proportionality), δl = 0.0304 mm

Yield load = 161 kN

Maximum load = 242 kN

Length of specimen at fracture = 249 mm

Determine :

(i) Young's modulus of elasticy *(ii) Yield point stress*

(iii) Ultimate stress *(iv) Percentage elongation.*

Sol. (*i*) **Young's modulus of elasticity E :**

Stress, $$\sigma = \frac{P}{A} = \frac{40}{\frac{\pi}{4} \times (0.04)^2} = 3.18 \times 10^4 \text{ kN/m}^2$$

Strain, $$e = \frac{\delta l}{l} = \frac{0.0304}{200} = 0.000152$$

$$\therefore \quad E = \frac{\text{stress}}{\text{strain}} = \frac{3.18 \times 10^4}{0.000152} = \mathbf{2.09 \times 10^8 \text{ kN/m}^2}. \quad \textbf{(Ans.)}$$

(*ii*) **Yield point stress :**

$$\text{Yield point stress} = \frac{\text{yield point load}}{\text{area}}$$

$$= \frac{161}{\frac{\pi}{4} \times (0.04)^2} = \mathbf{12.8 \times 10^4 \text{ kN/m}^2}. \quad \textbf{(Ans.)}$$

(*iii*) **Ultimate stress :**

$$\text{Ultimate stress} = \frac{\text{maximum load}}{\text{area}}$$

$$= \frac{242}{\frac{\pi}{4} \times (0.04)^2} = \mathbf{19.2 \times 10^4 \text{ kN/m}^2}. \quad \textbf{(Ans.)}$$

(*iv*) **Percentage elongation :**

$$\text{Percentage elongation} = \frac{\text{length of specimen at fracture} - \text{original length}}{\text{original length}}$$

$$= \frac{249 - 200}{200} = 0.245 = \mathbf{24.5\%}. \quad \textbf{(Ans.)}$$

Example 12.4. *A steel wire 2 m long and 3 mm in diameter is extended by 0.75 mm when a weight W is suspended from the wire. If the same weight is suspended from a brass wire, 2.5 m long and 2 mm in diameter, it is elongated by 4.64 mm. Determine the modulus of elasticity of brass if that of steel be 2.0 × 10^5 N/mm^2.*

Sol. *Given :* l_s = 2 m, d_s = 3 mm, δl_s = 0.75 mm ; E_s = 2.0 × 10^5 N/mm^2 ;

l_b = 2.5 m ; d_b = 2 mm ; δl_b = 4.64 m.

Modulus of elasticity of brass, E_b :

From Hooke's law, we know $\delta l = \dfrac{Wl}{AE}$

where, δl = extension, l = length, A = cross-sectional area, and

E = modulus of elasticity.

Case I : *For steel wire :*

$$\delta l_s = \frac{Wl_s}{A_s E_s}$$

or
$$0.75 = \frac{W \times (2 \times 1000)}{\left(\frac{\pi}{4} \times 3^2\right) \times 2.0 \times 10^5}$$

or
$$W = 0.75 \times \left(\frac{\pi}{4} \times 3^3\right) \times 2.0 \times 10^5 \times \frac{1}{2000} \qquad ...(i)$$

Case II : *For brass wire :*

$$\delta l_b = \frac{Wl_b}{A_b E_b}$$

$$4.64 = \frac{W \times (2.5 \times 1000)}{\left(\frac{\pi}{4} \times 2^2\right) \times E_b}$$

or
$$W = 4.64 \times \left(\frac{\pi}{4} \times 2^2\right) \times E_b \times \frac{1}{2500} \qquad ...(ii)$$

Equating eqns. (*i*) and (*ii*), we get

$$0.75 \times \left(\frac{\pi}{4} \times 3^2\right) \times 2.0 \times 10^5 \times \frac{1}{2000} = 4.64 \times \left(\frac{\pi}{4} \times 2^2\right) \times E_b \times \frac{1}{2500}$$

or
$$\mathbf{E_b = 0.909 \times 10^5\ N/mm^2.} \quad \textbf{(Ans.)}$$

Example 12.5. *A steel bar is 900 mm long ; its two ends are 40 mm and 30 mm in diameter and the length of each rod is 200 mm. The middle portion of the bar is 15 mm in diameter and 500 mm long. If the bar is subjected to an axial tensile load of 15 kN, find its total extension.*

Take E = 200 GN/m² (G stands for giga and 1 G = 10⁹)

Sol. Refer to Fig. 12.14.

Load, P = 15 kN

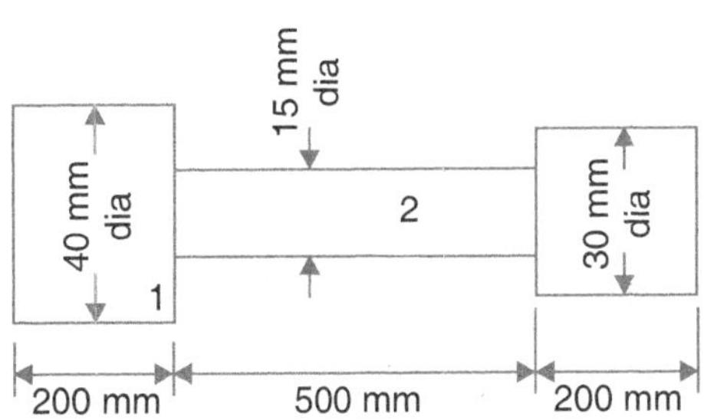

Fig. 12.14

Area, $A_1 = \dfrac{\pi}{4} \times 40^2$

$= 1256.6\ \text{mm}^2 = 0.001256\ \text{m}^2$

Area, $A_2 = \dfrac{\pi}{4} \times 15^2$

$= 176.7\ \text{mm}^2 = 0.0001767\ \text{m}^2$

Area, $A_3 = \dfrac{\pi}{4} \times 30^2$

$= 706.8\ \text{mm}^2 = 0.0007068\ \text{m}^2$

Lengths : l_1 = 200 mm = 0.2 m, l_2 = 500 mm = 0.5 m and l_3 = 200 mm = 0.2 m

Total extension of the bar :

Let δl_1, δl_2 and δl_3 be the extensions in the parts 1, 2 and 3 of the steel bar respectively.

Then, $$\delta l_1 = \frac{Pl_1}{A_1E}, \delta l_2 = \frac{Pl_2}{A_2E}, \delta l_3 = \frac{Pl_3}{A_3E} \quad \left[\because \; E = \frac{\sigma}{e} = \frac{P/A}{\delta l/l} = \frac{P.l}{A.\delta} \text{ or } \delta l = \frac{Pl}{AE}\right]$$

Total extension of the bar,

$$\delta l = \delta l_1 + \delta l_2 + \delta l_3$$

$$= \frac{Pl_1}{A_1E} + \frac{Pl_2}{A_2E} + \frac{Pl_3}{A_3E} = \frac{P}{E}\left[\frac{l_1}{A_1} + \frac{l_2}{A_2} + \frac{l_3}{A_3}\right]$$

$$= \frac{15 \times 10^3}{200 \times 10^9}\left[\frac{0.20}{0.001256} + \frac{0.50}{0.0001767} + \frac{0.20}{0.0007068}\right]$$

$$= 0.0002454 \text{ m} = 0.2454 \text{ mm}$$

Hence total extension of the steel bar = 0.2454 mm. (Ans.)

Example 12.6. *The bar shown in Fig. 12.15 is subjected to a tensile load of 50 kN. Find the diameter of the middle portion if the stress is limited to 130 MN/m². Find also the length of the middle portion if the total elongation of the bar is 0.15 mm. Take E = 200 GN/m².*

Sol. Magnitude of tensile load, $P = 50$ kN

Stress in the middle portion, $\sigma = 130$ MN/m²

Total elongation of the bar, $\delta l = 0.15$ mm $= 0.15 \times 10^{-3}$ m

Modulus of elasticity, $E = 200$ GN/m²

Diameter of the middle portion, d :

Now, stress in the middle portion, $$\sigma = \frac{P}{A} = \frac{50 \times 1000}{(\pi/4)\, d^2} = 130 \times 10^6$$

$$\therefore \quad d = \left[\frac{50 \times 1000}{\pi/4 \times 130 \times 10^6}\right]^{1/2} = 0.0221 \text{ m or } 22.1 \text{ mm}$$

Hence diameter of the middle portion = 22.1 mm. (Ans.)

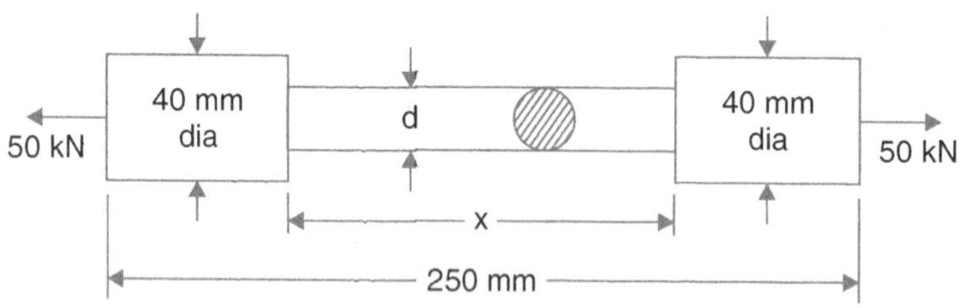

Fig. 12.15

Length of the middle portion :

Let the length of the middle portion = x metre

Stress in the end portions, $$\sigma' = \frac{50 \times 1000}{\pi/4 \times \left(\frac{40}{1000}\right)^2} = 39.79 \times 10^6 \text{ N/m}^2$$

$$\therefore \text{ Elongation of the end portion} = \sigma' \times \frac{(0.25 - x)}{E}$$

Also, elongation of the end portion + extension of the middle portion = 0.15×10^{-3}

$$\frac{39.79 \times 10^6 \times (0.25 - x)}{200 \times 10^9} + \frac{130 \times 10^6 \times x}{200 \times 10^9} = 0.15 \times 10^{-3}$$

$$39.79 \times 10^6 \times (0.25 - x) + 130 \times 10^6 \times x = 200 \times 10^9 \times 0.15 \times 10^{-3}$$

Dividing both sides by 39.79×10^6, we get

$$0.25 - x + 3.267\, x = 0.754$$

$$\therefore \quad x = 0.222 \text{ m or } 222 \text{ mm}$$

Hence length of the middle portion = 222 mm. (Ans.)

Example 12.7. *A steel tie rod 50 mm in diameter and 2.5 m long is subjected to a pull of 100 kN. To what length the rod should be bored centrally so that the total extension will increase by 15 percent under the same pull, the bore being 25 mm diameter ? Take E = 200 GN/m².*

Sol. Refer to Fig. 12.16 (*a*), (*b*)

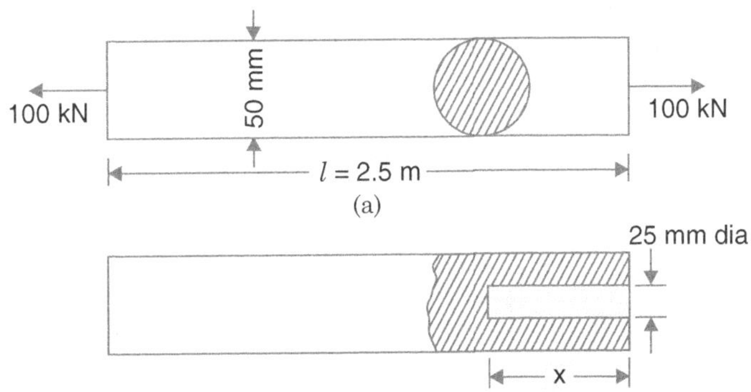

Fig. 12.16

Diameter of the steel tie rod = 50 mm = 0.05 m

Length of the steel rod, $l = 2.5$ m

Magnitude of the pull, $P = 100$ kN

Diameter of the bore = 25 mm = 0.025 m

Modulus of elasticity, $E = 200 \times 10^9$ N/m²

Length of the bore, x :

Stress in the solid rod, $$\sigma = \frac{P}{A} = \frac{100 \times 1000}{\pi/4 \times (0.05)^2} = 50.92 \times 10^6 \text{ N/m}^2$$

Elongation of the solid rod, $$\delta l = \frac{\sigma l}{E} = \frac{50.92 \times 10^6 \times 2.5}{200 \times 10^9} = 0.000636 \text{ m or } 0.636 \text{ mm}$$

Elongation after the rod is bored = $1.15 \times 0.636 = 0.731$ mm

Area at the reduced section $= \pi/4\,(0.05^2 - 0.025^2) = 0.001472$ m²

Stress in the reduced section, $$\sigma' = \frac{100 \times 1000}{0.001472} = 67.93 \times 10^6 \text{ N/m}^2$$

$\therefore$ Elongation of the rod $$= \frac{\sigma\,(2.5 - x)}{E} + \frac{\sigma\,.\,x}{E} = 0.731 \times 10^{-3}$$

$$= \frac{50.92 \times 10^6\,(2.5 - x)}{200 \times 10^9} + \frac{67.9 \times 10^6 x}{200 \times 10^9} = 0.731 \times 10^{-3}$$

$$= 50.92 \times 10^6\,(2.5 - x) + 67.9 \times 10^6 x$$

$$= 200 \times 10^9 \times 0.731 \times 10^{-3}$$

$$= (2.5 - x) + 1.33\,x = 2.87$$

or $$x = 1.12 \text{ m}$$

Hence length of the bore = 1.12 m. (Ans.)

Example 12.8. *A brass bar having cross-sectional area of 1000 mm² is subjected to axial forces shown in Fig. 12.17. Find the total elongation of the bar. Modulus of elasticity of brass = 100 GN/m².*

Sol. Refer to Fig. 12.17. The loading of three different portions *LM*, *MN* and *NP* is shown separately in the Fig. 12.18. It may be noted that portion *LM* is under the tensile force 50 kN to the left, and to the right of it, there is again an effective force 50 kN which is a resultant of three forces to its right

i.e., (80 – 20 – 10) = 50 kN.

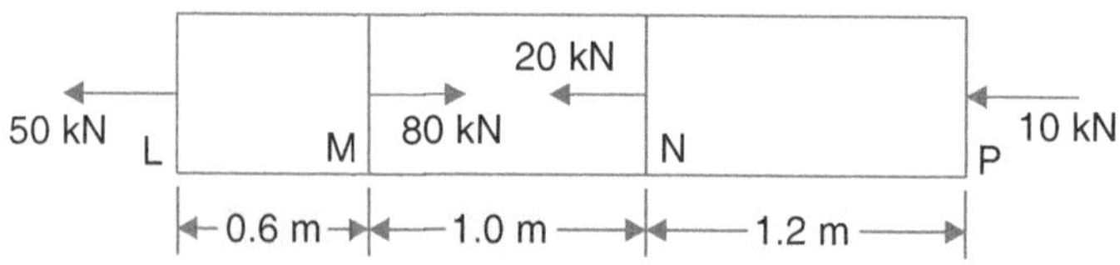

Fig. 12.17

Similarly in portion *MN*, the compressive force on the left is 30 kN (*i.e.*, 80 – 50) and 30 kN on the right (*i.e.*, 20 + 10). In *NP*, the compressive load is 10 kN, (*i.e.*, 80 – 50 – 20) and on the right, there is already a compressive load of 10 kN. So we observe that the bar is in equilibrium under the action of these forces.

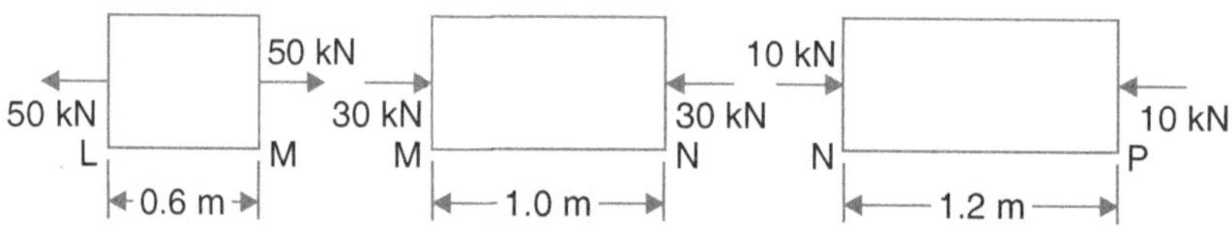

Fig. 12.18

Total elongation of the bar :

Let δl_1, δl_2 and δl_3 be the changes in length *LM*, *MN* and *NP* respectively.

Then, $$\delta l_1 = \frac{P_1 l_1}{AE} \text{ ...... increase (+)}$$

$$\delta l_2 = \frac{P_2 l_2}{AE} \text{ ...... decrease (–)}$$

$$\delta l_3 = \frac{P_3 l_3}{AE} \text{ ...... decrease (–)}$$

$\therefore$ Net change in length, $\delta l = \delta l_1 - \delta l_2 - \delta l_3$

$$= \frac{P_1 l_1}{AE} - \frac{P_2 l_2}{AE} - \frac{P_3 l_3}{AE} = \frac{1}{AE}(P_1 l_1 - P_2 l_2 - P_3 l_3)$$

$$= \frac{10^3}{1000 \times 10^{-6} \times 100 \times 10^9}(50 \times 0.6 - 30 \times 1 - 10 \times 1.2)$$

$$= \frac{1}{10^5}(30 - 30 - 12) = \frac{1}{10^5} \times (-12)$$

$$= -0.00012 \text{ m} = -0.12 \text{ mm}$$

Negative sign indicates that the bar is **shortened by 0.12 mm. (Ans.)**

Example 12.9. *In Fig. 12.19 is shown a steel bar of cross-sectional area 250 mm² held firmly by the end supports and loaded by an axial force of 25 kN.*

Determine :

(*i*) *Reactions at L and M.* (*ii*) *Extention of the left portion.*

$E = 200$ *GN/m²*.

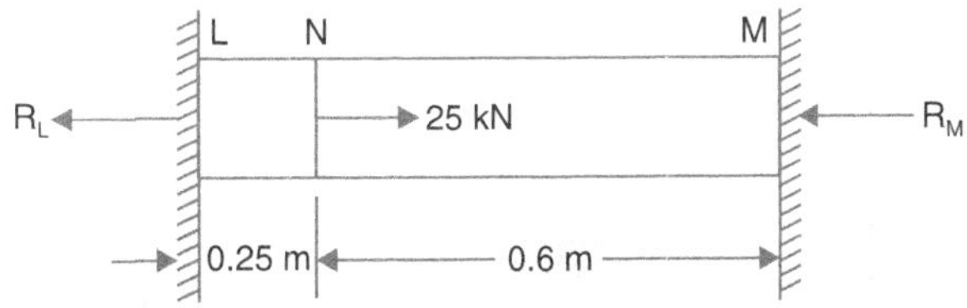

Fig. 12.19

Sol. Refer to Fig. 12.19.

(*i*) **Reactions at L and M :**

As the bar is in equilibrium

$\therefore \quad R_L + R_M = 25$ kN ...(*i*)

Also, since total length of the bar remains unchanced,

$\therefore$ Extension in *LN* = contraction in *MN*

$$\frac{R_L \times 0.25}{A \times E} = \frac{R_M \times 0.6}{A \times E}$$

$$R_L \times 0.25 = R_M \times 0.6$$

$$R_L = \frac{R_M \times 0.6}{0.25} = 2.4\, R_M$$

Substituting the value of R_L in (*i*), we get $2.4\, R_M + R_M = 25$

From which $\mathbf{R_M = 7.353}$ **kN. (Ans.)**

$\therefore$ $\mathbf{R_L}$ = 25 – 7.353 = **17.647 kN. (Ans.)**

(*ii*) **Extention of left portion :**

Extention of left portion $= \dfrac{R_L \times 0.25}{A \times E} = \dfrac{17.647 \times 10^3 \times 0.25}{250 \times 10^{-6} \times 200 \times 10^9} = 0.0000882$ m

= **0.0882 mm. (Ans.)**

Example 12.10. *Two parallel steel wires 6 m long, 10 mm diameter are hung vertically 70 mm apart and support a horizontal bar at their lower ends. When a load of 9 kN is attached to one of the wires, it is observed that the bar is 24° to the horizontal. Find 'E' for wire.*

Sol. Refer to Fig. 12.20.

Two wires *LM* and *ST* made of steel, each 6 m long and 10 mm diameter are fixed at the supports and a load of 9 kN is applied on wire *ST*. Let the inclination of the bar after the application of the load be θ.

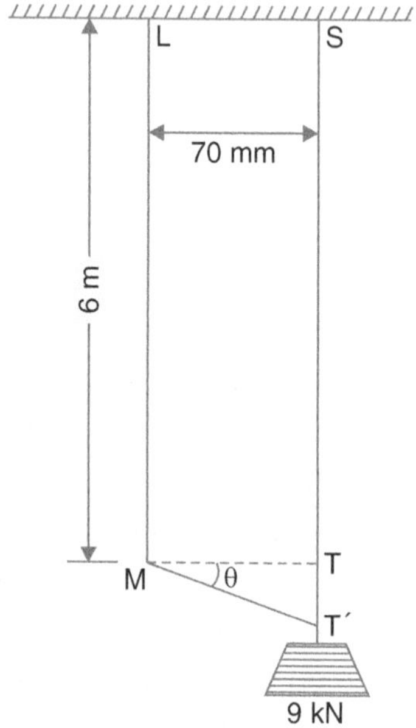

Fig. 12.20

The extension in the length of steel wire *ST*,

$$\delta l = 70 \tan \theta = 70 \times \tan 2.4° = 70 \times 0.0419 = 2.933 \text{ mm}$$
$$= 0.00293 \text{ m}$$

$\therefore$ Strain in the wire, $e = \dfrac{\delta l}{l} = \dfrac{0.00293}{6} = 0.000488$

and stress in the wire

$$\sigma = \frac{P}{A} = \frac{9000}{\frac{\pi}{4} \times \left(\frac{10}{1000}\right)^2} = 11.46 \times 10^7 \text{ N/m}^2$$

Young's modulus, $\mathbf{E} = \dfrac{\sigma}{e} = \dfrac{11.46 \times 10^7}{0.000488} = 235 \times 10^9 \text{ N/m}^2$

$= \mathbf{235 \ GN/m^2}$. **(Ans.)**

Example 12.11. *A concrete cylinder of diameter 150 mm and length 300 mm when subjected to an axial compressive load of 240 kN resulted in an increase of diameter by 0.127 mm and a decrease in length of 0.28 mm. Compute the value of Poisson's ratio* $\mu \left(= \frac{1}{m}\right)$ *and modulus of elasticity E.*

Sol. Diameter of the cylinder, $d = 150$ mm

Length of the cylinder, $l = 300$ mm

Increase in diameter, $\delta d = 0.127$ mm (+)

Decrease in length, $l = 0.28$ mm (–)

Axial compressive load, $P = 240$ kN

Poisson's ratio, μ :

We know that,

Linear strain $= \dfrac{\delta l}{l} = \dfrac{0.28}{300} = 0.000933$

and, lateral strain $= \dfrac{\delta d}{d} = \dfrac{0.127}{150} = 0.000846$

$\therefore$ Poisson's ratio, $\mu = \dfrac{\text{lateral strain}}{\text{linear strain}} = \dfrac{0.000846}{0.000933} = 0.907$

Modulus of elasticity, E :

Using the relation, $$E = \frac{\text{stress}}{\text{strain (linear)}} = \frac{P/A}{\delta l/l}$$

$$E = \frac{240/(\pi/4 \times 0.15^2)}{\left(\frac{0.00028}{0.3}\right)} = \frac{240 \times 4 \times 0.3}{\pi \times 0.15^2 \times 0.00028}$$

$$= 14.55 \times 10^6 \text{ kN/m}^2 = 14.55 \text{ GN/m}^2$$

∴ **Young's modulus, E = 14.44 GN/m². (Ans.)**

Example 12.12. *For a given material, Young's modulus is 110 GN/m² and shear modulus is 42 GN/m². Find the bulk modulus and lateral contraction of a round bar of 37.5 mm diameter and 2.4 m long when stretched 2.5 mm.*

Sol. Young's modulus, $E = 110 \text{ GN/m}^2$

Shear modulus, $C = 42 \text{ GN/m}^2$

Diameter of round bar, $d = 37.5 \text{ mm} = 0.0375 \text{ m}$

Length of round bar, $l = 2.4 \text{ m}$

Extension of bar, $\delta l = 2.5 \text{ mm} = 0.0025 \text{ m}$

Bulk modulus, K :

We know that, $$E = 2C\left(1 + \frac{1}{m}\right)$$

$$110 \times 10^9 = 2 \times 42 \times 10^9 \left(1 + \frac{1}{m}\right)$$

$$\frac{1}{m} = 1.31 - 1 = 0.31 \text{ or } m = \frac{1}{0.31} = 3.22$$

Substituting this value of m in the equation

$$K = \frac{mE}{3(m-2)} \quad K = \frac{3.22 \times 110 \times 10^9}{3(3.22 - 2)} = \mathbf{96.77 \text{ GN/m}^2}. \quad \textbf{(Ans.)}$$

Lateral contraction, δd :

Longitudinal strain, $$\frac{\delta l}{l} = \frac{0.0025}{2.4} = 0.00104$$

and lateral strain $$= 0.00104 \times \frac{1}{m} = 0.00104 \times \frac{1}{3.22} = 0.000323$$

∴ Lateral contraction, $\delta d = 0.000323\, d = 0.000323 \times 37.5 =$ **0.0121 mm. (Ans.)**

Example 12.13. *The following data relate to a bar subjected to a tensile test :*

Diameter of the bar, $d = 30 \text{ mm } (= 0.03 \text{ m})$

Tensile load, $P = 54 \text{ kN}$

Gauge length, $l = 300 \text{ mm } (= 0.3 \text{ m})$

Extension of the bar, $\delta l = 0.112 \text{ mm}$

Change in diameter, $\delta l = 0.00366 \text{ mm}$

Calculate : (i) Poisson's ratio (ii) The values of three moduli.

Sol. (*i*) **Poisson's ratio $\frac{1}{m}$ or (μ) :**

Stress, $\sigma = \frac{P}{A} = \frac{54}{\pi/4d^2} = \frac{54}{\frac{\pi}{4} \times (0.03)^2} = 76394 \text{ kN/m}^2 = 76.4 \text{ MN/m}^2$

Linear strain $= \frac{\delta l}{l} = \frac{0.112}{30} = 3.73 \times 10^{-4}$

Lateral strain $= \frac{\delta d}{d} = \frac{0.00366}{30} = 1.22 \times 10^{-4}$

$\therefore$ Poisson's ratio, $\mu = \frac{1}{m} = \frac{\text{lateral strain}}{\text{linear strain}} = \frac{1.22 \times 10^{-4}}{3.73 \times 10^{-4}} = \mathbf{0.327}$. **(Ans.)**

(*ii*) **The values of three moduli, E, C and K :**

We know that, $\mathbf{E} = \frac{\text{stress}}{\text{strain}} = \frac{76.4}{3.73 \times 10^{-4}} = \mathbf{2.05 \times 10^5 \ MN/m^2}$. **(Ans.)**

Also, $E = 2C\left[1 + \frac{1}{m}\right]$ [Eqn. 12.7]

or $\mathbf{C} = \frac{E}{2\left[1 + \frac{1}{m}\right]} = \frac{2.05 \times 10^5}{2\,(1 + 0.327)} = \mathbf{0.77 \times 10^5 \ MN/m^2}$. **(Ans.)**

Again $E = 3K\left[1 - \frac{2}{m}\right]$

$\therefore$ $\mathbf{K} = \frac{E}{3\left[1 - \frac{2}{m}\right]} = \frac{2.05 \times 10^5}{3\,(1 - 2 \times 0.327)} = \mathbf{1.97 \times 10^5 \ MN/m^2}$. **(Ans.)**

Example 12.14. *A concrete column of cross-sectional area 400 mm × 400 mm is reinforced by four longitudinal 50 mm diameter round steel bars placed at each corner. If the column carries a comprehensive load of 300 kN, determine :*

(*i*) *Loads carried ;*

(*ii*) *The compressive stress produced in the concrete and steel bars.*

Young's modulus of elasticity of steel is 15 times that of concrete.

Sol. Refer to Fig. 12.21.

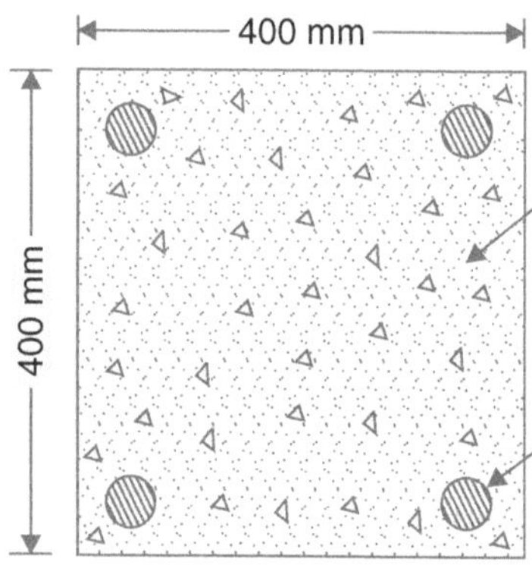

Fig. 12.21

Cross-sectional area of the column

$= 0.4 \times 0.4 = 0.16 \text{ m}^2$

Area of steel bars,

$A_s = 4 \times \pi/4 \times (0.05)^2 = 0.00785 \text{ m}^2$

$\therefore$ Area of concrete,

$A_c = 0.16 - 0.00785 = 0.1521 \text{ m}^2$

Since the steel bars and concrete shorten by the same amount under the compressive load,

$\therefore$ Strain in steel bars = strain in concrete

or $e_s = e_c$

$$\frac{\sigma_s}{E_s} = \frac{\sigma_c}{E_c}$$

or $\sigma_s = \sigma_c \cdot \frac{E_s}{E_c} = 15\,\sigma_c$ $(\because \; E_s = 15E_c)$

Also load shared by steel bars + load shared by concrete = 300000 N

$$P_s + P_c = 300000 \text{ N}$$

or $$\sigma_s \times A_s + \sigma_c \times A_c = 300000$$

$$15\sigma_c \times 0.00785 + \sigma_c \times 0.1521 = 300000$$

$$\sigma_c (15 \times 0.00785 + 0.1521) = 300000$$

$$\sigma_c = 1.11 \times 10^6 \text{ N/m}^2 = 1.11 \text{ MN/m}^2$$

$$\sigma_s = 15\, \sigma_c = 15 \times 1.11 \times 10^6$$

$$= 16.65 \times 10^6 \text{ N/m}^2 = 16.65 \text{ MN/m}^2$$

Load carried by steel bars, $P_s = \sigma_s \times A_s = 16.65 \times 10^6 \times 0.00785$

$= 130.7 \text{ kN} \simeq$ **131 kN. (Ans.)**

Load carried by concrete, $P_c = 1.11 \times 10^6 \times 0.1521 = 168.9 \simeq$ **169 kN. (Ans.)**

Example 12.15. *A copper rod of 40 mm diameter is surrounded tightly by a cast iron tube of 80 mm external diameter, the ends being firmly fastened together. When put to a compressive load of 30 kN, what load will be shared by each ? Also determine the amount by which the compound bar shortens if it is 2 m long.*

Take $E_{c.i.} = 175 \text{ GN/m}^2$ *and* $E_c = 75 \text{ GN/m}^2$.

Sol. Refer to Fig. 12.22.

Diameter of the copper rod = 0.04 m

∴ Area of copper rod,

$$A_c = \frac{\pi}{4} \times 0.04^2 = 0.0004\, \pi \text{ m}^2$$

External diameter of cast iron tube = 0.08 m

∴ Area of cast iron tube,

$$A_{c.i.} = \frac{\pi}{4}(0.08^2 - 0.04^2) = 0.0012\, \pi \text{ m}^2$$

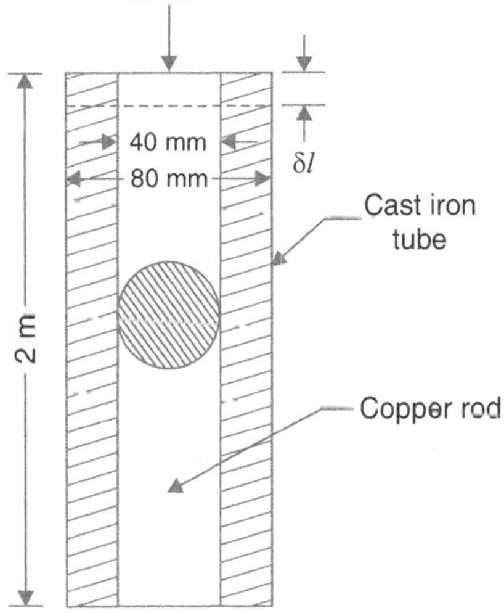

Fig. 12.22

We know that,

Strain in cast iron tube = strain in the copper rod

$$\frac{\sigma_{c.i.}}{E_{c.i.}} = \frac{\sigma_c}{E_c}$$

or $$\frac{\sigma_{c.i.}}{\sigma_c} = \frac{E_{c.i.}}{E_c} = \frac{175 \times 10^9}{75 \times 10^9} = 2.33$$

∴ $$\sigma_{c.i.} = 2.33\, \sigma_c \quad ...(i)$$

Also, total load = load shared by cast iron tube

+ load shared by copper rod.

or $$P = P_{c.i} + P_c$$

$$30 = \sigma_{c.i.} \,.\, A_{c.i.} + \sigma_c \,.\, A_c$$

or $$30 = \sigma_{c.i.} \times 0.0012\, \pi + \sigma_c \times 0.0004\, \pi \quad ...(ii)$$

Substituting the value of $\sigma_{c.i.}$ from (*i*) in (*ii*), we get

$$30 = 2.33\ \sigma_c \times 0.0012\ \pi + \sigma_c \times 0.0004\ \pi$$

$$= \sigma_c\ (0.008785 + 0.001257) = 0.010042\ \sigma_c$$

$$\sigma_c = \frac{30}{0.010042} = 2987.5 \text{ kN/m}^2$$

And from equation (*i*),

$$\sigma_{c.i} = 2.33\sigma_c = 2.33 \times 2987.5 = 6960.8 \text{ kN/m}^2$$

Load shared by the copper rod,

$$P_c = \sigma_c A_c = 2987.5 \times 0.0004\ \pi = \mathbf{3.75\ kN.} \quad \textbf{(Ans.)}$$

Load shared by cast iron tube,

$$\mathbf{P_{c.i.}} = 30 - 3.75 = \mathbf{26.25\ kN.} \quad \textbf{(Ans.)}$$

$$\text{Strain} = \frac{\sigma_c}{E_c} \text{ or } \frac{\sigma_c}{E_c} = \frac{\text{decrease in length}}{\text{original length}} = \frac{\delta l}{l}$$

$$\therefore \quad \delta l = \frac{\sigma_c}{E_c} \times l = \frac{2987.5 \times 10^3}{75 \times 10^9} \times 2 = 0.0000796 \text{ m or } 0.0796 \text{ mm}$$

Hence decrease in length = 0.0796 mm. (Ans.)

Example 12.16. *A solid steel cylinder 500 mm long and 70 mm diameter is placed inside an aluminium cylinder having 75 mm inside diameter and 100 mm outside diameter. The aluminium cylinder is 0.16 mm longer than the steel cylinder. An axial load of 500 kN is applied to the bar and cylinder through rigid cover plates as shown in Fig. 12.23. Find the stresses developed in the steel cylinder and aluminium tube. Assume for steel, E = 220 GN/m² and for aluminium E = 70 GN/m².*

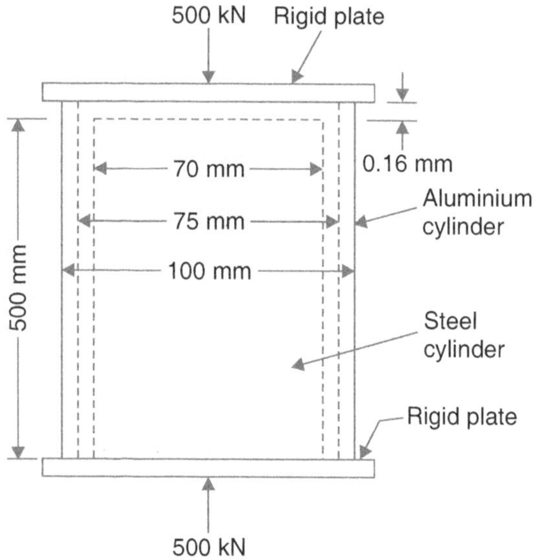

Fig. 12.23

Sol. Refer to Fig. 12.23.

Since the aluminium cylinder is 0.16 mm longer than the steel cylinder, the load required to compress this cylinder by 0.16 mm will be found as follows :

$$E = \frac{\text{stress}}{\text{strain}} = \frac{P/A}{\delta l/l} = \frac{Pl}{A\,.\,\delta l}$$

or

$$P = \frac{E\,.\,A\,.\,\delta l}{l}$$

$$= \frac{70 \times 10^9 \times \pi/4(0.1^2 - 0.075)^2 \times 0.00016}{0.50016}$$

$$= 76944 \text{ N}$$

When the aluminium cylinder is compressed by its extra length 0.16 mm, the load then shared by both aluminium as well as steel cylinder will be,

$$500000 - 76944 = 423056 \text{ N}$$

Let e_s = strain in steel cylinder

e_a = strain in aluminium cylinder

σ_s = stress produced in steel cylinder

σ_a = stress produced in aluminium cylinder

E_s = 220 GN/m^2

E_a = 70 GN/m^2

As both the cylinders are of the same length and are compressed by the same amount,

$$\therefore \quad e_s = e_a$$

or
$$\frac{\sigma_s}{E_s} = \frac{\sigma_a}{E_a} \quad \text{or} \quad \sigma_s = \frac{E_a}{E_s} \cdot \sigma_a = \frac{220 \times 10^9}{70 \times 10^9} \times \sigma_a = \frac{22}{7}\sigma_a$$

Also
$$P_s + P_a = P$$

or
$$\sigma_s \cdot A_s + \sigma_a \cdot A_a = 423056$$

$$\frac{22}{7}\sigma_a \times A_s + \sigma_a \times A_a = 423056$$

$$\therefore \quad \sigma_a = \frac{423056}{\frac{22}{7}A_s + A_a} = \frac{423056}{\frac{22}{7} \times \frac{\pi}{4} \times 0.07^2 + \frac{\pi}{4}(0.1^2 - 0.075^2)}$$

$$= \frac{423056}{0.012095 + 0.003436} = 27.24 \times 10^6 \text{ N/m}^2$$

$$= 27.24 \text{ MN/m}^2$$

and
$$\sigma_s = \frac{22}{7} \times 27.24 = 85.61 \text{ MN/m}^2$$

Stress in the aluminium cylinder due to load 76944 N

$$= \frac{76944}{\frac{\pi}{4}(0.1^2 - 0.075^2)} = 23.39 \times 10^9 \text{ N/m}^2 = 22.39 \text{ MN/m}^2$$

$\therefore$ Total stress in **aluminium cylinder**

$$= 27.24 + 22.39 = \mathbf{49.63\ MN/m^2}. \quad \textbf{(Ans.)}$$

and stress in **steel cylinder = 85.61 MN/m². (Ans.)**

Example 12.17. *A beam weighing 450 N is held in a horizontal position by three vertical wires, one attached to each end of the beam, one to the middle of its length. The outer wires are of brass of diameter 1.25 mm and the central wire is of diameter 0.625 mm. If the beam is rigid and wires of the same length and unstressed before the beam is attached, estimate the stresses induced in the wires. Take Young's modulus for brass as 86 GN/m² and for steel 210 GN/m².*

Sol. Refer to Fig. 12.24.

Let P_b = load taken by the brass wire, P_s = load taken by the steel wire.

Then
$$2P_b + P_s = P \qquad ...(i)$$

Since the beam is horizontal, all wires will extend by the same amount.

i.e.,
$$e_b = e_s \qquad (\because \text{ length of each wire is same})$$

where e_b = strain in brass wire, and

e_s = strain in steel wire.

$$\frac{\sigma_b}{E_b} = \frac{\sigma_s}{E_s}$$

$$\frac{P_b}{A_b \cdot E_b} = \frac{P_s}{A_s \cdot E_s} \quad \text{or} \quad P_s = \frac{P_b A_s \cdot E_s}{A_b \cdot E_b}$$

$$= \frac{P_b \times \frac{\pi}{4} \times (0.625 \times 10^{-3})^2 \times 210 \times 10^9}{\frac{\pi}{4}(1.25 \times 10^{-3})^2 \times 86 \times 10^9}$$

or $$P_s = 0.61\, P_b \qquad ...(ii)$$

Substituting the value of P_s in equation (*i*), we get

$$2P_b + 0.61\, P_b = P$$

$$2.61\, P_b = 450$$

$$P_b = 172.4 \text{ N}$$

and $$P_s = 0.61 \times 172.4 = 105.2 \text{ N}$$

Now, *stress, induced in the brass wire,*

$$\sigma_b = \frac{P_b}{A_b} = \frac{172.4}{\frac{\pi}{4}(1.25 \times 10^{-3})^2} = 1.40 \times 10^8 \text{ N/m}^2$$

$$= \mathbf{140 \text{ MN/m}^2. \text{ (Ans.)}}$$

and *stress induced in a steel wire,*

$$\sigma_s = \frac{105.2}{\frac{\pi}{4}(0.625 \times 10^{-3})^2} = 3.429 \times 10^8 \text{ N/m}^2$$

$$= \mathbf{342.9 \text{ MN/m}^2. \text{ (Ans.)}}$$

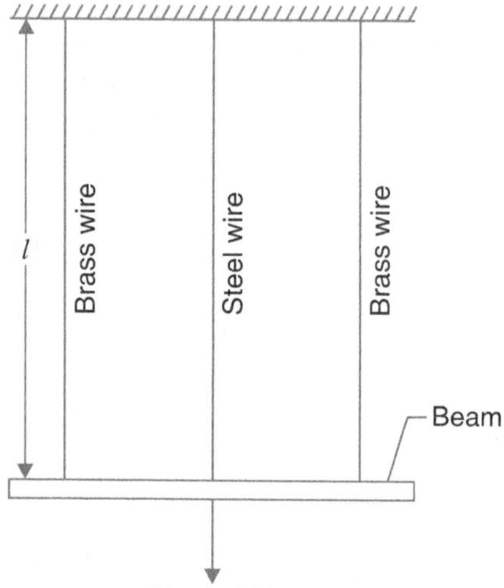

Fig. 12.24

Example 12.18. *A steel rod 15 m long is at a temperature of 15°C. Find the free expansion of the length when the temperature is raised to 65°C. Find the temperature stress produced when :*

(*i*) *The expansion of the rod is prevented ;*

(*ii*) *The rod is permitted to expand by 6 mm.*

Take $\alpha = 12 \times 10^{-6}$ *per°C and* $E = 200$ *GN/m*^{2}.

Sol. Free expansion of the rod $= l\,\alpha\,(t_2 - t_1) = 15 \times 12 \times 10^{-6}\,(65 - 15)$

$= 0.009$ m $= 9$ mm

(*i*) **When the expansion is fully prevented :**

Temperature stress $= \alpha\,(t_2 - t_1)\,E = 12 \times 10^{-6}\,(65 - 15) \times 200 \times 10^9$

$= 120 \times 10^6$ N/m^2 = **120 MN/m^2. (Ans.)**

(*ii*) **When the rod is permitted to expand by 6 mm :**

In this case, the amount of expansion prevented = 9 – 6 = 3 mm

$$\therefore \qquad \text{Strain, } e = \frac{\text{Expansion prevented}}{\text{Original length}} = \frac{3}{15 \times 1000} = 0.0002$$

∴ Temperature stress $= e \times E = 0.0002 \times 200 \times 10^9 = 40 \times 10^6$ N/m^2

= **40 MN/m^2. (Ans.)**

Example 12.19. *A copper flat measuring 60 mm × 30 mm is brazed to another 60 mm × 60 mm mild steel flat as shown in Fig. 12.25. If the combination is heated through 120°C, determine :*

(*i*) *The stress produced in each of the bar ;*

(*ii*) *Shear force which tends to rupture the brazing ; and*

(*iii*) *Shear stress.*

Take, $\alpha_c = 18.5 \times 10^{-6}$ *per °C*

$\alpha_s = 12 \times 10^{-6}$ *per °C*

$E_c = 110\ GN/m^2$

$E_s = 220\ GN/m^2$

Length of each flat = 400 mm.

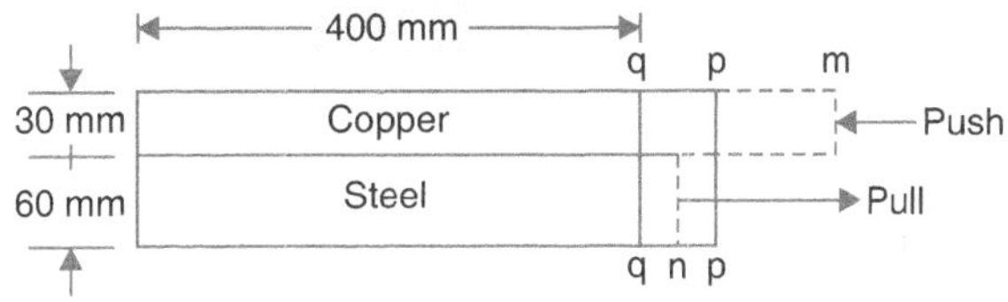

Fig. 12.25

Sol. Refer to Fig. 12.25.

(*i*) **Stress produced in each bar :**

As α_c is greater than α_s, therefore extension of copper flat will be more than that of steel flat, but since they are brazed together, *the former will try to pull the latter and the latter to push the former ; finally, however, they will become stable at certain position pp after compromise.*

Increase in length of copper flat, $qm = \alpha_c (t_2 - t_1)\, l$

Increase in length of steel flat, $qn = \alpha_s (t_2 - t_1)\, l$

Now compressive strain in copper flat,

$$e_c = \frac{pm}{l} = \frac{qm - qp}{l} \quad \text{or} \quad e_c = \frac{qm}{l} - \frac{qp}{l}$$

$$= \alpha_c (t_2 - t_1) - e \qquad ...(i)$$

where $\frac{qp}{l} = e$ = common strain

Tensile strain in steel, $e_s = \frac{np}{l} = \frac{qp - qn}{l} = e - \alpha_s (t_2 - t_1)$ ...(*ii*)

Adding (*i*) and (*ii*), we get

$$e_c + e_s = \alpha_c (t_2 - t_1) - \alpha_s (t_2 - t_1) = (\alpha_c - \alpha_s)(t_2 - t_1)$$

But, $$e_c = \frac{\sigma_c}{E_c},\ e_s = \frac{\sigma_s}{E_s} \quad \text{or} \quad \frac{\sigma_c}{E_c} + \frac{\sigma_s}{E_s} = (\alpha_c - \alpha_s)(t_2 - t_1)$$

or $$\frac{\sigma_c}{110 \times 10^9} + \frac{\sigma_s}{220 \times 10^9} = (18.5 \times 10^{-6} - 12 \times 10^{-6}) \times 120$$

or $$2\sigma_c + \sigma_s = 220 \times 10^9 \times 120\,(18.5 \times 10^{-6} - 12 \times 10^{-6})$$

or $$2\sigma_c + \sigma_s = 171.6 \times 10^6 \qquad ...(iii)$$

At position *pp*,

Pull on steel = push on copper

$$\sigma_s . A_s = \sigma_c . A_c$$

$$\sigma_s \times \frac{60}{1000} \times \frac{60}{1000} = \sigma_c \times \frac{60}{1000} \times \frac{30}{1000}$$

$$\sigma_s = \frac{\sigma_c}{2} = 0.5\ \sigma_c$$

Putting this value of σ_s in (*iii*), we get

$$2\sigma_c + 0.5\ \sigma_c = 171.6 \times 10^6$$

$\sigma_c = 68.64 \times 10^6\ \text{N/m}^2$ = **68.64 MN/m². (Ans.)**

and $\sigma_s = 0.5\ \sigma_c$ = **34.32 MN/m². (Ans.)**

(*ii*) **Shear force :**

Shear force $= \sigma_s \cdot A_s = (\sigma_c \cdot A_c) = 34.32 \times 10^6 \times \frac{60}{1000} \times \frac{60}{1000}$

$= 123552$ or **123.552 kN. (Ans.)**

(*iii*) **Shear stress :**

Shear stress $= \frac{\text{shear force}}{\text{shear area}} = \frac{123552}{\frac{400}{1000} \times \frac{60}{1000}}$

$= 5.148 \times 10^6$ N/m² = **5.148 MN/m². (Ans.)**

HIGHLIGHTS

1. A *load* may be defined as the combined effect of external forces acting on a body.
   The loads may classified as :
   (*i*) Dead loads, (*ii*) live or fluctuating loads, (*iii*) inertia loads or forces and (*iv*) centrifugal loads or forces. *Or* (*i*) Tensile loads, (*ii*) compressive loads, (*iii*) Torsional or twisting loads, (*iv*) Bending loads and (*v*) shearing loads, Or (*i*) Point loads, (*ii*) Distributed loads.
2. The internal resistance which the body offers to meet with the load is called *stress*. When any type of *simple* stress develops, we can calculate the magnitude of the stress by

$$f = \frac{W}{A}$$

where f = stress, W = load,

and A = area over which stress develops.

3. The strain (e) is the deformation produced by the stress.

Tensile strain, $e_t = \frac{\delta l}{l} = \frac{\text{Increase in length}}{\text{Original length}}$

Compressive strain, $e_c = \frac{\delta l}{l} = \frac{\text{Decrease in length}}{\text{Original length}}$

Shear strain, $e_s = \tan \phi = \phi$ radian. [Since ϕ is very small]

Volumetric strain, $e_v = \frac{\delta V}{V}$.

4. *Hooke's law* states that within elastic limit, stress varies directly as strain

i.e., Stress $\propto$ strain

or $\frac{\text{Stress}}{\text{Strain}} = \text{constant}$

Modulus of elasticity, $E = \frac{f}{e}$

Modulus of rigidity,

$$C \text{ (or } N \text{ or } G) = \frac{f_s}{e_s}$$

Bulk modulus of elasticity $= \frac{f_n}{e_v}$.

5. The ratio of lateral strain to linear strain is known as *Poisson's ratio*

i.e., Poisson's ratio = $\dfrac{\text{Lateral or secondary strain}}{\text{Linear or primary strain}} = \dfrac{1}{m}$

where m is a constant and its value varies between 3 and 4 for different metals.

6. Relations between elastic moduli :

$$E = 2C\left(1 + \frac{1}{m}\right) \qquad ...(i)$$

$$E = 3K\left(1 - \frac{2}{m}\right) \qquad ...(ii)$$

$$E = \frac{9KC}{3K + C} \qquad ...(iii)$$

7. If the temperature of a body is lowered or raised, its dimensions will decrease or increase correspondingly. If these changes, however, are checked, the stresses thus developed in the body are called *temperature stresses and* corresponding strains are called *temperature strains*.

Temperature strain $= \alpha\,(t_2 - t_1)$ ...(Compressive)

Temperature stress developed $= \alpha\,(t_2 - t_1)\,E$ ... (Compressive)

OBJECTIVE TYPE QUESTIONS

Fill in the blanks:

1. A .... may be defined as the combined effect of external forces acting on a body.
2. A .... load is one which is considered to act at a point.
3. A .... load is one which is spread in some manner over the length of the beam.
4. Bending and torsion stresses are classified as .... stresses.
5. Simple stress is often called .... stress.
6. When any type of simple stress develops, we can calculate the magnitude of the stress by, $f = \dfrac{W}{.......}$
7. The .... is the deformation produced by the body.
8. The shear strain is measured by the .... through which the body distorts.
9. .... strain is defined as the ratio between change in volume and original volume of the body.
10. The strains which disappear with the removal of load are termed as .... strains.
11. The body which regains its original position on the removal of the force is called an .... body.
12. The body is said to be .... if the strains exist even after the removal of external force.
13. Robert Hooke discovered experimentally that within elastic limit .... varies directly as ....
14. .... modulus is the ratio between tensile stress and tensile strain or compressive stress and compressive strain.
15. .... modulus of elasticity may be defined as the ratio of normal stress (on each face of a solid cube) to volumetric strain and is denoted by the letter K.
16. .... of a metal is its ability to withstand various forces to which it is subjected during a test or in service.
17. Young's modulus is indicative of the property called ....

18. .... is the property that enables the formation of a permanent deformation in a material.

19. .... is made use of in stamping images or coins and ornamental work.

20. .... is the ability of a metal to withstand elongation or bending.

21. .... is the property by virtue of which a material may be hammered or rolled into thin sheets without rupture.

22. .... is the strength with which the material opposes rupture.

23. The area under the stress-strain curve indicates the ....

24. Lack of ductility is ....

25. .... is usually defined as resistance of a material to penetration.

26. The converse of hardness is known as ....

27. .... limit is the maximum stress at which stress remains directly proportional to strain.

28. .... limit is the maximum stress which the material can withstand without causing permanent deformation which remains after removal of stress.

29. .... strength is the stress at which a material exhibits a specified limiting permanent set.

30. .... point is the stress at which there first occurs a marked increase in strain without an increase in stress.

31. .... strength is calculated by dividing the maximum load carried by the specimen during a tension test by the original cross-sectional area of the specimen.

32. .... strength is determined by dividing the load at the time of fracture by the original cross-sectional area.

33. Percentage .... $= \dfrac{\text{Final length} - \text{Original length}}{\text{Original length}} \times 100.$

34. A tensile ultimate stress is called ....

35. The miximum (or greatest) calculated stress to which a part of machine or structure is ever subjected to is called the .... stress.

36. The ratio of Lateral strain to linear strain is known as ....

37. $E = 2C(1 + \text{...........})$.

38. $E = \dfrac{9\ldots}{3K + C}$

Answers

1. load	**2.** point	**3.** distributed	**4.** indirect
5. direct	**6.** *A*	**7.** strain	**8.** angle
9. Volumetric	**10.** elastic	**11.** elastic	**12.** plastic
13. stress, strain	**14.** Youngs	**15.** Bulk	**16.** Strength
17. stiffness	**18.** Plasticity	**19.** Plasticity	**20.** Ductility
21. Malleability	**22.** Toughness	**23.** toughness	**24.** brittleness
25. Hardness	**26.** softness	**27.** Proportional	**28.** Elastic
29. Yield	**30.** Yield	**31.** Ultimate	**32.** Rupture
33. elongation	**34.** tenacity	**35.** working	**36.** Poisson's ratio
37. $\dfrac{1}{m}$	**38.** *KC*.		

Theoretical Questions

1. Define the term '*load*'. How are the loads classified ?
2. What is the difference between a point load and a distributed load ?
3. What is a simple stress?
4. Define the term '*strain*' and explain the following:
   (*a*) Tensile strain;
   (*b*) Compressive strain;
   (*c*) Shear strain.
5. State and explain Hooke's law.
6. Define the following:
   (*a*) Young's modulus (*b*) Modulus of rigidity
   (*c*) Bulk or volume modulus of elasticity.
7. Explain briefly any three of the following mechanical properties of metals:
   (*i*) Strength; (*ii*) Elasticity;
   (*iii*) Plasticity; (*iv*) Ductility;
   (*v*) Malleability; (*vi*) Tenacity (or toughness);
   (*vii*) Brittleness.
8. What is the significance of a tensile test?
9. Draw a stress-strain curve for a mild steel specimen and explain the following:
   (*i*) Proportional limit
   (*ii*) Elastic limit
   (*iii*) Yield strength
   (*iv*) Yield point
   (*v*) Ultimate strength
   (*vi*) Rupture strength.
10. Define the term '*Factor of safety*' and state its significance.
11. Write a short note on '*strain hardening*'.
12. Explain the term 'Poisson's ratio'.
13. Derive relations between E and C and E and K.
14. What are the temperature stresses and temperature strains?

Unsolved Examples

Stress, Strain and Young's Modulus :

1. A bar 0.3 m long is 50 mm square in section for 120 mm of its length, 25 mm diameter for 80 mm and of 40 mm diameter for the remaining length. If a tensile force of 100 kN is applied to the bar, calculate the maximum and minimum stresses produced in it, and the total elongation, Take E = 200 GN/m^2 and assume uniform distribution of load over the cross-sections. [**Ans.** 204 MN/m^2, 40 MN/m^2, 0.1453 mm]
2. A brass rod 20 mm diameter was subjected to a tensile load of 40 kN. The extension of the rod was found to be 254 divisions in the 200 mm extension metre. If each division is equal to 0.001 mm, find the elastic modulus of brass. [**Ans.** 100.25 GN/m^2]

3. Find the total extension of the bar, shown in Fig. 12.26 for an axial pull of 40 kN. [**Ans.** 0.225 mm]

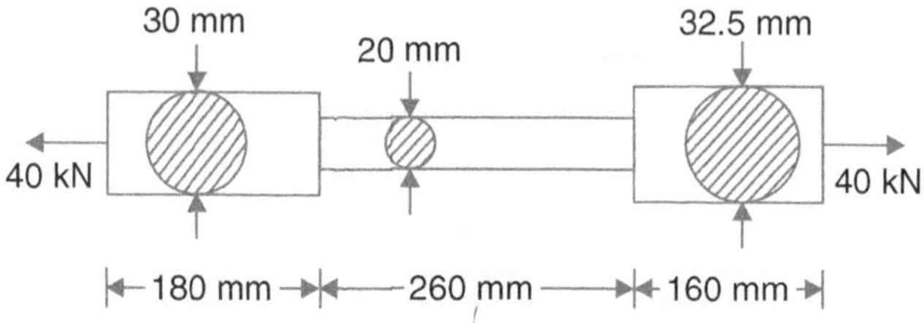

Fig. 12.26

4. A brass bar having a cross-sectional area of 1000 mm^2 is subjected to axial forces as shown in Fig. 12.27. Find the total change in length of the bar. Take E_s = 105 GN/m^2. [**Ans.** 0.1143 m (decrease)]

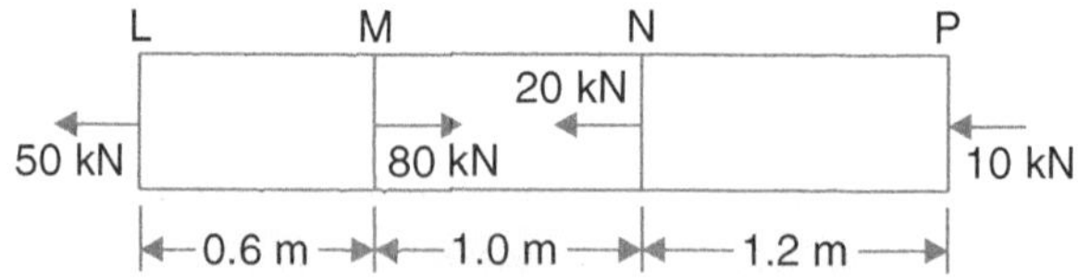

Fig. 12.27

5. A tie bar has enlarged ends of square section 60 mm × 60 mm as shown in Fig. 12.28. If the middle portion of the bar is also of square section, find the size and length of the middle portion, if the stress there is 140 MN/m^2, the total extension of the bar is 0.14 mm.

Take E = 200 GN/m^2. [**Ans.** 25 mm × 25 mm, 1.79 m]

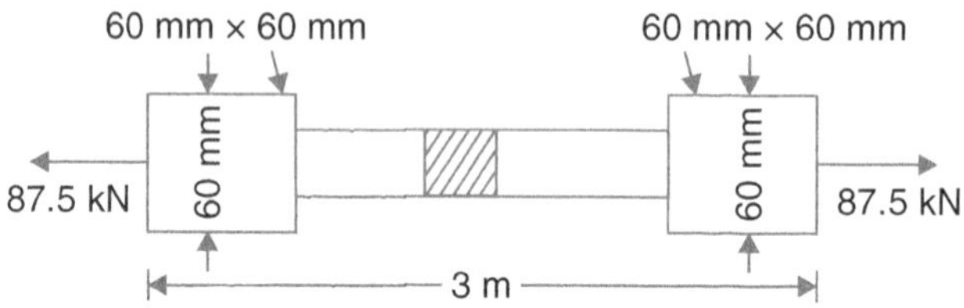

Fig. 12.28

6. A circular rod 0.2 m long, tapers from 20 mm diameter at one end to 10 mm diameter at the other. On applying an axial pull of 6 kN, it was found to extend by 0.068 mm. Find the Young's modulus of the material of the rod. [**Ans.** 112.3 GN/m^2]

7. A vertical rod 3 m long is rigidly fixed at upper end and carries an axial tensile load of 50 kN force. The rod tapers uniformly from a diameter of 50 mm at the top to 30 mm at the bottom. Calculate the total extension of the bar. Take density of material = 1 × 10^5 kg/m^3 and E = 210 GN/m^2. [**Ans.** 0.6217 mm]

Composite Sections :

8. A solid steel bar, 40 mm diameter, 2 m long passes centrally through a copper tube of internal diameter 40 m, thickness of metal 5 mm and length 2 m. The ends of the bar and tube are brazed together and a tensile load of 150 kN is applied axially to the compound bar. Find the stresses in the steel and copper, and extension of the compound bar. Assume E_c = 100 GN/m^2 and E_s = 200 GN/m^2.

[**Ans.** 93.1 MN/m^2, 46.55 MN/m^2, 0.931 mm]

9. A weight of 100 kN is suspended from a roof by two steel rods each 5 m long, joined to form an included angle of 100°. If the working stress is 56 MN/m^2, find the diameter of the rods and the amount each is stretched by the weight. Neglect weight of rods and take E = 210 GN/m^2. [**Ans.** 42.1 mm, 1.33 mm]

10. Three parallel wires in the same vertical plane jointly support a load of 15 kN. The middle wire is of steel and is 1 m long, while the outer ones of brass, the length of each being 1.05 m. The area of cross-section of each wire is 200 mm^2. After the wires have been so adjusted as to each carry 1/3 of the load, a further load of 35 kN is added. Find the stress in each wire and fraction of the whole load carried by the steel wire. Take E_s = 200 GN/m^2 and E_b = 80 GN/m^2. [**Ans.** σ_b = 62.8 MN/m^2, σ_s = 124.32 MN/m^2, 49.7%]

11. A reinforced concrete column is 300 mm × 300 mm in section. The column is provided with 8 bars each of 20 mm diameter. The column carries a load of 360 kN. Find the stresses in concrete and the steel bars. Take E_s = 210 GN/m^2 and E_c = 14 GN/m^2. [**Ans.** 2.87 MN/m^2, 43.05 MN/m^2]

12. A reinforced concrete column is 300 mm in diameter and has 4 steel bars each of 12 mm diameter embedded in it. If the allowable stresses in steel and concrete are 65.0 MN/m^2 and 4.0 MN/m^2 respectively, calculate the safe axial load which the column can carry. Take E_s = 15 E_c. [**Ans.** 308.078 kN]

13. A compound tube consists of a steel tube 170 mm external diameter and 10 mm thickness and an outer brass tube 190 mm external diameter and 10 mm thickness. The two tubes are of the same length. The compound tube carries an axial load of 1 MN. Find the stresses and the load carried by each tube and the amount by which it shortens. Length of each tube is 0.15 m. Take E_s = 200 GN/m^2 and E_b = 100 GN/m^2. [**Ans.** σ_s = 127.34 MN/m^2, σ_b = 63.67 MN/m^2, P_s = 0.64 MN, P_b = 0.36 MN ; 0.096 mm]

14. Two steel plates each 25 mm thick, are held together with the help of 20 mm diameter steel bolts. If the pitch of the threads is 2.5 mm, find the increase in stress in the shank of the bolt when the nut is turned through $\frac{1}{200}$ of a turn with respect to the head of the bolt, assuming that the plates do not deform. Take E_s = 205 GN/m^2. [**Ans.** 51.25 MN/m^2]

$$\left[\textbf{Hint.} \text{ Stretch in 50 mm length of the shank, } \delta l = 2.5 \times \frac{1}{200} = 0.0125 \text{ mm} \quad \therefore \quad \text{strain, } e = \frac{0.0125}{50}\right]$$

15. A rigid cross bar is supported horizontally by two vertical bars, L and M, of equal length and hanging from their tops. The bars L and M are 0.6 m apart. The cross bars stays horizontal even after a vertical force of 6 kN is applied to it at a point 0.4 m from M. If the stress in L is 200 MN/m^2, find the stress in M and the areas of cross-section of the two rods.
Take E_L = 200 GN/m^2 and E_M = 130 GN/m^2. [**Ans.** 130 MN/m^2, 20.0 mm^2, 15.385 mm^2]

16. A 28 mm diameter steel bar, 400 mm long is placed centrally within a brass tube having an inside diameter of 30 mm and outside diameter of 40 mm. The bar is shorter in length than the tube by 0.12 mm. While the bar and tube are held vertically on a rigid horizontal plateform, a compressive force of 60 kN is applied at the top of the tube through a rigid plate. Determine the stresses induced in both the bar and the tube. Take E_s = 200 GN/m^2 and E_b = 100 GN/m^2.
[**Ans.** 48.85 MN/m^2 (Comp.) ; 54.42 MN/m^2 (Comp.)]

17. A vertical tie of uniform strength is 18 m long. If the area of the bar at the lower end is 500 mm^2, find the area at the upper end when the tie is to carry a load of 700 kN. The material of the tie weighs 8 × 10^5 N/mm^3. [**Ans.** 500.5 mm^2]

18. A 25 mm diameter rod of steel passes centrally through a copper tube 63 mm external diameter, 40 mm internal diameter, and 1.30 m long. The tube is closed by rigid washers of negligible thickness and nuts threaded on the rod. Find the stresses in each when the nuts are tightened until the tube is reduced in length by 0.13 mm. Find the increase in the stresses if one nut is tightened by 1/2 of a turn relative to the other, there being 0.4 threads per mm. Take E_s = 200 GN/m^2 and E_c = 100 GN/m^2.
[**Ans.** σ_c = 10 MN/m^2 (comp.), σ_s = 37.9 MN/m^2 (tensile)
Increase in σ_c = 33.2 MN/m^2 (comp.) Increase in σ_s = 125.8 MN/m^2 (tensile)]

19. A steel rod 18 mm in diameter passes centrally through a steel tube 30 mm in external diameter and 2.5 mm thickness. The tube is 0.75 m long and is closed by rigid washers of negligible thickness which are fastened by nuts threaded on the rod. The nuts are tightened until the compressive load on the tube is 30 kN. Calculate the stresses in the tube and the rod.

[**Ans.** σ_{st} = 90.15 MN/m^2 (Comp.) ; σ_{sr} = 76.52 MN/m^2 (tensile)]

20. A compound bar consists of a central steel strip 25 mm wide and 6.4 mm thick placed between two strips of brass each 25 mm wide and x mm thick. The strips are firmly fixed together to form a compound bar of retangular section 25 mm wide and $(2x + 6.4)$ mm thick. Determine :

(*i*) The thickness of the brass strips which will make the apparent modulus of elasticity of compound bar 157 GN/m^2.

(*ii*) The maximum axial pull the bar can then carry if the stress is not to exceed 157 MN/m^2, in either the brass or the steel. Take E_s = 207 GN/m^2 and E_b = 114 GN/m^2.

[**Ans.** (*i*) x = 3.726 mm, (*ii*) 41.226 kN]

21. A weight of 50 kN is hanging from three wires of equal length, middle one is of steel and the other two are of copper. If 300 mm^2 is the cross-sectional area of each wire, then find out the load shared by each. Take E_s = 210 GN/m^2 and E_b = 120 GN/m^2. [**Ans.** σ_c = 13.6 kN, σ_s = 22.8 kN]

Poisson's Ratio :

22. A steel bar of rectangular cross-section 20 mm × 10 mm is subjected to a pull of 20 kN in the direction of its length. Taking E = 204 GN/m^2 and m = 10/3, find the percentage decrease of cross-section.

[**Ans.** 0.0294%]

23. A vertical circular bar 20 mm diameter, 3 m long carries a tensile load of 150 kN. Calculate :

(*i*) Elongation ; (*ii*) Decrease in diameter ; and

(*iii*) Volumetric strain. [**Ans.** (*i*) 15.9 mm, (*ii*) 0.0265 mm, (*iii*) 0.00265]

24. A steel bar 300 mm long, 50 mm wide and 12 mm thick is subjected to an axial pull of 84 kN. Find the change in length, width, thickness and volume of the bar. Take E = 200 GN/m^2 and Poisson's ratio = 0.32.

[**Ans.** δl = 0.21 mm (increase), δ_b = 0.0112 mm (decrease), δt = 0.0027 mm (decrease), δV = 45.36 mm^3 (increase)]

Thermal Stresses and Strains :

25. A railway line is laid so that there is no stress in the rails at 60°C. Calculate the stress in the rails at 20°C, if

(*i*) no allowance is made for contraction ;

(*ii*) there is an allowance of 5 mm contraction per rail

The rails are 30 mm long. E = 210 GN/m^2, α = 0.000012 per °C.

[**Ans.** (*i*) 100.8 MN/m^2, (*ii*) 65.8 MN/m^2]

26. A steel rod 30 mm diameter and 300 mm long is subjected to tensile force P acting axially. The temperature of the rod is then raised through 80°C and the total extension measured as 0.35 mm. Calculate the value of P.

Take E_s = 200 GN/m^2 and α_s = 12 × 10^{-6} per °C. [**Ans.** 6.29 kN]

27. A steel bar is placed between two copper bars each having the same area and length as the steel bar at 15°C. At this stage, they are rigidly connected together at both the ends. When the temperature is raised to 315°C, the length of the bars increases by 1.5 mm. Determine the original length and final stresses in the bar.

Take E_s = 210 GN/m^2, E_c = 110 GN/m^2, α_s = 0.000012 per °C, α_c = 0.0000175 per °C.

[**Ans.** 0.34 m, σ_c = 84.5 MN/m^2 (comp.), σ_s = 169 MN/m^2 (tensile)]

Part II
EXPERIMENTS

Experiment No. 1

Object. To verify triangle law of forces and parallelogram law of forces with the help of Gravesand's apparatus.

Apparatus. Gravesand's apparatus, paper sheet, weights, thread, drawing pine, mirror strip, three scale pans, set square, pencil etc.

Theory. The *"triangle law of forces"* states that if three coplanar forces acting on a particle can be represented in magnitude and direction by the three sides of a triangle taken in order, the forces will be in equilibrium. *This law can also be stated as:* If two forces acting on a particle are represented in magnitude and direction by the two sides of triangle taken in order then their resultant will be given by the third side of a triangle taken in opposite direction.

"Parallelogram of forces" states that if a particle is acted by two forces represented in magnitude and direction by the two sides of a parallelogram drawn from a point then the resultant is completely represented by the diagonal passing through the same point.

Procedure. Refer to Fig. 1.1

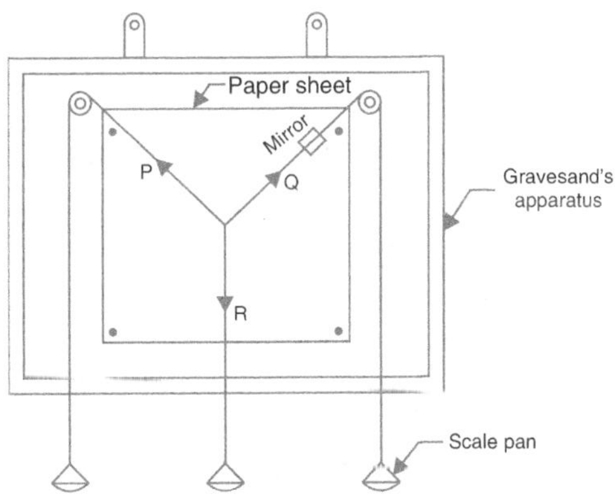

Fig. 1.1

1. Fix the paper sheet with drawing pins on the board set in a vertical plane.

2. Pass one thread over the pulleys carrying a pan (for adding weight during performance of the experiment) at its each end.

3. Take a second thread and tie its one end at the middle of the first thread and tie a pan at its other end.

4. Add weights in the pans in such a manner that the small knot comes approximately in the centre.

5. Displace slightly the pans from their position of equilibrium and note if they come to their original position of rest. This will ensure free movement of pulleys.

6. Take the mirror strip and place under each thread lengthwise and mark two points keeping the eye, the thread and its image in the same line without disturbing the system.

7. Mark lines of forces and write the magnitude of forces.

8. Remove the paper from the board and produce the lines to meet at O.

9. Use Bow's notation to name the forces P, Q and R as AB, BC and CA.

10. Select a suitable scale and draw the line ab parallel to force P and cut it equal to the magnitude of P. From b draw the line bc parallel to force Q and cut it equal to the magnitude of Q (Fig. 1.2). Calculate the magnitude of ca *i.e.,* R_1 which will be equal to the third force R which proves the triangle law of forces.

If R_1 differs from original magnitude of R, then percentage error is found as follows :

$$\text{Percentage error} = \frac{R - R_1}{R} \times 100$$

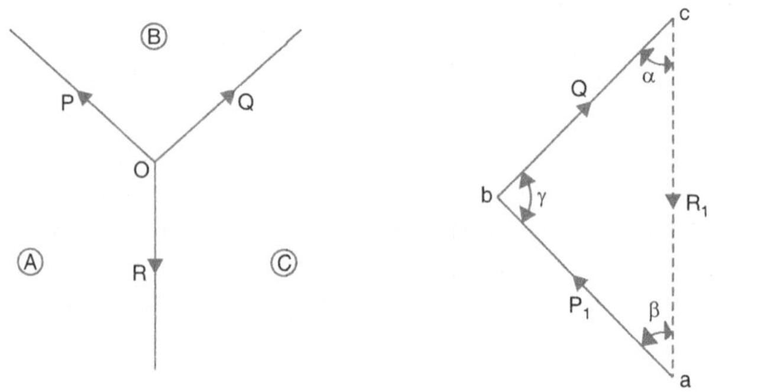

(*a*) Space diagram (*b*) Vector diagram

Fig. 1.2

Mathematically. Measure angles α, β and γ and by using Lami's theorem check the following relation :

$$\frac{P}{\sin \alpha} = \frac{Q}{\sin \beta} = \frac{R_1}{\sin \gamma}$$

Parallelogram law of forces. Refer to Fig. 1.3.

Cut $OA = P$ and $OB = Q$. From A draw AC' parallel to OB and BC' parallel to OA. The resultant of forces P and Q will be shown by R_1 which will be equal to R.

Note that R and R_1 are in the opposite direction.

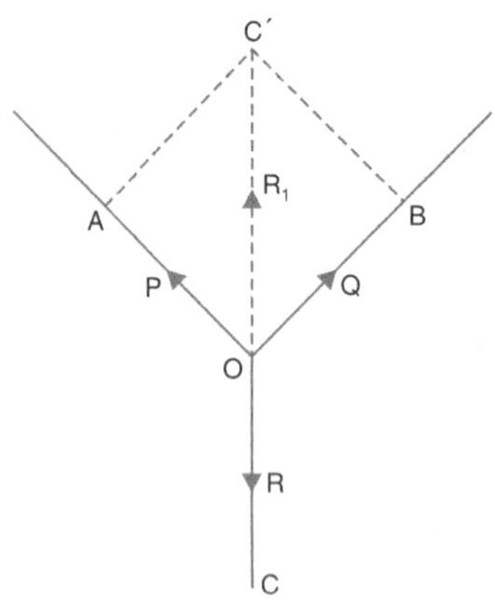

Fig. 1.3

Observations:

Scale.........

S. No.	*Total weight of pan P*	*Total weight of pan Q*	*Total weight of pan R*	*Calculate resultant* R_1	*%age error* $= \frac{R - R_1}{R} \times 100$

Precautions :

1. Pans/weights should not touch the vertical board.
2. There should be only one central knot on the thread which should be small.
3. While calculating the total force in each case the weight of the pan should be added to the weights put into the pan.
4. Make sure that all the pans are at rest when the lines of action of forces are marked.
5. All the pulleys should be free from friction.

Experiment No. 2

Object. To verify the polygon law of forces.

Apparatus. Gravesand's apparatus, paper sheet, weight box, thread, drawing pins, mirror strip, pans, set squares, pencil etc.

Theory. "*Polygon law of forces*" states that if a number of forces acting on a particle are represented in magnitude and direction by sides of a polygon taken in *same order*, then their resultant is represented in magnitude and direction by the closing side of the polygon taken in the *opposite direction*.

Procedure. Refer to Fig. 2.1

1. Set the board in a vertical plane and fix the paper sheet with drawing pins.
2. Pass a thread over two pulleys.
3. Take a second thread and tie the middle of this thread to the middle of first thread.
4. Pass the ends of the second thread over the other set of two pulleys.
5. Take a third thread and tie its one end to the point of first two threads.
6. Attach pans to the free ends of the threads as shown in Fig. 2.1.
7. Place the weights in the pans in such a manner that the knot comes approximately in the centre of the paper.

8. Take the mirror strip and place it under the threads turn by turn and mark the points by keeping the eye, the thread and its image in the same line without disturbing the system.

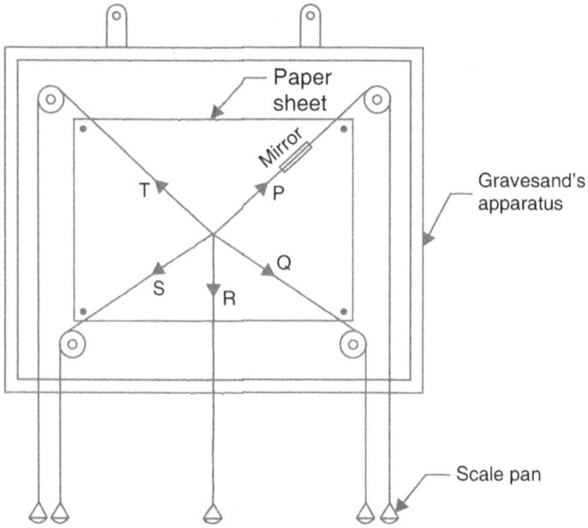

Fig. 2.1

9. Mark the lines of forces and write down the magnitude of forces.

10. Remove the paper from the board and produce the lines to meet at O.

11. Select a suitable scale draw the vector diagram (Fig. 2.2) by moving in one direction (*i.e.* clockwise or anticlockwise). Draw ab parallel to AB and cut it equal to force P; draw bc parallel to BC and cut it equal to Q ; draw cd parallel to CD and cut it equal to force R; draw de paralle to DE and cut it equal to force S. Vector ae will be the resultant force T_1 taken in opposite direction and should be equal to force T which proves the law of polygon of forces. If ae is not equal to T then percentage error is found as follows.

$$\text{Percentage error} = \frac{T - T_1}{T} \times 100$$

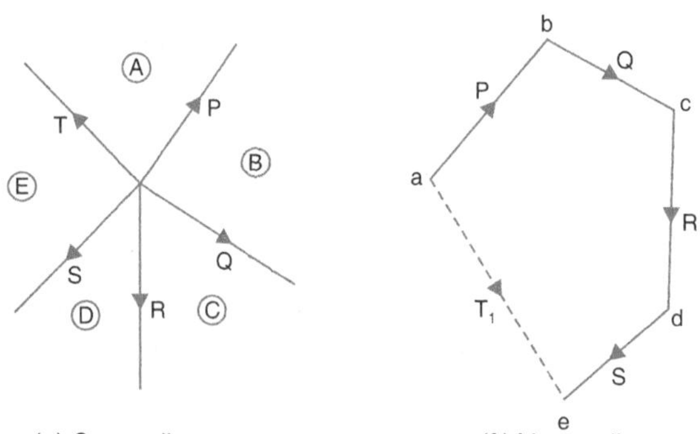

(*a*) Space diagram (*b*) Vector diagram

Fig. 2.2

Observations :

Scale.........

S.No	*Forces (Total weights of pans)*					*Calculated resultant* T_1	*Percentage error*
	P	*Q*	*R*	*S*	*T*		$= \frac{T - T_1}{100} \times 100$

Precautions :

1. Pans/weights should not touch the board.
2. There should be only central knot on the thread which should be small.
3. While calculating the total force in each case the weight of the pan should be added to the weights put into the pan.
4. Make sure all pans are at rest when the lines of action of forces are marked.
5. All the pulleys should be free from friction.

Experiment No. 3

Object. To verify the forces in the members of a jib crane.

Apparatus. Jib crane apparatus, weights, metre rod etc.

Theory. The experiment on jib crane is based on the *triangle law of forces* which states that if two forces acting upon a body can be represented in magnitude and direction by the two sides of a triangle taken in order, then their resultant is given by the third side of the triangle, taken in opposite direction.

Thus, the forces in jib and tie will be found out graphically with the help of triangle law of forces. The forces can also be calculated by applying sine formula. These calculated forces will be compared with the observed forces noted from the spring balances attached with jib and tie.

The jib crane apparatus (Fig. 3.1) consists of an inclined wooden rod called the *jib*. At the bottom of the jib end is fitted a compression balance which is pivoted and a hook is attached to the other end from which a chain hangs to carry the weight. A vertical tubular rod is fixed at the base and it carries hooks at different levels. One end of the tie rod is fitted to the jib and the second is attached to one of the hooks of the tubular rod. A spring balance is attached in the tie. The vertical tubular rod is fixed on a heavy base to avoid tilting and is supported for rigidity as shown in Fig. 3.1.

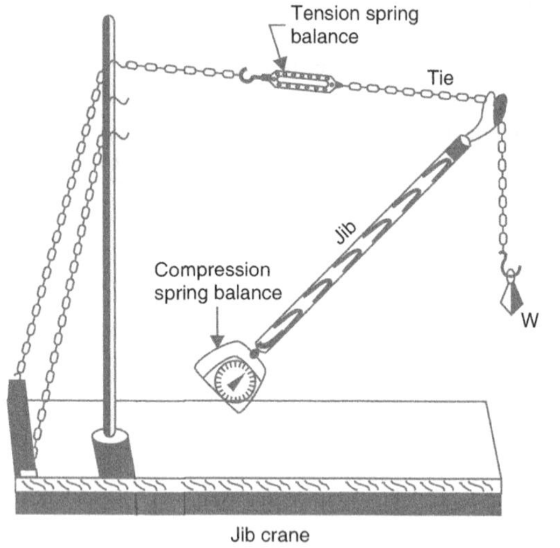

Fig. 3.1

Procedure :

1. Note down, the zero error (initial reading) in the compression and tension spring balances separately.

2. Attach the known weight, W with the chain as shown in Fig. 3.1.

3. Note down the final readings of compression and tension spring balances.

4. From these two readings, find the observed value by subtracting the initial reading from the final reading.

5. Measure the lengths of vertical post, tie and jib.

6. From these dimension draw the outline diagram LMN of jib crane.

7. Select a suitable scale and draw pq parallel to ML (direction of weight W) and cut equal to W ; draw qr parallel to LN and pr parallel to MN meeting at r. Then vectors pr and qr parallel to LN and pr parallel to MN meeting at r. Then vectors pr and qr represent forces in tie (tension) and jib (compression) respectively. Refer to Fig. 3.2.

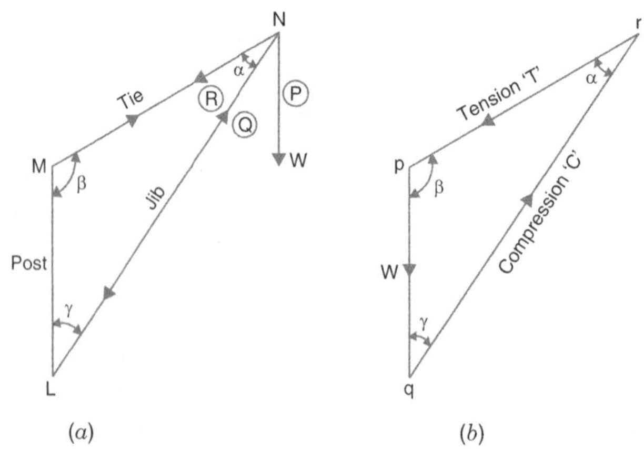

Fig. 3.2

Mathematically

In ΔLMN, measure α, β and γ.

Using sine law, we get

$$\frac{W}{\sin\alpha} = \frac{C}{\sin\beta} = \frac{T}{\sin\gamma}$$

Using sine law the value of C and T can be calculated. The percentage error can be calculated in the observed and calculated values of forces in jib and tie.

Precautions :

1. Measure the lengths accurately.
2. The weights should be suspended gently without any jerk.
3. The jib and the tie spring balances must be properly oiled for free movement.
4. Note down the initial and final readings of spring balance carefully.

Experiment No. 4

Object. To verify the reactions at the support of a simply supported beam.

Apparatus. A graduated wooden beam, two compression spring balances, weights, sliding hooks etc. Refer to Fig. 4.1.

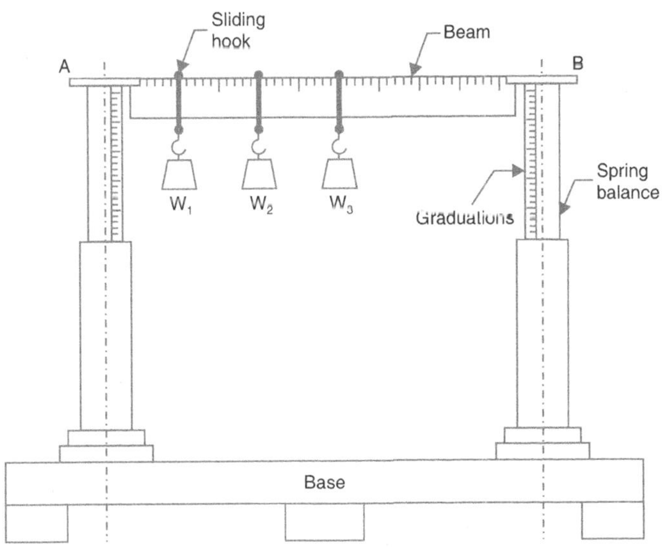

Fig. 4.1

Observations.

Scale ....

S.No.	W (kgf)	Length of members			Initial readings on balances (kgf)		Final readings on balances (kgf)		Observed forces from spring balances (kgf)		Calculated forces (kgf)		Percentage error	
		LM	MN	NL	Jib	Tie	Jib	Tie	Jib = C_1	Tie = T_1	Jib = C	Tie = T	Jib = $\frac{C_1 - C}{C} \times 100$	Tie = $\frac{T_1 - T}{T} \times 100$

Theory. This experiment is based on '*principle of moments*' which states that if a body is in equilibrium under the action of a number of coplanar forces then the algebraic sum of all the forces and their moments about any point in their plane are zero.

Mathematically. The body will be in equilibrium, if

$\Sigma H = 0$ *i.e.*, the algebraic sum of horizontal forces is zero.

$\Sigma V = 0$ *i.e.*, the algebraic sum of vertical forces is zero.

$\Sigma M = 0$ *i.e.*, the algebraic sum of moments about a point is zero.

Procedure :

1. Place the graduated beam on the compression spring balances.
2. Take the initial reading of the spring balances at both the ends A and B.
3. Now suspend weight at different points on the beam.
4. Note down the final reading of the spring balances. The difference between the final and initial readings of the balances give the observed values of the reactions at the end A and B.
5. Measure the distance of each weight from one support.
6. Take the moments about the support to calculate the reaction. The reaction at the other support may be found out by subtracting first reaction from total vertical load. If there is difference in the observed and the calculated reactions then calculate the percentage error.
7. Repeat the experiments by changing the loads at different positions.

Observations : *On next page*

Calculations : Ref. to Fig. 4.2

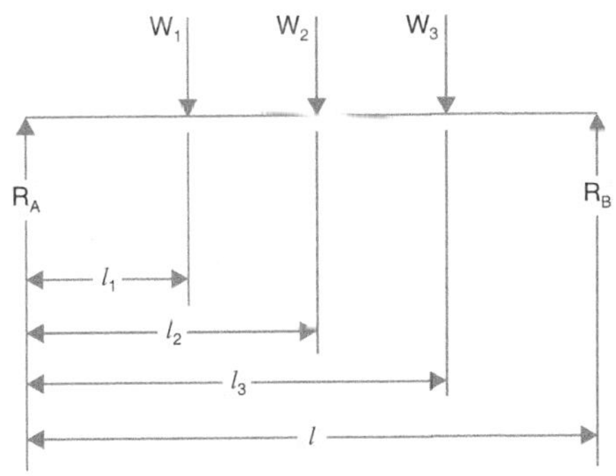

Fig. 4.2

Taking moments about A

$$W_1l_1 + W_2l_2 + W_3l_3 - R_B \times l = 0$$

or
$$R_B \times l = W_1l_1 + W_2l_2 + W_3l_3$$

or
$$R_B = \frac{W_1l_1 + W_2l_2 + W_3l_3}{l}$$

Also
$$R_A + R_B = W_1 + W_2 + W_3$$
$$R_A = (W_1 + W_2 + W_3) - R_B$$

Observations :

S.No.		
Initial readings of the balances at	A	
	B	
Final readings of the balances at	A	
	B	
Observed value of reactions (Final readings –initial readings)	A	
	B	
Weights suspended	W_1	
	W_2	
	W_3	
Distances of loads from left end (A) support.	l_1	
	l_2	
	l_3	
Moments of loads from left end support.	$W_1 \times l$	
	$W_2 \times l_2$	
	$W_3 \times l_3$	
Sum of moments	$W_1l_1 + W_2l_2 + W_3l_3$	
Calculated values of reactions	R'_A	
	R'_A	
Percentage error in reactions at	A	
	B	
Mean value :......		

$\therefore$ Percentage error in reaction R_A at A

$$= \frac{R_A - R'_A}{R_A} \times 100$$

$\therefore$ Percentage error in reaction R_B at B

$$= \frac{R_B - R'_B}{R_B} \times 100$$

Precautions :

1. Measure the distances accurately.
2. The weights should be suspended gently from hooks.
3. The initial and final readings of the spring balances should be noted carefully.
4. Before noting down the final readings, the beam should be slightly pressed downwards, so as to avoid any friction at the support.

Experiment No. 5

Object. To find co-efficient of friction between two surfaces, angle of response, mechanical advantage and efficiency of an inclined plane.

Apparatus. Inclined plane apparatus (Fig. 5.1), slider, weight box, pan, thread etc.

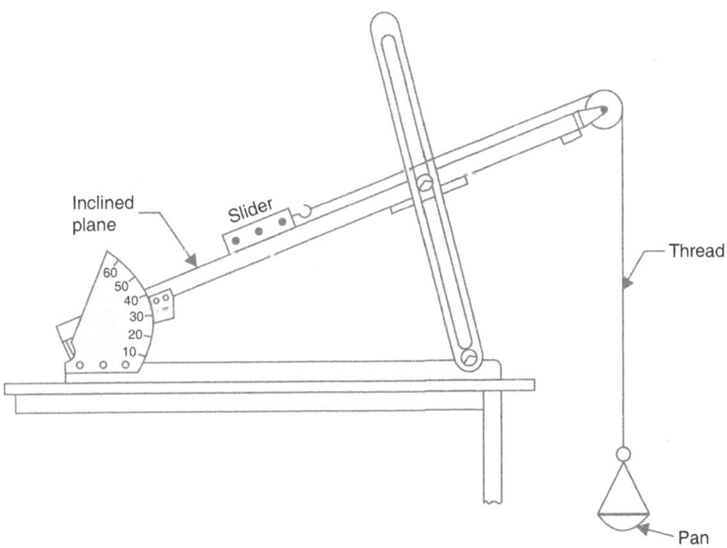

Fig. 5.1. Inclined plane apparatus

Theory. When a body slides upon another body, the property by virtue of which the motion on one relative to the other is retarded is called *friction*. The frictional force is directly proportional to the normal reaction 'N' *i.e.*,

$$F \propto N$$

or $$F = \mu N \text{ or } \mu = \frac{F}{N}$$

Suppose a body of weight W is to be lifted by an inclined plane and this requires effort P. When this load just moves upwards a frictional force F acts downwards which opposes its motion.

Refer to Fig. 5.2

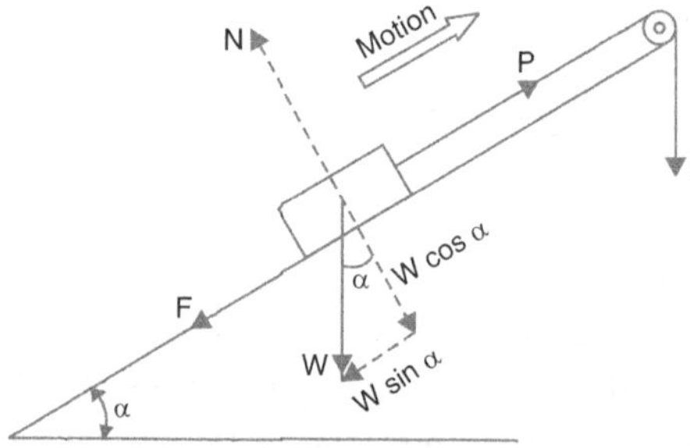

Fig. 5.2

Component of load W parallel to the plane = $W \sin \alpha$

Component of load W perpendicular to the plane = $W \cos \alpha$

Considering equilibrium parallel to the plane

$$P = F + W \sin \alpha$$

or,
$$F = P - W \sin \alpha \qquad ...(i)$$

Considering equilibrium perpendicular to the plane

$$N = W \cos \alpha \qquad ...(ii)$$

From (*i*) and (*ii*),

Co-efficient of friction,

$$\mu = \frac{F}{N} = \frac{P - W \sin\alpha}{W \cos\alpha}$$

Mechanical advantage (M.A) $= \dfrac{W}{P}$

Velocity ratio (V.R) $= \dfrac{\text{distance moved by effort}}{\text{distance moved by load}}$

Let effort P comes down through one centimetre, movement of the load along the plane = 1 cm

Vertical uplift of load $= 1 \times \sin \alpha$

$$\text{V.R.} = \frac{1}{1 \times \sin \alpha} = \text{cosec } \alpha$$

$$\% \text{ efficiency} = \frac{\text{M.A}}{\text{V.R.}} \times 100$$

Procedure :

1. Take the inclined plane apparatus and keep it first horizontal and put the slider on it.
2. Increase the inclination of the inclined board gradually till the slider just begins to slide downwards on it.

Note the angle in this position. This is called *angle of repose*.

3. Place the slider on the plane with the desired angle α.
4. Tie the slider to the pan with the help of a thread passing over the pulley.

5. Put the weights in the pan till the slider just starts moving. Note down the weights.
6. Measure the angle of inclination from the scale provided and find the value of μ.
7. Calculate M.A., V.R. and % efficiency.

Observations :

S.No.	*Total weight of the slider W*	*Weight of Pan + weights in the pan P*	$\mu = \dfrac{P - W \sin \alpha}{W \cos \alpha}$	$M.A. = \dfrac{W}{P}$	$V.R. = \text{cosec}\ \alpha$	$\% \eta = \dfrac{M.A.}{V.R.} \times 100$

Precautions :

1. The plane should be clean and smooth.
2. The guide pulley should move freely. It should be lubricated to make it frictionless.
3. Weight should be added gently in the pan.
4. The slider should just begin to move slowly, it should not move abruptly.
5. The direction of thread should be parallel to the inclined plane.

Experiment No. 6

Object. To find the mechanical advantage, velocity ratio and efficiency of a simple screw jack.

Apparatus. Screw jack apparatus (Fig. 6.1), weights, string, vernier calliper, outside calliper, metre rod etc.

Theory. Screw jack works on the principle of screw and nut. It is used for raising heavy loads through small efforts.

Let p = pitch of the screw ;

D = diameter of flanged table

W = load to be lifted

P = effort required to lift the load.

(Pitch is axial distance between the corresponding points on two consecutive threads.)

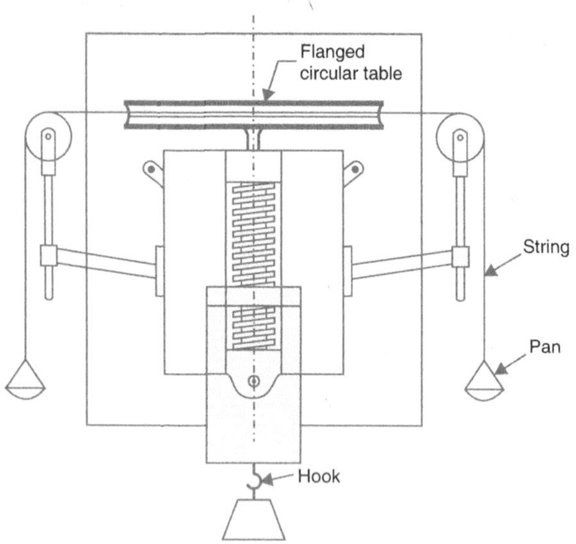

Fig. 6.1. Simple screw jack apparatus.

Now, $$\text{V.R.} = \frac{\text{distance moved by the effort in one revolution}}{\text{distance moved by the load in one revolution}}$$

$$= \frac{\pi D}{p}$$

and $$\text{M.A.} = \frac{\text{load lifted}}{\text{effort applied}} = \frac{W}{P}$$

∴ $$\%\ \eta = \frac{\text{M.A.}}{\text{V.R.}} \times 100$$

Procedure :

1. Measure the circumference of the flanged circular table with the help of an inextensible thread and metre rod. Or measure the diameter of the table with an outside calliper.

2. Measure the pitch of the screw with help of a vernier calliper.

3. Wrap the string round the circumference of the flanged table and pass it over one pulley. Similarly wrap another string over the circumference of flanged table and take it over the second pulley. The free ends of both the strings be tied to two pans in which the weights are to be placed.

4. Hang a known weight (W) on the hook (as shown in Fig. 6.1) and some weights in the pans so that the load W is just lifted, the effort P is equal to the sum of weights placed in both the pans.

5. Calculate the M.A., V.R. and % η.

6. Repeat the experiment three or four times.

Observations :

S.No.	Circumference of the table $= \pi D$	Pitch p	$V.R. = \dfrac{\pi D}{p}$	Load on table W	Effort $P = P_1 + P_2$	$M.A. = \dfrac{W}{P}$	$\%\eta = \dfrac{M.A.}{V.R.} \times 100$

Precautions:

1. To avoid the side thrust use both the pulleys to find the value of effort P.
2. The screw should be well-lubricated to reduce the friction.
3. The pulleys should be free from friction.
4. The load and effort should move slowly.
5. The strings should not overlap.
6. The string should be free from any knot.

Experiment No. 7

Object. To find the mechanical advantage, velocity ratio and efficiency of a worm and worm wheel.

Apparatus. Worm and worm wheel apparatus (Fig. 7.1) weights, string, metre rod, outside calliper, pan etc.

Theory. The worm and worm wheel apparatus consists of a toothed wheel (known as worm wheel) fixed with a drum on it. A string is wound round the drum carrying at its one end load W to be lifted. The worm wheel meshes with a worm which is fixed on a metallic spindle. The spindle carries a pulley from which a string hangs for application of effort.

Let d = diameter of drum fixed on the toothed or worm wheel

D = diameter of pulley attached to worm

T = number of teeth on worm wheel.

When one revolution is given to the pulley, only one tooth of worm wheel moves if the worm threads are of single start.

Distance moved by the effort $= \pi D$

Distance moved by the load $= \dfrac{\pi d}{T}$

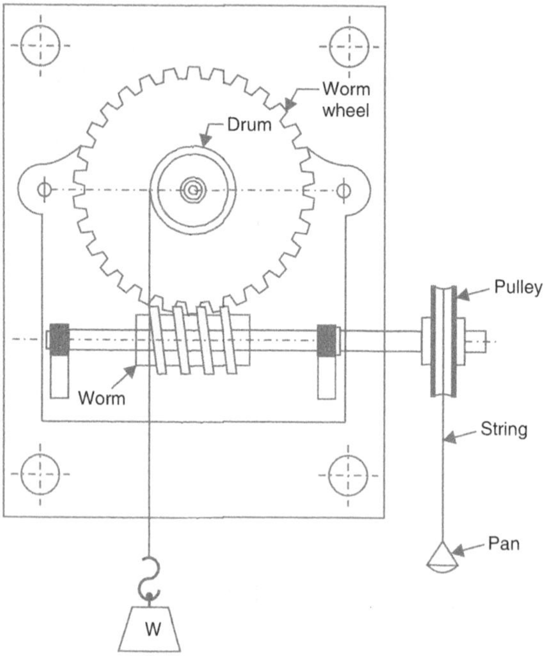

Fig. 7.1 Worm and worm wheel apparatus.

$$\text{V.R.} = \frac{\text{distance moved by the effort}}{\text{distance moved by the load}} = \frac{\pi D}{\frac{\pi d}{T}}$$

$$= \frac{TD}{d} \text{ (Neglecting thickness of string)}$$

$$\text{M.A.} = \frac{\text{load lifted}}{\text{effort applied}} = \frac{W}{P}$$

$$\therefore \quad \%\eta = \frac{\text{M.A.}}{\text{V.R.}} \times 100 = \frac{W/P}{TD/d} \times 100$$

$$= \frac{W \times d}{P \times T \times D} \times 100$$

Procedure :

1. Measure the circumference of drum and the pulley with the help of outside calliper.
2. Wrap the string round the drum and attach a load W.
3. Wrap another string round the pulley and attach a pan as shown in Fig. 7.1
4. Go on adding weights in the pan till the load just starts moving upwards.
5. Note down the weights in the effort pan.
6. Calculate M.A., V.R. and %η.
7. Repeat the experiment with different loads.

Observations :

Weight of pan =
Diameter of the pulley, D =
Diameter of the load drum, d =
Number of teeth on the worm wheel, T =

S. No.	*Load suspended* W	*Total effort P = wt. of pan + wts. in pan*	$M.A. = \frac{W}{P}$	$V.R. = \frac{T.D.}{d}$	$\%\eta = \frac{M.A.}{V.R.} \times 100$

Precautions :

1. Lubricate well the bearings of worm and teeth of worm wheel to decrease the friction.
2. Add the weights in the pan gently
3. The pan carrying the weights should not touch the wall.
4. The load and effort should move slowly.
5. The string should not overlap on the drum or the pulley.

Experiment No. 8

Object. To find mechanical advantage, velocity ratio and efficiency of a single purchase crab winch.

Apparatus. Single purchase crab winch apparatus, weights, pan, string etc.

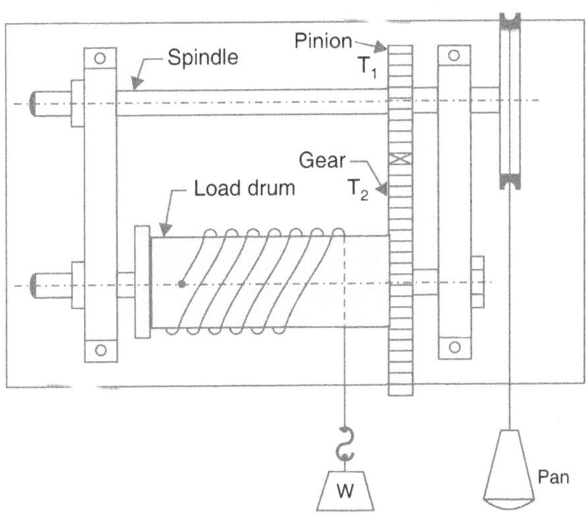

Fig. 8.1. Single purchase crab winch apparatus.

Theory. The single purchase crab winch apparatus consists of two toothed wheels, one small and the other large one. Smaller wheel is known as a *pinion* while the larger one is called *gear*. These wheels are mounted on two spindles as shown in Fig. 8.1. The gear is attached to the load drum. A string/rope is wound round the drum which carries the load W to be lifted. Another string/rope passes round the pulley to which is attached a scale pan in which weights are added to find the value of effort 'P' required to lift the load 'W'.

Let T_1 = number of teeth on the pinion ;

T_2 = number of teeth on the gear ;

D = diameter of the pulley ;

d = diameter of the load drum.

Let the pulley revolve through one revolution, then

Distance moved by the effort = πD

Number of revolutions of gear or the load drum = $\dfrac{T_1}{T_2}$

Distance moved by the load = $\pi d \times \dfrac{T_1}{T_2}$

$$\text{V.R.} = \frac{\text{distance moved by the effort}}{\text{distance moved by the load}}$$

$$= \frac{\pi D}{\pi d \times \dfrac{T_1}{T_2}} = \frac{D}{d} \times \frac{T_2}{T_1} \qquad \text{[Neglecting thickness of string/rope]}$$

$$\text{M.A.} = \frac{\text{load lifted}}{\text{effort applied}} = \frac{W}{P}$$

$$\%\eta = \frac{\text{M.A.}}{\text{V.R.}} \times 100.$$

Procedure :

1. Measure the circumference of the pulley and load drum with a string and metre rod or measure the diameter with an outside calliper.
2. Count the number of teeth on the pinion and the wheel.
3. Wrap a string/rope round the effort pulley and attach a scale pan at its free end.
4. Wrap another string/rope round the drum to carry load W in such a manner so that as the effort is applied, the load is lifted up.
5. Hang a load W on the string/rope of the load drum and put the weights in the effort pan so that load starts moving up gradually.
6. Note down the values of W and P and calculate the M.A., V.R. and efficiency.
7. Repeat the experiment with different loads.

Observations :

Number of teeth of pinion $= T_1$

Number of teeth of gear $= T_2$

Diameter of the pulley $= D$

Diameter of the load drum $= d$

S. No.	Load W	Effort P	M.A. = W/P	$V.R. = \frac{D}{d} \times \frac{T_2}{T_1}$	$\% \eta = \frac{M.A.}{V.R.} \times 100$

Precautions :

1. All the moving parts should be lubricated to reduce friction.
2. Weights should be put gently in the effort pan.
3. Add weight of the pan in the total effort.
4. The strings on the pulley and drum should not overlap.
5. Strings should be knot-free.

Experiment No. 9

Object. To determine the mechanical advantage, velocity ratio and efficiency of the first system of pulleys.

Apparatus. Set of pulleys, scale pan, weights, strings, hooks etc.

Theory. Fig. 9.1 shows the pulleys arranged in the first system of pulleys. The upper pulley is fixed to the wooden frame whereas the lower pulleys are movable. The weight to be lifted is hanged from the lowest pulley.

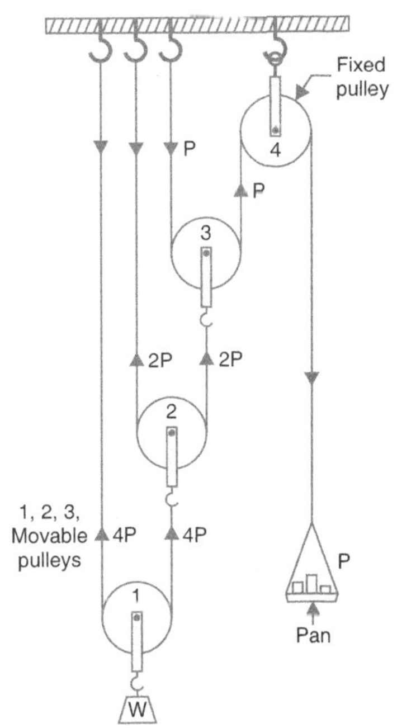

Fig. 9.1. First system of pulleys

When the effort P moves downwards by a distance y, then

Pulley 3 moves by a distance $= \frac{y}{2}$

Pulley 2 moves by a distance $= \frac{1}{2} \times \frac{y}{2} = \frac{y}{2^2}$

Pulley 1 moves by a distance $= \frac{1}{2} \times \frac{y}{2^2} = \frac{y}{2^3}$

$\therefore$ The load 'W' attached to the pulley 1 moves by a distance

$$= \frac{y}{2^3}$$

If there are 'n' number of movable pulleys, then distance moved by the load

$$= \frac{y}{2^n}$$

$\therefore$ Velocity ratio,

$$\text{V.R.} = \frac{\text{distance moved by the effort}}{\text{distance moved by the load}}$$

$$= \frac{y}{\frac{y}{2^n}} = 2^n$$

and $$\text{M.A.} = \frac{W}{P}$$

Hence $$\%\eta = \frac{\text{M.A.}}{\text{V.R.}} \times 100 = \frac{W}{P \times 2^n} \times 100$$

Procedure :

1. Fix one end of the string passing round each pulley to the hook and the second end to the block of next pulley.
2. Keep all the pulleys movable except the last pulley which is fixed and from which the effort is to be applied.
3. Hang a load 'W' from the lowest pulley.
4. Add weights in the pan till the load just starts moving up.
5. Note down the effort applied and calculate M.A.
6. Note down the distance moved by the effort and distance moved by the load and find V.R.
7. Verify the velocity with the formula (V.R. = 2^n).
8. Find out the % efficiency.
9. Repeat the experiment with different loads.

Observations :

S. No.	Load W (kgf)	Effort = wt of the pan + wt in the pan P (kgf)	M.A. = $\frac{W}{P}$	Distance moved by the effort = y	Distance moved by the load = x	V.R. = $\frac{y}{x}$	%η = $\frac{M.A.}{V.R.} \times 100$

Precautions :

1. The string should be light in weight and inextensible.
2. The pulleys should be well lubricated to reduce the friction.
3. The weights should be added gently in the pan.
4. The pulleys should be parallel to one another.
5. The load or effort should not touch any object.
6. While calculating total effort the weight of the pan should be added to the weights placed in the pan.

Experiment No.10

Object. To calculate personal horse power on suitable apparatus in laboratory.

Apparatus. Personal horse power apparatus (Fig. 10.1) consisting of a pulley fitted on a frame and a handle, spring balance, ropes, weights, metre rod, revolution counter, stop watch etc.

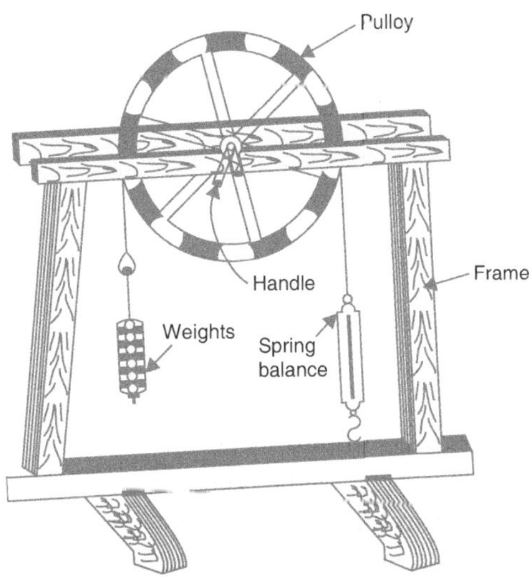

Fig. 10.1. Personal horse power apparatus.

Theory. Power is defined as the rate of doing work. An agent is said to be possess one H.P. if it is capable of doing 75 mkgf of work per second or 4500 mkgf of work per minute.

With the help of this apparatus it can be known at which rate an individual can do work.

Procedure :

1. Measure the diameter of the pulley.
2. Pass the rope on the pulley hanging vertically downwards.
3. Attach one end of the rope to a spring balance and hang weight '*W*' from the other end.
4. Rotate the wheel against the load (*i.e.,* from load side to the spring balance side) at uniform speed for one minute which can be checked from the spring balance which should show a constant reading.
5. Note down the number of revolutions from the revolution counter.
6. Note down the reading of spring balance and dead load *W*.
7. Calculate the horse power.

Calculations :

Let W = dead load

S = spring balance reading

D = diameter of the pulley

W = number of revolutions per minute.

Then, Torque, $$T = (W - S)\,\frac{D}{2}$$

and $$\text{H.P.} = \frac{2\pi NT}{4500}$$

Observations :

S. No.	*Dai. of pulley = D (m)*	*Dead load = W (kgf)*	*N(r.p.m.)*	*Initial reading of spring balance (kgf)*	*Final reading of spring balance (kgf)*	*Observed reading of spring = (Final- Initial) (kgf)*	*Torque,* $T = (W - S)\,\frac{D}{2}$	$H.P. = \frac{2\pi NT}{4500}$

Precautions :

1. The wheel should be rotated at uniform speed.
2. The apparatus should be rigidly fixed at the base.
3. r.p.m. should be counted carefully.
4. While rotating the pulley hand should not be changed.